Resource, Mobility, and Security Management in Wireless Networks and Mobile Communications

OTHER TELECOMMUNICATIONS BOOKS FROM AUERBACH

Resource, Mobility, and Security Management in Wireless Networks and Mobile Communications

Edited by
Yan Zhang
Honglin Hu
Masayuki Fujise

Auerbach Publications
Taylor & Francis Group
Boca Raton   New York

Auerbach Publications is an imprint of the
Taylor & Francis Group, an informa business

Auerbach Publications
Taylor & Francis Group
6000 Broken Sound Parkway NW, Suite 300
Boca Raton, FL 33487-2742

© 2007 by Taylor & Francis Group, LLC
Auerbach is an imprint of Taylor & Francis Group, an Informa business

No claim to original U.S. Government works
Printed in the United States of America on acid-free paper
10 9 8 7 6 5 4 3 2 1

International Standard Book Number-10: 0-8493-8036-7 (Hardcover)
International Standard Book Number-13: 978-0-8493-8036-5 (Hardcover)

Library of Congress Cataloging-in-Publication Data

Resource, mobility, and security management in wireless networks and mobile communications / editors, Yan Zhang, Honglin Hu, Masayuki Fujise.
    p. cm.
    Includes bibliographical references and index.
    ISBN 0-8493-8036-7 (alk. paper)
    1. Wireless communication systems--Security measures. 2. Wireless communication systems--Quality control. 3. Cellular telephone systems. 4. Wireless LANs. 5. Mobile communication systems. 6. Resource allocation. 7. Radio frequency allocation. I. Zhang, Yan, 1977- II. Hu, Honglin, 1975- III. Fujise, Masayuki, 1950-.

TK5103.2.R395 2006
621.384--dc22   2006047729

Visit the Taylor & Francis Web site at
http://www.taylorandfrancis.com

and the Auerbach Web site at
http://www.auerbach-publications.com

Contents

v

About the Editors

Yan Zhang received his Ph.D. from the School of Electrical and Electronics Engineering, Nanyang Technological University, Singapore. From August 2004 through May 2006 he worked at the National Institute of Information and Communications Technology (NICT), Singapore. Presently, he is working at the Simula Research Laboratory, Lysaker, Norway. He is on the editorial board of the *International Journal of Network Security*, is the series editor for the Auerbach Publications series "Wireless Networks and Mobile Communications," and has served as co-editor for the following books: *Resource, Mobility and Security Management in Wireless Networks and Mobile Communications; Wireless Mesh Networking: Architectures, Protocols, and Standards; Millimeter-Wave Technology in Wireless PAN, LAN, and MAN; Distributed Antenna Systems: Open Architecture for Future Wireless Communications; Security in Wireless Mesh Networks; Wireless Area Networks; Wireless Quality of Service: Techniques, Standards and Applications*; and *Handbook of Research on Wireless Security*. His research interests include resource, mobility, energy, and security management in wireless networks and mobile computing. He is a member of IEEE and IEEE Com Soc. His e-mail address is yanzhang@ieee.org

Honglin Hu received his Ph.D. degree in communications and information systems in January 2004 from the University of Science and Technology of China (USTC), in Hefei, China. From July 2004 through January 2006, he was with Future Radio, Siemens AG Communications in Munich, Germany. Presently he is working at the Shanghai Research Center for Wireless Communications (SHRCWC), which is also known as the International Center for Wireless Collaborative Research (Wireless CoRe). He also serves as an associate professor at the Shanghai Institute of Microsystem and Information Technology (SIMIT), Chinese Academy of Science (CAS). He is mainly working for international standardization and other collaborative activities. He is a member of IEEE, IEEE Com Soc, and IEEE TCPC. In addition, he serves as a member of the Technical Program Committee for IEEE WirelessCom 2005, IEEE ICC 2006, IEEE IWCMC 2006, IEEE ICC 2007, and IEEE/ACM Q2Swinet 2006. Since June 2006, he has served on the editorial board of Wireless Communications and Mobile Computing for John Wiley & Sons. His e-mail address is hlhu@ieee.org

Masayuki Fujise received his B.S., M.S., and Dr. Eng. degrees in communication engineering from Kyushu University, Fukuoka, Japan, in 1973, 1975, and 1987, respectively, and he received his M. Eng. degree in electrical engineering from Cornell University, Ithaca, New York, in 1980. In 1975, he joined KDD (Kokusai Denshin Denwa Co. Ltd.) and was with their Research and Development Laboratories. In 1990, he became department head at ATR (Advanced Telecommunications Research Institute International) Optical and Radio Communications Research Laboratories in Kyoto, Japan. In 1997, he joined CRL (Communications Research Laboratory) Ministry of Posts and Telecommunications. In April 2004 the name was changed to NICT (National Institute of Information and Communications Technology), which is an incorporated administrative agency. Presently, he is the director of the Singapore Wireless Communications Laboratory of NICT. He is also a guest professor at Yokohama National University. His interests include radio-on-fiber transmission technology, software-defined radio technology, and ad hoc wireless network technology. His e-mail address is fujise@nict.go.jp

Contributors

Avesh K. Agarwal
Department of Electrical and
    Computer Engineering
North Carolina State University
Raleigh, North Carolina

Attahiru Sule Alfa
Department of Electrical and
    Computer Engineering
University of Manitoba
Winnipeg, Manitoba, Canada

Boris Bellalta
Universitat Pompeu Fabra
Departament de Tecnologia
Barcelona, Spain

Sunghyun Choi
School of Electrical Engineering and
    Computer Science
Seoul National University
Seoul, Korea

Kihwan Choi
School of Electrical Engineering and
    Computer Science
Seoul National University
Seoul, Korea

Felipe A. Cruz-Pérez
Communication Section
CINVESTAV-IPN
Mexico City, Mexico

Christos Douligeris
Department of Informatics
University of Piraeus
Piraeus, Greece

Yuguang Fang
Department of Electrical and
    Computer Engineering
University of Florida
Gainesville, Florida

Masayuki Fujise
NICT Singapore
National Institute of Information and
    Communication Technology
Singapore

Hossam S. Hassanein
School of Computing
Queen's University
Kingston, Ontario, Canada

Honglin Hu
Siemens AG
Munich, Germany

Yoji Kawamoto
DRM Development Department
Technology Development Group
Sony Corporation
Tokyo, Japan

Geir M. Køien
Telenor R&D
Mobile Systems Group
Grimstad, Norway

Victor C.M. Leung
Department of Electrical and
   Computer Engineering
The University of British Columbia
Vancouver, British Columbia, Canada

Jie Li
Department of Computer Science
Graduate School of Systems and
   Information Engineering
University of Tsukuba, Tsukuba
   Science City, Japan

Wei Li
Department of Electrical Engineering
   and Computer Science
The University of Toledo
Toledo, Ohio

Wei Liang
Department of Electrical and
   Computer Engineering
North Carolina State University
Raleigh, North Carolina

Xiaokang Lin
Department of Electronics
   Engineering
Tsinghua University
Beijing, P.R. China

Jijun Luo
Siemens AG
Munich, Germany

Maode Ma
School of Electrical and Electronics
   Engineering
Nanyang Technological University
Singapore

Zuji Mao
Lucent Technologies, Inc.
Westford, Massachusetts

Shirley Mayadewi
Department of Electrical and
   Computer Engineering
University of Manitoba
Winnipeg, Manitoba, Canada

Michela Meo
Dipartimento di Elettronica
Politecnico di Torino
Torino, Italy

Jelena Mišić
Department of Computer Science
University of Manitoba
Winnipeg, Manitoba, Canada

Vojislav B. Mišić
Department of Computer Science
University of Manitoba
Winnipeg, Manitoba, Canada

Hussein T. Mouftah
School of Information Technology
   and Engineering
University of Ottawa
Ottawa, Ontario, Canada

Nidal Nasser
Department of Computing and
   Information Science
Telecom Technology Toronto
University of Guelph
Guelph, Ontario, Canada

Lauro Ortigoza-Guerrero
Wireless Facilities Inc. (WFI)
San Diego, California

Chang Woo Pyo
NICT Japan
National Institute of Information and
Communication Technology
Yokosuka, Japan

Eranga Perera
School of Electrical Engineering and
Telecommunications
The University of New South Wales
Sydney, Australia

G.R. Reddy
Department of Computer Science
University of Manitoba
Winnipeg, Manitoba, Canada

Aruna Seneviratne
National ICT Australia Ltd.
Locomotive Work-Shop
Australian Technology Park
Everleigh, NSW, Australia

Vijay Sivaraman
School of Electrical Engineering &
Telecommunications
The University of New South Wales
Sydney, Australia

Joo-Han Song
4G System Laboratory
Telecommunication R&D Center
Samsung Electronics Co. Ltd.
Suwon, Korea

Abd-Elhamid M. Taha
Electrical and Computer Engineering
Department
Queen's University
Kingston, Ontario, Canada

Dimitrios D. Vergados
Department of Information and Com-
munication Systems Engineering
University of the Aegean
Karlovassi, Samos, Greece

Qian Wang
Department of Electrical & Computer
Engineering
University of Manitoba
Winnipeg, Manitoba, Canada

Wenye Wang
Department of Electrical and Com-
puter Engineering
North Carolina State University
Raleigh, North Carolina

Vincent W.S. Wong
Department of Electrical and Com-
puter Engineering
The University of British Columbia
Vancouver, British Columbia, Canada

Shaoqiu Xiao
NICT Singapore
National Institute of Information and
Communication Technology
Singapore

Yang Xiao
Department of Computer Science
The University of Alabama
Tuscaloosa, Alabama

Yongqiang Xiong
Microsoft Research Asia
Beijing, P.R. China

Shouyi Yin
Department of Electronics
Engineering
Tsinghua University
Beijing, P.R. China

Qian Zhang
Department of Computer Science
Hong Kong University of Science
    and Technology
Kowloon, Hong Kong

Yan Zhang
NICT Singapore
National Institute of Information and
    Communication Technology
Singapore

Yanchao Zhang
Department of Electrical and
    Computer Engineering
University of Florida
Gainesville, Florida

Jialing Zheng
School of Electrical and Electronics
    Engineering
Nanyang Technological University
Singapore

Mingtuo Zhou
NICT Singapore
National Institute of Information and
    Communication Technology
Singapore

Yun Zhou
Department of Electrical and
    Computer Engineering
University of Florida
Gainesville, Florida

Preface

Wireless networks have experienced exponential growth during the past few years and are gradually emerging as a new discipline. Motivated by the demand for efficient algorithms in utilizing the scarce bandwidth, the desire for seamless roaming, and the requirement for solid security, all the mobile communications systems, including GSM/CDMA/3G/4G/802.11x/802.16, have to address resource, mobility, and security management issues. Hence, these three issues are common challenges in all wireless networks and mobile computing scenarios.

The book deals with resource, mobility, and security management for the particular standard in 802.11x/3G/4G and also the generic techniques applicable in all wireless networks. Consequently, this book is organized into three parts:

> Part I: Resource Management
> Part II: Mobility Management
> Part III: Security Management

The book covers the related key challenges and solutions in mobile ad hoc networks, wireless sensor networks, Bluetooth, Quality-of-Service (QoS), wireless local area network (WLAN), 3G, and heterogeneous wireless networks. The topics include call admission control (CAC), routing, multicast, medium access control (MAC), scheduling, bandwidth adaptation, handoff management, location management, network mobility, secure routing, key management, authentication, security, privacy, performance simulation, and analysis, etc. It can serve as a useful reference for students, educators, faculties, telecom service providers, research strategists, scientists, researchers, and engineers in the field of wireless networks.

This book has the following salient features:

It offers comprehensive, self-contained information on resource, mobility, and security management in wireless networks.

It serves as an easy cross-reference owing to the broad coverage on resource, mobility, and security management in a mobile computing environment.

It details the particular techniques in efficiently eliminating bandwidth insufficiency, increasing location management performance, and decreasing the associated authentication traffic.

It presents the interaction and coupling among the three components in wireless networks.

It provides background, application, and standard protocols.

It identifies the direction of future research.

We would like to acknowledge the effort and time invested by all contributors for their excellent work. All of them are extremely professional and cooperative. Special thanks go to Richard O'Hanley, Kimberly Hackett, and Karen Schober of Taylor & Francis Group for their support, patience, and professionalism from the beginning until the final stage of production. Last but not least, a special thank you goes to our families and friends for their constant encouragement, patience, and understanding throughout this project.

Yan Zhang, Honglin Hu, and Masayuki Fujise

RESOURCE MANAGEMENT

Chapter 1

Call Admission Control in Wireless LANs

Boris Bellalta and Michela Meo

Contents

1.1 Introduction

The increasing popularity of wireless local area networks (WLANs) based on the IEEE 802.11 technology, due to the ease in installing access facilities and to the affordable price of equipment, is pushing operators to deploy WiFi WLANs as access networks to their services. However, the limitations of this technology, such as the still limited radio resources, the poor channel quality depending on relative position of mobile nodes (MNs), the interference from hidden terminals, and the anomaly observed when MNs transmit at different speeds, make it difficult to cope with the need to provide a variety of services with different characteristics in terms of QoS (Quality-of-Service) requirements. Indeed, current WLAN networks provide best-effort services without any QoS guarantees. Due to the higher bandwidth provided, multimedia services such as video and voice streaming can perform well under low load conditions; but when traffic intensity increases, the delay and bandwidth of streaming flows are severely affected, thus degrading the received quality of service.

Solutions are therefore needed for service differentiation at the access and for providing QoS guarantees. In particular, services can be distinguished in two classes based on the mechanisms employed at the transport layer: (1) *elastic* flows, usually adopted to deliver data services such as file transfer applications, correspond to traffic carried by Transmission Control Protocol (TCP) and TCP-like protocols, that adapt the traffic generation rate to the network working conditions, attempting in this way to reduce network congestion; (2) *streaming* flows, adopted by multimedia applications, tend to generate traffic unaware and independently of the network conditions.

Based on this differentiation, some possible solutions for guaranteeing acceptable QoS levels consist of performing call admission control (CAC).

Flows can be accepted as long as the total traffic is below a given threshold. By properly choosing the threshold, QoS levels for accepted flows can be reasonably guaranteed. Moreover, service differentiation can be achieved by applying the scheme to streaming or elastic flows, or both.

This chapter first proposes a model of the IEEE 802.11 medium access control (MAC) layer and investigates the performance perceived by streaming and elastic flows in a hot-spot scenario. It then uses the model as the core of a CAC scheme. Indeed, the scheme is based on a bidimensional Markov chain describing the dynamics of flow arrivals and completions for each class of traffic. The transition rates in the chain are set according to the MAC layer model.

1.2   Related Work

The IEEE 802.11 MAC and the physical (PHY) layer specifications for the 802.11b standard in the 2.4 GHz band are explained in [1]. For other PHY specifications, refer to the amendments [2] (802.11a) or [3] (802.11g). These posterior PHY specifications allow the system to achieve higher data rates by improving the modulation and coding techniques, the MAC specification remaining unchanged. The current version of the MAC protocol does not implement any QoS mechanism, which is solved with the IEEE 802.11e standard [4].

Since the first specifications of the IEEE 802.11 appeared, a large amount of research has been done to analytically model the IEEE 802.11 MAC protocol. We remark the seminal articles of Bianchi [5] and Tay and Chua [6], referred to in most posterior works. In these two articles, the authors address the performance analysis of the MAC protocol assuming a finite number of saturated sources (i.e., sources always have a packet ready to be transmitted) that compete for the use of the shared channel. In both articles, the authors find the value of the MAC parameters that maximize the network throughput. The decoupling assumption (i.e., nodes attempt to transmit to the channel independently of each other) is applied to simplify the protocol analysis, which can be solved through a fixed point procedure. The presented results can be used to understand how the MAC protocol performs or to obtain measures such as the maximum throughput given a number of nodes in the system. These articles are well complemented by the delay analysis of the MAC protocol presented in [7], where Carvalho and Garcia-Luna-Aceves obtain the first and second moment of the service time, still in saturation conditions. It is remarkable that the authors find closed form expressions by linearizing the fixed point equations presented in the previously mentioned articles.

Models of the saturated system are unable to provide more detailed information about the behavior of the network (actual load, network utilization, etc.) or of each node individually (transmission attempt rate, collision probability, queueing delays, user throughput, etc.), which are of crucial importance to compute the grade of service the traffic flows receive from the network. Furthermore, protocol enhancements by means of dynamic tuning of backoff parameters or other MAC parameters have marginal effects when the network is unsaturated and, therefore, the real gain is also marginal. In recent years, several articles have addressed the modeling task of the IEEE 802.11 MAC performance under non-saturation conditions. Two groups of articles can be found, based on a Markovian stochastic analysis, which are extensions of the model presented by Bianchi [5]. In [8], the authors introduce new states in the Markov chain that describe the backoff algorithm, modeling the time in which the mobile nodes are empty (i.e., has no packet ready to be transmitted). Another group of articles is based on the observation that the attempt rate of a node is a regenerative process and it can be computed using the renewal-reward theorem. This approach is used in [9], [10], and [11]. Surprisingly, the decoupling assumption works well also under nonsaturated conditions, and results obtained by both type of models are accurate. One of the benefits of the unsaturated system models is that they allow one to analyze the system performance when nodes have different (heterogeneous) traffic profiles. For example, in [8], the authors analyze the system under the presence of streaming and elastic flows, respectively carried by UDP and TCP transport layer protocols.

Today, WLANs are basically used to access the Internet and download Web pages or other types of information. Therefore, most of the data is carried by the TCP from the Internet to the end user (i.e., most of the traffic flows in the downlink direction). The first work that analytically addressed the performance of a WLAN with TCP traffic is [12]. The authors consider both the data traffic in the downlink direction and the feedback traffic in the uplink direction due to TCP ACKs, modeling the access point and the mobile nodes as saturated sources. To catch the effect that the number of backlogged nodes is not constant and depends on the number of TCP connections, the authors propose the use of a discrete-time Markov chain to obtain the probability that n nodes have an ACK ready to be transmitted, and thus compete with the AP (access point) to transmit on the channel. The average system throughput is obtained by means of a time-scale decomposition. A similar approach is used in [13], including the effect of delayed ACK techniques and the presence of short-lived TCP connections. More recently, in [14] this problem is also considered under heterogeneous radio conditions. Similar results are obtained in the three articles. Finally, in [15], the authors analyze the presence of downlink/uplink and bidirectional TCP flows but under the assumption that the TCP advertisement window is equal to one.

The introduction of VoIP (Voice-over-IP) services over WLANs is going to further increase the use of WLANs, and it is a hot topic in the research community. In articles such as [11], [16], [17], or [18], a capacity analysis is presented. Similar conclusions are presented in all of them, remarking that the AP is the bottleneck of the network. To avoid this problem and increase the capacity, several solutions can be considered from the IEEE 802.11e specifications [4], for example, using the TXOP (Transmission Opportunity) option [17].

Therefore, motivated by the expected growth of multimedia traffic in future WLANs, a lot of work has been done in resource allocation strategies and admission control with the goal to provide mechanisms that ensure differentiated services. To differentiate among traffic flows, three main strategies are used in the literature, based on the specifications of [4]: (1) different DIFS values for each QoS level, (2) setting a different CW_{min} value to each flow [19,20], and (3) using the TXOP [17,21]. The first one assures rigid flow differentiation, but the second and third allow for more fine tuning of the service received by each flow. In these schemes, admission control is mandatory to ensure that the system remains always in a stable state, blocking new flows if necessary. Two groups of admission control schemes exist: (1) model-based schemes [20,22,23], which estimate the future system status using mathematical models of the system; and (2) measure-based schemes [24], which predict the future system status from current measures. Anyway, this is a soft classification because the major part of CAC schemes uses both models and measure information to achieve its goal.

One of the main difficulties in admission control schemes is to predict the future system state using actual system information such as the nonlinear behavior of the IEEE 802.11 MAC protocol parameters (conditional collision probability, transmission probability, mobile node queue utilization, etc.) with the number of flows and their traffic characteristics. For example, in Pong and Moors [22], to compute the effect of the introduction of a new flow, the admission control assigns to the new flow the characteristics obtained from a similar throughput flow.

Banchs et al. [20] propose an admission control scheme and a parameter tuning algorithm for the CW_{min} parameter assuming that $CW_{max} = CW_{min}$, or that the parameter $m = 0$. The rationale behind the assumption is that if the admission control provides the optimal CW_{min}, the fact that $m > 0$ can tend to suboptimal situations. The CAC is also based on the model of [5]. For each new request, the CAC estimates the collision probability by assuming that all nodes are saturated (thus, the conditional collision probability is the same for all flows) and computes the system achievable throughput. Under the assumption that the transmission probability is proportional to the throughput requested by each flow, the CAC computes the individual achievable throughput. It is remarkable that the derivation of optimal CW_{min} values allows the system to maximize the number of active flows.

In [24], the HARMONICA architecture for admission control and parameter tuning is presented. It uses LQI (Link Quality Indicator) to catch metrics such as packet dropping, link end-to-end delay, and throughput, which are used to decide if a new flow can be admitted and the optimum parameters can be assigned to each flow to maximize the network utilization.

Finally, in [23], an admission control scheme for both streaming and elastic flows is presented. The admission control admits or rejects streaming flows and adjusts the transmitting rate of elastic flows to avoid their interfering with streaming flows. The authors observe that the business ratio (fraction of time the channel is not empty) is practically equal to the channel utilization (fraction of time the channel is transmitting successful frames), independently of the number of users. One of the conclusions of the article was that, by maintaining the business ratio close to a certain threshold, the system throughput is maximized at the same time that delay and delay variation are minimized. The authors use the normalized throughput, which is linear with the business ratio until the selected threshold is reached (which is about 0.9 of the channel utilization for both RTS/CTS and Basic Access (BA) access schemes). For the rate control operation, it is assumed that a traffic-shaping procedure implemented at the nodes and the access point regulates the TCP traffic offered to the network.

1.3   A Hot-Spot Wireless Scenario

A wireless cell (or hot-spot) is the coverage area provided by a single access point (in [1] it is referred to as a Basic Service Set, BSS). The coverage area is the geographical area where both the AP and the mobile stations can communicate using the radio channel with an acceptable minimum quality; this quality can be measured in terms of SNR (signal-to-noise ratio) and other derived metrics such as the frame error ratio (FER). An Extended Service Set (ESS) contains multiple access points and their coverage areas. All or part of these coverage areas can overlap, so that a mobile station can select the AP to use; we call these areas reassociation or handoff areas.

Typical scenarios with this configuration are found in public areas (such as cafeterias, parks, airports) where users can access the Internet from their notebooks or PDAs; company buildings where workers use WLANs to communicate through the e-mail service, message applications, or VoIP; individual users at their homes, etc. In all these scenarios, the WLAN technology provides a certain grade of mobility and broadband access to the Internet at a very low cost.

This work considers a single BSS with an AP providing access to a fixed network to n MNs. Each node has a traffic profile specifying its basic configuration parameters, (i.e., bandwidth, packet arrival rate, expected frame length, etc.). The MNs and the AP use the DCF (Distributed Coordination

Function) of the IEEE 802.11 MAC and the DSSS PHY specifications in the 2.4 GHz band [1].

1.3.1 MAC Protocol Description

The IEEE 802.11 MAC is based on a distributed CSMA/CA protocol [1]. According to the basic access (BA) mechanism, when a node has no packets to transmit and receives a packet from the network layer, the node starts to sense the channel to determine its state, which can be either *busy* or *free*. If the channel is detected busy, the node waits until the channel is released. When the channel is detected free for a period of time larger than the DIFS (distributed inter-frame spacing) duration, a new backoff instance is generated. A backoff instance consists of a counter set to a random value each time it is generated. The random value is picked from a uniform distribution in the range $CW(k) = [0, min(2^k CW_{min} - 1, 2^m CW_{min} - 1)]$, where k is the current attempt to transmit the packet, CW_{min} is the minimum size of the contention window, and m defines the maximum size of the window. For each packet to be transmitted, k is initially set to 0 and it is increased by one at each failed transmission until a maximum number of retransmissions, called the Retry Limit, is reached, and the packet is dropped. The counter is decreased by one for each time-slot σ in which the channel is sensed free, and, when the countdown reaches zero, the node starts the packet transmission on the channel. If during the backoff countdown the channel is sensed busy, the backoff is suspended until the channel is detected free again.

A collision occurs if two nodes transmit at the same time; that is, the backoff instances from both nodes reach 0 at the same time. After the data packet is transmitted by the sender, the receiver waits for an SIFS (short inter-frame spacing) time and sends a MAC layer ACK to acknowledge the correct reception of the data packet. In case the sender does not receive the ACK frame, it starts the retransmission procedure. After discarding or successfully transmitting a packet, if more packets are ready to be transmitted, the node starts the transmission procedure again. Otherwise, it waits for a new packet from the network layer. Figure 1.1 plots an example of the BA mechanism with three mobile stations contending to transmit a packet.

Alternative to the BA mechanism, nodes can employ a RTS/CTS protocol to access the channel, so as to reduce the hidden terminal effect.

1.3.2 System Parameters

The system parameters are reported in Table 1.1, the overhead introduced by upper layers are listed in Table 1.2. We assume ideal channel conditions;

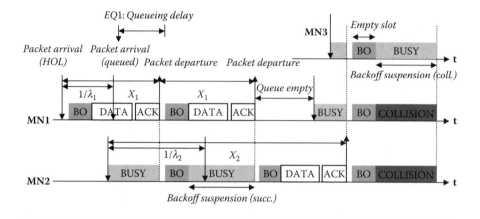

Figure 1.1   Example of the basic access mechanism.

that is, no packet is lost due to channel errors or the hidden terminal phenomenon. Figure 1.2 presents a sketch of the considered network. The fixed network is modeled by a simple 100 Mbps full-duplex link with a propagation delay of 2 ms in both directions. This link is used to interconnect a fixed node (server) where one endpoint of the traffic flows resides. The other endpoints are in the mobile nodes, which are linked to the server through the AP.

1.3.3   Frame Durations

When a node transmits a frame, two possible events can occur: a collision or a successful transmission. The duration of both events depends on the employed access mechanism. The successful transmission duration for the

Table 1.1   System Parameters of the IEEE 802.11b Specification [1]

Parameter	Value	Parameter	Value
R_{data}	2 Mbps	R_{basic}	1 Mbps
DIFS	50 μs	CW_{min}	32
SIFS	10 μs	CW_{max}	1024
SLOT (σ)	20 μs	m	5
EIFS	364 μs	ACK	112 bits @ R_{basic}
RTS	160 bits @ R_{basic}	CTS	112 bits @ R_{basic}
MAC header	240 bits @ R_{data}	MAC FCS	32 bits @ R_{data}
PLCP preamble	144 bits @ R_{basic}	PLCP header	48 bits @ R_{basic}
Retry Limit (R)	7	Q (queue length)	50 packets

Table 1.2 Protocol Overheads from Upper Layers

Parameter	Value
RTP header	12 bytes
TCP header	20 bytes
UDP header	8 bytes
IP header	20 bytes

BA and the RTS/CTS mechanisms, are, respectively, given by:

$$T_s^{ba} = \frac{PHY_H}{R_{basic}} + \frac{MAC_H + L_{data} + MAC_{FCS}}{R_{data}} + SIFS + \frac{PHY_H}{R_{basic}} + \frac{L_{ACK}}{R_{basic}} + DIFS \tag{1.1}$$

$$T_s^{rts} = O_{rts} + \frac{PHY_H}{R_{basic}} + \frac{MAC_H + L_{data} + MAC_{FCS}}{R_{data}} + SIFS + \frac{PHY_H}{R_{basic}} + \frac{L_{ACK}}{R_{basic}} + DIFS \tag{1.2}$$

where

$$O_{rts} = \frac{PHY_H}{R_{basic}} + \frac{RTS}{R_{basic}} + SIFS + \frac{PHY_H}{R_{basic}} + \frac{CTS}{R_{basic}} + SIFS \tag{1.3}$$

$$PHY_H = PLCP\ preamble + PLCP\ header \tag{1.4}$$

For the BA mechanism, the duration of a collision is equal to the maximum successful transmission duration of the colliding frames; but for the

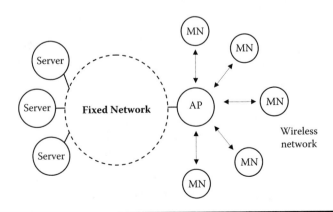

Figure 1.2 Sketch of the considered network.

RTS/CTS mechanism, the duration of a collision is constant and equal to:

$$T_c^{rts} = \frac{PHY_H}{R_{basic}} + \frac{RTS}{R_{basic}} + EIFS \tag{1.5}$$

The additional overhead of RTS/CTS access compensates for a low collision duration.

1.4   A Model of the IEEE 802.11 MAC Layer

This section presents a user-centric model of the DCF function of the 802.11 MAC layer. We approximate each MN by a finite-length queue with network-dependent service time.

1.4.1   A Mobile Node

Packets with average length L_i arrive at node i with rate λ_i. Both the time between packet arrivals and the service time are assumed to be exponentially distributed. Therefore, an MN (the AP included) is modeled by an $M/M/1/Q_i$ queue with Q_i as the queue length measured in packets.

The offered traffic load to the MAC layer and the queue utilization for node i are $v_i = \lambda_i X_i$ and $\rho_i = \lambda_i(1 - P_{b,i})X_i$, respectively, where X_i and $P_{b,i}$ are the mean service time and the packet blocking probability. The node throughput is $S_i = \rho_i L_i / X_i$.

By modeling each MN using an $M/M/1/Q_i$ queue, one can obtain simple expressions to measure the quality of the service observed by a node in terms of blocking probability, average queue length, and average transmission delay (including the service time):

$$P_{b,i} = \frac{v_i^{Q_i}}{\sum_{j=0}^{Q_i} v_i^j} \qquad EQ_i = \frac{\sum_{j=0}^{Q_i} j v_i^j}{\sum_{j=0}^{Q_i} v_i^j} \qquad ED_i = \frac{EQ_i}{\lambda_i(1 - P_{b,i})} \tag{1.6}$$

Finally, the probability to lose a packet is the probability that the packet is discarded at the queue entrance due to overflow, or dropped at the MAC layer because the number of retransmissions has exceeded the retry limit R_i. Then, a packet loss occurs with probability:

$$P_{L,i} = P_{b,i} + (1 - P_{b,i})P_{d,i} \tag{1.7}$$

where $P_{d,i}$ is the probability that a packet is dropped at the MAC layer.

1.4.2 The MAC Protocol

A node with a packet ready to transmit starts a backoff instance. Letting EB_i be the average number of slots selected by node i at each transmission attempt, the steady-state probability that the node transmits in a random slot given that a packet is ready in its transmission queue can be computed from:

$$\tau_i = \frac{E[Pr(Q_i(t) > 0)]}{EB_i + 1} = \frac{\rho_i}{EB_i + 1} \tag{1.8}$$

Node i transmission collides if any other node also transmits in the same slot. Then, the conditional collision probability for node i is:

$$p_i = 1 - \prod_{j \neq i} (1 - \tau_j) \tag{1.9}$$

To compute EB_i, two different approaches are found in the literature: (1) a stochastic (Markovian) approach [5] and (2) an average analysis [6]. Expressions found in both articles are different but numerically equal. For simplicity, we choose to use the expression of [6]; then EB_i is computed as:

$$EB_i = \frac{1 - p_i - p_i(2p_i)^{m_i}}{1 - 2p_i} \frac{CW_{min}}{2} - \frac{1}{2} \tag{1.10}$$

The effect of the *Retry Limit* R_i is considered in [25]. However, for operative values of $p_i < 0.4$, the effect of R_i on the average backoff time at each attempt is almost negligible. Using the conditional collision probability, the dropping probability at the MAC layer is given by the probability that a packet collides R_i times: $P_{d,i} = p_i^{R_i}$.

The service time (i.e., the time interval from the instant at which a packet enters into service until it is completely transmitted or discarded) is given by,

$$X_i = (M - 1)\left(EB_i \alpha_i + ET_{c,i}^{ba\|rts} \right) + EB_i \alpha_i + T_{s,i} \tag{1.11}$$

where M is the average number of transmissions, α_i is the average slot duration, and $ET_{c,i}$ is the average duration of a collision of node i. We approximate the value of $ET_{c,i}$ by:

$$\begin{cases} ET_{c,i}^{ba} \approx \dfrac{\sum_{j \neq i} \tau_j \max(T_{s,i}, T_{s,j})}{\sum_{j \neq i} \tau_j} \\ ET_{c,i}^{rts} = T_c^{rts} \end{cases} \tag{1.12}$$

where one neglects the fact that more than two packets collide simultaneously. Note that if the RTS/CTS access scheme is used, $ET_{c,i}^{rts}$ is constant and equal for all nodes. The average number of transmissions that a packet undergoes is computed under the decoupling assumption as:

$$M = \frac{1 - p_i^{R_i+1}}{1 - p_i} \tag{1.13}$$

A node freezes its backoff counter every time the channel is sensed busy and releases it after the channel is sensed free for a DIFS period. Therefore, the time between two backoff counter decrements is a random variable that depends on the behavior of the other nodes. By letting α_i be the average time between two backoff counter decrements — or equivalently, the average slot duration — one obtains:

$$\alpha_i = p_{e,i}\sigma + p_{s,i}\left(ET_{s,i}^{ba\|rts,*} + \sigma\right) + p_{c,i}\left(ET_{c,i}^{ba\|rts,*} + \sigma\right) \tag{1.14}$$

where $ET_{s,i}^{ba\|rts,*}$ and $ET_{c,i}^{ba\|rts,*}$ are the average durations of an observed successful transmission and a collision for node i when it is performing a backoff instance, respectively. To compute $ET_{c,i}^{ba\|rts,*}$, consider that the probability that more than two stations collide can be neglected; then:

$$\begin{cases} ET_{c,i}^{ba,*} \approx \dfrac{\sum_{j\neq i}\sum_{k>j,k\neq i}\max(T_{s,j},T_{s,k})\left(\tau_j\tau_k\prod_{u\neq\{j,k,i\}}(1-\tau_u)\right)}{\sum_{j\neq i}\sum_{k>j,k\neq i}\left(\tau_j\tau_k\prod_{u\neq\{j,k,i\}}(1-\tau_u)\right)} \\ ET_{c,i}^{rts,*} = T_c^{rts} \end{cases} \tag{1.15}$$

and

$$ET_{s,i}^* \approx \frac{\sum_{j\neq i}T_{s,j}\left(\tau_j\prod_{u\neq\{i,j\}}(1-\tau_u)\right)}{\sum_{j\neq i}\left(\tau_j\prod_{u\neq\{i,j\}}(1-\tau_u)\right)} \tag{1.16}$$

The probabilities $p_{e,i}$, $p_{s,i}$, and $p_{c,i}$ are related to the channel status in a given slot when a node is in backoff: $p_{e,i}$ is the probability that a slot is observed empty, $p_{s,i}$ is the probability that in a slot a successful transmission occurs, and $p_{c,i}$ is the probability that a collision occurs. Note that at the end of a successful transmission or a collision, one adds the duration of an empty slot, because the backoff counter is only decreased after the channel is sensed empty for the full duration of a slot. These channel probabilities can be computed as:

$$p_{e,i} = \prod_{j\neq i}(1-\tau_j) \qquad p_{s,i} = \sum_{z\neq i}\tau_z \prod_{j\neq z\neq i}(1-\tau_j) \qquad p_{c,i} = 1 - p_{e,i} - p_{s,i}$$

$$\tag{1.17}$$

Table 1.3   Model Validation: Traffic Profiles

Traffic Flow	Bandwidth	Frame Length	Retry Limit
Elastic (*E*1)	*Max. available*	1500 bytes	7
Streaming type 1 (*S*1)	100 kbps	400 bytes	7
Streaming type 2 (*S*2)	250 kbps	700 bytes	7

Due to the dependence of previous expressions on the queue utilization of each node, ρ_i, and the fact that Equations 1.8 and 1.9 form a set of nonlinear equations, we have to use iterative numerical techniques to solve the model.

1.4.3   Model Validation

To validate the model and analyze the performance of the IEEE 802.11 MAC protocol, consider a single-hop scenario with three different types of flows whose characteristics are summarized in Table 1.3. The network comprises $n+1$ nodes, including the AP, and each node uses the BA access scheme and carries a single traffic flow. (We refer to streaming type 1 (streaming type 2) flows with $S1$ ($S2$) and we use $E1$ to refer to elastic flows.)

Analytical results are compared against simulations performed using the ns2 package [26]. However, we have also built a detailed simulator of the IEEE 802.11 MAC protocol using the COST (Component Oriented Simulation Toolkit) simulation package [27] and verified that it provides equivalent results with respect to ns2, but allowing a higher flexibility to monitor the dynamics of the MAC parameters.

1.4.3.1   Homogeneous Traffic Flows

A first validation is done considering that all nodes in the network have the same traffic profile ($S1$, $S2$, or $E1$). Figure 1.3 shows predicted and simulated aggregate throughput against the number of flows for the three traffic classes specified in Table 1.3. The analytical and simulation models bring very close results, thus showing the accuracy of the model.

As the elastic results are obtained from saturated nodes, equivalent results are obtained in [5,6]. For the unsaturated traffic flows, Tables 1.4 and 1.5 report the values of other parameters, such as the conditional collision probability p_i, the queue utilization ρ_i, the average queueing delay ED_i, and packet losses $P_{L,i}$. The model captures the nonlinear dynamics of these parameters, especially the complex transition from unsaturated to saturated conditions. Note that under saturation conditions, as one would expect, parameters such as the conditional collision probability are equal and independent of the traffic load. Differences between the model and

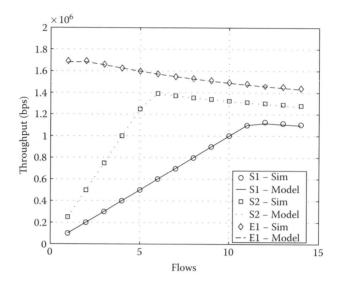

Figure 1.3 Homogeneous case: aggregate throughput for *S1*, *S2*, and *E1* traffic profiles.

the simulations (the model is pessimistic) are primarily motivated by the assumption of an exponential packet length distribution in the model that is constant in simulation.

1.4.3.2 Heterogeneous Traffic Flows

Having evaluated the model in the homogeneous scenario, one can now define a heterogeneous scenario. We chose a configuration in which two types of streaming flows (*S1* and *S2*) compete for the channel in the presence of elastic flows *E*1.

Table 1.4 *S1* — Homogeneous Case: Model Validation for Several Performance Parameters

		Model				*Simulation*		
$n_{s,1}$	p_i	ρ_i	ED_i	$P_{L,i}$	p_i	ρ_i	ED_i	$P_{L,i}$
1	0.0000	0.0813	0.0028	0.0000	0.0000	0.0817	0.0027	0.0000
2	0.0052	0.0877	0.0030	0.0000	0.0009	0.0862	0.0028	0.0000
4	0.0185	0.1044	0.0037	0.0000	0.0054	0.0960	0.0032	0.0000
6	0.0372	0.1295	0.0047	0.0000	0.0135	0.1122	0.0038	0.0000
8	0.0663	0.1731	0.0066	0.0000	0.0305	0.1388	0.0050	0.0000
10	0.1227	0.2760	0.0122	0.0000	0.0706	0.2075	0.0086	0.0000
12	0.3188	0.9979	1.2965	0.0697	0.3021	0.9400	0.7747	0.0587

Table 1.5   *S2* — Homogeneous Case: Model Validation for Several Performance Parameters

	Model				Simulation			
$n_{s,2}$	p_i	ρ_i	ED_i	$P_{L,i}$	p_i	ρ_i	ED_i	$P_{L,i}$
1	0.0000	0.1697	0.0045	0.0000	0.0000	0.1703	0.0042	0.0000
2	0.0120	0.2019	0.0056	0.0000	0.0038	0.1886	0.0048	0.0000
4	0.0560	0.3337	0.0112	0.0000	0.0266	0.2754	0.0079	0.0000
6	0.2066	0.9981	0.9176	0.0718	0.2042	0.9901	0.7777	0.0680
8	0.2534	1.0000	1.5862	0.3234	0.2511	0.9998	1.5498	0.3217
10	0.2897	1.0000	2.0729	0.4718	0.2843	0.9999	2.0441	0.4686
12	0.3191	1.0000	2.5593	0.5690	0.3097	0.9999	2.5247	0.5652

Two basic scenarios are considered:

1. Scenario 1: a variable number of *S1* flows ($n_{s,1}$) and a fixed number of nodes with *S2* flows ($n_{s,2} = 2$).
2. Scenario 2: a variable number of *S1* flows ($n_{s,1}$), a fixed number of nodes with *S2* flows ($n_{s,2} = 2$), and a fixed number of nodes with *E1* flows ($n_{e,1} = 2$).

Figure 1.4 and Figure 1.5 report the throughput for the three types of traffic flows in both scenarios. It is important to note that the model

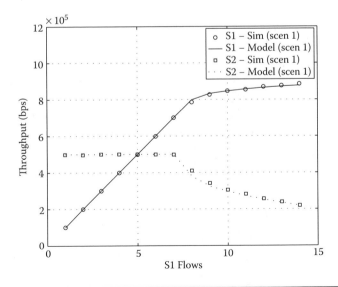

Figure 1.4   Heterogeneous case: aggregate throughput for *S1* and *S2*, nodes in scenario 1.

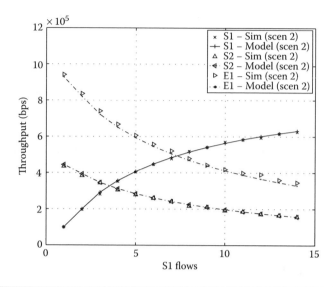

Figure 1.5 Heterogeneous case: aggregate throughput for *S1*, *S2*, and *E1* nodes in scenario 2.

can capture the point where both $S1$ and $S2$ flows fail to achieve their bandwidth requirements. Table 1.6 compares the queue utilization of a node with simulation results (scenario 1). Note that the model provides pessimistic values, but matches the dynamics of the queue utilization.

Table 1.7 shows the conditional collision probability, the expected number of slots of the backoff instance before a transmission attempt, and the channel probabilities observed by an $S1$ flow.

The first column of Table 1.8 includes the queue occupation for $S1$ nodes in the homogeneous scenario. Note how the introduction of $n_{s,2} = 2$ $S2$ flows causes an increment of the queue utilization for $S1$ flows. Therefore, a clear interaction from $S2$ flows exists and is added to the interaction between $S1$ flows. The total interaction, which is nonlinear with the

Table 1.6 *S1, S2* — Queue Utilization, Scenario 1

$n_{s,1}$	Model		Simulation	
	ρ_i (S1)	ρ_i (S2)	ρ_i (S1)	ρ_i (S2)
1	0.1308	0.2190	0.1130	0.2005
2	0.1471	0.2406	0.1210	0.2118
4	0.1983	0.3082	0.1559	0.2563
6	0.3255	0.4732	0.2441	0.3634
8	0.9304	1.0000	0.8743	0.9954
10	0.9999	1.0000	0.9892	0.9996

Table 1.7 Conditional Collision Probability, Expected Number of Slots of Each Backoff Instance, and Channel Probabilities (*S1*), Scenario 2

	Model					Simulation				
$n_{s,1}$	p_i	EB_i	$p_{e,i}$	$p_{s,i}$	$p_{c,i}$	p_i	EB_i	$p_{e,i}$	$p_{s,i}$	$p_{c,i}$
1	0.1785	19.91	0.8214	0.1656	0.0129	0.1789	19.84	0.8149	0.1643	0.0130
2	0.2039	20.95	0.7960	0.1858	0.0181	0.2037	20.96	0.7892	0.1839	0.0179
4	0.2525	23.39	0.7474	0.2221	0.0303	0.2477	23.91	0.7439	0.2163	0.0292
6	0.2890	25.75	0.7109	0.2472	0.0418	0.2812	26.55	0.7085	0.2409	0.0398
8	0.3185	28.07	0.6814	0.2659	0.0528	0.3088	28.91	0.6833	0.2570	0.0491
10	0.3431	30.33	0.6568	0.2805	0.0625	0.3305	31.37	0.6609	0.2710	0.0576

number of nodes, makes the queue utilization of *S1* nodes saturate more rapidly. At the same time, for a fixed number of *S2* nodes ($n_{s,2} = 2$), one observes how their queue utilization is also correlated with the queue utilization of *S1* flows.

For admission control purposes, it is worth noting that in scenario 1, with $n_{s,1} = 7$ flows, if another *S1* flow is accepted, the new accepted flow will perform correctly while the *S2* flows will perform poorly. Therefore, if the *S2* service degradation is not acceptable, this new *S1* should be rejected. Notice also that one cannot evaluate independently the two types of flows because the maximum number of flows for each type must be related to the presence of the other type of flows.

1.4.4 Model Applications

One of the most interesting features of the model is its flexibility to reproduce several situations of interest:

■ *Complex scenarios.* Because each node can be configured independently, it is easy to model nodes with different functions in the

Table 1.8 *S1, S2* — Mutual Interactions among Flows, Scenario 1

	Homogeneous		Heterogeneous			
$n_{s,1}$	ρ_i (S1)	p_i (S1)	ρ_i (S1)	ρ_i (S2)	p_i (S1)	p_i (S2)
1	0.0813	0.0000	0.1308	0.2190	0.0258	0.0206
2	0.0877	0.0052	0.1471	0.2406	0.0364	0.0309
4	0.1044	0.0185	0.1983	0.3082	0.0667	0.0607
6	0.1295	0.0372	0.3255	0.4732	0.1270	0.1198
8	0.1731	0.0663	0.9304	1.0000	0.2817	0.2795
10	0.2760	0.1227	0.9999	1.0000	0.3191	0.3191

same network (AP, MNs, etc.). This flexibility was difficult to achieve using saturated models due to their reduced parametrization.

■ *Multirate capabilities.* The extension of the model to multirate networks is rather straightforward, as one can assign to each flow/node a different value of R_{data}.

■ *Admission control and resource scheduling.* Because the model catches the relationships between flows, it can be used to evaluate different resource strategies, such as those based on setting the CW value, and test the overall effect of this setting on network performance.

1.5 User-Level Performance in Hot-Spot WLANs

The model presented in Section 1.4 is applied to a real scenario: an infrastructured WLAN network (BSS). First, using the ns2 package [26] as a simulation tool, one can investigate the performance of a basic WLAN cell with two types of traffic: (1) TCP-like traffic (i.e., persistent connections using the RENO version of the TCP protocol), and (2) VoIP CBR sources. Both simulation and analytical results are derived considering the use of the RTS/CTS mechanism. Performances are evaluated in terms of MAC layer throughput, which includes the upper layers overhead. To clarify the notation, we refer to every parameter related to the AP with the subscript d (downlink), and with the subscript u (uplink), we refer to the parameters of the MNs. The number of elastic flows in the downlink (uplink) is denoted by $n_{e,d}$ ($n_{e,u}$) and the number of VoIP calls by n_s.

Currently, some articles address the issue of analytically modeling the TCP throughput performance in WLAN networks. Basically, for the downlink direction, Bruno et al. [12], Miorandi et al. [13], and Lebeugle and Proutiere [14] present several models to compute WLAN throughput but with similar theoretical basis. In the uplink direction, a model is presented by Leith and Clifford [28] with also similar applicability. For what concerns TCP flows simultaneously in both directions, Pilosof et al. [29] explain the major observed phenomena but, to the best of our knowledge, the only work that treats it analytically is presented by Bruno et al. [15] under the assumption that the TCP advertisement window is equal to one.

1.5.1 A General View of TCP Performance

WLAN cells (or hot-spots) based on the IEEE 802.11 technology are deployed massively in business or public areas. From [30], more than 90 percent of the total traffic is TCP based. Nowadays, most traffic is due to HTTP transactions; however, it is worth mentioning that peer-to-peer (P2P) traffic reaches significant levels, higher than that obtained by e-mail or File Transfer Protocol (FTP) services. Another interesting observation is the asymmetry of the traffic flows, where the 85 percent goes from the fixed network to mobile nodes (downlink) and the remaining 15 percent, which is a significant value, goes from mobile nodes to the fixed network (uplink).

One of the most important issues for hot-spot operators is to know the grade of service that a user receives. This grade of service, in the case of elastic traffic, can be measured in terms of delay to visualize the data object requested (a Web page, a photograph, etc.), the time spent to receive or send an e-mail with an attached file, to transfer some files to another computer using FTP, a P2P transfer, etc. This delay is directly related to the bandwidth used by each flow. Therefore, knowledge of system performance can be used to provide service guarantees to users with different traffic profiles. For example, in the presence of users with a minimum bandwidth requirement and a set of users with a pure best-effort policy, admission and rate control strategies should be used to meet the requirements.

To compute the performance of the wireless cell, TCP protocol behavior should be considered. However, a detailed model of the TCP is a difficult and complex task and is outside the scope of this chapter. We therefore suggest a simple analysis based on the assumption that a node with a TCP flow can be modeled as a saturated queue. This assumption, as we show, allows one to obtain accurate results in terms of steady-state performance.

A first simulation result consisting of only downlink $S_{e,d}^{tcp}$ and uplink $S_{e,u}^{tcp}$ TCP flows is shown in Table 1.9. The maximum TCP window size has been fixed to $W = 1$ and $W = 42$ (as is commonly used in the operative TCP versions [29]). Two basic system throughput tendencies can be underlined:

1. The increment in the number of downlink TCP flows does not reduce the aggregated throughput due to the fact that TCP reduces the

Table 1.9 $S_{e,d}^{tcp}$ **and** $S_{e,u}^{tcp}$ **— Aggregate Throughput for Persistent TCP Flows (Mbps),** $L_{tcp} = 1500$ **bytes (including the TCP header)**

Flows	$S_{e,d}^{tcp}$ (W = 1)	$S_{e,u}^{tcp}$ (W = 1)	$S_{e,d}^{tcp}$ (W = 42)	$S_{e,u}^{tcp}$ (W = 42)
1	0.896	0.896	1.285	1.286
2	1.189	1.265	1.285	1.372
4	1.282	1.275	1.284	1.455
6	1.267	1.273	1.284	1.454
8	1.277	1.267	1.280	1.453
10	1.265	1.270	1.269	1.455

channel contention [12] (the average number of backlogged nodes with feedback traffic is less than the number of TCP flows).

2. As the number of TCP flows in the uplink increases, the aggregate throughput increases because the TCP window of MNs reaches its maximum value (for $W > 1$) despite packet losses and the starvation of the downlink ACK flow [28, 29].

However, previously mentioned works also show that unfairness exists among TCP flows in both the uplink and downlink directions. For the sake of simplicity, we assume that all flows share fairly the aggregate throughput; then each flow receives $S_{e,z}^{tcp}/n_{e,z}$ bps, with $z = \{d, u\}$.

1.5.1.1 Downlink TCP Flows

In the downlink, TCP flows compete, through the AP, with their own feedback traffic sent by the MNs.

A single downlink TCP flow, with maximum window size equal to $W = 1$, has a throughput proportional to L_{tcp}/RTT, where RTT is the TCP *round-trip time*. Then, the throughput is computed from:

$$S_{e,d=1}^{tcp}(W=1) = \frac{L_{tcp}}{\frac{L_{tcp}}{100e6} + X_d(L_{tcp}) + \frac{L_{ack}}{100e6} + X_u(L_{ack}) + 2\delta} \tag{1.18}$$

where $X_d(L_{tcp})$ ($X_u(L_{ack})$) is the service time over the WLAN for a TCP (ACK) packet and δ is the signal propagation delay. Simulation ($S_{e,d=1}^{tcp}(W=1)$ = 0.896 Mbps) and analytical results ($S_{e,d=1}^{tcp}(W=1)$ = 0.879 Mbps) show, as one expects, a good match.

With $W = 1$ and a single TCP flow, there is no competition to access the channel between the AP and the MNs because the two nodes never simultaneously have a packet ready to be transmitted at the MAC queue. From Table 1.9, as the number of simultaneous TCP flows increases, despite keeping $W = 1$, the AP queue tends to always have a packet ready to be transmitted, which justifies the assumption to model the access point as a saturated queue [12–14], which is obviously confirmed for values of $W > 1$.

For each received packet, an MN sends the correspondent ACK (we have not considered the delayed ACK technique as in [13]). Therefore, the number of ACK packets sent by an MN will be $1/n_{e,d}$ per packet emitted by the AP.

To analyze this situation, several approximations can be used:

■ *All sources are saturated (Model A_d).* In this model, both the AP and the MNs are considered saturated with data and ACK packets, respectively. This approximation provides pessimistic results because

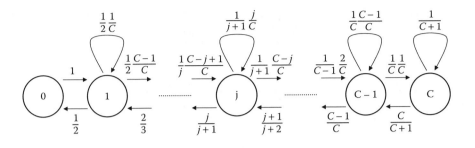

Figure 1.6 **Discrete Markov chain describing the evolution of the number of backlogged nodes.**

the level of contention suffered by data packets is very high, as the AP must compete with all MNs.

- *A time-scale decomposition (Model B_d)* [12,13]. Introduced in [12] and used also in [13], this model computes the distribution of backlogged nodes ($n^b_{e,d}$); that is, the probability that $n^b_{e,d}$ of the $n_{e,d}$ MNs are backlogged. The system throughput is computed averaging the throughput obtained with $n^b_{e,d}$ saturated nodes for $n^b_{e,d} = 0 \ldots n_{e,d}$. We suggest a novel variant of this model, where transitions between states are done after any successful transmission in the channel and not only after a successful transmission of the AP as in [12]. Figure 1.6 shows the DTMC (discrete-time Markov chain), which governs the number of backlogged nodes as a function of the number of downlink TCP flows C. Note that the DTMC changes its state after any successful transmission over the channel, independent of whether it was done by the AP or an MN. The probability to move from state $j - 1$ to state j depends on the probability that the AP transmits, $1/j$, and the probability that the packet was sent to a non-backlogged MN, $(C-j+1)/C$. The probability to remain in the same state j is the probability that the AP transmits a packet, $1/(j + 1)$, which is sent to a backlogged node with probability j/C. Finally, the probability to move from state j to state $j - 1$ is the probability that a backlogged MN transmits, $j/(j + 1)$. Note that a single ACK is stored in each MN queue.
- *Uncorrelated ACKs (Model C_d)*. Finally, by simply applying the MAC model described in Section 1.4, a good approximation can also be obtained. In this case, we configure the arrival rate of the MNs as $\lambda_{e,u} = \lambda_{e,d}/n_{e,d}$, where $\lambda_{e,d}$ is computed as the maximum arrival rate of the AP under the condition that $v_d = 1$ and assuming that the TCP packets are distributed uniformly among the destination nodes. It is worth noting that we can model the delayed ACKs technique by simply dividing the value of $\lambda_{e,u}$ by the delayed ACK factor γ, $\lambda_{e,u} = \lambda_{e,d}/(\gamma \cdot n_{e,d})$.

Table 1.10 $S_{e,d}^{tcp}$ — **Comparison of TCP Downlink Models (Mbps),** $L_{tcp} = 1500$ **bytes (including the TCP header)**

Flows	Simulation W = 42	Model A_d	Model B_d W >> 1	Model C_d
1	1.285	1.271	1.357	1.272
2	1.285	1.082	1.302	1.257
4	1.284	0.828	1.260	1.250
6	1.284	0.673	1.243	1.248
8	1.280	0.556	1.234	1.247
10	1.269	0.476	1.228	1.246

In Table 1.10 the downlink throughput obtains by simulation and comparison with the outcomes of the three models previously described. Note how model A_d clearly overestimates the negative effects of the feedback traffic and, thus, the throughput obtained is less than in simulation. Model B_d provides a very good approximation. Finally, model C_d also shows very accurate results, despite the assumption of Poisson arrivals for the ACK packets, which corresponds to assuming that there is no correlation with the reception of TCP data packets. These results allow us to validate our MAC model in this new scenario.

1.5.1.2 Uplink TCP Flows

TCP flows in the uplink compete among themselves and with the feedback traffic from the AP ACKs. Leith and Clifford [28] show the existing unfairness among competing uplink TCP flows. They also propose an analytical model for the uplink TCP throughput, addressing an ACK prioritization at the AP using the EDCF (Enhancement Distributed Coordination Function [4]) to reduce the inherent asymmetry of the WLAN. The asymmetry is due to the fact that MNs gain the $n_{e,u}\rho_u/(n_{e,u}\rho_u + \rho_d)$ transmission opportunities to transmit TCP data packets and the AP only gains $\rho_d/(n_{e,u}\rho_u + \rho_d)$ for the ACK packets. Their model assumes that MNs are saturated and the AP is not. To compute the transmission probability of the AP, they use the fact that all TCP data packets are answered by a single ACK packet. Then, by analogy with the model presented in this chapter, the transmission probability of the AP is the probability that the AP observes a successful transmission in the channel; that is, $\tau_d = p_{s,d}$, with $p_{s,d}$ computed as in Equation 1.17. This first model is called Model A_u.

Table 1.11 compares the results obtained by simulation for $W = 42$ with the results obtained by previous Model A_u and an additional model that assumes that the AP is also saturated (Model B_u). Both models provide good accuracy.

Table 1.11 $S_{e,u}^{tcp}$ — Comparison of TCP Uplink Performance Models (Mbps), $L_{tcp} = 1500$ bytes (including the TCP header)

Flows	Simulation W = 42	Model A_u W >> 1	Model B_u
1	1.286	1.271	1.271
2	1.372	1.293	1.418
4	1.455	1.298	1.500
6	1.454	1.297	1.526
8	1.453	1.295	1.538
10	1.455	1.293	1.544

1.5.1.3 Simultaneous Downlink and Uplink TCP Flows

Finally, when there are multiple TCP flows in both directions, $n_{e,d}$ in the downlink and $n_{e,u}$ in the uplink, the performance of downlink flows is severely affected, as one can see in Table 1.12. In this configuration, the AP queue is shared by both ACKs and data packets, while MNs only send either TCP data packets or ACKs (in the system, there are $n_{e,u}$ nodes sending TCP data packets and $n_{e,d}$ nodes sending ACKs). The results confirm those in [29] about the different behavior of the TCP window for the uplink and downlink flows. Pilosof et al. [29] argue that the TCP window for uplink senders reaches the maximum value, even with high ACK losses at the AP buffer, while downlink flows struggle with low window values (0–2 packets) caused by frequent timeouts due to data packet drops.

For $W = 1$ and a low number of upstream and downstream flows, the RTT is relatively independent of whether the TCP data packet is sent by an MN or a server in the fixed network, and, thus, the uplink and downlink performances are the same. Beyond $n_{e,d} = n_{e,u} = 6$, differences

Table 1.12 $S_{e,d}^{tcp}, S_{e,u}^{tcp}$ — Comparison of Simultaneous TCP Downlink/Uplink Flows (Mbps), $L_{tcp} = 1500$ bytes (including the TCP header)

Flows (in each direction)	Downlink W = 1	Uplink	Downlink W = 42	Uplink	Model A_b Downlink W >> 1	Uplink
1	0.582	0.660	0.200	1.084	0.234	0.907
2	0.631	0.625	0.003	1.364	0.149	1.158
4	0.633	0.630	0.000	1.424	0.086	1.340
6	0.629	0.630	0.000	1.425	0.060	1.412
8	0.626	0.634	0.000	1.452	0.046	1.449
10	0.606	0.651	0.000	1.423	0.037	1.471

between the streams become perceptible, and the tendency of the downlink throughput $S_{e,d}^{tcp}$ to decrease becomes clearly visible while the $S_{e,u}^{tcp}$ throughput continues growing. However, when $W > 1$, the results show that the uplink TCP flows achieve much higher throughput than the downlink flows.

Modeling this situation is very complex due to the interaction in the AP queue of the two types of packets: ACKs and data packets. We suggest a simple approximation (Model A_b) that captures the main tendencies observed in the simulations:

- The downlink queue is always saturated. The average packet length transmitted by the AP is computed from $EL_d = \phi_d L_{tcp} + \phi_u L_{ack}$, where ϕ_d and ϕ_u are the probabilities that a packet sent by the AP is a data or an ACK packet, respectively.
- These probabilities are computed from: $\phi_d = 1 - \phi_u$ and $\phi_u = n_{e,u}/(n_{e,u} + n_{e,d})$. Note that if $n_{e,d} = 0$, this model is equivalent to the model used for the uplink (Model B_u); and if $n_{e,u} = 0$, the model is equivalent to the model used for the downlink (Model C_d).
- MNs with uplink data packets are always saturated.
- Nodes with uplink ACKs have $\lambda_{ack} = \lambda_{e,d}/n_{e,d}$, where $\lambda_{e,d} = \phi_d/X_d(EL_d)$.

At the access point queue, ACKs controlled by the transmission opportunities of MNs compete with data packets of the downlink flows. As the number of uplink flows increases, because MNs have more transmission opportunities than the AP, the number of ACKs in the AP increases. Thus, the downlink flows suffer from both contending for buffer space in the AP and for contending on the access to the channel. Downlink TCP flows tend to starve, as can be noticed from results in Table 1.12.

1.5.2 Voice-over-IP (VoIP)

Voice communication using WLAN technology as the access network could be a promising alternative to traditional cellular networks (2G, 3G). Currently, roaming problems between WLAN coverage areas must be solved to provide continuous service to the user. However, novel proposals to interconnect and manage WLAN cells using common fixed infrastructure operators remain a promising reality [31]. Moreover, three major technological issues of the IEEE 802.11 MAC protocol itself must be solved or improved to achieve efficient use of transmission resources:

1. High protocol overheads
2. Unfairness between uplink and downlink streams
3. Fast VoIP degradation in the presence of TCP flows

Table 1.13   Typical Values of Most Used Codecs in VoIP

Codec	G.711	G.723.1	G.726–32	G.729
Bit Rate (kbps)	64	5.3/6.3	32	8
Framing Interval (ms)	20	30	20	2x10
Payload (bytes)	160	20/24	80	10
Packets/s	50	33	50	50

A criterion to determine the maximum number of VoIP calls that can be transported by a network (also called the VoIP capacity), given the desired voice quality in terms of bandwidth, delay, losses, can be found in [32]. For good quality, the average delay must be less than 150 ms, with losses less than 3 percent. Medium quality is achieved with delays between 150 and 400 ms and packet losses less than 7 percent. Finally, poor voice quality corresponds to delays greater than 400 ms and losses greater than 7 percent. Considering that the WLAN is only one hop of the whole path between the two endpoints, for a conservative good design, the quality target should be set to at least one third of the maximum recommended values.

Table 1.13 summarizes the basic characteristics of the most frequently used voice codecs for VoIP. The average throughput is plotted in Figure 1.7 for the *G.*711 and *G.*729 voice codecs. The AP is the bottleneck of the system that limits the VoIP capacity.

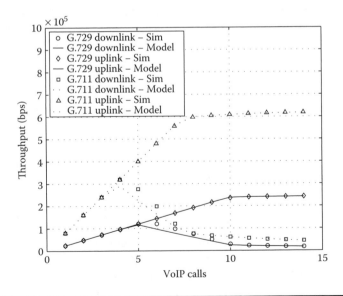

Figure 1.7   VoIP: throughput (AP and mobile nodes) for two voice codecs (*G.*711 and *G.*729).

Table 1.14 VoIP Efficiency over WLAN

Codec	G.711	G.723.1	G.726–32	G.729
Max. number of calls: C_{voip} (no contention)	4	9/9	5	6
Efficiency (no contention: η, $\eta = C_{voip} \cdot B_{voice}/R_{DATA}$)	12.8%	2.37%/2.85%	8%	2.4%
Max. number of calls (contention)	4	7/7	4	5

1.5.2.1 Protocol Overheads

Taking into account the parameters defined by the IEEE 802.11b standard [1], as summarized in Table 1.1, one can compute the maximum number of voice calls without contention; that is, the channel is ideally shared among voice calls, and with contention; see results in Table 1.14. From the results, one can conclude that the contention to access the channel reduces the maximum number of calls, but it is not the primary limiting factor. It is clear that the main problem is the large overhead introduced by the higher-layer protocols. A technique to solve this situation is header compression, such as ROHC (RObust Header Compression, RFC 3243).

1.5.2.2 Unfairness

Because the AP carries the same data as the whole set of MNs, it must attempt to transmit n times more than each MN. Therefore, it is desirable that the transmission attempts of the AP are n times greater than the transmission attempts of a single MN (or equivalently, $\tau_d = n\tau_u$) to achieve fair access to the channel (each node access to the channel is proportional to the traffic volume it must send).

As a measure of the system fairness, we compare:

$$w_d = \frac{\tau_d}{n_u\tau_u + \tau_d} \qquad w_u = \frac{n_u\tau_u}{n_u\tau_u + \tau_d} \qquad (1.19)$$

Considering the G.729 voice codec, the first two columns of Table 1.15 show w_d and w_u versus the number of mobile nodes. As the number of voice calls (or mobile nodes) increases, the AP unfairness grows, resulting in a fast saturation of the AP queue that limits the system capacity. To solve this problem, a simple solution consists of updating the CW_{min} value of MNs each time a new call arrives at the system. By computing the value of CW^*_{min} for each MN, assuming $m = 0$, one obtains

$$CW^*_{min} = \frac{n_s\rho_u(EB_d + 1)}{\rho_d} - 1 \qquad (1.20)$$

Table 1.15 Bandwidth Share between Uplink and Downlink VoIP Flows

Voice Calls (G.729)	AP (w_d)	Mobile Nodes (w_u)	AP (w_d)	Mobile Nodes (w_u)
	CW_{min}		CW_{min}^*	
1	0.500	0.500	0.500	0.500
2	0.482	0.517	0.501	0.498
3	0.466	0.533	0.504	0.495
4	0.450	0.549	0.506	0.493
5	0.428	0.571	0.510	0.489
6	0.352	0.647	0.514	0.485
7	0.217	0.782	0.515	0.484
8	0.210	0.789	0.516	0.483

The fairness obtained by applying this solution is shown in the two rightmost columns of Table 1.15. Note that the AP gets equal or more transmission opportunities than the uplink MNs.

This solution can also be combined with the one presented in [17], where the $TXOP$ mechanism of EDCF [4] is used to provide fairness. However, in both cases, a limited gain is obtained in terms of capacity increment. Finally, another solution, similar to the use of $TXOP$, is presented in [18]. In this case, several voice packets are encapsulated in only one multicast packet, which is sent to all MNs, where each one gets its own data.

1.5.2.3 Interaction with TCP Flows

Because the AP queue is shared by all downlink streams, the VoIP packets must compete for the buffer space with all the other flows, which can be streaming User Datagram Protocol (UDP) or elastic (TCP) flows (both data and ACKs destined to an MN). In the previous section, we concluded that TCP traffic tends to saturate the MAC queue and cause high losses if it is shared with VoIP packets. However, uplink flows reduce the transmission opportunities gained by the AP, especially if they are uplink TCP flows that also tend to saturate the MNs queue.

Table 1.16 clearly shows the negative influence of TCP traffic. Note the fast degradation of the VoIP throughput with TCP downlink flows and the inoperability of any VoIP call with just a single TCP uplink flow. It is also interesting to observe that, when the AP queue is saturated with VoIP traffic, the interaction with TCP traffic is reduced due to the starvation of TCP flows.

Therefore, the presence of TCP traffic in both the downlink (buffer losses) and the uplink (AP starvation) leads to low performance of VoIP calls. These problems must be solved in order to deploy a successful VoIP service over 2 WLAN.

In the downlink, a simple classification/prioritization scheme can be used (e.g., the dual queue proposed in [33]) where the TCP and UDP

Table 1.16 Downlink Throughput for VoIP Calls in Presence of TCP Flows (Mbps)

VoIP calls		TCP Downlink			TCP Uplink		
(G.729)	No TCP Flows	1	5	10	1	5	10
1	0.024	0.023	0.023	0.022	0.018	0.008	0.006
2	0.048	0.047	0.045	0.043	0.026	0.014	0.014
3	0.072	0.071	0.066	0.062	0.029	0.023	0.021
4	0.096	0.094	0.088	0.081	0.030	0.029	0.031
5	0.120	0.114	0.106	0.102	0.029	0.030	0.028
6	0.127	0.126	0.126	0.118	0.127	0.027	0.016

packets occupy separated buffers. However, the main problem is with TCP uplink flows because each node acts independently of the others. The only possible solution is to set different MAC parameters (such as CW_{min}, R, m, $TXOP$) for each mobile node to reduce the interaction of these TCP flows on VoIP calls.

1.6 Call Admission Control

With the goal to solve the performance problems previously exposed about the interaction between TCP and UDP traffic, we propose an admission control scheme that is capable of differentiating between *streaming* and *elastic* traffic and, at the same time, between *uplink* and *downlink* flows, providing an acceptable grade of service for streaming flows (VoIP).

1.6.1 CAC Architecture

The CAC entity is located at the AP. When an application wants to use the cell resources, it sends a request packet (e.g., it could be similar to the Add Traffic Specification, ADDTS, packet [4]) to the AP with the traffic profile required by the flow. Using the information provided by the application, the CAC decides if the new state of the network is feasible. If so, it sends a positive response to the request. Otherwise, it sends a negative response and the new flow is rejected, preserving the grade of service of the active flows already in the system.

To differentiate between TCP and UDP downstream flows, the AP uses a dual queue [33], wherein UDP packets are isolated from TCP packets and with a service prioritization for the UDP queue. Moreover, the upstream flows are differentiated by setting different MAC layer parameters (CW_{min}) for each flow.

1.6.1.1 A Dual Queue Scheme at the Access Point

Similar to the case in [33], we propose a dual queue strategy to differentiate between downstream TCP and UDP packets. We refer with $\rho_{s,d}$ to the UDP queue utilization and with $\rho_{e,d}$ to the TCP queue utilization, respectively. TCP packets are served only when the UDP queue is empty; then the probability of transmitting a downstream TCP packet is $1 - \rho_{s,d}$. Note that, because the upstream feedback traffic is proportional to the downstream TCP traffic, one can assume minimal impact of uplink TCP ACKs over UDP packets.

Without considering the presence of TCP uplink flows at the moment, the downlink TCP throughput can be measured as a function of the UDP queue utilization as $ES_{e,d} = (1 - \rho_{s,d})S_{e,d}^{tcp}$, using the model presented in the previous section (Model C_d). Clearly, the elastic flows can suffer starvation when $\rho_{s,d} \approx 1$. However, this can be solved by setting a maximum $\rho_{s,d}^{tb}$ value (e.g., $\rho_{s,d}^{tb} = 0.8$) which ensures that at least $1 - \rho_{s,d}^{tb}$ of the transmissions are for TCP downstream flows.

1.6.1.2 Uplink Differentiation via Different Values of CW_{min}

To differentiate VoIP flows with respect to TCP uplink flows (which cause a major performance degradation), we propose to use different CW_{min} values that prioritize streaming flows over elastic flows. Let Ψ_{CW} be the set of all possible CW_{min} values that can be used by uplink elastic flows, with $\Psi_{CW} = \{32, 64, 128, 256, 1024\}$. When the CAC receives a new request of a TCP uplink flow, it computes the suitable CW_{min} value for the new and all the already active uplink elastic flows, and broadcasts the new CW_{min} values. If there are no VoIP flows in the system, we assume that all nodes and the AP use the standard value of CW_{min}.

1.6.2 Proposed Algorithm

We propose an algorithm based on the estimates provided by the model presented in the previous section to predict the network behavior when a new flow request is received. Notice that the estimates can be either provided by another model or based on measurements. However, we believe that the proposed model is a good trade-off between simplicity and accuracy. Each request is configured to provide at least the following information: (1) flow type (F), which can be elastic or streaming; (2) requested bandwidth (B); and (3) average frame length (L). Using these pieces of information, the suggested algorithm operates as follows:

1. A new request is received by the admission control with parameters (F, B, L).
2. Using the current system information plus the new flow request, the new system state is estimated:
   a. If the new request is for a downlink elastic flow, it is accepted if the number of downlink elastic flows is less than the threshold $N_{e,d}^{tb}$. The elastic downlink throughput is auto-regulated by the dual-queue mechanism and TCP dynamics.

    b. If the new request is for an uplink elastic flow, it is accepted if the number of uplink elastic flows is less than the threshold $N_{e,u}^{tb}$ and if the new state is feasible. If it is not feasible, it tests from the set of CW_{min} values; if using another CW_{min} value for the elastic flows, the new state is possible.

    c. If the new request is for a downlink streaming flow, the CAC evaluates the queue utilization of the downlink UDP queue. If $\rho_{s,d}^* < \rho_{s,d}^{tb}$, the new flow is accepted. The $\rho_{s,d}^*$ parameter is estimated considering also the presence of the new flow. After accepting the new flow, the current $\rho_{s,d}$ is also updated with the information of the new flow.

    d. If the new request is for an uplink streaming flow, the CAC evaluates the system state. If the new state is feasible, it accepts the flow; if not, it tests if another combination of CW_{min} for elastic flows can make the new state feasible.

    e. If the new request is for a bidirectional streaming flow, the system evaluates if both the uplink and downlink flows can be accepted using the previous explanations.

3. If no state is feasible, reject the new flow.

1.6.3   Performance Evaluation

1.6.3.1   Model of a Cell

Four types of flows are considered: downlink/uplink streaming flows and downlink/uplink elastic flows. Under the assumption of exponential distributions of flow arrivals and departures, the system can be described by a continuous-time Markov chain (CTMC). If one also assumes that all VoIP calls comprise one uplink and one downlink flow and uses the same codec, then the state of the CTMC is given by the vector $(n_{e,d}, n_{e,u}, n_s)$, where $n_{e,d}$, $n_{e,u}$, and n_s denote the number of elastic flows (downlink and uplink) and VoIP calls that are active in the cell. To solve this CTMC, we suggest breaking the three-dimensional CTMC in two bidimensional CTMCs. First, CTMC ($CTMC_A$) comprises the situation where the VoIP calls compete with uplink TCP flows and, second, CTMC ($CTMC_B$) comprises the situation where downlink TCP flows compete with uplink TCP flows. The partial results of both CTMC can be averaged using the approximation that with probability $\rho_{s,d}$ the system works in the situation described by $CTMC_A$ and with probability $1 - \rho_{s,d}$ the system behavior can be modeled by $CTMC_B$.

While the number of elastic flows can grow to infinity, the maximum number of streaming flows is limited by the bandwidth requirements of the voice calls to N_{voip}^{tb}. To solve these infinite bidimensional CTMC, we need to truncate the state space. Without loss of generality, we introduce a realistic minimum bandwidth $B_{e,min}$ required for an elastic flow, which gives a maximum number of $N_{e,u}^{tb}$ ($N_{e,d}^{tb}$) uplink (downlink) elastic flows. The CTMC state space is described by:

$$CTMC_A : S_A = \{(n_{e,u}, n_s)| \ S_{e,u}^{tcp}(n_{e,u}, n_s)/n_{e,u} \geq B_{e,min},$$
$$S_{s,d}^{voip}(n_{e,u}, n_s) > 0.97 \cdot n_s B_s\}$$
$$CTMC_B : S_B = \{(n_{e,u}, n_{e,d})| \ S_{e,u}^{tcp}(n_{e,u}, n_{e,d})/n_{e,u} \geq B_{e,min}$$
$$S_{e,d}^{tcp}(n_{e,u}, n_{e,d})/n_{e,d} \geq B_{e,min}\}$$

(1.21)

For both voice and elastic flows, the user population is considered infinite with steady-state arrival rates $\lambda_{e,u}$ and $\lambda_{e,d}$ for elastic flows and λ_s for VoIP calls. The elastic flow duration is a function of the bandwidth observed by the elastic flows and the flow length (amount of data to transmit) FL_e, with the departure rate equal to $\mu_{e,x} = S_{e,x}^{tcp}(..)/(n_{e,x}FL_e)$, and μ_s for streaming flows, which have a fixed average duration.

1.6.3.2 Parameters

The goal of this subsection is to evaluate the effect of elastic uplink flows on the capacity of VoIP calls. The parameters considered to test our CAC algorithm are the following:

1. Voice calls:
   a. Voice codec: *G.729* with $L = 20$ bytes and rate equal to 8 kbps.
   b. We assume that the voice call requests follow a Poisson process with rate $\lambda_s = 0.0083$ calls/s (one call every two minutes), and that the duration of a call is exponentially distributed with mean $1/\mu_s = 240$ s (four minutes).
2. Elastic flows:
   a. The arrival process of elastic flows is also assumed to be Poisson with rates $\lambda_{e,d} = 1$ and $\lambda_{e,u}$ flows/s. Because we wish to evaluate the impact of uplink elastic flows over VoIP calls, the parameter $\lambda_{e,u}$ will vary.
   b. Two elastic flow lengths are considered: $FL_e = 1$ Mbits and $FL_e = 2$ Mbits, with exponential distribution (equal values for the downlink and the uplink).
   c. For the sake of simplicity, $B_{e,min}$ is computed to allow a maximum number $N_{e,u}^{tb} = N_{e,d}^{tb} = 10$ of active elastic flows in the system.

1.6.3.3 Considered metrics

To evaluate the proposed mechanism, we compute the following performance metrics:

1. Voice calls:
   a. Blocking probability: BP_s

2. Elastic flows:
    a. Uplink (downlink) blocking probability: $BP_{e,u}$ ($BP_{e,d}$)
    b. Uplink (downlink) elastic throughput: $ES_{e,u}$ ($ES_{e,d}$)

The analysis of the previous metrics is straightforward from the CTMC, which defines the system state. The averaged downlink throughput is measured as:

$$ES_{e,d} = (1 - \rho_{s,d}) E\left[S_{e,d}^{tcp}(n_{e,u}, n_{e,d})\right] \qquad (1.22)$$

where $E[S_{e,d}^{tcp}(n_{e,u}, n_{e,d})]$ is the averaged downlink throughput computed from $CTMC_B$. The averaged uplink elastic throughput can be approximated by:

$$ES_{e,u} = \rho_{s,d} E\left[S_{e,u}^{tcp}(n_{e,u}, n_s)\right] + (1 - \rho_{s,d}) E\left[S_{e,u}^{tcp}(n_{e,u}, n_{e,d})\right] \qquad (1.23)$$

where $E[S_{e,u}^{tcp}(n_{e,u}, n_s)]$ and $E[S_{e,u}^{tcp}(n_{e,u}, n_{e,d})]$ are computed, respectively, from $CTMC_A$ and $CTMC_B$. The values $S_{e,d}^{tcp}(n_{e,u}, n_{e,d})$, $S_{e,u}^{tcp}(n_{e,u}, n_s)$, and $S_{e,u}^{tcp}(n_{e,u}, n_{e,d})$ are computed using the models of the previous section. A similar approximation is applied to compute the blocking probability, assuming that all downlink TCP flows are blocked if some VoIP conversations are active.

1.6.4 Numerical Results

Here we present some numerical results to investigate the performance of the proposed CAC scheme. Results are shown as the comparison of the performance achieved by the WLAN using a nonadaptive CAC called *simple CAC*, according to which all flows always use a fixed value of CW_{min} equal to 32 and using the proposed CAC scheme, called *adaptive CAC*.

Figure 1.8 shows the voice call blocking probability BP_s for two different elastic flow lengths, with the *simple CAC* and the *adaptive CAC*. For both CAC schemes, as we expect, the blocking probability increases as the traffic intensity of uplink elastic flows grows. However, under the *adaptive CAC*, the blocking probability exhibits substantially lower values, due to growth in the number of feasible states deriving from the reduction of the transmission rate of elastic flows (due to the increment of the value of their CW_{min} parameter), which decreases the possibility that the arrival of a new voice call is blocked. Observe that for values of $\lambda_{e,u}$ greater than $\lambda_{e,u} = 1.5$ ($FL_e = 2$ Mbits) ($\lambda_{e,u} = 3.25$ ($FL_e = 1$ Mbits)), and using the *simple CAC* scheme, all voice calls are blocked; while using the *adaptive CAC* scheme, this probability remains rather constant at about 0.2.

Previous results show the goodness of the *adaptive CAC* to reduce the blocking probability of voice calls. However, it has negative effects on

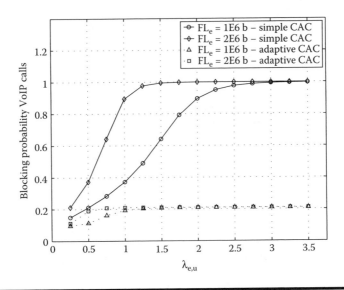

Figure 1.8 BP_s: blocking probability for VoIP calls.

the elastic flows, increasing the blocking probability of them and reducing their throughput. Figure 1.9 reports the blocking probability for uplink elastic flows. Two regions can be observed: (1) the blocking probability of elastic flows is mainly caused by voice calls ($\lambda_{e,u} \leq 0.5$, $FL_e = 2$ Mbits) or ($\lambda_{e,u} \leq 1.5$, $FL_e = 1$ Mbit), and (2) the blocking probability of elastic

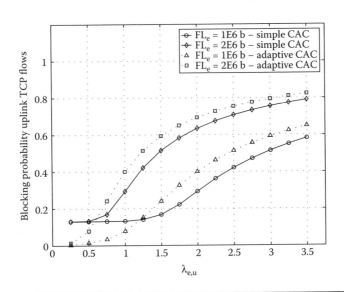

Figure 1.9 $BP_{e,u}$: blocking probability for uplink elastic flows.

flows is caused by their own traffic. In the first region, for low traffic intensity of elastic flows, the blocking probability using the *adaptive CAC* decreases as the number of uplink elastic flows that can coexist with the VoIP calls grows. However, in the second region, as the average number of active elastic flows is higher, the blocking probability of voice calls is near 1 (see Figure 1.8); thus, the presence of active voice calls is negligible, and the blocking probability of elastic flows is due mainly to its own traffic. Using the *adaptive CAC*, which preempts the elastic flows (reducing their transmission rate), the blocking probability of elastic flows increases as the output rates for uplink elastic flows decrease, which results in values higher than the *simple CAC*. As a conclusion, the *adaptive CAC* is significantly better for both VoIP calls and elastic flows than the *simple CAC* when the uplink elastic flow intensity is not enough to saturate the system. Otherwise, for high elastic flow intensity, the scheme maintains low the blocking probability of VoIP calls while the performance of elastic flows is not severely reduced.

The higher number of accepted VoIP calls using the *adaptive CAC* results in an increment of the VoIP throughput and then a higher utilization ($\rho_{s,d}$) of the downlink streaming queue at the AP, thereby decreasing the transmission opportunities of the TCP downlink traffic. This situation entails an increase in the blocking probability of downlink TCP flows (Figure 1.10).

Finally, observe the uplink/downlink elastic throughput using the *simple* and the *adaptive CAC* schemes (Figure 1.11 and Figure 1.12), which show coherent results with previous explanations.

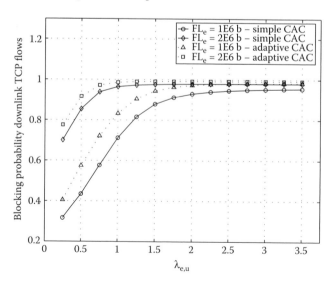

Figure 1.10 $BP_{e,d}$: **blocking probability for downlink elastic flows.**

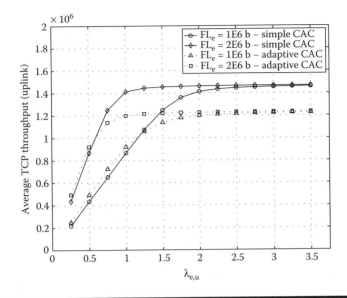

Figure 1.11 $ES_{e,u}$: average elastic throughput uplink.

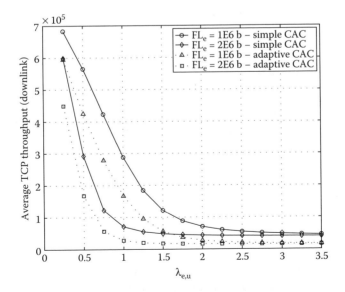

Figure 1.12 $ES_{e,d}$: average elastic throughput downlink.

From the presented numerical results, we can conclude that if a CAC is implemented for WLANs to provide QoS to multimedia traffic flows, it must modify the MAC parameters of MNs to optimize the overall WLAN performance.

1.7 Conclusions

In this chapter we discussed some issues related to the provision of integrated services through WiFi access networks. The popularity of WiFi access and the common use of a variety of different services is making this topic more and more crucial for operators in the field.

In particular, the performance of the IEEE 802.11 MAC protocol (DCF) was investigated by means of a novel user-centric model that was validated against simulation results. The model was used to analyze the case of heterogeneous traffic scenarios, in which elastic and streaming traffic, representing, respectively, TCP-based applications and VoIP service, share the common radio resources. The interaction between these two classes of traffic was highlighted.

The analysis suggested that, in order to provide good QoS to traffic flows whose requirements are so different from each other, CAC is needed. A novel scheme was then proposed based on both the use of the network state and a smart setting of the MAC layer parameters.

Acknowledgments

This work was partially supported by the Catalonian Government under the i2CAT (Advanced Internet in Catalonia) project, the Spanish Government under project TIC2003-09279-C02-01, and by the European Commission under NEWCOM Network of Excellence.

References

[1] IEEE Std 802.11. Wireless LAN Medium Access Control (MAC) and Physical Layer (PHY) Specifications. ANSI/IEEE Std 802.11, 1999 Edition (Revised 2003).

[2] IEEE Std 802.11a. Wireless LAN Medium Access Control (MAC) and Physical Layer (PHY) Specifications; Amendment: High-Speed Physical Layer in the 5 GHz Band. IEEE Std 802.11a, 1999.

[3] IEEE Std 802.11g. Wireless LAN Medium Access Control (MAC) and Physical Layer (PHY) Specifications; Amendment: Further Higher Data Rate Extension in the 2.4 GHz band. IEEE Std 802.11g, 2003.

[4] IEEE Std 802.11e. Wireless LAN Medium Access Control (MAC) and Physical Layer (PHY) Specifications; Amendment: Medium Access Control (MAC) Quality of Service Enhancements. IEEE Std 802.11e, 2005.

[5] Giuseppe Bianchi. Performance Analysis of the IEEE 802.11 Distributed Coordination Function. *IEEE Journal on Selected Areas in Communications*, 18(3), 535–547, March 2000.

[6] Y.C. Tay and K.C. Chua. A Capacity Analysis for the IEEE 802.11 MAC Protocol. *Wireless Networks*, 7, 159–171, March 2001.

[7] Marcelo M. Carvalho and J.J. Garcia-Luna-Aceves. Delay Analysis of IEEE 802.11 in Single Hop Networks. *11th IEEE International Conference on Network Protocols (ICNP'03)*, Atlanta, GA, November 2003.

[8] David Malone, Ken Dury, and Douglas J. Leith. Modeling the 802.11 Distributed Coordination Function with Heterogeneous Finite Load. *Workshop on Resource Allocation in Wireless Networks*, Trento, Italy, April 2005.

[9] Omesh Tickoo and Biplab Sikdar. Queuing Analysis and Delay Mitigation in IEEE 802.11 Random Access MAC based Wireless Networks. *IEEE INFOCOM 2004*, Hong Kong, China, March 2004.

[10] Kamesh Medepalli and Fouad. A. Tobagi. System Centric and User Centric Queuing Models for IEEE 802.11 based Wireless LANs. *IEEE Broadnets 2005*, Boston, MA, October 2005.

[11] Nidhi Hegde, Alexandre Proutiere, and James Roberts. Evaluating the Voice Capacity of 802.11 WLAN under Distributed Control. *IEEE LAN/MAN 2005*, Megalo Arsenali, Chania, Greece, September 2005.

[12] Raaele Bruno, Marco Conti, and Enrico Gregori. Analytical Modeling of TCP Clients in Wi-Fi Hot Spot Networks. *Networking 2004, LNCS 3042*, pp. 626–637, May 2004.

[13] Daniele Miorandi, Arzad A. Kherani, and Eitan Altman. A Queuing Model for HTTP Traffic over IEEE 802.11 WLANs. *ITC Specialist Seminar on Performance Evaluation of Wireless and Mobile Systems*, Antwerp, Belgium, September 2004.

[14] F. Lebeugle and A. Proutiere. User-Level Performance in WLAN Hotspots. *ITC Specialist Seminar on Performance Challenges for Ecient Next Generation Networks*, China, 2005.

[15] R. Bruno, M. Conti, and E. Gregori. Stochastic Models of TCP Flows over 802.11 WLANs. Technical Report IIT TR-11/2005, CNR-IIT, June 2005.

[16] David P. Hole and Fouad A. Tobagi. Capacity of an IEEE 802.11b Wireless LAN Supporting VoIP. *IEEE International Conference on Communications (ICC'04)*, Paris, France, June 2004.

[17] Peter Clifford, Ken Duffy, Douglas Leith, and David Malone. On Improving Voice Capacity in 802.11 Infrastructure Networks. *IEEE Wireless Networks, Communications, and Mobile Computing (Wirelesscom'05)*, Maui, Hawaii, June 2005.

[18] Wei Wang, Soung Chang Liew, and Victor O.K. Li. Solutions to Performance Problems in VoIP over a 802.11 Wireless LAN. *IEEE Transactions on Vehicular Technology*, 54(1), 366–385, January 2005.

[19] Albert Banchs, Xavier Perez, Markus Radimirsch, and Heinrich J. Stuttgen. Service Differentiation Extensions for Elastic and Real-Time Trac in 802.11

Wireless LAN. *IEEE Workshop on High Performance Switching and Routing (HPSR 2001)*, Dallas, TX, May 2001.

[20]   Albert Banchs, Xavier Perez-Costa, and Daji Qiao. Providing Throughput Guarantees in IEEE 802.11e Wireless LANs. *ITC Specialist on Providing QoS in Heterogeneous Environments Seminar*, Berlin, Germany, September 2003.

[21]   A. Ksentini, A. Gueroui, and M. Naimi. Adaptive Transmission Opportunity with Admission Control for IEEE 802.11e Networks. *ACM/IEEE MSWIM 2005*, Montreal, Quebec, Canada, October 2005.

[22]   Dennis Pong and Tim Moors. Call Admission Control for IEEE 802.11 Contention Access Mechanism. *IEEE Globecom 2003*, San Francisco, CA, December 2003.

[23]   Hongqiang Zhai, Xiang Chen, and Yuguang Fang. A Call Admission and Rate Control Scheme for Multimedia Support over IEEE 802.11 Wireless LANs. *Quality of Service in Heterogeneous Wired/Wireless Networks (QShine'04)*, Dallas, TX, October 2004.

[24]   Liqiang Zhang and Sherali Zeadally. HARMONICA: Enhanced QoS Support with Admission Control for IEEE 802.11 Contention-based Access. *10th IEEE Real-Time and Embedded Technology and Applications Symposium (RTAS'04)*, Toronto, Ontario, Canada, May 2004.

[25]   Haitao Wu, Yang Peng, Keping Long, Shiduan Cheng, and Jian Ma. Performance of Reliable Transport Protocol over IEEE 802.11 Wireless LAN: Analysis and Enhancement. *IEEE INFOCOM 2002*, New York, June 2002.

[26]   NS2. Network Simulator. http://www.isi.edu/nsnam/ns/, February 2005 (release 2.28).

[27]   Gilbert (Gang) Chen. Component Oriented Simulation Toolkit. http://www.cs.rpi.edu/scheng3/, 2004.

[28]   D.J. Leith and P. Clifford. Using the 802.11e EDCF to Achieve TCP Upload Fairness over WLAN Links. *Modeling and Optimization in Mobile, Ad Hoc, and Wireless Networks (WiOpt'05)*, Trentino, Italy, April 2005.

[29]   Saar Pilosof, Ramachandran Ramjee, Danny Raz, Yuval Shavitt, and Prasun Sinha. Understanding TCP Fairness over Wireless LAN. *IEEE INFOCOM 2003*, San Francisco, CA, March 2003.

[30]   Chen Na, Jeremy K. Chen, and Theodore S. Rappaport. Hotspot Traffic Statistics and Throughput Models for Several Applications. *IEEE Globecom*, Dallas, TX, November 2004.

[31]   J. Barcelo, A. Sfairoupoulou, J. Infante, M. Oliver, and C. Macian. Barcelona's Open Access Network Pilot. *IEEE/Create-Net Conference on Testbeds and Research Infrastructures for the Development of Networks and Communities (TrindentCom'06)*, Barcelona, Spain, July 2006.

[32]   C. Casetti and C.-F. Chiasserini. Improving Fairness and Throughput for Voice Traffic in 802.11e EDCA. *IEEE PIMRC 2004*, Barcelona, Spain, September 2004.

[33]   Jeonggyun Yu, Sunghyun Choi, and Jaehwan Lee. Enhancement of VoIP over IEEE 802.11 WLAN via Dual Queue Strategy. *IEEE International Conference on Communications (ICC'04)*, Paris, France, June 2004.

Chapter 2

Activity Scheduling in Bluetooth Sensor Networks

Jelena Mišić, G.R. Reddy, and Vojislav B. Mišić

Contents

2.1   Overview

Sensor networks are finding widespread use in diverse application areas such as environmental monitoring, health care, logistics, surveillance, and others. In most cases, the sensor network must satisfy the two simultaneous, yet often conflicting goals of maintaining a desired packet arrival rate at the network sink and maximizing the network lifetime. The problem is further compounded by the fact that most of those networks operate on battery power with as little human intervention as possible. This chapter considers a Bluetooth scatternet operating as the sensor network with a medium data rate sensing application. In this setup, we present an approach to activity management in Bluetooth sensor networks that utilizes cross-layer adaptive sleep management on a per-piconet basis. The effects of finite buffers in individual nodes are also considered. The proposed approach is shown to be computationally simple yet effective.

2.2   Introduction

A wireless sensor network consists of a number of wireless sensor nodes, spread across a given area, that are used for event detection and reporting [1]. In the sensor network, a node (sink) issues queries that request data from the sensing node; a group of such nodes that can provide the requested data (known as the source) sends it to the sink [3]. Because sensor networks are primarily used for event detection tasks, the rate at which data is propagated from source node(s) to the sink must be high enough to obtain the desired reliability R, which is commonly defined as the number of data packets required per second for reliable event detection at the sink [2]. At the same time, sensor networks frequently operate on battery power, which means that energy efficiency must be maintained.

Reliable event detection using minimal energy resources requires simultaneous achievement of several subgoals. The packet loss along the path from the source to the sink must be minimized. At the physical (PHY) layer, packets can be lost due to noise and interference, while at the medium access control (MAC) layer, losses may be incurred by collisions. Because sensors are continuously monitoring the environment and sending data, retransmission of lost packets is not necessary; it would use up the bandwidth to send stale data, and thus impair the performance of the sensor network in qualitative terms.

Packet delays must be minimized as well, including queueing delays experienced in various devices along the data path and also delays due to congestion in the network. (Queueing delays are incurred at the MAC layer, while congestion detection and control are performed at the transport layer.)

Finally, packet propagation should take place along the shortest paths, while avoiding congested nodes and paths; this is the responsibility of the network layer.

The sensor nodes are generally battery operated and have limited computational capabilities, which in turn means that the protocol stack must be as simple as possible. As a result, the simultaneous goals of minimizing packet losses and maximizing the efficiency (and, by extension, maximizing the lifetime of the network as well) necessitate that some of the aforementioned functions of the different layers are performed together. That is, cross-layer optimization of network protocol operation is needed; the feasibility of this optimization is determined by the communication technology used to implement the network.

This chapter describes a Bluetooth sensor network operating under the scheme that integrates congestion control with reliability and energy management. We investigate the performance of the proposed solution and the ways in which its operation can be optimized. We start by discussing the suitability of Bluetooth technology for use in sensor networks, and present the most important among the existing solutions for congestion control and energy management problems in sensor networks. Then we develop two algorithms that schedule the sleep of individual slaves: the first one maintains fixed event reliability at the sink and the second one keeps the satisfactory event reliability by avoiding congestion. The performance of the proposed algorithms is analyzed in detail. A note on our simulation setup and a brief summary conclude the chapter.

2.3   Bluetooth and Sensor Networks

Bluetooth was originally intended as a simple communication technology for cable replacement [6]. However, its use has been steadily growing in a diverse set of applications [2]. Bluetooth operates in the Industrial, Scientific and Medical band at 2.4 GHz using the Frequency Hopping Spread Spectrum (FHSS) technique, which makes it highly resilient to the noise and interference from other networks operating in the same band, such as IEEE 802.11b and IEEE 802.15.4 [8]. The raw data rate of 1 Mbps (or 3 Mbps, if the recent version 2.0 of the standard is used) and the default transmission range of 10 to 100 meters [6] make Bluetooth networks suitable for medium-rate wireless personal area networks (WPANs). These same qualities mean that Bluetooth is suitable for the construction of low-cost sensor networks, offering coverage of sensing areas with diameters of several tens to several hundred meters [2].

2.3.1   Piconet Operation

Bluetooth devices are organized into piconets, small networks with up to 8 active nodes and up to 255 inactive ones [6]. Bluetooth uses a TDMA/TDD polling protocol where all communications are performed under the control of the piconet master. Bluetooth uses a set of RF frequencies (79 or 23, in some countries) in the ISM band at about 2.4 GHz. The FHSS technique is utilized to combat interference. Each piconet hops through the available RF frequencies in a pseudo-random manner. The hopping sequence, which is determined from the Bluetooth device address of the piconet master, is known as the channel [6]. Each channel is divided into time slots of $T = 625 \ \mu s$, which are synchronized to the clock of the piconet master. In each time slot, a different frequency is used.

All communications in the piconet take place under the control of the piconet master. All slaves listen to downlink transmissions from the master. The slave can reply with an uplink transmission if and only if addressed explicitly by the master, and only immediately after being addressed by the master. Data is transmitted in packets, which take one, three, or five slots; link management packets also take one slot each. The RF frequency does not change during the transmission of the packet. However, once the packet is sent, the transmission in the next time slot uses the next frequency from the original hopping sequence (i.e., the two or four frequencies from the original sequence are simply skipped). By default, all master transmissions start in even-numbered slots, while all slave transmissions start in odd-numbered slots. A downlink packet and the subsequent uplink packet are commonly referred to as a frame. Therefore, the master and the addressed slave use the same communication channel, albeit not at the same time. This communication mechanism, known as Time Division Duplex (TDD), is schematically shown in Figure 2.1. This approach is collision-free and,

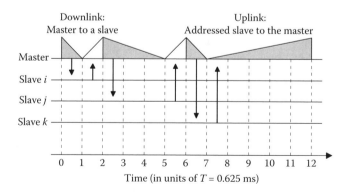

Figure 2.1   TDD master-slave communication in Bluetooth. Gray triangles denote data packets; white triangles denote empty (POLL and NULL) packets.

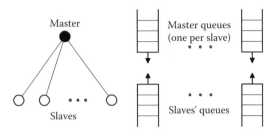

Figure 2.2   The Bluetooth piconet and its queueing model.

consequently, more energy efficient than the collision-based MACs used in IEEE 802.11 and IEEE 802.15.4 [11].

Because of the TDD communication mechanism, all communications in the piconet must be routed through the master. Each slave will maintain (operate) a queue where the packets to be sent out are stored. The master, on the other hand, operates several such queues, one for each active slave in the piconet. The piconet and the corresponding queueing model is shown in Figure 2.2. We note that these queues may not physically exist, for example, all downlink packets might be stored in a single queue; however, the queueing model provides a convenient modeling framework that facilitates the performance analysis of Bluetooth networks.

2.3.2   Intra-Piconet Polling

The master polls the slave by sending the data packet from the head of the corresponding downlink queue. The slave responds by sending the data packet from the head of its uplink queue. When there is no data packet to be sent, single-slot packets with zero payload are sent — POLL packets in the downlink and NULL packets in the uplink direction [6]. As the process of polling the slaves is actually embedded in the data transmission mechanism, we use the term "polling" for every downlink transmission from the master to a slave.

Because packets must wait at the slave or at the master before they can be delivered to their destinations, the delays they experience are mainly queueing delays. Therefore, the intra-piconet polling scheme is obviously the main determinant of performance of Bluetooth piconets, and one of the main determinants of performance of Bluetooth scatternets. As usual, the main performance indicator is the end-to-end packet delay, with lower delays being considered as better performance. There are, however, at least two other requirements to satisfy. First, the piconet master should try to maintain fairness among the slaves, so that all slaves in the piconet receive equal attention in some shorter or longer time frame. (Of course, their traffic load should be taken into account.) Second, Bluetooth devices are, by default, low-power devices, and the polling scheme should be sufficiently simple in terms of computational and memory requirements.

The polling schemes can roughly be classified according to the following criteria:

■ The number of frames exchanged during a single visit to the slave can differ; it can be set beforehand to a fixed value, or it can be dynamically adjusted on the basis of current and historical traffic information.

■ Different slaves can receive different portions of the bandwidth; again, the allocation can be done beforehand, or it can be dynamically adapted to varying traffic conditions. The latter approach is probably preferable in Bluetooth piconets, which are ad hoc networks formed by mobile users, and the traffic can exhibit considerable variability. In fact, due to users' mobility, even the topology of the piconet can change on short notice. However, the fairness of polling might be more difficult to maintain under dynamic bandwidth allocation.

■ Finally, the sequence in which slaves are visited can be set beforehand, or it can change from one piconet cycle to another, depending on the traffic information. In either case, slaves that had no traffic in the previous cycle(s) can be skipped for one or more cycles, but the polling scheme must ensure that the fairness is maintained.

The current Bluetooth specification does not specifically require or prescribe any specific polling scheme [6]. This may not seem to be too big a problem, because optimal polling schemes for a number of similar single-server, multiple-input queueing systems are well known [17,18]. However, the communication mechanisms used in Bluetooth are rather specific and the existing results cannot be applied. It should come as no surprise, then, that a number of polling schemes have been proposed and analyzed [7,13]. Many of the proposed schemes are simply variations of the well-known limited and exhaustive service scheduling [24], but several improved adaptive schemes have been described as well [13,15].

In our work, we have chosen the so-called *E-limited service* polling scheme in which the master stays with a slave for a fixed number M of frames ($M > 1$), or until there are no more packets to exchange, whichever comes first. Packets that arrive during the visit are allowed to enter the uplink queue at the slave and can be serviced — provided the limit of M frames is not exceeded [24]. This scheme has been found to offer better performance than either limited or exhaustive service, and the value of M can be chosen to achieve minimum delays for given traffic burstiness [20].

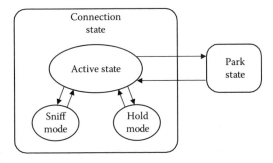

Figure 2.3   Connection states and modes.

2.3.3   Power-Saving Operation Modes

An active slave is assigned a three-bit active mode address AM_ADDR that is unique within the piconet [6]. However, the slave need not listen to master transmissions all the time; it may choose to switch to one of the so-called modes in which it can detach itself from the piconet for prolonged periods without having to surrender its piconet address; these modes are known as HOLD and SNIFF. The slave can also switch to a parked state, in which it releases its active mode address; this switch can be initiated by either the master or the slave itself. Broadcast as well as unicast messages can target active or parked slaves only, as appropriate. The connection states and modes are schematically shown in Figure 2.3.

An active slave can temporarily detach itself from the piconet by entering the so-called HOLD mode, the operation of which is shown in Figure 2.4. In this mode, the master will not poll the slave for a specified time interval, referred to as the *holdTO*, or hold timeout. The inactivity period due to the HOLD mode affects only ACL links established between the master and the slave; SCO and eSCO links, if any, remain operational even when the slave is in the HOLD mode. During the HOLD mode, the slave can engage in other activities such as scanning, paging, or joining another piconet. The slave can also enter a low-power mode to conserve energy.

The actual duration of the HOLD mode is negotiated between the master and the slave; the negotiation process can be initiated by either the master or the slave itself. The initiating party proposes the switch to the HOLD mode as well as the hold timeout; the responding party can accept it, or respond with a counterproposal of its own.

Another mode that can be entered from the active connection state is the SNIFF mode, the operation of which is shown in Figure 2.5. In this mode, the slave is absent from the piconet for a specified time, during which the master will not poll it. The slave periodically joins the piconet to listen to master transmissions. If no transmission is initiated, or even detected,

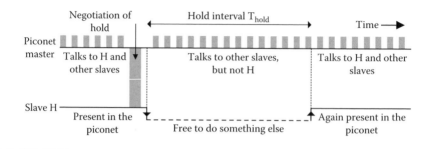

Figure 2.4 The operation of the HOLD mode.

during the predefined window, the slave again detaches itself from the piconet for another interval of absence. During this time interval, the slave can engage in other activities, similar to the HOLD mode described above. Again, the inactivity due to SNIFF mode affects only the ACL link or links that may be set between the master and the slave in question, but not the SCO or eSCO ones, if any.

As is the case with the HOLD mode, the SNIFF mode and its parameters are negotiated between the master and the slave. The negotiation process can be initiated by either party; the initiating Link Manager (LM) proposes the SNIFF mode and its parameters to the corresponding LM of the other participant. Once the switch and the parameters are accepted, the slave can start alternating between active and SNIFF mode. Unlike the HOLD mode, which is a one-off event, the SNIFF mode lasts until one of the participants explicitly requests its termination.

A connection can break down for different reasons, including power failure, user movement, or severe interference. To detect the loss of connection, the traffic on each link must be monitored on both the master and the slave side. This is accomplished through the so-called supervision

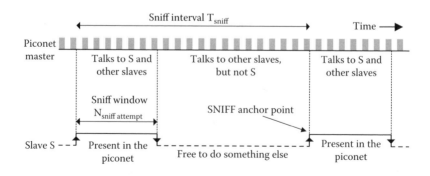

Figure 2.5 The operation of the SNIFF mode.

timer, $T_{supervision}$, which is reset to zero every time a valid packet is received on the associated physical link. If the timer reaches the *supervisionTO* timeout, the value of which is negotiated by the LM, the link is considered lost, and the associated active piconet member address can be reassigned to another device. The value of the supervision timeout should be longer than negotiated HOLD and SNIFF periods. The same link supervision timer is used for all logical transports carried over the same physical link.

2.3.4   *Bluetooth Scatternets*

A group of independent piconets interconnected through shared devices, or bridges, forms a scatternet; this work focuses on slave-slave bridges that act as slaves in each of the piconets they visit. As most Bluetooth devices — bridges included — have only one radio interface, the bridge must visit adjacent piconets in different time periods. Consequently, both the intra-piconet polling scheme and the inter-piconet (bridge) scheduling scheme are important factors that determine the performance of a Bluetooth scatternet [21].

Most of the schemes are based on the concept of rendezvous points: time instants at which the bridge should be present in the piconet to exchange data with its master [14]. These time instants can be fixed before the actual data transfer, maybe even for the entire lifetime of the scatternet, or they can be negotiated as necessary between the piconet master(s) and the bridge(s). The rationale for the existence of a predefined rendezvous point is to have both participants join the exchange simultaneously. If this is not the case, the participant that switches earlier would have to wait idle and thus waste time and, ultimately, bandwidth.

The schedule of rendezvous points can be fixed beforehand or adaptive. The former case may be suitable for sensor networks that have comparatively well-known traffic requirements; it certainly is unsuitable for sensor networks that feature activity management in which sensors are going to sleep for prolonged periods of time. This case seems easier to handle using the latter approach with adaptive scheduling of rendezvous points. But in either case, the main problem with rendezvous-based bridge scheduling remains: the overhead incurred by the construction and maintenance of the schedule of rendezvous points.

This overhead can be avoided if the bridge (or bridges) could operate without such a schedule. It turns out that such an approach, which will be referred to as *walk-in bridge scheduling*, is indeed feasible [22]. Under walk-in scheduling, the bridges can switch between piconets at will, without any prior arrangement. The piconet masters will poll their slaves as determined by the chosen intra-piconet polling scheme, which includes the bridge as well as other slaves. The master will, therefore, poll the bridge in each piconet cycle, and the exchange will start only if the bridge is found to be present.

The main advantage of the walk-in scheme lies in the absence of rendezvous points, which means that any given piconet can accommodate several bridge devices simultaneously, and any given bridge can visit several piconets in sequence. Walk-in bridge scheduling can thus be applied with ease in scatternets of arbitrary size, and there is no performance penalty due to the construction and subsequent maintenance of the schedule of rendezvous points. Neither of these features can be achieved with rendezvous-based scheduling.

2.4 Related Work

A number of schemes have been proposed for event detection and data transmission in wireless sensor networks, most notably the following.

Directed Diffusion (DD) has been proposed for event detection and reliable data transfer in wireless sensor networks [12]. In this scheme, a node requests data by sending an interest query for named data; once a match for the required data is found, the results are transferred to the querying node. In this process, intermediate nodes can aggregate the obtained data, store them in their cache, and redirect them to their neighboring nodes. However, to store the interest queries and the resulting datasets, the DD scheme assumes that all nodes are roughly equivalent in terms of computational and memory capabilities. This might cause significant overhead for sensor networks, which generally have serious power and processing limitations [9]. Furthermore, the guaranteed end-to-end data delivery (which DD supports) is not required for event detection, due to the fact that correlated data flows from several source nodes are loss tolerant as long as event features are reliably detected [2,4].

Pump Slowly Fetch Quickly (PSFQ) scheme is based on propagation of data from the source node by injecting data at relatively low speed and allowing nodes that experience data loss to fetch any missing data packets from immediate neighbors by requesting retransmission [26]. The main source of packet loss in this scheme is the poor quality of wireless links and the resulting transmission errors, while traffic congestion and resulting packet blocking due to buffer overflows at various stages in the network are not considered [2]. This is not a realistic assumption for sensor networks, especially in view of the fact that some packet loss may be acceptable due to correlation of sensed data.

Event-to-Sink Reliable Transport (ESRT) has been designed for reliable event detection [2]. ESRT utilizes the notion of event-to-sink (multipoint-to-point) reliability, rather than the more common end-to-end (point-to-point) reliable transport protocols. The philosophy of ESRT is to prevent the nodes from sending extra packets but, at the same time, from optimizing the number of nodes required to sense the data for a given task. Also, ESRT transfers raw data to the sink without any kind of in-network processing [10,25].

In the energy-efficient **CODA transport protocol** [25], the proposed congestion handling technique uses heuristic mechanisms for monitoring network operations to avoid congestion. These mechanisms include receiver-based congestion detection, open-loop hop-by-hop back-pressure, and closed-loop multisource regulation. In receiver-based congestion detection, CODA uses combinations of present and past channel loading conditions and current buffer levels to predict congestion occurrence in the network. A simple technique is used for monitoring the message queue. Although CODA achieves congestion control, the messaging overhead required in controlling congestion leads to higher energy consumption. Also, congestion control is not directly connected to the application reliability at the sink.

The aforementioned approaches can be roughly classified into those that achieve individual packet reliability using packet retransmission, and those that try to obtain a sufficient number of packets at the sink using some kind of feedback to inform the sensing nodes to decrease the reporting rate. Neither of them considers the effects of finite buffer limitations of sensing and bridging devices.

Furthermore, all these approaches either do not consider the impact of the MAC protocol at all or assume the use of collision-based (and thus generally inefficient) protocols such as CSMA-CA. This was the motivation that led us to investigate the possibility of implementing wireless sensor networks using Bluetooth and its collision-free MAC protocol.

The suitability of Bluetooth as the platform to implement sensor networks has been investigated by building a Bluetooth protocol stack for the TinyOS operating system [16]. The experiments were conducted on actual Bluetooth devices, known as BTnodes, which were developed at ETH Zurich [5]. The BTnodes were equipped with two radios to enable multihop networking. The network was then tested for throughput and energy consumption. The results suggest that Bluetooth-based sensor networks could be appropriate for event-driven applications that exchange bursts of data for a limited time period.

One possible limitation for the use of Bluetooth to implement sensor networks is the limited number of slaves. As mentioned above, a Bluetooth piconet can have, at most, seven active slaves at any given time, while up to 255 others can be parked [6]. Consequently, sensor networks with a large number of nodes must be implemented either as Bluetooth scatternets, or perhaps by combining Bluetooth with other communication technologies such as IEEE 802.15.4 [11]. This is a promising area for further research.

2.5 Congestion Control in Sensing Scatternets

Let us consider a Bluetooth sensor network implemented as a scatternet such as the one shown in Figure 2.6. (In an earlier work, that same topology was used to assess the impact of finite buffer size on scatternet performance [21].) In this setup, one of the nodes in piconet P_1 acts as a sink, while all other nodes act as sources. Each piconet represents a cluster of sensor nodes that is controlled and coordinated by the piconet master. Each slave maintains an uplink queue toward the master, and for each slave, the master maintains a downlink queue. The master also maintains outgoing downlink queues for each bridge. Each bridge has one incoming queue and several outgoing queues, one per each piconet it visits. When the sink needs to acquire some information from the network, it injects a query that is propagated through the network. Once the query reaches the source nodes, they take actions to respond to the query: they collect the data and send it back to the sink.

The traffic model depends on the sensing application. Consider a relatively high-bandwidth, low-cost, surveillance-based sensing application

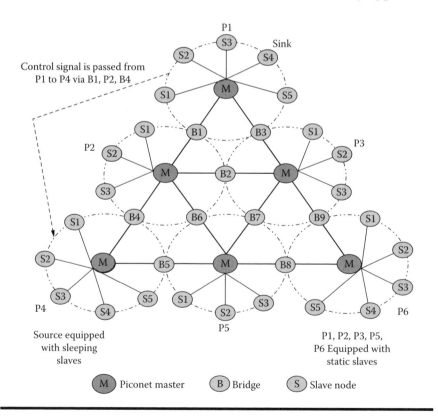

Figure 2.6 Wireless sensor network with triangular topology.

where compressed still images are taken as a result of event detection and sent to the sink. This setup can be used in applications such as road traffic control or asset protection. It is unlikely that the sensed data will fit in a single Bluetooth packet — even with the Enhanced Data Rate facility defined in Bluetooth v2.0, the largest packet size is still only 1023 bytes [6] — and, therefore, the traffic must consist of bursts of several packets. We assume that the packet burst size is geometrically distributed with mean burst size $\overline{B}$ (in this case, $\overline{B} = 3$), and each packet has a length of five Bluetooth time slots T. The packet burst arrival rate λ to each sensing node is presented as the number of packet burst arrivals per time slot. For example, values used in our model are 0.002 to 0.005 packet bursts (images) per time slot, which translates into approximately 3.2 to 8 packet bursts (images) per second. The packet burst arrival rate and the probability of traffic locality are uniform for all the slaves.

Furthermore, we assume that the nodes within a piconet can exchange some other data with the master as required by the sleep management algorithm. Therefore, the locality probability (i.e., the probability that the traffic generated by the slave will have destinations in the same piconet) is set to some small value P_l; the complementary probability that the destination is in another piconet (sink) is $1 - P_l$. When the traffic is generated by a slave for another piconet, the packets are routed through the intermediate piconets, via bridges and piconet master(s), to the destination piconet by taking the shortest path. For simplicity, we assume that neither the masters nor the bridges generate any traffic. We also assume that intra-piconet polling uses the E-limited scheduling scheme, while bridge scheduling is performed using the walk-in approach [19].

Considering event reliability, we can distinguish between no less than three related, yet quite distinct concepts:

1. Absolute event reliability corresponds to the number of packets received per second at the sink from all source piconets. We can also introduce absolute event reliability per source piconet, which is the number of packets per second received by the sink from the given source piconets.
2. Relative reliability is defined as the ratio of the number of received packets from the source piconet at the sink and the number of transmitted packets by that piconet.
3. Finally, the desired reliability is the number of data packets required for reliable event detection at the sink. This number is determined by the requirements of the sensing application.

Our initial exploration focused on the relative reliability per source piconet. When the packet arrival rate (and, consequently, the traffic load) increases, the reliability will increase but only up to a certain point, which can be explained by the lack of congestion control. Namely, the increase in traffic toward the sink will, at some point, overload the bridge buffers along the way. As a result, bridge buffers will begin to drop packets, which leads to a reduction in relative reliability for traffic from a given piconet.

One approach to minimizing packet losses at the bridge buffers and maximizing relative reliability at the sink would be to control the traffic load; this can be accomplished by controlling the number of active slaves in all the source piconets. Because the operation of a piconet is entirely controlled by the piconet master, it is the piconet master that needs to instruct slaves to temporarily suspend their activities; this is performed at the request of the network sink. (Should reliability fall below a predefined limit, the sink can request the master to increase the number of active slaves.) Activation and deactivation can be accomplished by unparking some parked slaves and parking previously active slaves.

An alternative (and much faster) procedure is to put active slaves in one of the possible power-saving modes, such as SNIFF or HOLD [6]. In both of these, the slave in question retains its network address, although the master will not try to poll it. In our experiments, we have assumed that the slave will enter a low-power mode and thus conserve energy. Upon returning to active state, the slave again begins to listen to the master's transmissions, while the master is free to poll the slave at will.

While both SNIFF and HOLD modes could, in theory, be utilized to implement the power-saving mechanism, the HOLD mode has a distinct advantage. Namely, the duration of each HOLD interval is negotiated anew between the master and the slave in question, which opens the possibility for adjustment to any desired time interval. The SNIFF mode, on the other hand, entails distinct procedures for initiation and termination, which makes it less suitable for our purposes. Overall, the use of the HOLD mode gives us both the effectiveness and flexibility of the procedure, which is why we have chosen to implement the activation and deactivation of slaves using the HOLD mode.

2.6 Maintaining Fixed Reliability at the Sink

Let us assume that N_p piconets are reporting the sensing information to the master in the sink piconet. Piconets are indexed by index $i = 1 \ldots N_p$, and each piconet P_i has m_i ordinary slaves (i.e., slaves without the bridging function). The upper limit of reliability of sensed information is determined by the piconet capacity. If five slot packets are used and there is no downlink data traffic (which, in fact, is needed to carry control information), the maximum absolute reliability is $R_{max} = 1/(T + 5T) = 266$ packets per second, where $T = 625$ μs is the duration of Bluetooth time slot.

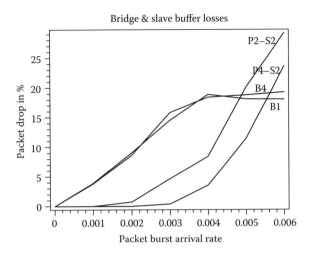

Figure 2.7 Blocking probability versus offered load at the slaves in P_2 and P_4 and the bridges B_1 and B_4.

In practice, the maximum achievable number will be lower, due to the losses at the bridges and presence of downlink traffic needed to send queries and control information. In many cases it will suffice to maintain the reliability at some application-defined level R; assuming uniform conditions, the absolute reliability contributed by each piconet is $R_i = R/N_p$.

We also need to estimate packet losses at the slave and bridge buffers. The bridge loss rate is a function of total piconet load, bridge load, bridge polling parameter M_b, slave polling parameter M_s, and bridge buffer size. In case the topology is fixed and the polling parameters are known, we can assume that the bridge loss rate depends on the bridge packet arrival rate and total piconet load. In this case, the bridge loss rate can be approximated with $P_{b,i} = K_i \overline{B} L \lambda_{b,i}$, where $\overline{B}$ is the average burst size, L is the packet size in slots, $\lambda_{b,i}$ is the burst arrival rate toward the bridge, and K_i is the proportionality constant [19]. Measured values of blocking probabilities are shown in Figure 2.7.

Therefore, the sink can calculate losses from source piconets and communicate them to source piconets to adjust the slaves' activities. Of course, these losses should not be too high — say, up to a few percent — otherwise, the network is operating in the congestion regime, in which case it is better to partition it into sections with separate (and different) sinks and thus avoid congestion. When losses along the path are known, the source piconet can compensate for the losses by scaling its absolute reliability to $R'_i = \dfrac{R_i}{\prod_{over\ path}(1-P_{b,i})}$. The absolute reliability must be transformed into the average number of active slaves per piconet. The mean number of packets contributed by the slave per second is $R_s = \lambda \bar{B}/T$, while the mean number of active slaves per piconet is $A_{s,i} = \frac{R_i'T}{\lambda B}$.

Procedure LONG : managing the long activity period

Data: a, b, m_i
Result: initial value of the short activity management counter C_s
begin
  if $a \leq b$ **then**
    put $m_i - 1$ most recently used slaves to HOLD mode for bT_u seconds;
    remaining slave should be active for aT_u seconds;
  else
    put $m_i - \lceil \frac{a}{b} \rceil$ most recently active slaves to sleep for bT_u;
    among $\lceil \frac{a}{b} \rceil$ remaining slaves, activate $\lfloor \frac{a}{b} \rfloor$ least recently used slaves for
    next bT_u seconds;
    the remaining slave S^* should be active for $(a \bmod b)T_u$ seconds;
  end
  $C_s = a \bmod b$,
end

The mean value of $A_{s,i}$ slaves at any given time can be obtained in the following manner. Assume that $A_{s,i}$ is a rational number: $A_{s,i} = \frac{a}{b}$, where a, b are integers. Further assume that the activity control process consists of basic time units T_u when the slave can be put in HOLD state. (Note that T_u should be much larger than the Bluetooth time slot; in this work, we assume that T_u is one second.) Then, a units of activity must be executed by the slaves over every b time unit. The values for a and b should be selected according to the desired level of granularity of the sleep control. Let us denote the long activity management period with bT_u, and the short activity management period with T_u.

Procedure SHORT : managing the long activity period

Data: C_s
begin
  if $C_s > 0$ **then**
    $C_s = C_s - 1$;
  else if $C_s = 0$ **then**
    $C_s = C_s - 1$;
    put slave S^* to HOLD for $(b - a \bmod b)T_u$ seconds;
end

Within the long activity management period, we try to minimize the number of slaves needed to accomplish this activity requirement. In effect, this is an attempt to minimize the protocol overhead because the slaves will sleep in the HOLD mode and this must be negotiated; the less negotiation we undertake, the more efficient the protocol becomes.

During the short management cycles, we will try to balance the utilization of various slaves in an effort to extend the battery life of each slave. Additionally, feedback from the sink can be communicated to the source piconets to slightly decrease or increase the average number of active slaves, which will result in decrementing or incrementing the value of a.

In this manner, we are able to maintain the reliability at the sink at the desired level. The entire procedure is shown in Algorithm 1.

Algorithm 1: Maintaining fixed reliability at the sink

Data: total event reliability at the sink R, scatternet topology, N_p, m_i,
$\quad\quad i = 1 \ldots N_p$, packet burst arrival rate λ per slave, mean burst size $\overline{B}$
begin
$\quad$ **for** *each piconet P_i* **do**
$\quad\quad$ estimate event reliability R_i;
$\quad\quad$ estimate load through outgoing bridges;
$\quad\quad$ estimate packet loss through each bridge;
$\quad\quad$ estimate total packet loss toward the sink;
$\quad\quad$ recalculate R'_i;
$\quad\quad$ find $A_{s,i}$, a, b;
$\quad\quad$ $C_0 = 0$;
$\quad\quad$ **after** *every T_u seconds* **do**
$\quad\quad\quad$ $C_0 = C_+1$;
$\quad\quad\quad$ *management of long activity period*;
$\quad\quad\quad$ **if** $(C_0 \bmod b == 0)$ **then**
$\quad\quad\quad\quad$ | call LONG;
$\quad\quad\quad$ **end**
$\quad\quad\quad$ *management of short activity period*;
$\quad\quad\quad$ call SHORT;
$\quad\quad$ **end**
$\quad$ **end**
end

To validate this algorithm, we performed simulation experiments with the required event reliability of 20 packets per second at the sink from all the piconets. The packet burst arrival rate for each slave, when active, was set to $\lambda = 0.001$. The reliability requirement was mapped into bridge packet burst arrival rates, and losses through the bridges were estimated as 3 percent for B_4 and 5 percent for B_1, respectively. Then the source piconet transmission rates were set to 4.3 packets per second for P_4, P_5, and P_6 and 4.21 packets per second from P_2 and P_3. The resulting activity of the

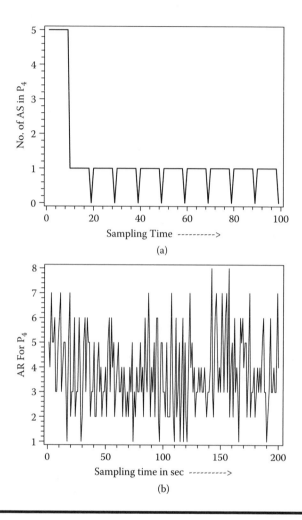

Figure 2.8 Mean number of active slaves and absolute reliability at the sink, for packets from slaves in P_4: (a) number of active slaves over time, and (b) absolute reliability from P_4 at the sink.

slaves in piconet P_4 and the reliability at the sink are shown in Figure 2.8; as can be seen, the algorithm manages to maintain the mean value of absolute reliability around the desired value, while the number of active slaves is minimized.

2.7 Optimizing Reliability at the Sink

While the algorithm described above manages to maintain the reliability at the desired level, it does so without respect for other considerations, in particular the congestion level and the losses due to finite buffers at the bridges and masters through which the packets must pass. Fortunately, a scheme can be devised to simultaneously perform sleep management *and* congestion control, and thus accomplish reliable event detection while minimizing energy consumption.

In some cases, it might be desirable to operate the network in the area of no congestion or mild congestion so as to have minimal losses and to extend the lifetime of the network. (This may be likely to happen when the process being observed changes very slowly over time.) To that end, let us define the relative reliability of the event from a piconet P_i as the ratio of the number of packets generated by the piconet and the number of those packets that are actually received by the sink: $RR_i = \prod_{overpath}(1 - P_{b,i})$. Relative reliability depends on the network load and can be used to detect congestion.

Using the scatternet topology from Figure 2.6 as an illustration, the algorithm, shown as Algorithm 2 on the next page, operates as follows. Initially, the exterior piconets operate with five active slaves, while the interior ones operate with only three, because of their higher carried load. The desired reliability is chosen by the user; the actual reliability is periodically calculated at the sink and communicated to the source piconet (i.e., to its master, which then manages individual slave activity).

Then, the piconet master is able to calculate the relative reliability over the period that is a multiple of the long activity management period. Given that the length of the long management period is bT_u, the length of period for estimating reliability is cbT_u. We have chosen $b = 10$, $T_u = 1\,s$, and $c = 6$; those values give $60\,s$ as the period for estimating reliability and $10\,s$ for the period for changing total slave activity. As in the example with fixed reliability, the total slave activity is calculated as a rational number $A_{s,i} = a/b$, where $b = 10$ for simplicity. Slave activity (expressed through the variable a, the current value of which is denoted as a_c) can be changed only at the boundaries of reliability estimation period. The algorithm also maintains the history of slave activity, as an exponentially weighted moving average a_h; it increases slave activity only if there is an increasing trend of activity and the relative reliability is below the threshold. (In our case, the smoothing constant is $\alpha = 0.5$.)

The number k regulates the step of algorithm progression. It can be set to any value between 1 and b that corresponds to exclusion of one slave. Higher values of k result in faster reactions with possible oscillations, while smaller values lead to slow adaptation to network conditions. We found that $k = b$ gives satisfactory behavior of the algorithm.

When the network experiences congestion, some of the data packets transferred from the source area to the sink are lost due to buffer overflow at the intermediate bridges. This overflow results in a sudden drop in the relative event reliability sensed by piconets, which is taken as a sign of congestion and low reliability in the network. At this point of time, the event reliability calculated at the source master (based on the measurement taken at the sink) will drop below the desired event reliability. Hence, the source master must increase the duration of the hold mode for its slaves. Once a slave is put into HOLD mode, it stops sending the sensed data to the source master, thereby conserving its battery power and reducing congestion in the network. Once the sleep time expires, the slave returns from the HOLD mode and starts collecting data and is ready to be polled.

Algorithm 2: Controlling relative reliability at the sink (algorithm is executed at each piconet)

Data: piconet index i, m_i, penalty k, measured reliability R_i
begin
> **while** *true* **do**
> > **after** *every cbT_u seconds*
> > > $C_0 = C_0 + 1$
> > > **if** $(C_0 \bmod b == 0)$ **then**
> > > > | call *LONG*;
> > > call *SHORT*;
> > **end**
> > receive measured reliability from the sink;
> > calculate RR_i, $a_c = a$;
> > **if** $RR_i > T_H$ **then**
> > > | $a = a_c - k$;
> > > **else if** $(RR_i < T_L)$ AND $(a_b > a_c)$ **then**
> > > > | $a = a_c + 1$;
> > > **else if** $(RR_i < T_L)$ AND $(a_b < a_c$ **then**
> > > > | $a = a_c - k$;
> > > **end**
> > > $a_b = \alpha a_b + (1 - \alpha)a_c$
> > **end**
> **end**
end

Table 2.1 shows a representative measured trace from the simulator to illustrate our sleep regulation technique. In this case, sleep management is applied to all piconets.

We note that the packet transmission rates will be affected by the packet loss caused by noise and interference. While this packet loss is indeed possible, its effects will be negligible because of the following:

■ Bluetooth uses FHSS, which makes it rather resilient to noise and interference [27].
■ Bluetooth packets can be protected using Forward Error Correction (FEC), at the expense of slight reduction of their information carrying capacity.

Table 2.1   Simulator Trace for P_4

Interval	Relative Reliability	Active slaves in P_4	Slaves Put on HOLD	Slaves Back from HOLD
1	55	5	0	0
2	64	5	2	0
3	75	3	2	0
4	88	1	0	0
5	86	2	1	1
6	85	3	2	2
7	90	3	2	2
8	91	2	1	1
9	88	3	2	2
10	89	3	2	2
11	91	2	1	1
12	88	3	2	2
13	92	3	2	0
14	88	1	0	1

■ Furthermore, the Bluetooth polling algorithm requires that polling be performed using full length packets (i.e., at least one time slot T), which allows the nodes to acknowledge proper packet reception (or lack thereof) without additional overhead.

■ Finally, it might be argued that this packet loss will cause the algorithms to mistakenly increase the number of active nodes. However, from the standpoint of the activity management algorithm, packet loss due to noise and interference does not differ — in qualitative terms — from the loss caused by congestion, that is, by buffer blocking at the intermediate nodes. As long as the thresholds and parameters of the algorithm are properly adjusted, the algorithms are able to maintain the received reliability within the desired limits.

2.8   Performance: Relative and Absolute Reliability

Simulations were carried out to explore the behavior of the relative reliability observed at the sink for each source piconet when the sleep management scheme is applied in the entire scatternet. Figure 2.9 presents the relative reliability observed at the sink for packets sent from slaves in P_4 (which is an exterior piconet) for packet arrival rates of 0.002, 0.003, and 0.004. All other parameters were set to the same values as before. Because all piconets operate under the sleep management scheme (except P_4, which is the subject of experiment), the relative reliability at the sink is

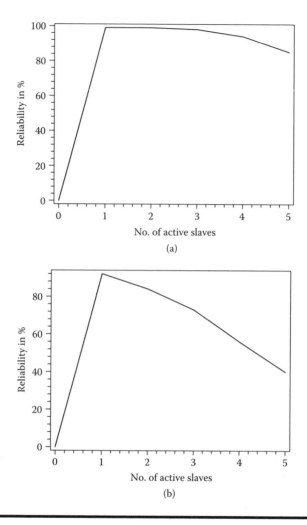

Figure 2.9 **Relative reliability at the sink for packets from the slaves in piconet P_4 versus the number of active slaves in P_4: packet burst arrival rates of (a) 0.002 packets per slot, and (b) 0.005 packets per slot.**

much higher than in the case when only one piconet uses the scheme; the peak value exceeds 95 percent under a wide range of packet arrival rates. Note that the event reliability for P_4 remains within limits T_L and T_H for one to two active slaves. The average number of active slaves for which the event reliability is within the limits decreases with the packet burst arrival rate, which is expected.

Figure 2.10 presents the analogous dependency, only this time the relative reliability corresponds to packets sent to the sink from slaves in P_2,

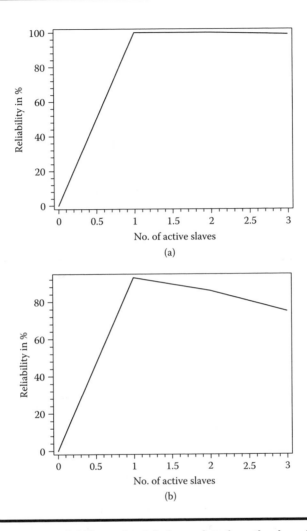

Figure 2.10 **Relative reliability at the sink, for packets from the slaves in piconet P_2 versus the number of active slaves in P_2: packet burst arrival rate (a) 0.002 packets per slot and (b) 0.005 packets per slot.**

which is an interior piconet. Because the same congestion control mechanism is used in all the piconets, the shape of the dependencies is almost identical to those from the previous set of diagrams.

To calculate the dispersion of the relative reliability, its mean, variance, and standard deviation are calculated for the data in Figures 2.9 and 2.10. We note that the increase in arrival rate leads to a decrease in mean value and an increase in variance, which can be used to indicate serious congestion. Figure 2.11 shows the development of the number of active slaves in P_4 over time, including the warm-up period of the simulator. The packet

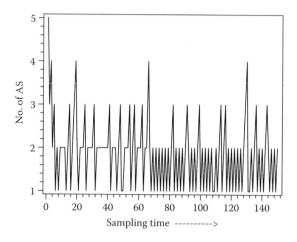

Figure 2.11   Fluctuation in the number of active slaves (AS) for P_4.

burst arrival rate was set to $\lambda = 0.002$ packet bursts per Bluetooth time slot. We note that the algorithm maintains average number of active slaves around 1.9.

Figure 2.12 and Figure 2.13 show absolute reliability observed at the sink for packets originating from slaves in piconets P_4 and P_2, respectively. At lower packet burst arrival rates, the absolute reliability is a monotonically increasing function of the number of slaves. While this result differs from the corresponding dependencies of relative reliability from Figures 2.9 and 2.10, keep in mind that the sleep management scheme was designed with the goal of maintaining the *relative* reliability, not its absolute counterpart, within certain limits. Of course, congestion control could be designed the other way around, that is, by specifying the desired absolute reliability and trying to achieve it with the highest possible relative reliability, as shown in Section 2.6.

2.9   Performance: Packet Loss at the Bridge Buffers

Another sign of congestion (and, by extension, a decrease in reliability) is the increase in packet loss rates at the bridge buffers. We have measured packet loss rates in our scatternet using the same setup as above: the sink was in piconet P_1, the desired reliability was set to 60 percent, the packet burst arrival rate was set to 0.005 (bursts per Bluetooth time slot), the bridge residence time was set to one piconet cycle, and polling parameters for slaves and bridges were $M_s = 3$ and $M_b = 12$, respectively. However,

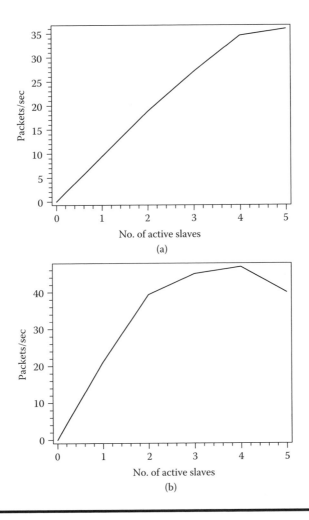

Figure 2.12   **Absolute reliability at the sink for packets from slaves in P_4 versus the number of active slaves in P_4: packet burst arrival rate of (a) 0.002 packets per slot and (b) 0.005 packets per slot.**

to get better insight, we varied the traffic locality probability in the range $P_l = 0.3 \ldots 0.8$ and the bridge buffer size in the range $8 \ldots 20$.

Packet losses at the buffers of bridges B_4, B_9, B_1, and B_3 are shown in Figure 2.14. Because of the symmetry of the network, B_4 and B_9 exhibit similar packet loss rates; we note that packet losses become significant for high inter-piconet traffic (i.e., with $P_l = 0.3$ and lower), which is characteristic of sensor networks. Similar conclusions hold for packet loss rates at the buffers of "interior" bridges B_1 and B_3 (which carry the data from

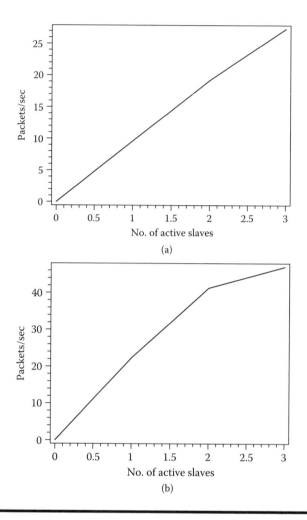

Figure 2.13   Absolute reliability at the sink for packets from slaves in P_2 versus the number of active slaves in P_2: packet burst arrival rate of (a) 0.002 packets per slot and (b) 0.005 packets per slot.

P_4, P_6, P_2, and P_3 to P_1), shown in Figures 2.14(c) and 2.14(d). Because B_1 and B_3 carry data packets from two piconets each, the loss of data packets at these bridges is greater when compared to B_4 and B_9. We observe that under a realistic locality probability of $P_l = 0.3$, buffer sizes of 12 packets or more suffice to keep the packet loss very low; this offers a substantial advantage over the value of 40 or more, which is necessary in the network without sleep management [21].

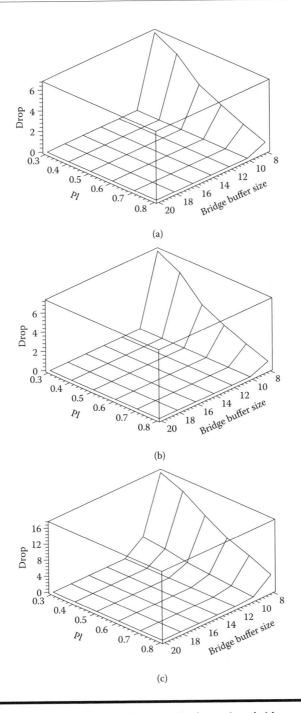

Figure 2.14   Bridge buffer drop rate (in percent) for various bridges at packet burst arrival rate of 0.005 packets per slot: (a) bridge B_4, (b) bridge B_9, (c) bridge B_1, and (d) bridge B_3.

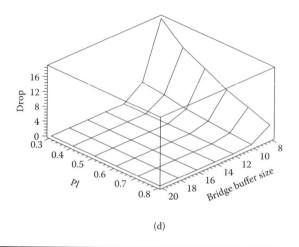

(d)

Figure 2.14   (Continued).

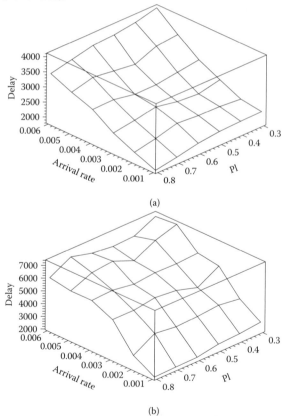

(a)

(b)

Figure 2.15   End-to-end packet delays: end-to-end delays for (a) traffic from P_6 to P_1 and (b) traffic from P_3 to P_1.

2.10   Performance: End-to-End Delay

Finally, end-to-end packet delays for traffic from P_6 to P_1 and from P_3 to P_1, are shown in Figures 2.15(a) and 2.15(b), respectively. Both queueing and transmission delays are taken into consideration to calculate end-to-end delays. In this case, P_l varied in the range $0.3 \dots 0.8$, and packet burst arrival rates were in the range $\lambda = 0.001 \dots 0.006$. Because the interior piconets have more bridges than the exterior ones, their carried load is higher and so are the delays.

A note on simulations. All simulation results presented were obtained with a custom-built Bluetooth simulator implemented using the Artifex object-oriented Petri net engine by RSoft Design, Inc. [23].

2.11   Conclusion

In this chapter we have developed and evaluated two congestion control algorithms for Bluetooth-based sensor networks. Both algorithms are based on sleep scheduling of Bluetooth slaves. The first algorithm maintains required (fixed) event reliability at the sink using the minimum slave activity. It uses precalculated activity values obtained from the analytical and simulation models of the network.

The second algorithm keeps the whole network within the acceptable range of packet losses while maintaining minimum slave activity. In this case, source piconets use the information measured at the sink to regulate the activity of their slaves. Simulation results confirm that this sleep management policy results in decreased bridge buffer loss rates in all downstream bridges toward the sink, which means that bridges can be designed with smaller buffer space.

References

[1] M. Achir and L. Ouvry. Power Consumption Prediction in Wireless Sensor Networks. In *ITC Specialist Seminar on Performance Evaluation of Wireless and Mobile Systems*, Antwerp, Belgium, August 2004.

[2] O.B. Akan and I.F. Akyildiz. ESRT: Event-to-Sink Reliable Transport in Wireless Sensor Networks. In *IEEE/ACM Transactions on Networking*, 13(5), 1003–1016, October 2005.

[3] I.F. Akyildiz, W. Su, Y. Sankarasubramaniam, and E. Cayirci. Wireless Sensor Networks: A Survey. *Computer Networks, (Elsevier) Journal*, 38:393–422, March 2002.

[4] I.F. Akyildiz, M.C. Vuran, and O.B. Akan. On Exploiting Spatial and Temporal Correlation in Wireless Sensor Networks. In *Proc. WiOpt'04: Modeling*

and Optimization in Mobile, Ad Hoc and Wireless Networks, pp. 71–80, March 2004.

[5] J. Beutel, O. Kasten, and M. Ringwald. BTnodes — A Distributed Platform for Sensor Nodes. In *Proc. 1st Intl. Conf. on Embedded Networked Sensor Systems (SenSys)*, pp. 292–293, November 2003.

[6] Bluetooth SIG. *Draft Specification of the Bluetooth System*. Version 2.0, November 2004.

[7] A. Capone, R. Kapoor, and M. Gerla. Efficient polling schemes for Bluetooth picocells. In *Proc. IEEE Int. Conf. on Communications ICC 2001*, Vol. 7, pp. 1990–1994, Helsinki, Finland, June 2001.

[8] N. Golmie, R.E. Van Dyck, and A. Soltanian. Interference of Bluetooth and IEEE 802.11: Simulation Modeling and Performance Evaluation. In *Proceedings 4th ACM International Workshop on Modeling, Analysis and Simulation of Wireless and Mobile Systems*, pp. 11–18, Rome, Italy, July 2001.

[9] B. Hong and V.K. Prasanna. Optimizing System Life Time for Data Gathering in Networked Sensor Systems. In *Algorithms for Wireless and Ad-hoc Networks (A-SWAN) (Held in conjunction with MobiQuitous)*, August 2004.

[10] B. Hong and V.K. Prasanna. Optimizing a Class of In-Network Processing Applications in Networked Sensor Systems. In *1st IEEE International Conference on Mobile Ad-hoc and Sensor Systems (MASS 2004)*, October 2004.

[11] Standard for Part 15.4: Wireless MAC and PHY Specifications for Low Rate WPAN. IEEE Std 802.15.4, IEEE, New York, NY, October 2003.

[12] C. Intanagonwiwat, R. Govindan, D. Estrin, J. Heidemann, and F. Silva. Directed Diffusion for Wireless Sensor Networking. *ACM/IEEE Transactions on Networking*, 11(1):2–16, February 2003.

[13] N. Johansson, U. Körner, and P. Johansson. Performance Evaluation of Scheduling Algorithms for Bluetooth. In *Proceedings of BC'99 IFIP TC 6 Fifth International Conference on Broadband Communications*, pp. 139–150, Hong Kong, November 1999.

[14] P. Johansson, R. Kapoor, M. Kazantzidis, and M. Gerla. Rendezvous Scheduling in Bluetooth Scatternets. In *Proc. IEEE Int. Conf. on Communications ICC 2002*, pp. 318–324, New York, April 2002.

[15] Y.-Z. Lee, R. Kapoor, and M. Gerla. An Efficient and Fair Polling Scheme for Bluetooth. In *Proceedings MILCOM 2002*, Vol. 2, pp. 1062–1068, 2002.

[16] M. Leopold, M. Dydensborg, and P. Bonnet. Bluetooth and Sensor Networks: A Reality Check. In *1st ACM Conference on Sensor Networks*, November 2003.

[17] H. Levy, M. Sidi, and O.J. Boxma. Dominance Relations in Polling Systems. *Queueing Systems Theory and Applications*, 6(2):155–171, 1990.

[18] Z. Liu, Ph. Nain, and D. Towsley. On Optimal Polling Policies. *Queueing Systems Theory and Applications*, 11(1–2):59–83, 1992.

[19] J. Mišić and V.B. Mišić. *Performance Modeling and Analysis of Bluetooth Networks: Polling, Scheduling, and Traffic Control*. CRC Press, Boca Raton, FL, 2005.

[20] J. Mišić, K.L. Chan, and V.B. Mišić. Performance of Bluetooth Piconets under E-Limited Scheduling. Tech. Report TR 03/03, Department of

Computer Science, University of Manitoba, Winnipeg, Manitoba, Canada, May 2003.

[21] J. Mišić, V.B. Mišić, and G.R. Reddy. On the Performance of Bluetooth Scatternets with Finite Buffers. In *Proc. WWAN2005 International Workshop on Wireless Ad Hoc Networking (ISCDS'05 Workshops)*, pp. 865–870, Columbus, OH, June 2005.

[22] V.B. Mišić, J. Mišić, and K.L. Chan. Walk-in Scheduling in Bluetooth Scatternets. *Cluster Computing*, 8(2/3):197–210, 2005.

[23] RSoft Design, Inc. *Artifex v.4.4.2.* San Jose, CA, 2003.

[24] Hideaki Takagi. *Queueing Analysis*, Vol. 1: Vacation and Priority Systems. North-Holland, Amsterdam, the Netherlands, 1991.

[25] C. Wan, S.B. Eisenman, and A.T. Campbell. CODA: Congestion Detection and Avoidance in Sensor Networks. In *Proceedings of the First International Conference on Embedded Networked Sensor Systems*, pp. 266–279. ACM Press, November 2003.

[26] C.Y. Wan, A.T. Campbell, and L. Krishnamurthy. PSFQ: A Reliable Transport Protocol for Wireless Sensor Networks. In *First ACM International Workshop on Wireless Sensor Networks and Applications (WSNA 2002)*, pp. 1–11, September 2002.

[27] S. Zürbes. Considerations on Link and System Throughput of Bluetooth Networks. In *Proceedings of the 11th IEEE International Symposium on Personal, Indoor and Mobile Radio Communications PIMRC 2000*, Vol. 2, pp. 1315–1319, London, U.K., September 2000.

Chapter 3

Traffic-Aware Routing for RTC in Wireless Multi-Hop Networks

Shouyi Yin, Yongqiang Xiong,
Qian Zhang and Xiaokang Lin

Contents

3.1 Introduction

Real-time communication (RTC), as one of the most popular applications, has drawn great attention in recent years, especially in wireless multi-hop networks [1], which is emerging due to its decentralized nature and the popularity of wireless devices. This chapter focuses on wireless multi-hop networks consisting of IEEE 802.11 [35] devices. Routing is the key issue for RTC over wireless multi-hop networks, because it determines whether the forthcoming RTC traffic[1] can be served on a high-quality path. There are two major challenges for routing RTC traffic over wireless multi-hop networks: (1) strong Quality-of-Service (QoS) provision and (2) severe interference in wireless networks. On the one hand, RTC applications have critical delay and bandwidth requirements; to ensure providing high-quality support for the forthcoming RTC traffic, the system needs to accurately predict the path quality in advance before the traffic really being accepted in order to select the best candidate path. On the other hand, in wireless multi-hop networks, interference is the key factor impacting path performance. To serve the forthcoming RTC traffic, in addition to the interference coming from the physical environment, two types of traffic also interfere with this RTC flow: one is neighboring traffic, including the traffic cross the same node and adjacent nodes, and the other is the RTC traffic itself (we call it self-traffic) along the path. Both types of interference should be considered when estimating path quality.

Figure 3.1 illustrates the impact of self-traffic. Suppose the RTC traffic is sending along node A to node D via node B and C sequentially. We focus on analyzing the performance of the link from A to B. First, this RTC flow sending from node B to node C will contend with the transmission on node A in the same channel, thus affecting the traffic on link A–B. Moreover, when node C begins forwarding this RTC traffic to D, node B cannot receive packets from A; thus, transmission on link A–B is also affected by the same traffic flow on link C–D. Thus, we can see that in addition to the neighboring traffic, self-traffic also affects the path quality, so we need to take the self-traffic effect into account when estimating the path quality for the forthcoming RTC traffic. However, traditional measurement-based routing schemes cannot get an accurate estimation, especially for large-volume RTC traffic, because when such a scheme performs probing and measurement at the routing selection stage, in which the forthcoming traffic is still not injected into the network. That is, the measurement results

---

[1] In this chapter, the terms "traffic" and "flow" are interchangeable.

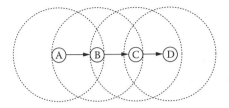

Figure 3.1   RTC traffic from node A to D via B and C. Traffic on link A-B will interfere with the same flow on link B-C and link C-D, respectively.

cannot account for self-traffic interference. This weakness will cause incorrect routing selection for large-volume RTC traffic.

Recently, people have tried to equip multiple radios with wireless nodes to improve the capacity of the wireless network. Routing in a multi-radio network is different than that in a traditional single-radio network. In a multi-radio network, two adjacent nodes or links can choose two noninterfering radios or channels. This means that a node can send and receive packets on two nonoverlapping radios or channels simultaneously, and two adjacent nodes or links can send packets at the same time without mutual interference. The interference caused by self-traffic and neighboring traffic will be affected by this channel diversity in multi-radio networks. Self-traffic interference is more severe in the multi-radio scenario.

It is challenging to take both neighboring traffic and self-traffic into account because self-traffic interference will not appear until the RTC traffic is injected into the path. To address this issue, we analyze how neighboring traffic and self-traffic affect path quality. Interestingly, we find that both interferences affect packet service time, in essence. Therefore, considering that the rate of RTC flow is well controlled (e.g., a video stream delivered with a 256-Kbps bit rate), then to obtain an accurate estimation of path quality, we analyze the expected RTC packet service time over IEEE 802.11 networks to account for the impact of both existing neighboring traffic and the nonexisting self-traffic. We derive the packet service time from the IEEE 802.11 MAC standard according to the neighboring traffic information, which can be measured, and self-traffic information, which can be obtained from the application layer; thus, we can predict the path quality accurately.

Based on the aforementioned idea, we propose a novel traffic-aware routing metric to address the quality prediction issue for RTC traffic. We use the expected end-to-end transmission time on the path as the quality metric, which can also reflect the path bandwidth [2]. The proposed traffic-aware metric, PPTT (Path Predicted Transmission Time), is the sum of the delay estimation on each link along the routing path, which further consists of packet service time and queue delay. In the PPTT scheme, when we model the impact of the interfering traffic and link condition, we classify them according to the different channels they are using. In this way, we can avoid selecting the same channel for adjacent nodes, which can lead to bad link quality. By explicitly considering the radio characteristic, our proposed routing metric, PPTT, is also a unified metric for both single-radio and multi-radio networks.

We implemented the PPTT routing scheme based on the Mesh Connectivity Layer (MCL) software [3] and studied its performance for both single-radio and multi-radio scenarios in a wireless multi-hop testbed consisting of 32 nodes equipped with two IEEE 802.11 a/b/g combo cards in our building. Experiment results showed that our proposed routing metric outperforms other non-traffic-aware routing schemes. We also conducted simulations in network simulator (ns2 [4]) for random topologies. Rather promising performance results were obtained.

The contributions of this chapter are as follows. First, by explicitly taking self-traffic into account, together with neighboring traffic, we offer a unified traffic-aware routing metric to predict the path quality in both single-radio and multi-radio multi-hop networks, which is then demonstrated to improve network performance through real implementation. Second, we estimate the transmission time based on the analysis of the 802.11 MAC behavior for the forthcoming RTC flow that needs to be served in a multi-hop wireless network. Having such a predicted RTC transmission time, we can offer more accurate guidance for QoS-related services.

The remainder of this chapter is organized as follows. Section 3.2 introduces related schemes for RTC in wireless multi-hop networks. Section 3.3 illustrates why we need a new routing metric for RTC. The details of our proposed new traffic-aware routing metric and the PPTT routing scheme are described in Section 3.4, along with the prediction of the packet service time. In Section 3.5 we discuss the issues of implementation of PPTT. Section 3.6 demonstrates the effectiveness of our proposed scheme with experiment results, both in a real testbed and in a network simulator. Finally, we draw some conclusions and future perspective in Section 3.7.

3.2   Related Work

To support RTC traffic in wireless multi-hop networks, in the literature there are numerous related proposals working on different layers. They mainly focus on the MAC and network layers. There are some transportation- or application-centric schemes for RTC. However, the most important and challenging issues are still with MAC scheduling and routing in wireless multi-hop networks.

On MAC layer, the proposals target serving RTC traffic with higher priority or reserving dedicated resources to access the channel. DBASE [5] lets RTC traffic have different MAC parameters, including contention window size, frame size, and inter-frame size, from non-RTC traffic to contend for the channel, and similar ideas were applied in IEEE 802.11e [6–8]. Some protocols [1,9,10] are designed for TDMA-based MAC to reserve dedicated timeslots for RTC traffic, and [11,12] focus on QoS-aware scheduling to avoid the cost of time synchronization in TDMA-based schemes. Although MAC-based schemes work well in single-hop wireless networks by controlling nodes in a one-hop neighborhood, they cannot guarantee the performance of RTC in a multi-hop scenario. By only enhancing the MAC functionality, the end-to-end QoS guarantee for RTC cannot be supported because, in a multi-hop scenario, neighbors several hops away may also affect the performance of the current node. Thus, guidance from routing layer, such as local delay or bandwidth indication, is quite essential to those enhanced MAC schemes.

Several QoS routing schemes [13,14] have been proposed for wireless multi-hop networks. They can be classified into reservation-oriented and reservation-less approaches. Reservation-less approaches [15,16] adopt ideas similar to those of the DiffServ framework for the Internet to offer a soft-QoS guarantee by serving different flows with different service classes on each node. These approaches, however, are another type of higher-level scheduling mechanism similar to MAC-based schemes, so they cannot provide QoS guidance such as delay or bandwidth budget to the underlying QoS-aware MAC in multi-hop networks. On the other hand, reservation-oriented approaches intend to offer a hard-QoS guarantee by reserving resources for each flow on every node (see [17–21]). To calculate the available resource for reservation, [22] proposed a formula to estimate the available bandwidth, and [24] estimates links according to the prediction of existing traffic and the location of mobile nodes. In summary, all these reservation-oriented protocols primarily focus on estimating the available bandwidth while not taking into account interference from all the traffic. However, without considering the self-traffic effect, all the above-mentioned solutions cannot offer an accurate prediction of end-to-end delay for the coming RTC traffic.

In addition to the work-specific target for QoS support, there are also many studies on generic routing protocols. Most traditional ad hoc routing schemes [25–28] adopt a simple metric called HOP-COUNT to find a shortest path from sender to receiver, but recent research results [29,30] show that this metric may work well in mobile scenarios where the topology changes dynamically, but results in poor performance in stationary mesh networks because it does not consider the link quality, which may lead to select long (in terms of distance) but error-prone routes. Recently, researchers proposed some link quality metrics such as "Per-hop Round Trip Time" (RTT) [31], "Per-hop Packet Pair Delay" (PktPair) [32], "Expected Transmission Count" (ETX) [30], and signal strength or signal-to-noise ratio (SNR) [33,34] to select a path with a good quality such as high bandwidth, low loss ratio, and short transmission time, in order to achieve higher network capacity.

Multi-radio networks have drawn significant attention in recent years. Broch et al. [39] proposed a multi-interface supported routing protocol that can be used in multi-radio networks. It uses HOP-COUNT as the routing metric. Draves et al. [2] proposed a link-quality path metric called Weighted Cumulative Expected Transmission Time (WCETT) for multi-radio wireless networks, in which the channel bandwidth and channel diversity are considered. WCETT combines each link's Expected Transmission Time (ETT) and explicitly accounts for interference among links that use the same channel, thus having good performance in multi-radio networks.

All these link-quality routing schemes, however, use a measurement-based scheme to probe the link conditions based on the current wireless condition and the existing traffic in the network, and thus they do not take into explicit consideration the self-traffic; therefore, they cannot obtain an accurate prediction of link quality due to the combination of interferences from the different traffic after the self-traffic is injected into the system.

3.3   Why a New Routing Metric for RTC?

Much prior research work has been conducted to propose different routing metrics, such as "HOP COUNTS" [25–28], ETX [30], RTT [31], PktPair [32], and WCETT [2], for wireless multi-hop networks. However, all these existing link-quality routing metrics cannot always guarantee the selection of a good path for RTC.

3.3.1   Single-Radio Scenario

Think about the single-radio scenario shown in Figure 3.2, where each node has an 802.11b radio with a fixed 2-Mbps link rate. Suppose there is a flow from node 6 to node 0 with 256-Kbps transmission rate of (flow-1). Now we inject another 384-Kbps RTC flow from node 3 to 4 (flow-2). There are two possible paths for flow-2. More specifically, path-1 is along nodes set (3, 5, 1, 4), where node 5 and node 1 are in the interference range of node 7 and node 8; path-2 is along nodes set (3, 10, 2, 9, 4), where no node along this path is interfered by flow-1.

We use the ETX metric to illustrate why we need a new routing metric because a recent report pointed out that ETX can achieve rather good performance in a single-radio scenario. The ETX metric measures the expected number of transmissions, including retransmissions, needed to send a unicast packet across a link. To derive ETX, each node broadcasts one probe packet every second. Its neighbors then calculate the loss rate of the

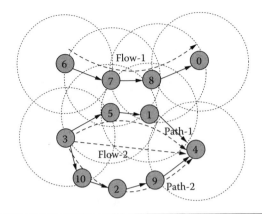

Figure 3.2   A scenario to show how self-traffic affects routing selection. The dotted circles denote the interference range.

probes on the links. Let p denote the probability that packet transmission fails. The ETX of this link then can be calculated by:

$$ETX = \frac{1}{1 - p} \qquad (3.1)$$

The path metric is the sum of the ETX value of each link in the path. The routing protocol then selects the path with minimal ETX. Specifically in this case, the measurement shows that the packet loss rates of the three links of path-1 are 13, 23, and 14 percent, respectively, due to neighboring traffic 6 to 0, and the packet loss rate of all links of path-2 is 0. Hence, the ETX of path-1 is $3.61(= 1/(1 - 13\%) + 1/(1 - 23\%) + 1/(1 - 13\%))$, while ETX of path-2 is $4(= 1/(1 - 0) + 1/(1 - 0) + 1/(1 - 0) + 1/(1 - 0))$. Therefore, ETX will select path-1 rather than path-2 for the new incoming RTC flow.

However, if this RTC flow is delivered through path-1, it will encounter higher contention and loss probability due to the joint inference from flow-1 and flow-2 itself. This is verified by simulation results. When we start flow-2 on path-1, the ETX value of path-1 is changed to $6.15(= 1/(1-48\%)+1/(1-59\%)+1/(1-44\%))$; and if we start flow-2 on path-2, ETX of path-2 will be changed to $5.92(= 1/(1 - 28\%) + 1/(1 - 44\%) + 1/(1 - 29\%) + 1/(1 - 25\%))$. So, contrary to ETX's selection, the performance of path-1 is worse than that of path-2 in terms of goodput to serve the RTC traffic: its goodput along path-1 is 318 Kbps, while the goodput along path-2 is 384 Kbps. This phenomenon results because ETX does not consider self-traffic interference. ETX selects path-1 according to the probing results. But the probing results do not include self-traffic interference because, at that stage, flow-2 is not injected into the network. However, the self-traffic interference cannot be neglected; after flow-2 is injected into path-1, the joint impact results in the performance of path-1 being worse than path-2. This weakness is universal for measurement-based routing schemes because they cannot measure the self-traffic interference before forthcoming traffic is injected. Thus, measurement-based routing schemes cannot reflect path quality accurately. The larger the RTC traffic volume, the more severe the problem is. When the RTC traffic load is light, the self-traffic interference is trivial. The lack of consideration for self-traffic may not cause performance loss. However, when the RTC traffic load is heavy, the performance will decrease significantly if self-traffic interference is ignored.

Furthermore, to study the self-traffic effect, we need to distinguish it according to the interference impact. Let us illustrate it with Figure 3.2. After selecting path-2 for flow-2, traffic on link (3,10) will be interfered by flow-2 itself on link (10,2) and (2,9), but these two interfering traffic flows have different impacts on link (3,10). Traffic on link (10,2) contends with link (3,10) and increases the contention probability. We call this kind of traffic — sent from nodes that are in the carrier sensing range of link's sender — *carrier sensing* (CS) *traffic*. For other interfering traffic on link (2,9), it is sent from hidden terminals for link (3,10), resulting in an increase in collision probability. We call this kind of traffic — from the sender that is in the carrier sensing range of link's receiver but not in the carrier sensing range of link's sender — *hidden terminal* (HT) *traffic*. HT traffic causes much more severe loss than CS traffic. This is also verified by simulation. For the 384-Kbps flow, the CS traffic on (10,2) causes a 2 percent loss ratio for link (3,10), while the HT traffic on (2,) causes a 30 percent loss ratio.

3.3.2   Multi-Radio Scenario

In a multi-radio network, the impact of self-traffic is even more severe than in a single-radio network. In a single-radio network, self-traffic will affect routing selection with neighboring traffic jointly. If there is no neighboring traffic, although self-traffic affects the path performance, it will not cause a mistake in routing selection. However, in a multi-radio network, even if there is no neighboring traffic, it may cause incorrect routing selection if self-traffic is not taken into account. Consider the scenario shown in Figure 3.3; there are two candidate multi-radio paths for a flow from node 1 to node 10. We vary the traffic rate of the flow and deliver it through these two paths, respectively. The delay and goodput of these two paths are compared in Figure 3.4. When the offered load is light, the self-traffic impact is not distinct, so that path-1 and path-2 achieve similar performances. When the offered load is heavy, the impact of self-traffic emerges. Because there is a hidden terminal in path-1 and no hidden terminal in path-2, the performance of path-1 is worse than that of path-2. This simple example shows that the self-traffic should be considered when selecting the high performance path for RTC in a multi-radio network.

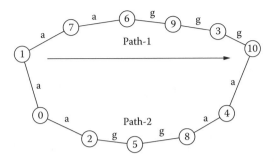

Figure 3.3   Multi-radio scenario. Each node has two radios (802.11a and 802.11g).

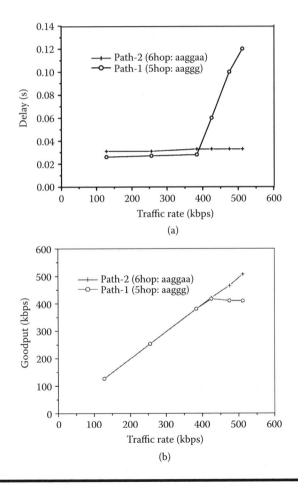

Figure 3.4   (a) Delay and (b) goodput comparison of path-1 and path-2.

From the above analysis for single-radio and multi-radio scenarios, we can conclude that:

1. Self-traffic interference should be considered explicitly to predict routing path quality.
2. Interfering traffic should be differentiated to reflect the different impacts on the same link.

Therefore, a prediction-based routing metric that considers self-traffic interference explicitly and can predict the whole path performance is required.

3.4 Traffic-Aware Routing Metric: PPTT

3.4.1 Basic Idea

In this section we present a new prediction-based routing metric to explicitly consider different types of interfering traffic (i.e., neighboring traffic and self-traffic) that interfere with the requested flow. We can classify this interfering traffic as CS and HT traffic, according to their relative positions. CS traffic is the cumulative traffic of all nodes that are in the carrier sensing range of the link's sender (traffic sender on this link). In 802.11 wireless networks, when the sender wants to transmit a packet across the link, it will compete with the nodes in CS range for channel access. Thus, a larger volume of CS traffic leads to a longer channel access time. HT traffic is the cumulative traffic of all nodes that are in the carrier sensing range of the link's receiver (traffic receiver on this link) but not in the carrier sensing range of the link's sender. A packet transmitted across the link can collide with the packets from hidden terminals. Thus, a larger volume of HT traffic causes more packet collisions, which results in longer retransmission times. Considering the different impacts of CS and HT traffic on link quality, we need to differentiate them by their locations. We use a 25-node grid topology network (Figure 3.5) to illustrate the differentiation of interfering traffic (in this topology, we set the interference range and transmission range of all nodes equal). There are four flows in the network. Flow a to e is delivered via links (a,b), (b,c), (c,d) and (d,e). It has two neighboring flows, f→g and j→m, because nodes f and j are in nodes b's and c's CS range, respectively. Considering link (b,c), CS traffic of this link includes flows a→b, c→d, and f→g, because they are all in the CS range of b; HT traffic includes flows d→e and j→k because they are in receiver c's CS range but out of sender b's CS range.

With these HT traffic and CS traffic concepts, we can further study the twofold impact of self-traffic interference. First, self-traffic will enlarge the CS traffic volume of each link in its delivery path. For example, traffic

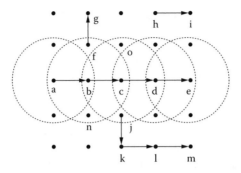

Figure 3.5   25-Node grid topology. Each node contributes CS traffic or HT traffic according to its relative position in the network.

of links (a,b) and (c,d) is the increment of CS traffic of link (b,c). As a result, self-traffic interference elongates the channel access time. Second, self-traffic will increase the HT traffic volume of some links in this path. For example, the traffic of link (d,e) is the increment of HT traffic of link (b,c). As a result, self-traffic interference causes more packet collisions, which in turn elongates the packet retransmission time.

In summary, the interfering traffic, whether it is neighboring traffic or self-traffic, acts as either CS traffic or HT traffic to impact the link quality, which results in the increment of packet transmission time. Therefore, by considering traffic interference, we propose a new time-based routing metric, called PPTT, to explicitly account for different types of traffic interference. PPTT tries to predict the end-to-end delay after the traffic starts to be delivered along the path.

To calculate PPTT, we predict the packet transmission time link by link, which we call link predicted transmission time (LPTT). Then, summing the LPTT of each link, we obtain the PPTT of the entire path. The LPTT is defined as the time from the instant the packet enters the queue of the link's sender to the instant it successfully reaches the receiver or is dropped, which comprises queueing delay and packet service time. Because RTC packets can queue up in the buffer for processing, we use the $M/M/1$ queueing model to calculate queueing delay. (Kim and Li [38] pointed out if the channel contains significantly less low-frequency energy than that of the arrival process, we can simply model the queue system by an exponential server.) The packet service time is the time span that MAC layer takes to send out the packet, and we calculate it based on the analysis of 802.11 MAC behavior.

As we can see, the packet transmission time on a certain link is related to its existing CS traffic, HT traffic, and self-traffic. Thus, the LPTT can be represented as $LPTT(\lambda_{cs}, \lambda_{ht}, \lambda)$, where λ is the traffic rate of this RTC flow, and λ_{cs} and λ_{ht} denote the average CS traffic rate and HT traffic rate, respectively. The current CS traffic and HT traffic of neighboring interfering traffic can be obtained by a simple one-hop signaling protocol. However, to calculate the LPTT, the CS and HT traffic obtained by exchanging the existing traffic information is not sufficient. The potential CS and HT traffic caused by self-traffic interference must be added to the current CS/HT traffic; therefore, we can use $(\lambda_{cs}, \lambda_{ht}, \lambda)$ to compute the accurate PPTT.

In the following we describe how to calculate the LPTT and PPTT in detail, as well as the packet service time for RTC traffic. Similar to other link-quality routing metrics, our proposed path metric is also the sum of individual link quality, so any routing protocols that use the PPTT metric will select the path with minimum path metric and, in return, can get the highest-quality path.

3.4.2 Path Predicted Transmission Time (PPTT) for 802.11

PPTT is calculated by summing each link's LPTT or, more precisely, $LPTT(\lambda_{cs}, \lambda_{ht}, \lambda)$. In this subsection we focus on calculating the potential CS traffic and HT traffic by considering the impact of self-traffic.

As mentioned above, CS traffic and HT traffic consist of two parts, neighboring traffic and self-traffic. The former part can be measured and collected periodically: along the corresponding routing path. To derive the latter part, we begin by characterizing the impact of self-traffic on each link in the path. The self-traffic along the path impacts each link by acting as CS and HT traffic. We use two parameters — carrier sensing factor (CSF) and hidden terminal factor (HTF) — to represent the self-traffic interference. The CSF of a link is the number of links in the path that are on the same channel and in the sender's CS range. The HTF of a link is the number of links in the path that are on the same channel and in the receiver's CS range, but not in the sender's range. For a link from node i to j, the increased CS traffic due to self-traffic interference is $CSF_{ij} \cdot \lambda$; and the increased HT traffic due to self-traffic interference is $HTF_{ij} \cdot \lambda$, where λ is the traffic rate along the path.

For link (i, j), considering self-traffic interference, the CS traffic becomes $\lambda_{cs} + CSF_{ij} \cdot \lambda$, and HT traffic becomes $\lambda_{ht} + HTF_{ij} \cdot \lambda$. Therefore, the LPTT of link i to j that has considered self-traffic interference is $LPTT(\lambda_{cs} + CSF_{ij} \cdot \lambda, \lambda_{ht} + HTF_{ij} \cdot \lambda, \lambda)$.

The CSF and HTF of a link are determined by link position and channel distribution of the path. Let us illustrate this with an example. Consider the four-hop path in Figure 3.6. First we consider the single-radio scenario (see Figure 3.6a). Because all nodes use the same radio (i.e., 802.11a), two adjacent links will interfere with each other and the one-hop away link must form the hidden terminal. Taking link (b,c) as an example, it has two CS links (a,b) and (c,d) and one HT link (d,e). Thus, CSF_{bc} is 2 and HTF_{bc} is 1. The CSF and HTF of other links can be calculated similarly. Second,

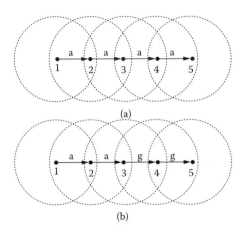

Figure 3.6   Illustration of CSF and HTF: (a) single-radio and (b) multi-radio.

we consider the multi-radio scenario (see Figure 3.6b). Each node has two radios, which can operate in 802.11a or 802.11g mode: links (a,b) and (b,c) use 802.11a radio, and links (c,d) and (d,e) use 802.11g radio. Because links on 802.11a mode and 802.11g do not interfere with each other, link (b,c) is only interfered by the self-traffic of link (a,b). Thus we get the value of CSF_{bc} (1) and HTF_{bc} (0), which are smaller than that in a single-radio case. Therefore, the channel diversity is reflected by the CSF and HTF of each link. In this sense, our proposed PPTT metric offers a unified way to take channel diversity into account. It can select the path with larger channel diversity that has smaller CSF and HTF, resulting in better performance.

For an n-hop path, we can obtain the CSF and HTF of each link according to its position in the path. Summing the LPTT of each link, we get the predicted transmission time of the entire path:

$$PPTT(\lambda) = \sum_{i=1}^{n} LPTT_i(\lambda_{cs}(i, i+1) + CSF_{i,i+1} \cdot \lambda,$$

$$\lambda_{ht}(i, i+1) + HTF_{i,i+1} \cdot \lambda, \lambda) \tag{3.2}$$

where λ is the average traffic rate of RTC.

3.4.3   Link Predicted Transmission Time (LPTT) for 802.11

LPTT is the prediction of packet transmission time across a link. A packet that will be transmitted over a link first enters the sender's queue to wait to be sent out. When the packet departs the queue, it enters the MAC layer. Therefore, the LPTT comprises the queueing delay and the MAC

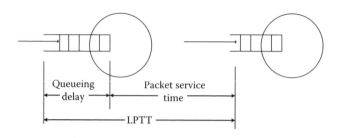

Figure 3.7 The link predicted transmission time (LPTT) comprises the queueing delay and packet service time.

layer processing time, which we also call *packet service time*. Figure 3.7 illustrates the component of the LPTT. Considering the traffic pattern of the RTC traffic, the queueing delay can be calculated according to a certain queueing model, that is, the $M/M/1$ model. The challenge in calculating the LPTT then becomes how to calculate packet service time.

There is some prior work [36,37] that studied the medium access control (MAC) layer processing time of the IEEE 802.11 protocol. All assume that every node in the network has the same packet collision probability. This is indeed the case in saturated networks. However, RTC traffic with the proper rate control mechanism will not saturate the networks. Thus, each node will have a different packet collision probability. To get an accurate estimation of the MAC layer processing time for the coming RTC flow and taking self-traffic interference into account, we introduce a new MAC model for RTC derived from the IEEE 802.11 protocol, in which we estimate the packet collision probability according to the given and its interfering traffic. As mentioned, self-traffic interference cannot be ignored for RTC. A packet may collide with the packets of HT traffic, which includes both existing neighboring traffic and coming self-traffic. We model the packet collision probability as a function of the HT traffic of each link. Thus, the packet collision probability can reflect the impact of both neighboring traffic and self-traffic.

The packet service time is calculated according to the specific MAC behavior and neighboring traffic conditions. An 802.11 node that wants to send a packet first waits for channel idle and performs a backoff period to access the channel and then sends control packets and data packets. If necessary, some control packets and data packets are retransmitted. Thus, the service time consists of channel access time, backoff time, control packets, and data packets transmission time. The channel access time is relative to CS traffic. The backoff time and packets transmission time are relative to HT traffic.

There are two access methods used under 802.11: (1) Distributed Coordination Function (DCF), which is more suitable for ad hoc networks, namely the basic access method; and (2) the Request-to-Send/Clear-to-Send

(RTS/CTS) access method. The basic access method uses only DATA and ACK packets. It has a well-known hidden-terminal problem. The DCF addresses this problem through the RTS/CTS access method. In the RTS/CTS access method, two more packets, RTS/CTS, are exchanged before transmitting DATA packets.

Before proceeding further, we need to introduce the following notations and assumptions. Considering a link from host i to j, we first define *contention link* and *hidden link*. We call a link a *contention link* of link (i, j) if the link's sender is in the carrier sensing range of node i. And we call a link a *hidden link* if the link's sender is the hidden terminal of node i. Let CL_{ij} be the set of contention links. If host i wants to send a packet, it will compete with the channel with links in CL_{ij}. Similarly, let HL_{ij} denote the set of hidden links. The packet sent from i to j may collide with the packets sent along the links in HL_{ij}. We use *DIFS, SIFS, EIFS*, and *slot* to denote the time interval of the DCF Inter-Frame Space, Short Inter-Frame Space, Extended Inter-Frame Space, and a slot [35], respectively. And we use S_D, S_A, S_R, and S_C to denote the size of DATA, ACK, RTS, and CTS packets, respectively. Let B_{ij} and CH_{ij} denote the bandwidth and channel number of link (i, j), respectively. In a 802.11 network, RTS, CTS, and ACK packets are transmitted at the basic rate; we use B_{basic} to denote this. And we use the notations *ACK, RTS*, and *CTS* for the required time periods of transmitting ACK, RTC, and CTS packets, respectively. In addition, we use $DATA_{ij}$ to denote the DATA packet transmission time along link (i, j). We assume that all the nodes in the network send packets exponentially. In our scheme, for simplicity of implementation, we use the Poisson traffic model. Despite the inaccuracies of this approximation, note that in Section 3.6, we can get rather good performance. In fact, our scheme is independent of the traffic model. We can use other traffic models, (e.g., MMPP [Markov Modulated Poisson Process] model) in our scheme. We use λ_{kl} to denote the packet transmitting rate along a certain link (k, l). We list all the notations in our analysis in Table 3.1.

Because both access methods are used in practical networks, we calculate the packet service time for these basic access methods in the following, and a similar derivation is given for the RTS/CTS access method in the Appendices (see Section 3.8.1).

In the basic access method, a node transmits a DATA packet if the channel is idle for a period of time that exceeds distributed interframe spacing (DIFS). If the channel is busy, it will wait until the end of the current transmission. It will further wait for an additional DIFS and a random backoff period determined by binary exponential backoff algorithm before transmission. The receiver replies with an ACK to the sender after successfully receiving the DATA packet. If the transmitter does not receive the ACK within a predefined time period, the entire process will be repeated.

Figure 3.8(a) shows the simplified state transmit diagram of the basic access method for transmitting a DATA packet from node i to node j. S0 is

Table 3.1   Notations in Our Analysis

(i, j)	Link from node i to node j
CL_{ij}	The set of contention links of link (i, j)
HL_{ij}	The set of hidden links of link (i, j)
DIFS, SIFS, EIFS and *slot*	DCF Inter-Frame Space, Short Inter-Frame Space, Extended Inter-Frame Space, and slot
S_D, S_A, S_R, and S_C	S_D represents the size of DATA packet, and others are likewise
B_{ij}	Bandwidth of link (i, j)
CH_{ij}	Channel used by link (i, j)
B_{basic}	Basic rate of 802.11 link
$DATA_{ij}$	DATA packet transmission time along link (i, j)
RTS, CTS, and *ACK*	*RTS* represents transmission time of RTS packet, and others are likewise
λ_{kl}	Packet transmitting rate along link (k, l)

the initial state, S1 is the state in which node i senses channel idle for DIFS and starts backoff with a random backoff time, S2 is the state in which the backoff timer of node i reaches zero and node i sends out the DATA packet, S3 is the state in which node j receives the DATA packet, and S4 is the state in which the retransmission time exceeds the LongRetryLimit (LRL).

When host i wants to send packet to host j, it first enters state S0. In this state, host i senses the channel, if the channel is idle for the DIFS period, it enters state S1, after delaying a random backoff time interval until

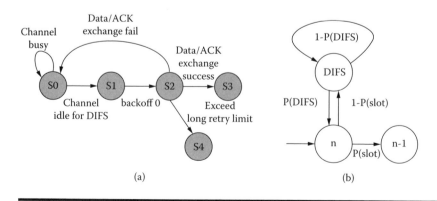

(a)

(b)

Figure 3.8   State transition of IEEE 802.11 MAC in the basic access method: (a) basic access and (b) backoff counter from n to $n-1$.

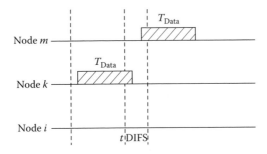

Figure 3.9 Two possible cases cause channel busy.

its backoff counter becomes 0, and then enters state S2. Otherwise, if the channel is busy in the DIFS period, it goes back to state S0. In state S2, host i sends a DATA packet to host j; if the DATA/ACK pair exchange successfully, it will enter state S3. If the exchange is failed, it will return to state S0. If the retransmission times exceed the LRL, it transits to state S4 and drops this packet. Thus, the average transition time from S0 to S3 and S4 is the service time of each packet.

For convenience, let P^i_{DIFS} and P^i_{slot} denote the probabilities of host i successes in sensing channel idle for the time interval DIFS and slot, respectively. And P^i_{DATA} is used to denote the probability of host i successfully sending the DATA packet.

There are two possible cases that node i will sense the channel busy in the DIFS period. These two cases are shown in Figure 3.9. Assume node i starts to sense channel at time t. First, if a neighbor node k sent packet in *DATA* period before t, the transmission will not be finished at t; thus, node i will sense the channel as busy. Second, if a neighbor node m starts to send packet in *DIFS* period after t, node i will also sense the channel as busy. Therefore, P^i_{DIFS} is equivalent to the probability that no link of CL_{ij} is transmitting packets in the time interval of *DATA + DIFS*. For a certain link (k, l), the probability of no packet transmitting along it in *DATA + DIFS* is:

$$\exp\left[-\left(\frac{S_{\text{D}}}{B_{kl}} + DIFS\right) \cdot \lambda_{kl}\right] \tag{3.3}$$

Thus,

$$P^i_{\text{DIFS}} = \prod_{(k,l)\in CL_{ij}} \exp\left[-\left(\frac{S_{\text{D}}}{B_{kl}} + DIFS\right) \cdot \lambda_{kl}\right]$$

$$= \exp\left[-\left(S_{\text{D}} \cdot \sum_{(k,l)\in CL_{ij}} \frac{\lambda_{kl}}{B_{kl}} + DIFS \cdot \sum_{(k,l)\in CL_{ij}} \lambda_{kl}\right)\right] \tag{3.4}$$

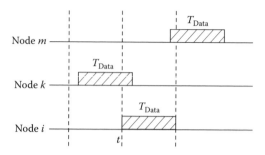

Figure 3.10 Two possible cases cause DATA packet collision.

Here, $\sum_{(k,l)\in CL_{ij}} \lambda_{kl}$ is the CS traffic of link (i, j), that is:

$$\lambda_{cs}(i, j) = \sum_{(k,l)\in CL_{ij}} \lambda_{kl} \tag{3.5}$$

Let

$$\lambda_{cs}^{norm}(i, j) = \sum_{(k,l)\in CL_{ij}} \frac{\lambda_{kl}}{B_{kl}} \tag{3.6}$$

We call $\lambda_{cs}^{norm}(i, j)$ the *nomalized CS traffic* of link (i, j).
Similarly, we get:

$$P_{slot}^{i} = \exp[-slot \cdot \lambda_{cs}(i, j)] \tag{3.7}$$

Two possible cases that cause DATA packet collision are shown in Figure 3.10. The DATA packet successful transmission probability P_{DATA}^{i} is equivalent to the probability that no link of HL_{ij} transmits a packet in the time interval of two DATA packets transmission.

$$
\begin{aligned}
P_{DATA}^{i} &= \prod_{(k,l)\in HL_{ij}} \exp\left[-\left(\frac{S_D}{B_{kl}} + \frac{S_D}{B_{ij}} \cdot \lambda_{kl}\right)\right] \\
&= \exp\left[-\left(S_D \cdot \sum_{(k,l)\in HL_{ij}} \frac{\lambda_{kl}}{B_{kl}} + \frac{S_D}{B_{ij}} \cdot \sum_{(k,l)\in HL_{ij}} \lambda_{kl}\right)\right]
\end{aligned} \tag{3.8}
$$

Here, $\sum_{(k,l)\in HL_{ij}} \lambda_{kl}$ is the HT traffic of link (i, j), that is:

$$\lambda_{ht}(i, j) = \sum_{(k,l)\in HL_{ij}} \lambda_{kl} \tag{3.9}$$

Let

$$\lambda_{\text{ht}}^{\text{norm}}(i, j) = \sum_{(k,l) \in HL_{ij}} \frac{\lambda_{kl}}{B_{kl}} \qquad (3.10)$$

We call $\lambda_{\text{ht}}^{\text{norm}}(i, j)$ the *normalized HT traffic* of link (i, j).

To obtain the packet service time, we consider the kth retransmission of the packet from node i to j. First, the node waits to ensure that the medium is idle for a DIFS period of time. This costs $\frac{DIFS}{P_{DIFS}^i}$ period of time. The backoff counter then selects a random number of backoff slots. Figure 3.8(b) shows the state transit diagram of a backoff counter.

At the state in which the backoff counter is n, if the channel is idle in the slot, it will transit to the next state in which the backoff counter is decreased by 1. If there are transmissions by other stations during the slot, then the station will freeze its backoff counter and will resume the count where it leaves off, after the DIFS interval in which channel is idle. Thus, considering the freezing of the backoff counter, the expected time duration of one backoff slot is:

$$\tau = \frac{slot}{P_{slot}^i} + \frac{1 - P_{slot}^i}{P_{slot}^i} \cdot \frac{DIFS}{P_{DIFS}^i} \qquad (3.11)$$

In 802.11, the backoff counter value is chosen randomly between 0 and the contention window CW. CW is an integer between CW_{min} and CW_{max}, with typical values being 31 and 1023, respectively. Initially, CW is equal to CW_{min}. Upon unsuccessful transmission, CW is doubled, until it reaches CW_{max}. After successful transmission, CW is again set to CW_{min}. Thus, the average number of backoff slots at the kth retransmission is $\frac{CW_{min}}{2} \cdot 2^{k-1}$.

If the transmission of a DATA packet is failed at the kth attempt, the time cost is:

$$t_k^f = \frac{DIFS}{P_{DIFS}^i} + \frac{CW_{min}}{2} \cdot 2^{k-1} \cdot \tau + DATA_{ij} + EIFS \qquad (3.12)$$

If the DATA packet is transmitted successfully at the kth attempt, the total spent time is:

$$t_k^s = \frac{DIFS}{P_{DIFS}^i} + \frac{CW_{min}}{2} \cdot 2^{k-1} \cdot \tau + DATA_{ij} + SIFS + ACK \qquad (3.13)$$

The probability that the kth retransmission is successful is $P_{DATA}^i \cdot (1 - P_{DATA}^i)^{k-1}$. Then the average packet service time of basic access method is:

$$T_{MAC} = \sum_{k=1}^{LRL} P_{DATA}^i (1 - P_{DATA}^i)^{k-1} \left(\sum_{i=1}^{k-1} t_i^f + t_k^s \right) + (1 - P_{DATA}^i)^{LRL} \sum_{k=1}^{LRL} t_k^f \qquad (3.14)$$

To obtain the queueing delay, we use the $M/M/1$ queueing model. Let λ denote the packet transmission rate across link (i, j), and the packet service rate is μ, where $\mu = 1/T_{MAC}$.

The queueing delay of the $M/M/1$ queue is:

$$T_{queue} = \frac{\lambda/\mu}{\mu - \lambda} \qquad (3.15)$$

Thus, substituting μ into Equation 3.15, we obtain:

$$T_{queue} = \frac{\lambda T_{MAC}^2}{1 - \lambda T_{MAC}} \qquad (3.16)$$

Thus, summing the queueing delay and packet service time, we obtain the final result for LPTT as:

$$LPTT(\lambda_{cs}, \lambda_{ht}, \lambda) = T_{queue} + T_{MAC} = \frac{T_{MAC}}{1 - \lambda T_{MAC}} \qquad (3.17)$$

Therefore, using Equations 3.2, 3.14 and 3.17, we can predict the transmission time along the path with the estimated traffic, including both neighbor traffic and self-traffic.

3.5 Implementation

In this section we discuss the implementation of the routing scheme with the new prediction-based routing metric, PPTT.

We have implemented our PPTT metric in an ad hoc routing framework called the MCL [3]. The MCL is a loadable Windows driver and is implemented as a 2.5 layer protocol. The MCL routes packets using the LQSR protocol, which is a link-state source routing protocol. We implemented our proposed PPTT metric in LQSR. We introduced a Link Info exchange scheme to exchange neighboring- and self-traffic information. The data structure of Link Info Message is shown in Figure 3.11(a). There are five fields in the Link Info Message. "Link Target Address" is a network address of the node where the link ends. That is, a Link Info Message contains some information on the link, which is from the node sending the Link Info Message to the "Link Target Address" node. In our implementation, we have considered the compatibility for both single-radio and multi-radio networks. In a multi-radio network, different links can use different channels and have different link bandwidths. We use the "Channel Number" and "Link BW" fields to deliver the channel number that the link used and the link bandwidth, respectively. "Tx Pkt Rate" is the packet sending rate

31	15	C
Link target addr		
Link BW	Channel num	
Tx Pkt rate	Rx Pkt rate	

(a)

31	15	C
Address n		
Link BW	Channel num	
$\lambda_{cs}(n, n+1)$	$\lambda_{cs}^{norm}(n, n+1)$	
$\lambda_{ht}(n, n+1)$	$\lambda_{ht}^{norm}(n, n+1)$	

(b)

Figure 3.11 Message format in the PPTT routing scheme: (a) link information message and (b) RREP message.

along this link, and "Rx Pkt Rate" is the packet receiving rate along this link.

Each node maintains two tables: "Local Link Info Table" (LLI Table) and "Neighboring Link Info Table" (NLI Table). We use a sample network shown in Figure 3.12 to illustrate these two tables. The LLI Table contains each link starting from the local node. For example, in the transmission range of node i, there are nodes h, j, a_1, a_2, a_3; thus, the LLI Table of node i looks like Table 3.2. Each node periodically broadcasts a Link Info Message that contains all its local links.

Each node can receive its neighbors' Link Info Message by gathering these Link Info Messages, the NLI Table can be constructed. For node i, it can receive the Link Info of nodes a_1, a_2, a_3, h, and j. Thus, the NLI Table of node i looks like Table 3.3. Based on this table, we can calculate the CS traffic and HT traffic of a link.

When a source node wants to send packets to a destination node, it generates and broadcasts an RREQ. When the destination node receives

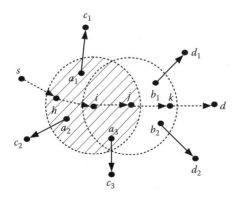

Figure 3.12 Sample network to illustrate LLI and NLI Tables.

Table 3.2   Local Link Info Table

Link	Bandwidth	Channel	Tx Pkt Rate	Rx Pkt Rate
(i,j)	B_{ij}	CH_{ij}	$\lambda_{TX}(i,j)$	$\lambda_{RX}(i,j)$
(i,h)	B_{ih}	CH_{ih}	$\lambda_{TX}(i,h)$	$\lambda_{RX}(i,h)$
(i,a_1)	B_{ia_1}	CH_{ia_1}	$\lambda_{TX}(i,a_1)$	$\lambda_{RX}(i,a_1)$
.	.	.	.	.
.	.	.	.	.
.	.	.	.	.

the RREQ, it generates and sends an RREP to the source node. The RREP contains the link info of each hop in the path. The data structure of RREP is shown in Figure 3.11(b). "Address n" is the address of nth node in the path; other fields are the traffic information of link $(n, n+1)$. When the RREP arrives at node n, node n calculates $\lambda_{cs}(n, n+1)$, $\lambda_{cs}^{\text{norm}}(n, n+1)$, $\lambda_{ht}(n, n+1)$, and $\lambda_{ht}^{\text{norm}}(n, n+1)$, and then fills these parameters into the RREP.

$\lambda_{cs}(n, n+1)$ and $\lambda_{cs}^{\text{norm}}(n, n+1)$ are calculated according to NLI Tables. Because all links in the NLI Table are contention links of node n, they can be calculated as follows:

$$\lambda_{cs}(n, n+1) = \sum_{(i,j)\in NLI_n, CH_{ij}=CH_{n,n+1}} \lambda_{TX}(i, j) \tag{3.18}$$

$$\lambda_{cs}^{\text{norm}}(n, n+1) = \sum_{(i,j)\in NLI_n, CH_{ij}=CH_{n,n+1}} \frac{\lambda_{TX}(i, j)}{B_{ij}} \tag{3.19}$$

where NLI_n is the set of links in the NLI Table of node n.

$\lambda_{ht}(n, n+1)$ is the HT traffic rate of link $(n, n+1)$. It is the sum of traffic rates of nodes that are neighbors of $n+1$, but not n. When RREP arrives at node n, it already contains $\lambda_{cs}(n+1, n+2)$. Therefore, by eliminating the traffic rate of nodes that are neighbors of both n and $n+1$, we can get $\lambda_{ht}(n, n+1)$. By the lookup NLI Table of node n, we can know which node is a neighbor of node $n+1$, because the NLI Table contains the link info

Table 3.3   Neighboring Link Info Table

Link	Bandwidth	Channel	Tx Pkt Rate	Rx Pkt Rate
(j,k)	B_{jk}	CH_{jk}	$\lambda_{TX}(j,k)$	$\lambda_{RX}(j,k)$
(h,s)	B_{hs}	CH_{hs}	$\lambda_{TX}(h,s)$	$\lambda_{RX}(h,s)$
(a_1,c_1)	$B_{a_1 c_1}$	$CH_{a_1 c_1}$	$\lambda_{TX}(a_1,c_1)$	$\lambda_{RX}(a_1,c_1)$
.	.	.	.	.
.	.	.	.	.
.	.	.	.	.

of node $n+1$. Thus, we can know which node is a neighbor of both n and $n+1$. We use A_n and A_{n+1} to denote the set of neighbor nodes of node n and $n+1$, respectively. Hence, we can calculate $\lambda_{ht}(n, n+1)$ as follows:

$$\lambda_{ht}(n, n+1) = \lambda_{cs}(n+1, n+2) - \sum_{i \in A_n \cap A_{n+1}, CH_{ij}=CH_{n,n+1}} \lambda_{TX}(i, j) \quad (3.20)$$

Similarly, $\lambda_{ht}^{norm}(n, n+1)$ can be calculated as follows:

$$\lambda_{ht}^{norm}(n, n+1) = \lambda_{cs}^{norm}(n+1, n+2) - \sum_{i \in A_n \cap A_{n+1}, CH_{ij}=CH_{n,n+1}} \frac{\lambda_{TX}(i, j)}{B_{ij}} \quad (3.21)$$

When a source node receives an RREP, it gets the CS and HT traffic information of each link in the path. In a multi-radio network, we also get the channel distribution of the entire path. We can calculate the CSF and HTF of each link according to the link position and channel distribution of the path. In the calculation of λ_{cs} and λ_{ht}, we only calculate those links that are on the same channel, because the links on different channels do not intefere with each other. Thus, the PPTT can be calculated based on this traffic information. The source node selects the min-PPTT path for coming traffic flow.

Moreover, in practice, wireless network interface cards (NICs) support the "autorate" feature. The NIC automatically selects the bandwidth for every packet. Thus we can get an accurate bandwidth of link only by measuring it empirically. In our implementation, we measure the bandwidth using the technique of packet pairs [32]. The accuracy of packet-pair measurement is studied in Draves et al. [2]. It can be concluded, despite some inaccuracies, that it is able to unambiguously distinguish between various channel bandwidths.

Because PPTT needs nodes to exchange traffic information between each other, this causes some overhead. In our implementation, each node broadcasts a Link Info Message periodically with the same frequency as that in WCETT. The size of each item in the Link Info Message is 12 bytes, and the complexity of broadcast is $O(n)$. Thus we can estimate the PPTT without incurring too much overhead.

3.6 Performance Evaluation

We evaluate the performance of the PPTT scheme in both single-radio and multi-radio scenarios. In this section we describe the results of our experiments on a real testbed and simulations by ns-2. First, we illustrate the accuracy of PPTT by simulation. Then we present experimental results of the PPTT scheme in our testbed and compare the performance of PPTT to ETX in the single-radio scenario. We evaluate PPTT in random topology

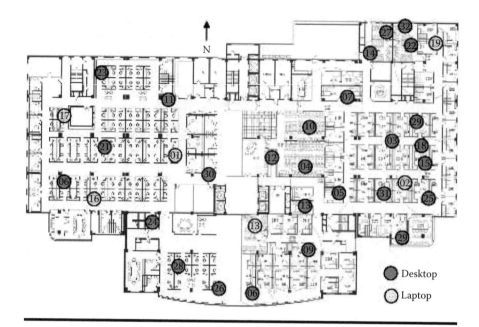

Figure 3.13 Our testbed, consisting of 32 nodes placed in fixed locations inside an office building.

through extensive simulations. Finally, we study the performance of PPTT in a multi-radio scenario by both experiments in testbed and simulations.

3.6.1 Testbed Experiments

We built a 32-node wireless testbed. Our testbed is located on one floor of a fairly typical office building. All nodes were placed in fixed locations and did not move during testing. The topology of the testbed is shown in Figure 3.13. The nodes are all DELL PCs equipped with a LinkSys Dual-Band Wireless A+G Adapter and an ORiNOCO 802.11abg ComboCard Gold card. In our experiments, each card runs on 802.11a and/or 802.11g radios. The cards all perform autorate selection and have RTS/CTS disabled.

We installed an MCL driver in which the PPTT metric is implemented on each PC in our testbed. The experimental data in this chapter is the result of measurements from our testbed.

3.6.2 Accuracy of PPTT

To verify the accuracy of PPTT, we conducted the following simulations. The topology is shown in Figure 3.14. There are two paths connected: 1 with 6 and 7 with 11, respectively. The transmission range and the

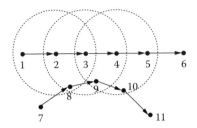

Figure 3.14 Two paths interfere with each other. Flow from node 1 to node 6 is background traffic, and we study the end-to-end delay of traffic from node 7 to node 11 by prediction and measurement.

interference range are equal. Nodes 8, 9, and 10 are interfered by the path 1 to 6. Each node has an 802.11b radio with 2-Mbps bandwidth.

We first only consider the self-traffic interference. We inject a flow from node 7 to node 11. There is no other neighboring traffic. Thus, only self-traffic will impact the end-to-end path delay. We vary the traffic sending rate and compare the end-to-end path delay with the computed PPTT. Figure 3.15(a) shows the comparison. It is shown that PPTT is well matched with the real end-to-end delay. We can also observe that as the traffic rate increases, the end-to-end delay increases exponentially. This is because as the traffic rate increases, the packet collision probability increases, which leads to the backoff time increasing exponentially.

Then we set the interfering traffic from 1 to 6 as 128 Kbps, and we also vary the traffic rate from node 7 to 11 and compare the prediction of path transmission time with the end-to-end path delay measured from the testbed. Figure 3.15(b) shows that PPTT is also well matched with real path delay.

3.6.3 Single-Radio Scenario

We first conduct some experiments to illustrate the self-traffic effect and PPTT performance in our testbed. We add two flows to our testbed: one is from node 26 to 10 and the other is from node 11 to 21 (Figure 3.16). The flow from node 11 to 21 starts earlier than flow from node 26 to 10 and acts as background traffic. From node 26 to 10, there are two paths: path1 (26, 30, 12, 10) and path2 (26, 8, 9, 13, 4, 10). Path1 is interfered by the background traffic. We begin by setting flow 11 to 21 as a fixed traffic rate, 2 Mbps, and varying the traffic rate from 26 to 10. The goodput comparison of PPTT to ETX is shown in Figure 3.17(a). We can see that when traffic rate from 11 to 21 is between 2.2 and 5 Mbps, PPTT can achieve larger goodput than ETX; and at other traffic rates, PPTT has comparable goodput with ETX.

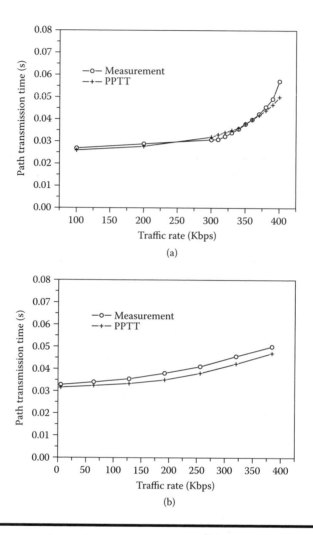

Figure 3.15 Comparison of PPTT with real end-to-end delay: (a) no neighboring traffic and (b) with neighboring traffic.

The better performance of PPTT derives from the consideration of both neighboring traffic interference and self-traffic interference. When the traffic rate is less than 2.2 Mbps, the self-interference is not serious. Although path1 is interfered by flow 11 to 21, it performs better than path1 because it is shorter. When the traffic rate is more than 2.2 Mbps, self-interference cannot be ignored. The performance of path2 is better than path1. ETX does not take self-interference into account, nor does it attempt to select path2. This is reflected in the fact that the goodput using ETX is lower than PPTT.

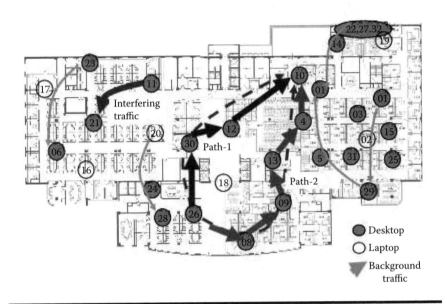

Figure 3.16   Our testbed, consisting of 32 nodes placed in fixed locations inside an office building.

Next we set the traffic rate from node 26 to 10 at a fixed 3.8 Mbps and vary the traffic rate from node 11 to 21. The comparison of goodput is shown in Figure 3.17(b).

The plots in Figure 3.17 show that PPTT outperforms ETX. When background traffic exceeds 512 Kbps, the cumulative interference of both neighboring traffic and self-traffic leads to a poorer performance of path1 than path2. Because PPTT takes both types of traffic interference into account, it can select the better one. This is reflected in the fact that the delay using PPTT is lower than in ETX and the goodput is better.

The main conclusion from these experiments is that PPTT performs better than ETX. The increase in performance is a result of the fact that PPTT takes both neighboring interference and self-interference into account. This sometimes leads it to select a longer path than ETX; however, these longer paths result in better performance.

Then we use the ns-2 simulator to evaluate PPTT in a random topology wireless network. The DCF of IEEE 802.11 is used as the MAC layer protocol. RTS/CTS are disabled. The bandwidth of each node is 2 Mbps. The carrier sensing range is 300 meters and transmission range is 250 meters.

We generate a random topology with 30 nodes in 1000 × 1000-meter rectangular field. All nodes are static. The source-destination pairs are spread randomly over the network. We establish 20 constant bit rate (CBR) connections by choosing source and destination randomly and ensure that there are at least two connections active at the same time. Each connection

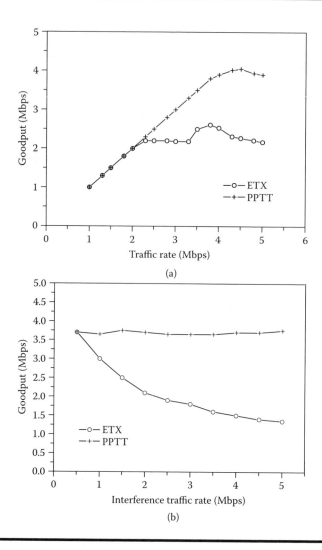

Figure 3.17   Goodput comparison of PPTT to ETX with varying (a) self-traffic and (b) neighboring traffic.

lasts for 20 seconds. The order in which the connections are established is randomized. The waiting time between the start of two successive connections is 5 seconds. Simulations last for 500 simulated seconds. To compare the performance under different circumstances of traffic load, we set traffic rate of each CBR connection as 128, 256, or 384 Kbps.

For each simulation, we calculate the average goodput and delay of all CBR connections. In Figures 3.18(a) and 3.18(b), compare goodput and end-to-end delay of ETX to PPTT. From these figures we can see that in the 128-Kbps case, ETX can achieve the same performance as PPTT.

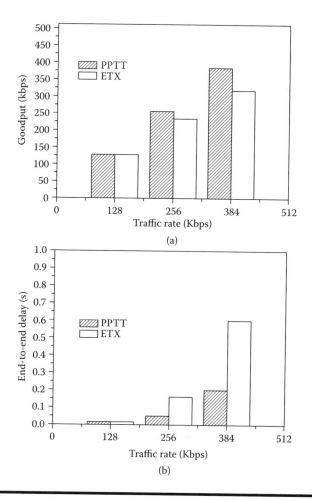

Figure 3.18   Comparison of (a) goodput and (b) delay of ETX to PPTT.

But in the 256-Kbps and 384-Kbps cases, the performance of PPTT is much better than that of ETX. The average goodput using the PPTT metric is up to 28 percent higher than for ETX. And the average end-to-end delay of each connection using PPTT metric is up to 52 percent less than for ETX.

The gain is a result of the fact that PPTT takes self-traffic interference into account as well as neighboring traffic interference. Two simultaneous connections possibly interfere with each other. Because ETX does not consider self-traffic interference, it may not select the high-quality path. The ability of PPTT to select good paths is illustrated in Figure 3.19. This figure shows the relationship between path length and throughput for ETX and PPTT. We can see that ETX might select paths in a sub-optimal manner. For example, considering the three-hop path in both plots, ETX sometimes selects

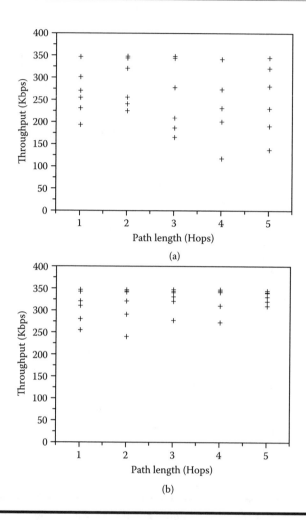

Figure 3.19   Relationship between path length and throughput of individual connections in a single-radio scenario: (a) ETX and (b) PPTT. Each dot represents a candidate path.

a low-throughput path. However, PPTT always selects higher-throughput paths.

3.6.4   Multi-Radio Scenario

Our proposed PPTT scheme has considered the channel diversity of a multi-radio network, so it can also serve as a multi-radio routing metric.

We first test the PPTT scheme in our testbed. Each node equips two wireless network cards: one operates on 802.11a mode and the other

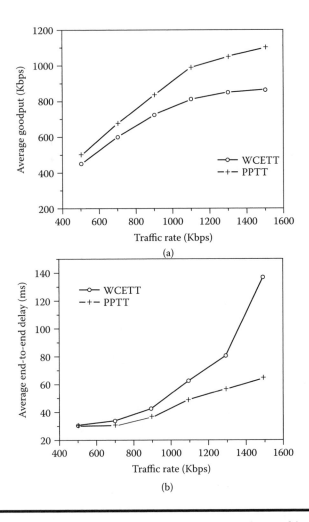

Figure 3.20   Performance comparison of PPTT and WCETT in a multi-radio testbed: (a) goodput and (b) delay.

operates on 802.11g mode. We randomly carry out User Datagram Protocol (UDP) connections in the testbed and vary the traffic rate of each connection to emulate different loaded cases, both for WCETT and PPTT. The comparison of goodput and delay with different traffic load are shown in Figures 3.20(a) and 3.20(b), respectively. The average goodput using PPTT is up to 30 percent higher than WCETT. And the average end-to-end delay using PPTT is up to 57 percent less than WCETT.

The ability of PPTT to select a high-performance path is illustrated in Figure 3.21. WCETT sometimes selects some low-throughput path, while PPTT always selects better paths. For example, let us think about an individual

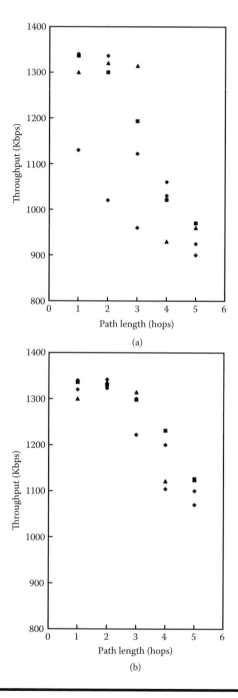

Figure 3.21 Relationship between path length and throughput of individual connections in a multi-radio scenario: (a) WCETT and (b) PPTT. Each dot represents a candidate path.

Figure 3.22   One flow from node 23 to node 26; two-candidate multi-radio path.

flow from node 23 to node 26. As Figure 3.22 shows, there are two candidate paths: Path-1 (23a, 17a, 8ag, 16g, 24ga, 28a, 26a)[2] and Path-2 (23g, 8g, 16g, 24ga, 28a, 26a). WCETT of path-2 is less than that of path-1; thus path-2 is selected by WCETT. However, there is a hidden terminal in path-2 while there is no hidden terminal in path-1. The self-traffic impact of path-2 is more serious than that of path-1. Therefore, the goodput of path-1 is 1.2 Mbps while that of path-2 is 900 Kbps. Because self-traffic is explicitly considered in the PPTT scheme, PPTT will select path-1 rather than path-2.

We then compare the performance of PPTT and WCETT by simulations. We generate a random topology with 30 static two-radio nodes in 1000 × 1000-meter rectangular field. We establish 20 CBR connections randomly over the network. We adjust the traffic rate of each CBR connection to compare the performance under different traffic-load cases. The simulation results are shown in Figure 3.23. The results show that PPTT achieves better performance in terms of goodput (Figure 3.23a) and delay (Figure 3.23b) than WCETT.

Both the experimental testbed and simulation results show that in light-loaded cases, PPTT can achieve performance similar to WCETT.

---

[2] 23a means the packet is incoming and outgoing both from 802.11a radio of node 23. 8ag means the packet is incoming from 802.11a radio of node 8 and outgoing from 802.11g radio of node 8. The others are likewise.

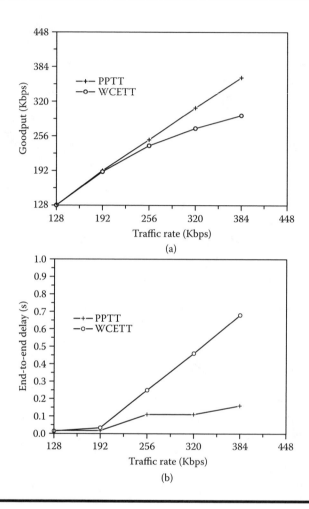

Figure 3.23 Performance comparison of PPTT and WCETT in a multi-radio scenario: (a) goodput and (b) delay.

In light-loaded cases, self-traffic interference is trivial; the path performance is determined by channel diversity. Because PPTT takes channel diversity into account, it can achieve comparable performance to WCETT. In heavy-loaded cases, PPTT can achieve lower delay and higher goodput than WCETT. This is because PPTT considers both channel diversity and self-traffic interference, while WCETT considers channel diversity only.

3.7 Conclusions

The self-traffic effect should be taken into account in the routing metric to get a more accurate estimation of the transmission time along a path, especially for the RTC flow, which has critical delay and bandwidth requirements. This chapter proposes a new prediction-based routing metric called PPTT to deduce the transmission time for the new given traffic before it is injected into the wireless mesh network, and tends to select the route with minimal PPTT. We analyze packet service time for rate-controlled RTC traffic to deduce the expected transmission time on the corresponding link. Based on this model, we estimate the LPTT according to the interfering traffic from neighbors, including both carrier sensing nodes and hidden terminal nodes. The calculation of LPTT reflects the effect of neighboring traffic, while the PPTT reflects the effect of self-traffic; thus, PPTT offers a unified traffic-aware routing metric for wireless mesh networks. Results from both real testbed and network simulator (ns2) experiments show that our PPTT algorithm out-performs other routing schemes without considering the self-traffic effect for RTC traffic. The average goodput improvement is about 28 percent, while the delay improvement is about 52 percent.

PPTT can also serve as a multi-radio routing metric because it calculates the interference effect for different channels accordingly, instead of putting them together. In our experiment, PPTT outperforms WCETT for RTC traffic in a multi-radio environment. As a next step, we are going to evaluate the performance of our proposed scheme in a larger wireless mesh network.

3.8 Appendices

3.8.1 *Packet Service Time under RTS/CTS Access Method*

In the RTS/CTS access method, the station that wants to send a DATA frame first transmits an RTS packet after the channel is available for a period longer than DIFS or the backoff time reaches zero. When the receiver receives the RTS, it transmits a CTS packet. If the CTS packet is not received within a predefined time interval, the sender retransmits the RTS packet. After successful reception of the CTS packet, the sender will send out the DATA packet.

Figure 3.24 shows a simplified diagram of host i attempting to send a packet to host j using the RTS/CTS access method.

Compared with the basic access method, the RTS/CTS access method uses a four-phase RTS-CTS-DATA-ACK handshake. When host i senses channel idle for a DIFS period and the backoff counter reaches zero, it enters into state S2 and sends RTS instead of DATA to node j. If the RTS/CTS exchange is successful, it will enter state S3. If the exchange is failed, it will return to state S0. We assume that if RTS/CTS exchanges successfully, the DATA will be sent successfully. Thus from S3, it will enter state S4. If the retransmission times exceed the SRL, it transits to state S5 and drops this

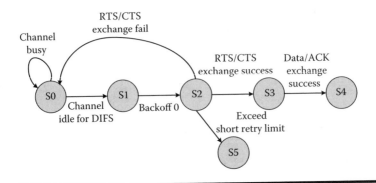

Figure 3.24 RTS/CTS access method of IEEE 802.11 MAC.

packet. The packet service time is the average transition time from S0 to S4 and S5.

P_{RTS}^i is used to denote the probability of host i sending the RTS packet successfully to host j. Similarly, this is equivalent to the probability that no packet is sent from the hosts of H_{ij} in the time of $DATA + RTS$. Thus:

$$P_{RTS}^i = \exp\left[-\left(S_D \cdot \lambda_{ht}^{norm}(i, j) + \frac{S_R}{B_{basic}} \cdot \lambda_{ht}(i, j)\right)\right] \qquad (3.22)$$

We consider the kth retransmission of the packet from node i to j. The channel access process and the backoff process are the same as in the basic access method. If the transmission of the RTS packet is failed at the kth attempt, the time spent for the kth attempt is:

$$t_k^f = \frac{DIFS}{P_{DIFS}^i} + \frac{CW_{min}}{2} \times 2^{k-1} \times \tau + RTS + EIFS \qquad (3.23)$$

If the RTS packet is transmitted successfully at the kth attempt, the spent time of the kth attempt is:

$$t_k^s = \frac{DIFS}{P_{DIFS}^i} + \frac{CW_{min}}{2} \times 2^{k-1} \times \tau + RTS + SIFS$$

$$+ CTS + SIFS + DATA_{ij} + SIFS + ACK \qquad (3.24)$$

The probability that the kth retransmission is successful is:

$$P_{RTS}^i \times \left(1 - P_{RTS}^i\right)^{k-1}$$

Thus, the average packet service time of RTS/CTS access method is:

$$T_{MAC} = \sum_{k=1}^{SRL} P_{RTS}^i \left(1 - P_{RTS}^i\right)^{k-1} \left(\sum_{i=1}^{k-1} t_i^f + t_k^s\right)$$
$$+ \left(1 - P_{RTS}^i\right)^{SRL} \sum_{i=1}^{SRL} t_i^f \qquad (3.25)$$

Acknowledgments

We would like to thank Richard Draves et al. for providing MCL code. This is the basis for implementing our proposed routing metric. We would also like to thank Yang Yang and Yunxin Liu for their help in the testbed experiments.

References

[1] T.-W. Chen, J.T. Tsai, and M. Gerla, QoS routing performance in multihop, multimedia, wireless networks, *Proceedings of IEEE ICUPC '97*, 1997.

[2] R. Draves, J. Padhye, and B. Zill, Routing in multi-radio, multi-hop wireless mesh networks, *Proceedings of the 10th Annual International Conference on Mobile Computing and Networking*, 2004:114–128, Philadelphia.

[3] Microsoft Mesh Connectivity Layer (MCL) Software, http://research. microsoft.com/mesh.

[4] The Network Simulator — ns-2, 2003, http://www.isi.edu/nsnam/ns.

[5] S.-T. Sheu and T.-F. Sheu, DBASE: A distributed bandwidth allocation/sharing/extension protocol for multimedia over IEEE 802.11 ad hoc wireless LAN, *Proceedings of INFOCOM*, 2001, 3(2001):1558–1567.

[6] S. Mangold, S. Choi, P. May, O. Klein, G. Hiertz, and L. Stibor, IEEE 802.11e Wireless LAN for Quality of Service, *Proceedings of European Wireless*, 2002.

[7] I. Aad and C. Castelluccia, Differentiation Mechanisms for IEEE 802.11. *Proceedings of INFOCOM*, 2001.

[8] J. Sheu, C. Liu, S. Wu, and Y. Tseng, A priority MAC protocol to support real-time traffic in ad hoc networks, *Wireless Networks*, 2004, 10(1):61–69.

[9] Y.-C. Hsu and T.-C. Tsai, Bandwidth routing in multihop packet radio environment, *Proceedings of the 3rd International Mobile Computing Workshop*, 1997.

[10] C.R. Lin and C.-C. Liu, An on-demand QoS routing protocol for mobile ad hoc networks, *Proceedings of IEEE Global Telecommunications Conference*, 2000.

[11] V. Kanodia, C. Li, A. Sabharwal, B. Sadeghi, and E. Knightly, Distributed multi-hop scheduling and medium access with delay and throughput constraints, *Proceedings of the Seventh Annual International Conference on Mobile Computing and Networking (MobiCom'01)*, Rome, Italy, 2001.

[12] H. Luo, S. Lu, V. Bharghavan, J. Cheng, and G. Zhong, A packet scheduling approach to QoS support in multihop wireless networks, *ACM Journal of Mobile Networks and Applications (MONET), Special Issue on QoS in Heterogeneous Wireless Networks*, 2002.

[13] C. Zhu and M.S. Corson, QoS routing for mobile ad hoc networks, *Proceedings of IEEE Infocom*, June 2001.

[14] C.R. Lin, On-demand QoS routing in multihop mobile networks, *Proceedings of IEEE Infocom*, April 2001.

[15] W. Liu and Y. Fang, Courtesy piggybacking: supporting differentiated services in multihop mobile ad hoc networks, *Proceddings of IEEE Infocom*, 2004.

[16] A. Veres, G. Ahn, A.T. Campbell, and L. Sun, SWAN: service differentiation in stateless wireless ad hoc networks, *Proceedings of IEEE Infocom*, June 2002.

[17] Q. Xue and A. Ganz, Ad hoc QoS on-demand routing (AQOR) in mobile ad hoc networks, *Journal of Parallel and Distributed Computing*, 2003, 63:154–165.

[18] C. Perkins and E. Belding-Royer, Quality of Service for Ad hoc On-Demand Distance Vector Routing (work in progress), October 2003. draft-perkins-manet-aodvqos-02.txt.

[19] K. Al Agha, G. Pujolle, H. Badis, and A. Munaretto, QoS for ad hoc networking based on multiple metrics: bandwidth and delay, *Proceedings of the Fifth IEEE International Conference on Mobile and Wireless Communications Networks (MWCN 2003)*, 2003.

[20] R. Sivakumar, P. Sinha, and V. Bharghavan, CEDAR: a core-extraction distributed ad hoc routing algorithm, *Proceedings of IEEE Infocom*, March 1999.

[21] X. Zhang, S.B. Lee, A. Gahng-Seop, and A.T. Campbell, INSIGNIA: An IP-based quality of service framework for mobile ad hoc networks, *Journal of Parallel and Distributed Computing (Special Issue on Wireless and Mobile Computing and Communications)*, 2000, 60(4):374–406.

[22] D.H. Cansever, A.M. Michelson, and A. H. Levesque, Quality of service support in mobile ad-hoc IP networks, *Proceedings of IEEE MILCOM*, 1999:30–34.

[23] S.H. Shah and K. Nahrstedt, Predictive location-based QoS routing in mobile ad hoc networks, *Proceedings of IEEE International Conference on Communications*, 2002.

[24] H. Sun and H.D. Hughes, Adaptive QoS routing based on prediction of local performance in ad hoc networks, *Proceedings of IEEE WCNC*, 2003.

[25] C.E. Perkins, E.M. Belding-Royer, and S. Das, Ad Hoc On Demand Distance Vector (AODV) Routing, IETF RFC 3561.

[26] D.B. Johnson, D.A. Maltz, and Y.-C. Hu, The Dynamic Source Routing Protocol for Mobile Ad Hoc Networks (DSR). Internet Draft (work in progress), IETF, April 2003. http://www.ietf.org/internet-drafts/draft-ietf-manet-dsr-09.txt

[27] V. Park and S. Corson, Temporally-Ordered Routing Algorithm (TORA) Version 1 Functional Specification, 2001, http://www.ietf.org/internet-drafts/draft-ietf-manet-tora-spec-04.txt.

[28] C.E. Perkins and P. Bhagwat, Highly dynamic destination-sequenced distance-vector routing (DSDV) for mobile computers, *Proceedings of the Conference on Communications Architectures, Protocols and Applications*, 1994:234–244, London, U.K.

[29] R. Draves, J. Padhye, and B. Zill, Comparison of routing metrics for static multi-hop wireless networks, *Proceedings of ACM SIGCOMM*, Portland, OR, August 2004.

[30] D.S.J. De Couto, D. Aguayo, J. Bicket, and R. Morris, A high-throughput path metric for multi-hop wireless routing, *Proceedings of the 9th Annual International Conference on Mobile Computing and Networking*, 134–146, San Diego, CA, 2003.

[31] A. Adya, P. Bahl, J. Padhye, A. Wolman, and L. Zhou, A multi-radio unification protocol for IEEE 802.11 wireless networks, *Proceedings of BroadNets*, 2004.

[32] S. Keshav, A control-theoretic approach to flow control, *Proceedings of SIGCOMM*, September 1991.

[33] Y.-C. Hu and D.B. Johnson, Design and demonstration of live audio and video over multihop wireless ad hoc networks, *Proceedings of the MILCOM*, 2002.

[34] T. Goff, N.B. Abu-Ghazaleh, D.S. Phatak, and R. Kahvecioglu, Preemptive routing in ad hoc networks, *Proceedings of the 7th Annual International Conference on Mobile Computing and Networking*, 2001:43–52, Rome, Italy.

[35] I.S. Committee, Wireless LAN Medium Access Control (MAC) and Physical Layer (PHY) Specifications, in IEEE 802.11 Standard, IEEE, ISBN 1-55937-935-9, 1997.

[36] G. Bianchi, Performance analysis of the IEEE 802.11 distributed coordinated function, *IEEE JSAC*, 2000, 18(3):535–547.

[37] N. Gupta and P.R. Kumar, A performance analysis of the IEEE 802.11 wireless LAN medium access control, *Communications in Information and Systems*, 2004, 3(4):279–304.

[38] Y.Y. Kim and S. Li, Modeling multipath fading channel dynamics for packet data performance analysis, *Wireless Networks*, 2000, 6(2000):481–492.

[39] J. Broch, D.A. Maltz, and D.B. Johnson, Supporting hierarchy and heterogeneous interfaces in multi-hop wireless ad hoc networks, *Proceedings of Parallel Architectures, Algorithms, and Networks*, 1999, 370–375.

Chapter 4

Reliable Multicast for Wireless LAN

Sunghyun Choi and Kihwan Choi

Contents

4.1   Overview

Prevailing IEEE 802.11 wireless local area network (WLAN) today does not support any medium access control (MAC) protocol for reliable multicast (RM). It only provides unreliable multicast without any feedback from receivers and reliable unicast with positive acknowledgment (ACK) from the receiver. This chapter delves into parity-based multicast protocols. To support RM over the 802.11 WLAN, two different RM MAC protocols — a polling-based feedback protocol and a contention-based feedback protocol — are developed. Both protocols improve the efficiency of multicast as well as the reliability by supporting the per-group feedback based on negative ACK (NAK). That is, combined with the reliable MAC protocols that support per-group feedback, parity-based loss recovery can improve multicast efficiency. Therefore, we investigate how the proposed RM MAC protocols can be combined with forward error correction (FEC) of the upper layer to achieve reliable and efficient multicast. For the performance evaluation, we consider multicasting scenarios using MAC-level reliable protocols and upper-layer-level FEC. The simulation results demonstrate that our multicast schemes substantially outperform the existing scheme based on per-packet feedback in terms of efficiency, and guarantee the RM, which the 802.11 cannot provide.

4.2   Introduction

RM is an efficient transmission scheme to deliver data from a sender to multiple receivers or "destinations." RM requires lower bandwidth and complexity than unicast to individual destinations in the network. Multicast can also be used by different applications, such as data distribution, video conferencing, shared whiteboards, online games, and distributed computing while guaranteeing the reliability.

A typical multicast scenario in wireless network is depicted in Figure 4.1. To distinguish between the original sender of the multicast data and the node transmitting multicast data to receivers in a single wireless cell, we refer to the former as the "original source" (e.g., a remote node multicasting stocked data) and to the latter as the "multicast source" (e.g., the access point of the cell). We consider the multicast environment, where a multicast source transmits multicast data to receivers in the same wireless link.

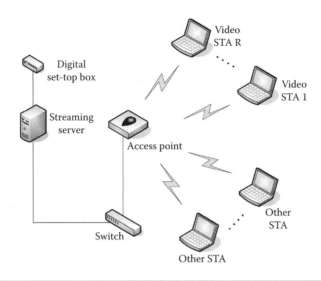

Figure 4.1   Network topology for reliable multicast.

We also assume that the speed of the link between the original source and multicast source is so high that the multicast source can be considered the original source.

In many wireless networks, including IEEE 802.11 WLAN, wireless nodes transmit data in broadcast manner due to their omni-directional antenna property. This transmission property is useful for accessing multiple nodes at one time, but prohibits collecting the response (e.g., ACK) from multiple nodes efficiently. Therefore, the main challenge in providing RM in a WLAN is the development of an efficient feedback scheme that is needed to provide the reliability.

A clear distinction between the multicasts in the WLAN and the traditional (wireline) network environment is that multiple nodes in the WLAN share a single medium so that a node can access other nodes in the same basic network in one-hop, while traditional networks often assume dedicated routes for each node. IEEE 802.11 WLAN, however, does not provide any MAC-layer protocol for RM. The IEEE 802.11 standard only defines unreliable multicast MAC protocol without any feedback and reliable unicast MAC protocol with ACK [1]. Unreliable multicast MAC protocol can exploit the omni-directional antenna property, but cannot guarantee the reliability of multicast. Reliable unicast MAC protocol supports reliability via automatic repeat request (ARQ) with ACK feedback.

To provide the reliability of multicast, we consider MAC-level multicast protocol with per-group feedback using NAK. Per-group feeback collects feedback information on the transmitted packets after transmission of a group of packets, while per-packet feedback collects feedback information on the transmitted packet after the transmission of each packet. The reason why we adopt NAK for feedback is that NAK facilitates forward error correction (FEC) of the upper layer, while drastically reducing the feedback congestion. Previous work [3,6] supports the combination of FEC and ARQ using NAK can improve the performance of RM by reducing bandwidth waste. Although FEC by itself cannot provide an efficient multicast, e.g., layered FEC in [3], FEC combined with ARQ has very high repair efficiency, thus substantially reducing the network bandwidth requirements for RM.

This chapter delves into two MAC-level ARQ protocols with per-group feedback using NAK for RM, and combines the protocols with the upper-layer FEC for efficient multicast. We propose two MAC protocols to avoid feedback collision: one uses polling-based NAK feedback and the other uses contention-based NAK feedback. First, with polling-based feedback, the multicast source calls each receiver so that the receiver transmits its own NAK to the source. After the multicast source receives a NAK from a receiver, the source transmits parity packets, where the number of additional parity packets is determined by the received NAK. The multicast source repeats the inquiry of NAK frames, and replies until the source recognizes that all receivers do not need to receive packets anymore. On the other hand, with contention-based feedback, the receivers transmit their NAK frames in a contentious manner based on carrier-sense multiple access with collision avoidance (CSMA/CA). When the multicast source successfully receives a NAK from a receiver, it transmits the corresponding number of packets to the receiver. With the proposed protocols, the multicast source can exploit the information of the received NAK, such as the number of additionally requested packets. Using the information, the multicast source prepares additional parity packets to satisfy the receivers. Because additional parity packets are transmitted based on the demands of receivers, we can avoid redundant medium access, which is a critical weakness of FEC.

We consider the multicast transmission from access point (AP) to stations in the infrastructure mode although the proposed protocols can be applied to the ad hoc mode with minor changes in the frame format. Therefore, the multicast source is always the AP and the receivers are the stations associated with the AP. This situation is a typical multicast environment, where AP is highly loaded for real-time multicast such as video broadcast.

4.3 Related Work

There is a large amount of literature on RM; some articles focus on the use of FEC. An iterative polling for feedback to enhance the efficiency of multicast is suggested in [9]. An early work [3] studied the use of FEC integrated with ARQ for RM. It assumed that the RM layer exists on the FEC layer or the RM layer can be integrated with the FEC layer. However, for WLAN, MAC-level RM can exploit the medium access property but IEEE 802.11 standard does not provide any MAC protocol for RM. In addition, MAC-level FEC is not rational because RM is not the only target service of WLAN and the cost of the implementation of FEC is too high.

Several RM MAC protocols for WLAN have been proposed [4,5,7]. To reserve medium for multicast and arrange the feedback order, RTS/CTS exchange is considered in [4]. Tone signaling to indicate the failure of multicast is proposed in [5]. In addition, a multicast protocol with leader-based feedback using intentional absence and collision of ACK is suggested to avoid feedback collision [7]. However, the feedback for each transmitted packet introduces considerable overhead, especially when the multicast source has a number of multicast packets. To reduce the feedback overhead, FEC and per-group feedback using NAK is considered for general networks [3]. They refer to the feedback system of ARQ-based multicast as per-packet feedback, and to the feedback system of FEC-based multicast as per-group feedback. Ideally, the source only needs to collect the maximum number of requested packets and respond with additional parity packets corresponding to the maximum number of requests. Moreover, per-group feedback can be equivalent to per-packet feedback when the transmission group consists of only a single packet.

4.4   Background

4.4.1   IEEE 802.11 MAC

IEEE 802.11 MAC [1] is based on logical functions called coordination functions, which determine when a station (STA) operating within a Basic Service Set (BSS)[1] is permitted to transmit and may be able to receive frames via the wireless medium. Two coordination functions are defined, namely, (1) the mandatory distributed coordination function (DCF), for a distributed, contention-based channel access, based on CSMA/CA, and (2) the optional point coordination function (PCF), for a centralized, contention-free channel access, based on the poll-and-response mechanism. Most of today's 802.11 devices operate in DCF mode only. We briefly explain in this section how the DCF and PCF work.

---

[1] There are two types of BSSs. An infrastructure BSS is composed of an access point (AP) and multiple STAs associated with the AP, where the AP works as a bridge between the wireless and wired domains; and an independent BSS (IBSS) is composed of multiple STAs.

4.4.1.1 Distributed Coordination Function (DCF)

The 802.11 DCF works with a single first-in-first-out (FIFO) transmission queue. The CSMA/CA constitutes a distributed MAC based on a local assessment of the channel status, that is, whether the channel is busy (i.e., somebody is transmitting a frame) or idle (i.e., no transmission). Basically, the CSMA/CA of the DCF works as follows.

When a frame arrives at the head of the transmission queue, and if the channel is busy, the MAC waits until the medium becomes idle and then defers for an extra time interval, called the DCF Distributed Inter-frame Space (DIFS). If the channel stays idle during the DIFS deference, the MAC then starts the backoff process by selecting a random backoff count. For each slot time interval during which the medium stays idle, the random backoff (BO) counter is decremented. When the counter reaches zero, the frame is transmitted. On the other hand, when a frame arrives at the head of the queue, if the MAC is in either the DIFS deference or the random BO process, the processes described above are applied again. That is, the frame is transmitted only when the random BO has finished successfully. When a frame arrives at an empty queue and the medium has been idle longer than the DIFS time interval, the frame is transmitted immediately.

Each STA maintains a contention window (CW), which is used to select the random BO count. The BO count is determined as a pseudo-random integer drawn from a uniform distribution over the interval [0,CW]. How to determine the CW value is further detailed below. If the channel becomes busy during a BO process, the BO is suspended. When the channel becomes idle again, and stays idle for an extra DIFS time interval, the BO process resumes with the latest BO counter value. The timing of DCF channel access is illustrated in Figure 4.2.

For each successful reception of a frame, the receiving STA immediately acknowledges the frame reception by sending an ACK frame. The ACK frame is transmitted after a short inter-frame space (SIFS), which is shorter than the DIFS. Other STAs resume the BO process after the DIFS idle time. Thanks to the SIFS interval between the data and ACK frames, the ACK

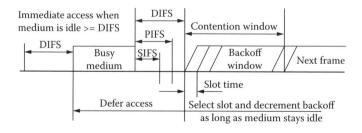

Figure 4.2   IEEE 802.11 DCF channel access.

frame transmission is protected from other STAs' contention. If an ACK frame is not received after the data transmission, the frame is retransmitted after another random BO.

The *CW* size is initially assigned CW_{min}, and increases when a transmission fails, that is, the transmitted data frame has not been acknowledged. After any unsuccessful transmission attempt, another BO is performed using a new CW value updated by:

$$CW := 2(CW + 1) - 1$$

with an upper bound of CW_{max}. This reduces the collision probability in case there are multiple STAs attempting to access the channel. After each successful transmission, the *CW* value is reset to CW_{min}, and the transmission-completing STA performs the DIFS deference and a random BO even if there is no other pending frame in the queue. This is often referred to as "post" BO, as this BO is done after, not before, a transmission. This post BO ensures that there is at least one BO interval between two consecutive MAC service data unit (MSDU) transmissions.

In WLAN environments, there may be hidden stations. Two stations, which can transmit to and receive from a common station while they cannot see each other, are hidden from each other. Because the DCF operates based on carrier sensing, the existence of such hidden stations can degrade network performance severely. To reduce the hidden station problem, IEEE 802.11 defines a Request-to-Send/Clear-to-Send (RTS/CTS) mechanism. That is, if the transmitting STA opts to use the RTS/CTS mechanism, before transmitting a data frame, the STA transmits a short RTS frame, followed by a CTS frame transmitted by the receiving STA. The RTS and CTS frames include information on how long it takes to transmit the subsequent data frame and the corresponding ACK response. Thus, other STAs hearing the transmitting STA and hidden STAs close to the receiving STA will not start any transmissions; their timer, called Network Allocation Vector (NAV), is set, and as long as the NAV value is non-zero, a STA does not contend for the medium. Between two consecutive frames in the sequence of RTS, CTS, data, and ACK frames, a SIFS is used. Figure 4.3 shows the timing diagram involved in an RTS/CTS frame exchange.

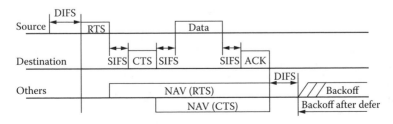

Figure 4.3   RTS-CTS frame exchange.

Table 4.1 MAC Parameters for 802.11a PHY

Parameters	SIFS (μsec)	DIFS (μsec)	Slot (μsec)	CW_{min}	CW_{max}
802.11a PHY	16	34	9	15	1023

All of the MAC parameters, including SIFS, DIFS, Slot Time, CW_{min}, and CW_{max}, depend on the underlying physical (PHY) layer. Table 4.1 shows these values for the 802.11a PHY [2]. Irrespective of the PHY, DIFS is determined by SIFS+2·SlotTime; and another important IFS, called the PCF IFS (PIFS), is determined by SIFS+SlotTime. With 802.11a, the transmission rate is up to 54 Mbps. There are other PHYs such as IEEE 802.11b PHY with rates of up to 11 Mbps. As we are discussing reliable multicast MAC in this chapter, our evaluation results in the following are relatively valid, irrespective of the underlying PHY.

4.4.1.2 Point Coordination Function (PCF)

To support time-bounded services, the IEEE 802.11 standard also optionally defines the PCF to let STAs have contention-free access to the wireless medium, coordinated by a point coordinator (PC), which is collocated within the AP. The PCF has higher priority than the DCF because the period during which the PCF is used is protected from the DCF contention via the NAV set. Under the PCF, the time axis is divided into repeated periods, called superframes, wherein each superframe is composed of a contention-free period (CFP) and a subsequent contention period (CP). During a CFP, the PCF is used for accessing the medium, while the DCF is used during a CP. It is mandatory that a superframe includes a CP of minimum length that allows at least one MSDU delivery under the DCF. See Figure 4.4 for the CFP and CP coexistence.

A superframe starts with a beacon frame, which is a management frame that maintains the synchronization of the local timers in the STAs and delivers protocol-related parameters. The AP generates beacon frames at regular beacon frame intervals; thus, every STA knows when the next beacon frame will arrive. This instance is called the target beacon transition time (TBTT), and is announced in every beacon frame. During a CFP, there is no contention among STAs; instead, STAs are polled. See Figure 4.4 for typical frame exchange sequences during a CFP. The PC polls an STA asking for a pending frame. If the PC itself has pending data for this STA, it uses a combined date and poll frame by piggybacking the CF-Poll frame into the data frame.

Upon being polled, the polled STA acknowledges the successful reception along with data. If the PC receives no response from a polled STA after

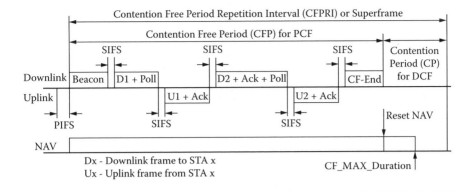

Figure 4.4   IEEE 802.11 PCF channel access during a CFP.

waiting for a PIFS interval, it polls the next STA or ends the CFP. Therefore, no idle period longer than PIFS occurs during CFP. The PC continues polling other STAs until the CFP expires. A specific control frame, called CF-End, is transmitted by the PC as the last frame within the CFP to signal the end of the CFP.

4.4.2   Upper-Layer FEC

FEC across packets is a well-known type of error control scheme to resist channel error causing packet erasures [3,8,13]. We consider the usage of *Reed-Solomon* (RS) codes for the upper-layer FEC. In general, an (n, k) RS code is used to encode k symbols of m bits into blocks composed of $n\,(=2^m - 1)$ symbols, where $m \geq 1$. Therefore, the encoding algorithm expands a block of k symbols to n symbols by adding $n - k$ redundant symbols.

An e-erasure-correcting RS code has the following parameters:

Block length : $n = 2^m - 1$ symbols

Message size : k symbols

Parity-checksize : $n - k = e$ symbols

Minimum distance : $d_{min} = n - k + 1$ symbols

By puncturing and shortening, (n, k) RS code can be reduced to (n', k') RS code with the same symbol size, where n' and k' are smaller than or equal to n and k, respectively. We assume punctured/shortened RS code with an 8-bit symbol (i.e., one byte) for our upper-layer FEC. For more details on coding theory, readers are encouraged to see Reference [15].

When the FEC is used at the upper layer, it is necessary to apply the RS coding across data packets. This is due to the fact that the nominal 802.11 MAC implementations discard the entire MAC frame in the event of

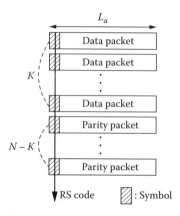

Figure 4.5   Upper-layer RS coding across packets.

an erroneous reception of the frame. The erroneous frame at the receiving MAC is never passed on to the higher layer. For this reason, if RS coding is applied within a single packet at the upper layer, the erroneous packet will not be available for error detection or correction at the upper layer.

Therefore, RS coding at the upper layer is applied across packets using an interleaver, that is, K data packets (of L_a length bytes) are buffered at the interleaver shown in Figure 4.5. The first symbol (byte) from each of the K data packets are sent through an (N, K) punctured RS coder, resulting in $(N-K)$ parity symbols each of which forms the first byte of the $(N-K)$ parity packets, respectively. This is repeated for the L_a bytes resulting in $(N-K)$ parity packets. Each packet of length L_a is generated by the RS encoder. Each data or parity packet is transmitted via a MAC frame; if this frame is discarded at the receiving MAC layer due to channel errors, it results in a symbol erasure at the RS decoder in the upper layer. Because the positions of lost packets are already known by counting the sequence numbers of the packets, we can apply RS erasure-correcting decode, which can correct up to $(N-K)$ packet losses (or erasures) out of N packets over which the RS coding was applied. The procedure of RS coding across data packets is depicted in Figure 4.5. As mentioned above, the block length, N, of the punctured RS code is smaller than or equal to 255 because the size of the RS code symbol is one byte.

4.5   MAC Protocol Description

As discussed in Section 4.2, per-group feedback collects feedback information after the transmission of a group of packets, while per-packet feedback collects feedback information after each packet transmission. For example, the standard DCF, which collects feedback information such as CTS or ACK from a receiver after transmission of a packet, as illustrated in Figure 4.3, is a typical per-packet feedback MAC protocol. In this section, we describe two types of multicast MAC protocols that use per-group feedback schemes. Both protocols adopt NAK feedback to reduce performance degradation caused by a large number of multicast receivers. However, each protocol exploits different feedback schemes to avoid excessive collision for NAK frame collection. One is based on iterative polling, while the other is based on contention to collect NAK frames from receivers without collision. From now on we refer to the former protocol as the polling-based feedback protocol (PBP), and to the latter protocol as the contention-based feedback protocol (CBP).

Before delving into our proposed MAC protocols, we define two terms: "packet" and "frame." We refer to a *packet* as the transmission unit at the layers above MAC. A packet from the upper layer can be seen as an MSDU by the MAC layer. On the other hand, we refer to a *frame* as the transmission unit at the MAC layer. A frame can be also referred to as an MPDU (MAC protocol data unit). Therefore, data packets are packetized into data frames, and parity packets are packetized into FEC frames by the MAC layer, respectively. In addition, we refer to a *multicast period* as the time period from when the multicast source starts to send a group of packets for multicast to when the multicast source finalizes the packet transmissions for multicast.

The proposed reliable multicast protocols work as follows. The upper layer of a multicast source puts down N packets to the MAC layer along with such additional information as the number of data packets, the number of parity packets, and the destinations for multicast. Then the MAC layer transforms packets to frames and transmits them to multiple destinations. To reconstruct data packets at the upper layer of each multicast receiver, the proposed MAC protocols continue transmitting frames until the number of frames received by each multicast receiver corresponds to the number of data packets informed at the beginning of multicast period.

A notable point is that the MAC of the multicast source does not transmit frames that have ever been transmitted. That is, a frame is transmitted once. After the multicast source transmits all N frames, there remains no FEC frame to transmit for upcoming NAK. If a receiver fails to receive more than or equal to K frames, it cannot overcome the packet erasure by using additional parity packets. To handle this severe frame erasure, we propose a possible solution based on retransmission Section 4.5.3.

The MAC frame formats used by our protocols are depicted in Figures 4.6 through 4.9. Figure 4.6 depicts the frame format of the RM frame, which is used to initiate a multicast period. The N field in the RM frame denotes the number of combined data and FEC frames, and the K field denotes the number of data frames. Because we assume that the length of

Octets:	2	2	32	6	2	1	1	4
	Frame control	Duration	RA bitmap	TA	Sequence control	N	K	FCS

Figure 4.6   MAC frame format of RM.

the RS code symbol is one byte, N and K are less than 255. We also assume that the number of multicast receivers is less than 256, which should be a large enough number for a single BSS (basic service set) of the typical WLAN. The length of the RA bitmap field in an RM frame is 256 bits, or 32 octets, each of which corresponds to the Association ID of the station. To inform the values of N and K, a 1-octet field is used for each, as shown in Figure 4.6. Figure 4.7 depicts the RNAK (request for NAK frames) frame, which is used for multicast source to request NAK frames from receivers. In an RNAK frame, the $\underline{N}$ field denotes the number of remaining packets out of N. Figure 4.8 describes the NAK frame format. There are two types of NAK frames: one contains a 1-octet "feedback" field and the other NAK contains a 32-octet "feedback" field. The 1-octet "feedback" field presents the number of necessary packets; the 32-octet "feedback" is used as the bitmap of the received packets when parity-based recovery turns out to have failed. We discuss the issue of using 32-octet "feedback" in Section 4.5.3. (Figure 4.9 describes the CF-END frame format.) The CF-END frame is originally used to announce the end of a CFP, in the standard PCF, as illustrated in Figure 4.4. In our protocols, the CF-END frame is used to finalize a multicast period.

4.5.1   Reliable Multicast MAC Protocol with Polling-Based Feedback

We now present a multicast MAC protocol with polling-based feedback. We assume that the multicast source is an AP and it tries to transmit at least K packets out of N packets to all receivers. Basically, according to this protocol, the AP asks the receivers in a round-robin fashion to find if they need additional parity packets. When an AP polls a receiver and responds to it by sending more parity packets, the other receivers are also able to receive these extra parity packets and exploit them to reconstruct

Octets:	2	2	6	6	1	1	4
	Frame control	Duration	RA	TA	CW	$\underline{N}$	FCS

Figure 4.7   MAC frame format of RNAK.

Octets:	2	2	6	6	1 or 32	4
	Frame control	Duration	RA	TA	Feedback	FCS

Figure 4.8   MAC frame format of NAK.

the original data packets. Each receiver replies to an RNAK from the AP with a NAK. Once an AP receives a NAK asking for no more packets from a specific receiver, the AP does not send an RNAK to the receivers again until the end of the multicast period. We refer to the NAK frame with zero of feedback field as "0-NAK." The reliable multicast MAC protocol with polling-based feedback, referred to as PBP, works as follows:

[A]   AP $\longrightarrow$ Receivers
    The AP sends an RM frame and K frames with SIFS interval.
[B]   AP $\longrightarrow$ Receiver
    The AP sends an RNAK to a receiver in the polling entry.
[C]   Receiver $\longrightarrow$ AP
    The receiver that received RNAK sends a NAK frame to the AP.
[D]   AP $\longrightarrow$ Receiver
    If the NAK is not a 0-NAK, the AP sends additional frames according to the received NAK, and goes to step [B]. If the NAK is a 0-NAK, the AP assigns the next receiver in the receiver entry to the polling entry, and step [B]. If there is no receiver for the AP to assign to the polling entry, the AP finalizes the multicast period by transmitting a CF-END frame.

Figure 4.10 shows an example of a multicast period using PBP. The AP can access the medium with PIFS (point inter-frame space) idle time, and initiate a multicast period by sending an RM frame. From the RM frame, receivers can collect the necessary information for multicast such as the number of data frames, the number of FEC frames, and the entry of multicast destinations. If a receiver is one of the multicast destinations, the receiver counts the number of received data and FEC frames, and updates its NAK for the feedback. When the AP asks the receiver to transmit its NAK by transmitting RNAK, the receiver sends a NAK frame and starts to receive additional FEC frames after SIFS. Because the AP transmits additional FEC

octets:	2	2	6	6	4
	Frame Control	Duration	RA	BSS ID	FCS

Figure 4.9   MAC frame format of CF-END.

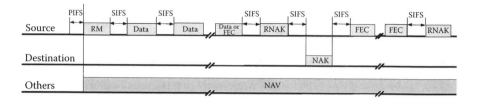

Figure 4.10 Reliable multicast with PBP.

frames with broadcast address, all of the receivers can receive the frames and update their own NAK frames for feedback.

Of course, control frames, such as NAK, RM, and RNAK, can suffer from channel error and hence be erroneously transmitted. To handle this problem, the protocol can include more a reliable response and retry process despite additional overhead. We describe this issue in Section 4.5.3.

4.5.2 Reliable Multicast MAC Protocol with Contention-Based Feedback

The polling-based protocol is simple and reliable but can introduce additional overhead, especially when channel error is scarce and the number of multicast receivers is large. In this section, we propose a contention-based feedback protocol to reduce the feedback overhead. The RM MAC protocol with contention-based feeback, referred to as CBP, works as follows:

[A] Main Transmission Period (AP ⟶ Receivers)
Send an RM frame and *K* data frames with SIFS interval. The RM contains the bitmap of the multicast destinations.

[B] Contention Round (AP ⟶ Receiver)
The AP sends an RNAK frame, which contains the contention window size. Each of the receivers sets its BO counter randomly using the contention window size specified in the RNAK. If there is no busy medium during a contention round, the AP finalizes the multicast period by transmission of a CF-END frame.

[C] NAK Transmission Round (Receiver ⟶ AP)
The receivers that need additional FEC frames try to transmit NAK frames using CSMA/CA. If a receiver's BO counter becomes zero, it sends a NAK to the AP. The NAK is transmitted once in a contention round even if the NAK does not reach the AP successfully.

[D] NAK Transmission Round (AP ⟶ Receivers)
The AP transmits additional frames with a broadcast MAC address corresponding to the number of requested frames in the received

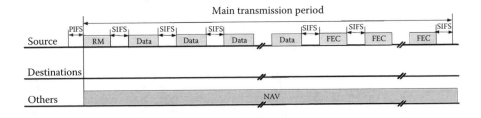

Figure 4.11 Frame diagram of main transmission period of CBP.

NAK. If the AP has counted down all its counters corresponding to contention window size, go to Step [B]. After the AP transmits additional FEC frames, receivers update their NAK frames, resume counting down their BO counter, and go to step [C].

Now, we explain each step in detail.

4.5.2.1 Main Transmission Period

The AP can start an RM period with PIFS access. The first part of an RM period is called the "main transmission period." In a main transmission period, the AP broadcasts an RM, K data frames with SIFS interval. At the beginning of a main transmission period, NAV is announced to reserve the medium for multicast period. All these are illustrated in Figure 4.11.

4.5.2.2 Contention Round

After a main transmission period, the AP can transmit additional FEC frames to support RM using parity-based recovery. First, the AP transmits an RNAK frame after SIFS from the main transmission to receive NAK frames from receivers. The receivers access the medium with contention to exchange NAK and FEC frames with the AP. We refer to this process as a "contention round," and the beginning of a contention round is depicted in Figure 4.12.

An RNAK frame contains the size of the contention window, which is used by receivers to set their BO counter for NAK transmissions in the contention round. After a receiver receives an RNAK, it randomly selects a slot from [0, $CW - 1$], where CW is announced via an RNAK. Then, it counts down its BO counter and transmits NAK based on CSMA/CA. The important difference between CBP and the standard DCF BO process is that CBP does not retransmit a NAK frame after a failure of NAK and FEC exchange. This difference is based on the fact that DCF can increase the contention window size, while CBP uses the announced contention

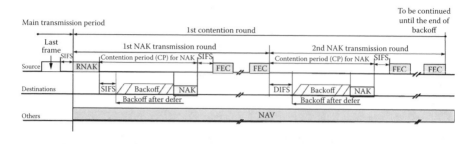

Figure 4.12   Frame diagram of the beginning of a contention round of CBP.

window size. We postpone how CBP determines the size of the contention window to Section 4.5.2.6.

4.5.2.3   NAK Transmission Round

Each contention round consists of several NAK transmission rounds. After SIFS time interval from receiving an RNAK, receivers start to transmit their NAK frames to the AP in a contentious manner, with the contention window size announced in the RNAK. This is called a "NAK transmission round." When a receiver transmits its NAK to the AP successfully, the AP in turn transmits FEC frames corresponding to the requested number of the received NAK. Not only the receiver, which transmitted the NAK, but all the multicast receivers receive and use these FEC frames from the AP because all the receivers have the identical RS decoder at the upper layer and the receiver address of FEC frames is the broadcast MAC address.

When the AP transmits additional FEC frames, the receivers should update their NAK frames considering the number of received FEC frames. If a receiver transmitted its NAK successfully but has not received enough additional FEC frames, the receiver should prepare another NAK for the next contention round, and then update it to reflect additionally received FEC frames that are initiated by the other receivers.

After the AP transmits FEC frames as a response to a NAK, receivers that have pending NAK frames resume BO and transmit their own updated NAK frames. This is another NAK transmission round, and it lasts until the next successful NAK transmission. NAK transmission rounds are repeated, exchanging NAK and FEC frames until the contention window, which is announced in the RNAK frame, is all counted down.

4.5.2.4   Last NAK Transmission Round

When the contention window is counted down to zero by the AP, it transmits the next RNAK, including a new contention window size to initiate another contention round as shown in Figure 4.13.

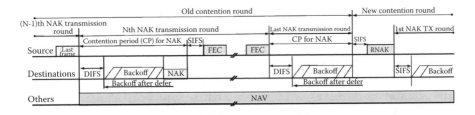

Figure 4.13   Frame diagram of consecutive contention rounds of CBP.

4.5.2.5   Last Contention Round

If the medium is idle during the entire contention round, then the AP recognizes that there is no station to transmit NAK and finalizes the RM session by transmitting a CF-END frame to reset NAV. That is, the last contention round consists of a single NAK transmission round with no busy medium. Figure 4.14 shows that the end of the last contention round is the end of the multicast period. The frame format of CF-END is depicted in the IEEE 802.11 standard [1].

4.5.2.6   Contention Window Size in RNAK

The contention window size for the contention-based feedback process should be announced at the beginning of each contention round. Therefore, we here develop an algorithm to determine the contention window size for each contention round. Although other algorithms can determine the optimal window size, the proposed algorithm is so simple that the multicast source determines the contention window size without complex calculation.

For simplicity of notation, we define that CW is the contention window size of the contention round, p_{access} is the probability that an arbitrary

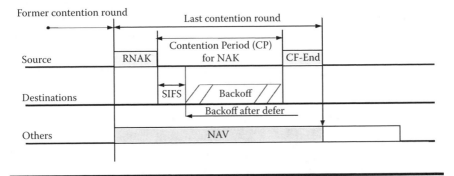

Figure 4.14   Frame diagram of the end of a contention round of CBP.

receiver accesses the medium in a slot time, n_{old} is the estimated number of receivers from the previous contention round, and n_{new} is the expected number of receivers in the next contention round, respectively.

Now we estimate the number of receivers through backtracking. During a contention round, three types of medium access are possible: (1) successful NAK transmission, (2) NAK collision, and (3) an idle slot time. Both NAK transmission and NAK collision result in a busy medium but the multicsat source transmits FEC frames only when a successful NAK-transmission occurs. For a simple estimation, we assume that there are only two receivers, which transmit NAK frames at each NAK-collision. For the last busy medium of the previous contention round was a NAK collision, let $n_{old} = 2$.

As tracking backward, for each NAK collision, n_{old} is updated by:

$$n_{old} = n_{old} + 2 \qquad (4.1)$$

and for each NAK transmission, n_{old} is updated by:

$$n_{old} = n_{old} \times 2 \qquad (4.2)$$

Because the probability that the number of packets a receiver needs is more than that of another receiver, it can be assumed to be 1/2, and Equation 4.2 makes sense.

Considering that the typical frame loss probability is a small number, we can expect that additional parity packets can be transmitted to receivers successfully with quite a large probability. From this assumption, we can estimate n_{new} as:

$$n_{new} = \left\lceil \frac{n_{old}}{2^m} \right\rceil \qquad (4.3)$$

where m is the number of NAK transmission rounds (i.e., successful NAK transmissions) in the previous contention round.

The expected number, n_{new}, of receivers should be in $[1, R]$, where R is the number of total multicast receivers (already known by the AP from the upper layer). Then we can determine CW for the next contention round as follows. First, we find p_{access}, which maximizes the probability for the success of NAK transmissions. By approximating the contention round to p-persistent CSMA, we can find $p_{access} : \max(n_{new} p_{access}(1 - p_{access})^{n_{new}-1})$ by differentiation.

The differentiation results in:

$$p_{access} \cdot n_{new} = 1 \qquad (4.4)$$

When the RNAK is announced by the AP, NAK pending stations schedule their own NAK frames randomly by selecting one out of $[0, CW - 1]$. This implies that $p_{access} = \frac{1}{CW}$. Therefore, we can determine the contention window size for the next contention round as:

$$CW = n_{new} \tag{4.5}$$

From the fact that n_{new} cannot exceed R, we assign $CW = R$ when $n_{new} \geq R$. In addition, we can assign R to the CW for the first contention round because of no previous contention round.

4.5.3 Discussion

4.5.3.1 Loss of Control Frame

Even short control frames can be corrupted for some receivers, especially when the group size of the receivers is large or the channel error rate is high. If a receiver did not receive an RM frame, the receiver cannot realize the initiated multicast session and it cannot prepare to receive packets from the source. To avoid this erroneous operation, an RM can be acknowledged similar to Group RTS (GRTS) [10]. That is, the multicast source sends an RM frame, including the order of CTS transmissions; then, all receivers included in the RM frame reply to the RM with CTS, respectively, with SIFS interval. Although the original idea of using GRTS was channel probing to measure the signal strength, we can adopt GRTS to establish a handshake for reliable multicast.

For PBP, all frames are exchanged with SIFS interval. There is no idle time longer than SIFS when PBP works. If an RNAK is erroneous and the polled receiver cannot send its NAK, the multicast source can transmit an RNAK again after PIFS idle time. If a NAK is erroneous, the multicast source can retransmit the RNAK after SIFS because the received erroneous frame might be a NAK responding to the previous RNAK.

However, CBP needs an additional action to prevent abnormal operations. First, idle time can be longer than DIFS, which can allow the medium access of other stations. Idle time longer than DIFS appears when the contention period contains a bulk of idle slots, and receivers resume their BO count with DIFS idle time. Second, the medium can be idle during the entire contention round because all receivers lost RNAK frames in CBP. For these cases, the multicast source might terminate the multicast session although several NAK frames remain for the reception of additional FEC frames. Therefore, additional RNAK transmissions are necessary to convince the multicast source to finalize the multicast period. Finally, an erroneous NAK might be perceived as a collision when the multicast source determines contention window size for CBP. Such a case can result in over-estimation of the number of the receivers, which implies excessively large contention window size and overhead.

4.5.3.2 Selective Retransmission

Because the number of data and FEC frames is finite, parity-based recovery can fail, especially when the error rate is high and the number, K, of data frames is relatively large compared with the number of FEC frames. To manage a parity-based recovery failure, selective retransmission, which transmits FEC frames more than once, can be considered. Assume that a receiver accepts an RNAK, which indicates that the remaining number of FEC frames is less than the number of frames that the receiver needs for the parity-based recovery. To reconstruct the original data packets, the receiver can request the multicast source to retransmit the lost frames. For the retransmission, the receiver can send a NAK with the bitmap of the received frames to inform whether each frame is received successfully. Because we assume that N and K are less than 256, the bitmap is 256 bits, or 32 octets, enough to indicate the reception information. If the multicast source receives a NAK with a bitmap from a receiver, it transmits all remaining FEC frames and retransmits the transmitted FEC frames referring to the bitmap of the NAK.

4.6 Performance Evaluation

In this section we evaluate the performance of the proposed and existing protocols in terms of multicast throughput. First we determine the idealized performance of retransmission-based multicast and parity-based multicast by mathematical analysis. Retransmission-based multicast adopts per-packet feedback for retransmission of packets, while parity-based multicast adpots per-group feedback for transmission of additional parity packets. We assume an ideal multicast environment in which the cost of feedback is zero. The more efficient this practical protocol we have proposed, the closer its performance reaches idealized performance. Then, simulation results of the proposed protocols will be compared with the retransmission-based multicast protocol using tone [5], which minimizes feedback overhead but requires additional hardware and bandwidth.

To evaluate performance, we exploit multicast throughput, which is defined as:

$$\xi = \frac{E[L_a]}{E[T_a]} \tag{4.6}$$

where L_a is the size of a packet and T_a is the elapsed time for an arbitrary data packet to be recovered by all receivers. The WLAN parameters used in the evaluation, such as transmission rate and SIFS, correspond to IEEE 802.11a [2], which supports basic transmission rate of 6 Mbps for broadcast.

For the evaluation, we use default parameter values respecting the typical WLAN environment; that is, the number of receivers is 10, the packet error rate (PER) is 0.1, the size of a packet is 1,000 bytes, and the environment of IEEE 802.11a is assumed. Unless specified otherwise, the parameters correspond to these default values. We use the term "PER" for the loss probability of data or parity frame. In other work, it is often referred to as "FER" (frame error rate). For simplicity, we assume that management frames such as RNAK, NAK, and RM are not corrupted by the channel errors. Each receiver suffers from the channel error independently. An independent channel condition for each receiver is common in a typical WLAN indoor environment [11].

4.6.1 Mathematical Analysis of Idealized Performance

If the cost of the feedback process is zero and $N = \infty$, then multicast using parity-based loss recovery can achieve its idealized performance. Here we derive multicast throughput mathematically assuming that the feedback cost can be ignored.

We first derive an unachievable lower bound to the expected number of packet transmissions required to transmit an arbitrary packet to all receivers. We assume all receivers experience independent homogeneous channel error. Let p be the packet error rate of receivers and L_r the number of additional packet transmissions required by a certain receiver. The distribution of L_r is:

$$P(L_r = 0) = (1 - p)^K \tag{4.7}$$

$$P(L_r = m) = \binom{K + m - 1}{K - 1} p^m (1 - p)^K, \quad m = 1, \cdots \tag{4.8}$$

Let L denote the maximum number of additional packets that are transmitted and R the number of multicast receivers. Its cumulative distribution is given by:

$$P(L \leq m) = [P(L_r \leq m)]^R, \quad m = 0, 1, \cdots \tag{4.9}$$

where

$$P(L_r \leq m) = \sum_{i=0}^{m} P(L_r = i), \quad m = 0, 1, \cdots \tag{4.10}$$

Finally, we define M as the number of transmission times an arbitrary data packet is transmitted before all receivers recover it accounting for parity packets. The mean number of transmission per arbitrary packet is [3]:

$$E[M] = (E[L] + K)/K \tag{4.11}$$

Therefore, idealized multicast throughput above MAC is given by:

$$\xi_{ideal} = \frac{L_a}{E[M] \cdot \left(\frac{PHY_{bdr} + MAC_{bdr} + L_a}{r} + SIFS \right)} \tag{4.12}$$

where L_a is the size of the packet, r denotes the packet transmission rate, PHY_{bdr} is the length of physical layer header size, MAC_{bdr} is the length of physical layer header size, and SIFS is a necessary interval between consecutive transmissions.

We can determine the idealized throughput of the retransmission-based multicast by setting $N = \infty$ and $K = 1$. Figure 4.15 shows the idealized throughput of retransmission-based multicast and parity-based multicast with varying PER. Parity-based multicast outperforms retransmission-based multicast across the entire PER range in Figure 4.15. One interesting case of this result is when PER is around 0.1, which corresponds to a typical channel error rate in WLANs; the gain of parity-based multicast protocol is significant. The parity-based multicast protocol outperforms the retransmission-based multicast protocol by up to 50 percent in terms of multicast throughput. The scalability of the idealized multicast is shown in Figure 4.16. As the number of data packets increases, the slope of the throughput degradation slows. In the next subsection, we compare the idealized multicast and our multicast protocols through simulation results.

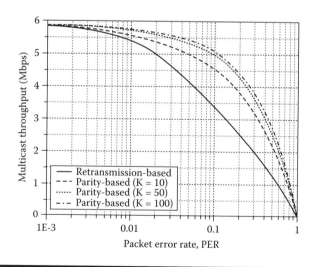

Figure 4.15   Idealized multicast throughput versus PER.

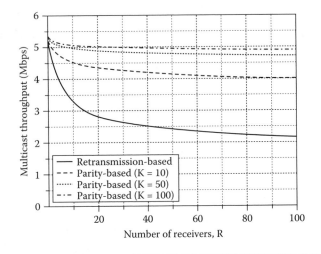

Figure 4.16   Idealized multicast throughput versus number of receivers.

4.6.2   *Simulation Results*

For practical reliable multicast protocols, the feedback from receivers is indispensable so that the overhead of feedback is added. The more efficient protocol collecting feedback is used, the closer to the upper bound the performance of the protocol comes. Here we evaluate our multicast protocols and also compare the proposed protocols with the existing retransmission-based multicast protocol with tone-based feedback [5] using NS-2 simulator [14]. We refer to the retransmission-based multicast protocol with tone-based feedback as TBP. An example of TBP operation is shown in Figure 4.17. First, the source transmits an RTS frame to receivers and waits for a tone signal during a specific time, that is, NRT (not-ready response time). If there is no tone signal for the RTS frame, the source transmits a data frame and waits for a tone signal during NRT. If there is a tone signal, the source retransmits the data frame until there is no tone signal. For details of the protocol, readers are encouraged to refer to Reference [5]. In this chapter we ignore the BO procedure of a protocol for a fair comparison. We simulate each protocol 10,000 times for each scenario under the 802.11a environment. We use parameters of the typical WLAN as mentioned at the beginning of this section.

Figure 4.18 reveals that PBP exceeds TBP over the entire range of error rates in terms of the multicast throughput, and is less than the idealized multicast throughput. Tone signaling time of TBP can be ignored, but the overhead of additional RTS and SIFS transmissions causes performance degradation. However, PBP achieves high performance because parity-based recovery improves the efficiency of multicast. With a large

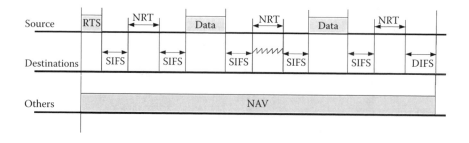

Figure 4.17   Reliable multicast with TBP.

number of data packets, the number of times exchanging NAK and FEC frames decreases so that the efficiency of feedback increases.

Figure 4.19 shows the scalability of PBP. As the number of multicast receivers increases, the AP should poll receivers many times so that the overhead for polling increases. On the other hand, as shown in Figure 4.17, tone signaling is not affected by the number of multicast receivers because overlapped tone signals are allowed. However, for PBP, the degradation from increased receivers is compensated by the scalability of parity packets. Parity-based multicast reduces unnecessary receptions such that it improves multicast efficiency. Moreover, in most cases, a WLAN infrastructure is composed of far fewer than 100 stations.

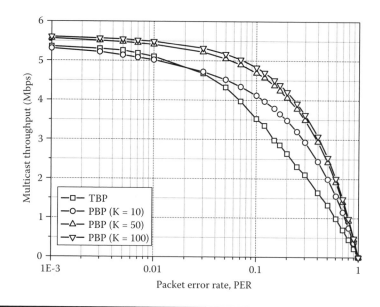

Figure 4.18   Multicast throughput of PBP versus PER.

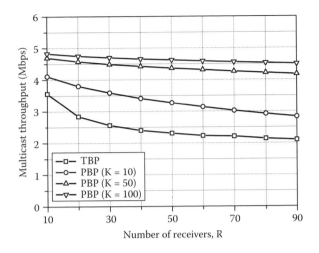

Figure 4.19   Multicast throughput of PBP versus the number of receivers.

Figure 4.20 and Figure 4.21 show the multicast performance of CBP. The main difference in the simulation results between PBP and CBP is that CBP is closer to idealized performance, especially in terms of scalability. We can conclude that the multicast throughput decreases more gently as the number of receivers grows by comparing the slopes of the lines in Figures 4.19 and 4.21.

We define *delay bound* as the time within which the data packets of receivers should be recovered. The delay bound can be interpreted as the

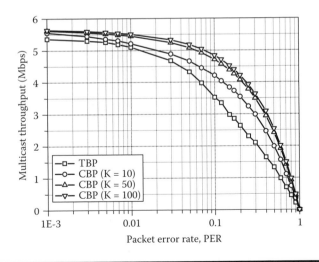

Figure 4.20   Multicast throughput of CBP versus PER.

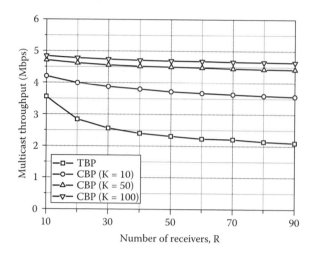

Figure 4.21   Multicast throughput of CBP versus number of receivers.

playout delay of chunks at the receiver [12]. With parity-based multicast protocols, the source exchanges blocks of packets and feedback with the receivers. The original packets that fail to be transmitted are recovered after the receiver accepts sufficient parity packets. Because of random packet error, the time from when an original packet is transmitted at the multicast source until it is recovered at the receiver can fluctuate from block to block. To remove this jitter, a playout delay can be used. The receiver attempts to play out each chunk, playout delay after the chunk is generated and starts to be transmitted.

The multicast source is fully loaded and tries to transmit packets within a given delay bound. For a longer delay bound, we can exploit per-group feedback more efficiently with a large number of data packets, K, to reduce the feedback cost. Therefore, the amount of available time for multicast leads the efficiency enhancement for per-group feedback. Figure 4.22 and Figure 4.23 show simulations with various delay bounds of multicast period. (Note that the x-axis of the delay bound is on a log scale in these figures.)

The usage of large RS code block requires sufficiently long delay bound. As shown in Figures 4.22 and 4.23, the delay bound should be longer than at least the transmission time of the data packets. With a long enough delay bound, we can achieve high multicast throughput. The proposed protocols can support 4.8 Mbps multicast throughput, while the existing multicast protocol can support only about 3.2 Mbps with 1000-millisecond delay bound. Because the number of receivers is 10 and PER is 0.1, the performance difference between PBP and CBP is not considerable, as seen by comparing Figures 4.22 and 4.23.

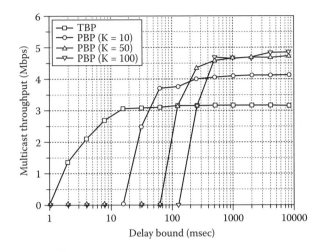

Figure 4.22   Multicast throughput of PBP versus delay bound.

Figure 4.24 and Figure 4.25 show the degradation of multicast throughput in the presence of heterogeneous receivers, that is, receivers with different loss probabilities. We consider a network composed of two types of heterogeneous receivers: (1) $R \cdot (1 - v)$ low-loss receivers with PER = 0.05 and (2) $R \cdot v$ high-loss receivers with PER = 0.25. This allows us to vary the fraction v of high-loss receivers among all receivers.

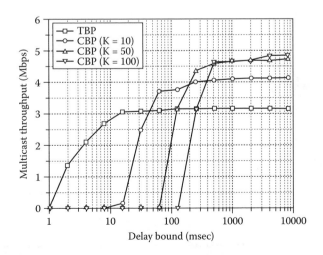

Figure 4.23   Multicast throughput of CBP versus delay bound.

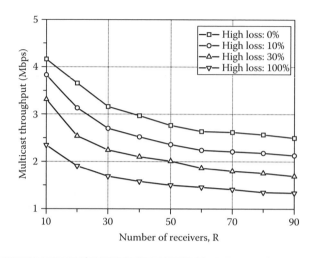

Figure 4.24 Multicast throughput of TBP for different fractions of high-loss receivers.

The results are quite different between TBP and PBP as shown in Figures 4.24 and 4.25. The total difference in multicast throughput between zero and pure high-loss receivers is about 1.5 Mbps for both TBP and PBP. However, PBP is more sensitive to the presence of high-loss receivers, as shown in Figure 4.25. When the ratio of high-loss receivers is 0.1, the throughput degradation of PBP is about 80 percent of the total difference

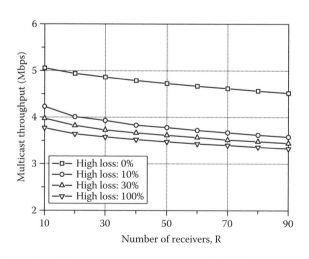

Figure 4.25 Multicast throughput of PBP ($K = 50$) for different fractions of high-loss receivers.

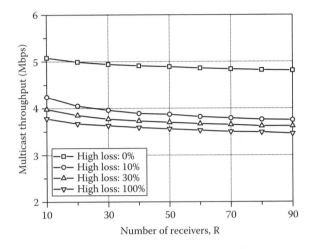

Figure 4.26   Multicast throughput of CBP ($K = 50$) for different fractions of high-loss receivers.

between zero and pure high-loss, while the degradation is about 25 percent of the total difference for TBP. The simulation results on CBP with heterogeneous error rates is shown in Figure 4.26. Compared with the PBP results, CBP is found to be more robust to performance degradation as network size increases. Although both PBP and CBP are robust in the high-loss environment, the proposed protocols are prone to the presence of heterogeneous loss as can be seen in Figures 4.25 and 4.26.

4.7   Summary

The omni-directional transmission property and unreliable multicast protocol of the IEEE 802.11 standard have created a strong demand for a reliable multicast protocol. We developed two MAC protocols that guarantee the reliability of multicast, and they can be combined with the upper-layer FEC. To improve the efficiency of multicast, our protocols transmit additional FEC frames based on the demands from receivers. To collect requests from receivers, we exploit polling-based feedback and contention-based feedback, and our protocol uses packet exchange, while the existing protocol requires additional hardware and bandwidth for tone signaling. The simulation results show that our multicast protocol, combined with upper-layer FEC, can achieve up to 50 percent multicast throughput gain over the existing retransmission-based multicast protocol in typical 802.11a WLAN environments.

References

[1] IEEE Standard 802.11-1999, Part 11: Wireless LAN Medium Access Control (MAC) and Physical Layer (PHY) Specifications, Reference number ISO/IEC 8802-11:1999(E), IEEE Standard 802.11, 1999 edition, 1999.

[2] IEEE 802.11a, Part 11: Wireless LAN Medium Access Control (MAC) and Physical Layer (PHY) Specifications, High-Speed Physical Layer in the 5 GHz Band, Supplement to IEEE 802.11 Standard, September 1999.

[3] Jorg Nonnenmacher, Ernst W. Biersack, and Don Towsley, "Parity-Based Loss Recovery for Reliable Multicast Transmission," *IEEE/ACM Transaction on Networking (TON)*, Vol. 6, No. 4, pp. 349–361, August 1998.

[4] Min-Te Sun, Lifei Huang, Anish Arora, and Ten-Hwang Lai, "Reliable MAC Layer Multicast in IEEE 802.11 Wireless Networks," in *Proc. International Conference on Parallel Processing (ICPP'02)*, 2002.

[5] Sandeep Gupta, Vikram Shankar, and Suresh Lalwani, "Reliable Multicast MAC Protocol for Wireless LANs," in *Proc. IEEE International Conference on Communications (ICC'03)*, May 2003.

[6] Don Towsley, Jim Kurose, and Sridhar Pingali, "A Comparison of Sender-Initiated and Receiver-Initiated Reliable Multicast Protocols," *IEEE Journal of Selected Areas in Communications (JSAC)*, Vol. 15, No. 3, pp. 398–406, April 1997.

[7] Joy Kuri and Sneha K. Kasera, "Reliable Multicast in Multi-access Wireless LANs," in *ACM/Kluwer Wireless Networks*, Vol. 6, No. 4, pp. 359–369, July 2001.

[8] C. Huitema, "The Case for Packet Level FEC," in *Proc. IFIP 5th International Workshop on Protocols for High Speed Networks (PfHSS'96)*, Sophia Antipolis, October 1996.

[9] Stijn van Langen, Sai Shankar N, and Warner ten Kate, "Performance Analysis of Iterative Polling Scheme for Real-Time Reliable Multicast," in *Proc. IEEE International Conference on Communications (ICC'03)*, May 2003.

[10] Zhengrong Ji, Yi Yang, Junlan Zhou, Mineo Takai, and Rajive Bagrodia, "Exploiting Medium Access Diversity in Rate Adaptive Wireless LANs," in *Proc. ACM International Conference on Mobile Computing and Networking (MOBICOM'04)*, Philadelphia, September 2004.

[11] European Telecommunications Standards Institute, "Channel Models for HIPERLAN/2 in Different Indoor Scenarios," *ETSI 3ERI085B*, 1998.

[12] James F. Kurose and Keith W. Ross, *Computer Networking: A Top-Down Approach Featuring the Internet,* 3rd ed., Addison Wesley, 2004.

[13] Mihaela van der Schaar, Santhana Krishnamachari, Sunghyun Choi, and Xiaofeng Xu, "Adaptive Cross-Layer Protection Strategies for Robust Scalable Video Transmission over 802.11 WLANs," *IEEE Journal of Selected Areas in Communications (JSAC)*, Vol. 21, No. 10, December 2003.

[14] "The Network Simulator — ns-2," http://www.isi.edu/nsnam/ns/, online link.

[15] Shu Lin and Daniel J. Costello, *Error Control Coding, Fundamentals and Applications,* 2nd ed. Englewood Cliffs, NJ: Prentice Hall, 2004.

Chapter 5

Wireless Network Tele-traffic Modeling with Lossy Link

*Yan Zhang, Wei Li, Yang Xiao, Mingtuo Zhou,
Shaoqiu Xiao, and Masayuki Fujise*

Contents

5.1 Overview

Tele-traffic modeling provides a significant and efficient manner in evaluating and optimizing wireless mobile network performance. Traditionally, wireless network performance is studied by considering the limited bandwidth in the radio interface, which may lead to a connection request rejection or an ongoing service termination (e.g., [17]). On the other hand, it is well known that wireless communication systems are also characterized by the highly unreliable wireless link. Consequently, it is reasonable and equally important to take into account the wireless link unreliability in investigating the wireless network's performance.

This chapter concentrates on the call completion characteristics to demonstrate the impact of inherently lossy link. The closed-form formula for the call completion probability is developed under the simple Gilbert-Elliot channel model and the generalized channel model with the relaxed call holding time distribution based on the complex theory and transform techniques (Laplace Transform and z-Transform). The specific results under the typical channel models and the commonly employed call holding times are presented for the sake of ready application and generalization demonstration. To reflect the unique erroneous characteristics of the wireless channel, we further extend to derive the call completion probability in the presence of the lossy link and the insufficient resource simultaneously. Illustrative examples indicate that the conventional result without considering link unreliability will greatly overestimate the realistic result; and that different call holding time distributions may lead to substantial performance discrepancy.

5.2 Introduction

A fact in the wireless network community is that a variety of wireless systems (e.g., IEEE 802.11a/b/g Wireless Local Area Network [1], IEEE 802.16 Wireless Metropolitan Area Networks [2], Global System for Mobile Communications (GSM) [3], General Packet Radio Service (GPRS) [3], Universal Mobile Telecommunication System (UMTS) [4], and Bluetooth [5]) have been standardized or are being drafted to satisfy the diverse requirements in the different scenarios with respect to the coverage, service categories, and data rate. Accordingly, intensive research and development activities for a range of relevant aspects are being carried out. As a highly desirable approach, tele-traffic modeling of wireless networks is becoming increasingly complicated to evaluate performance. Although wireless networks are rapidly evolving, two fundamental features constricting the capacity improvement and performance guarantee are undeniable: (1) the limited capacity of the radio interface and (2) the inherently unreliable wireless channel.

In a wireless network, the whole service coverage is partitioned into several continuous areas, each of which is called a cell. At the center of the cell, a base station (BS) is responsible for the relay communication between the mobile station (MS) and the core network. With rapidly increasing numbers of subscribers, the resource insufficiency becomes more serious. The MS initializing a connection request (new call) may be blocked when all the finite channels are occupied. Here, a connection request could be a voice call, a data session, a videoconference connection in next-generation wireless mobile networks, or a flow in the wireless IP environment. We use the general terminology "call" to represent such diverse requests. After successful initial access, the call continues its service in the cell. If the call is unable to normally complete in the serving cell, the MS will send a handoff request to the immediate neighboring cell to avoid connection termination. In case no resources are available in the target BS, the handoff call will be rejected and consequently the call connection is forced to terminate. It is believed that terminating an ongoing call is much more undesirable than blocking a new call. As a consequence, the limited bandwidth plays a significant role in wireless network performance as well as the user's service satisfaction. To efficiently share the finite bandwidth, a variety of studies have proposed radio resource management schemes to improve network performance or investigated the wireless network performance with generalized tele-traffic variables in stand-alone networks or the next-generation multi-tier wireless networks [17–23]. Examples include the guard channel [17], queueing priority scheme [18,19], subrating scheme [20], dynamic guard channel [21], new call bounding [26], directed retry [22], and measured-based scheme [23].

From the aforementioned studies, we are impressed that these works focus on the limited bandwidth effect on the call or network performance, and consequently another essential property of wireless systems that is, the wireless link unreliability is always ignored. However, it is well known that the wireless link is inherently time-varying and statistically unstable [44], and has a significant impact on system performance [45]. A seriously degraded wireless link may lead to physical link breakdown and hence call connection termination between the MS and BS. As a result, similar to insufficient resource, the unreliable wireless link also plays an equally important role in determining network performance [10,11].

It is observed that, in these studies, the call holding time is assumed to follow exponential distribution. Motivated by the multimedia traffic diversity in future generation wireless systems and facts based on empirical data, validation of the exponential assumption for call holding time has been extensively examined [24,27,29,30,33,35,36]. Lin and Chlamtac [24] studied the call completion characteristics caused by the effect of limited resource in the BS. The performance metrics call completion probability is derived

146 ■ *Resource, Mobility, and Security Management*

under Erlang call holding time. Fang et al. [27] further studied the call pattern with the relaxed call holding time assumption. Yeo and Jun [29] studied the hierarchical cellular system with hyper-Erlang [43] call holding time. The authors observed that the application of exponential or hyper-Erlang call holding time may lead to significant performance discrepancy. Ashtiani et al. [30] presented hyper-Erlang distribution to model cell residence time. Bolotin [33] claimed that the exponential assumption may be invalid for modern telephone services and proposed the lognormal distribution to approximate the wireline call holding time. Orlik and Rappaport [35] employed the sum of hyper-exponential (SOHYP) distribution, which has the advantage of approximating the behavior of any positive random variable and preservation of the Markovian property, to model the call holding time. Marsan et al. [36] model the call holding time as a hyper-exponential distribution with the application of the memory-less property to analyze the performance of hierarchical cellular networks, one of the highly promising network architectures in next-generation wireless networks.

In this chapter we relax the call holding time assumption and wireless channel model to derive the closed-form formula for the call completion probability based on the complex theory and transform techniques (including Laplace Transform and z-Transform). The remainder of the chapter is organized as follows. In Section 5.3, we describe typical call trajectory with the simple two-state Gilbert-Elliot channel model. The call completion probability is formulated on the basis of the proposed system model. In Section 5.4, based on a similar system model, the call completion probability is extended into the situation of the generalized channel model. In Section 5.5, we further derive the closed-form formula for call completion probability in the presence of link unreliability and also resource insufficiency. Numerical results are given in Section 5.6, followed by concluding remarks in Section 5.7.

5.3 Call Completion Probability with Gilbert-Elliot Model

5.3.1 Gilbert-Elliot Channel Model

For the time-varying channel, the Gilbert-Elliot model [6,7] has been widely used to capture the periods of signal degradation [12–14]. In this two-state Markov chain model, the state space of the wireless channel consists of $\Omega = \{ good, \ bad\}$. The channel stays in the *good* state for an exponentially distributed duration with mean $1/\mu_g$. It then transmits into the *bad* state and stays for an exponentially distributed period with average $1/\mu_b$. It is assumed that calls in the *good* channel state can communicate error-free and those in the *bad* channel state may, but not definitely, fail to complete

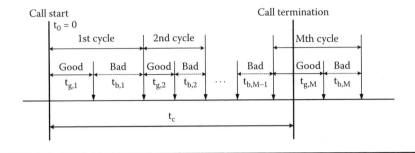

Figure 5.1   The time diagram for a typical call in mobile network over lossy link.

the conversation owing to the broken link. Without loss of generality, we denote the time when a call is initialized as $t_0 = 0$. The call experiences the wireless channel *good* state and *bad* state alternately. Define a *cycle* as the continuous *good* state and its next consecutive *bad* state. In Figure 5.1, the notation t_c represents the call holding time representing the requested period for a call connection to complete a communication. Denote the probability density function (p.d.f.) and cumulative distribution function (c.d.f.) of t_c as $f_{t_c}(t)$ and $F_{t_c}(t)$, respectively. Denote the Laplace Transform (LT) of the p.d.f. $f_{t_c}(t)$ as $f_{t_c}^*(s)$. The time duration of the *good* state during the ith cycle is denoted as $t_{g,i}$, where the notation g represents the *good* state; and the interval in the *bad* state during the ith cycle is denoted as $t_{b,i}$, where the notation b stands for the *bad* state. Then, $t_{g,i}(i = 1, 2\ldots)$ are exponentially independent and identical distributed (i.i.d.) random variables with the service rate μ_g, and $t_{b,i}(i = 1, 2\cdots)$ are the i.i.d. exponentially distributed random variables with the service rate μ_b. Hence, the p.d.f. of $t_{g,i}$ and $t_{b,i}$ $(i = 1, 2\ldots)$ is given by:

$$f(t_{g,i}) = \mu_g e^{-\mu_g t_{g,i}}; \quad t_{g,i} \geq 0, i = 1, 2\ldots \tag{5.1}$$

$$f(t_{b,i}) = \mu_b e^{-\mu_b t_{b,i}}; \quad t_{b,i} \geq 0, i = 1, 2\ldots \tag{5.2}$$

5.3.2   Link Reestablishment Procedure

The functionality of error handling is one of the key modules in BS software, wherein the component to handle the unreliable wireless channel is the core consideration. As the wireless channel associated with an ongoing call becomes worse, the BS should be able to monitor such degradation and notify the relevant module to deal with this accident (i.e., try to reestablish the impaired channel). Upon the time when the channel turns from the *good* state into the *bad* state, the call may not be dropped immediately because of the link layer error protection scheme. Usually there is a timer, called a monitoring channel timer (MCT) T_{mc}, in the link layer to monitor the channel state. As the channel becomes worse and no messages are exchanged between the MS and BS during the duration T_{mc}, a LINK-ReESTABLISH-REQUEST message is sent from the MS to BS, or in reverse, in an attempt to reestablish the physical link. Upon sending the message LINK-ReESTABLISH-REQUEST, the timer TIMER-LER with length T_{LER} is started to monitor the message transmission; if no acknowledgment LINK-ReESTABLISH-ACK is received during T_{LER}, the TIME-LER times out. Then, the message LINK-ReESTABLISH-REQUEST is sent out again. If after N_{LER} times attempts and no message LINK-ReESTABLISH-ACK is received, the previously allocated resource for this connection is released and can be utilized by other users. Practically, T_{LER} is very small because its value is normally determined under a *good* environment. In particular, in a GSM system, the Radio Link Protocol (RLP) uses a fixed retransmission timer recommended as 480 ms (full rate) or 780 ms (half rate), and the maximum number of retransmissions recommended as 6 [15]. In UMTS, the Radio Resource Control (RRC) protocol recommends the length of the retransmission timer T300 as 1000 ms and the maximum number of retransmission N300 = 3 [16].

5.3.3 Link Reestablishment Successful Probability

After the expiration of the timer MCT when the channel becomes worse and transmits into the *bad* state, the BS detects the link broken state and consequently a number of consecutive timer TIMER-LERs are probably started to try relinking with the MS. To present the attempt of the link reestablishment procedure, we define the link reestablishment successful probability as the probability that the physical link is successfully reestablished during the channel *bad* state. Due to the i.i.d. property of random variables $t_{ig}(i = 1, 2 \ldots)$ and $t_{b,i}(i = 1, 2 \ldots)$, the link reestablishment successful probabilities are the same during the channel *bad* state in any cycle, we focus on the first cycle to derive the link reestablishment successful probability P_s. A physical link can fail to be reestablished in case all reestablishment attempts are failed. Namely, the wireless channel remains in the *bad* state after the maximum number of link reestablishment attempts. Hence, we have:

$$P_s = 1 - Pr(T_{mc} + N_{LER}T_{LER} < t_{b,1}) \qquad (5.3)$$

Note that the timer length T_{LER} for monitoring the message LINK-ReESTABLISH-REQUEST transmission has no essential difference from any other expiration verification which is used to monitor the message between the BS and MS. In terms of implementation, T_{LER} can be regarded as a constant in BS software. The maximum number of retransmitting message

LINK-ReESTABLISH-REQUEST N_{LER} is also a fixed constant. The assumption that N_{LER} and T_{LER} are fixed is reasonable from an implementation point of view because, in practical BS software, it is convenient to set these parameters as constants although it may be inflexible.

For the variable T_{mc}, we first consider the situation when T_{mc} is assumed to be a fixed value. Let $\Delta = T_{mc} + N_{LER} T_{LER}$. Then,

$$Pr(t_{b,1} > T_{mc} + N_{LER} T_{LER}) = e^{-\mu_b \Delta}$$

This result yields the probability

$$P_s = 1 - e^{-\mu_b \Delta} \tag{5.4}$$

For a more complicated scenario, T_{mc} is assumed to be a random variable with p.d.f. $f_{T_{mc}}(t)$ and LT with p.d.f. $f^*_{T_{mc}}(s)$. On the other hand, the link reestablishment procedure is also related to the impatience characteristics of the users because too long a reestablishment will make the user impatient and thus press the "Cancel" button on the phone. Hence, the result of the link reestablishment procedure depends on the system configuration as well as on the user's behavior. Let T_u be the user's impatience time. Denote its p.d.f. and LT of p.d.f. as $f_{T_u}(t)$ and $f^*_{T_u}(s)$, respectively. In such case, a physical link may fail to reestablish due to one of the following reasons:

■ Event $\mathcal{B}_1$: user's impatience time T_u is less than $t_{b,1}$.
■ Event $\mathcal{B}_2$: all reestablishment attempts are failed. That is, the wireless channel remains in the *bad* state after the maximum number of link reestablishment attempts.

Hence, we have:

$$P_s = 1 - Pr(\mathcal{B}_1 \text{ or } \mathcal{B}_2)$$

$$= 1 - [Pr(\mathcal{B}_1) + Pr(\mathcal{B}_2) - Pr(\mathcal{B}_1\mathcal{B}_2)] \tag{5.5}$$

where:

$$Pr(\mathcal{B}_1) = Pr(T_u < t_{b,1}) = f^*_{T_u}(\mu_b) \tag{5.6}$$

$$Pr(\mathcal{B}_2) = Pr(T_{mc} + N_{LER} T_{LER} < t_{b,1}) \tag{5.7}$$

$$Pr(\mathcal{B}_1\mathcal{B}_2) = Pr(T_u < t_{b,1}, T_{mc} + N_{LER} T_{LER} < t_{b,1}) \tag{5.8}$$

Then, the probability for the event $\mathcal{B}_2$ is given by:

$$Pr(\mathcal{B}_2) = Pr(t_{b,1} > T_{mc} + N_{LER}T_{LER})$$

$$= f^*_{T_{mc}}(\mu_b)e^{-\mu_b N_{LER}T_{LER}} \qquad (5.9)$$

The probability for the simultaneous occurrence of the events $\mathcal{B}_1$ and $\mathcal{B}_2$ is given by:

$$
\begin{aligned}
Pr(\mathcal{B}_1\mathcal{B}_2) &= Pr(T_u < t_{b,1},\ T_{mc} + N_{LER}T_{LER} < t_{b,1}) \\
&= \int_0^\infty Pr(T_u < t)\, Pr(T_{mc} + N_{LER}T_{LER} < t)\mu_b e^{-\mu_b t}\,dt \\
&= \int_{N_{LER}T_{LER}}^\infty Pr(T_u < t)\, Pr(T_{mc} + N_{LER}T_{LER} < t)\mu_b e^{-\mu_b t}\,dt \\
&= \int_{N_{LER}T_{LER}}^\infty \int_0^t f_{T_u}(x)dx \int_0^{t-N_{LER}T_{LER}} f_{T_{mc}}(y)dy\mu_b e^{-\mu_b t}\,dt \qquad (5.10)
\end{aligned}
$$

where we have put the probability $Pr(T_{mc} + N_{LER}T_{LER} < t) = 0$ as $0 \le t \le N_{LER}T_{LER}$.

Substituting Equations 5.6, 5.9, and 5.10 into Equation 5.5, we obtain

$$
\begin{aligned}
P_s = {}& 1 - f^*_{T_u}(\mu_b) - f^*_{T_{mc}}(\mu_b)e^{-\mu_b N_{LER}T_{LER}} \\
& + \int_{N_{LER}T_{LER}}^\infty \int_0^t f_{T_u}(x)dx \int_0^{t-N_{LER}T_{LER}} f_{T_{mc}}(y)dy\mu_b e^{-\mu_b t}\,dt \qquad (5.11)
\end{aligned}
$$

Specifically, we assume that the timer length T_{mc} follows an exponential distribution with average $1/\mu_{mc}$. The user's impatience duration is assumed to be an exponential distribution with mean $1/\mu_u$, which has been popularly used in mobile network performance evaluations [32,34]. In this case, the probability for successful link reestablishment becomes:

$$P_s = \frac{\mu_b}{\mu_b + \mu_u} + \mu_b e^{-(\mu_b+\mu_u)N_{LER}T_{LER}} \left[\frac{1}{\mu_{mc} + \mu_u + \mu_b} - \frac{1}{\mu_b + \mu_u} \right]$$

5.3.4 Call Completion Probability

The average duration elapsed from the time the wireless channel turns into the *good* state to the time the user continues the conversation (if the link is reestablished successfully) is $T_{LER}/2$. Because this interval is very short compared with the call holding time, we ignore its effect and assume that if the call is not forced to block by the impaired channel, the user can continue his communication at the moment the channel becomes the *good* state. Let M denote the number of cycles needed for a user to complete the communication successfully without termination. Note that no messages are exchanged between the MS and BS during the channel *bad* state represented by the dotted line in Figure 5.1. The time diagram in Figure 5.1 shows that $M = 1$ if and only if the call holding time t_c is shorter than the wireless channel *good* state time duration $t_{g,1}$ in the first cycle; $M = 2$ if and only if the call holding time t_c is larger than $t_{g,1}$, but not longer than the summation of the channel good states during the first and second cycles, that is, $t_{g,1} + t_{g,2}$, etc. As a consequence, we develop the necessary and sufficient condition for $M = m$:

$$\sum_{i=1}^{m-1} t_{g,i} < t_c \leq \sum_{i=1}^{m} t_{g,i} \tag{5.12}$$

As stated, we *cannot* use the expression below to determine $M = m$ because the MS cannot exchange messages with the BS during channel *bad* state.

$$\sum_{i=1}^{m-1} t_{g,i} + \sum_{i=1}^{m-1} t_{b,i} < t_c \leq \sum_{i=1}^{m} t_{g,i} + \sum_{i=1}^{m-1} t_{b,i} \tag{5.13}$$

Define ξ_m as the summation of the number of m wireless channel *good* state time duration; that is:

$$\xi_m = (t_{g,1} + \cdots + t_{g,(m-1)} + t_{g,m}) \cdot \mathbf{1}_{m \geq 1} \tag{5.14}$$

where $\mathbf{1}_A$ equals 1 when the event A is true and zero otherwise. Let $f_{\xi_m}(t)$, $F_{\xi_m}(t)$ and $f^*_{\xi_m}(s)$ denote the p.d.f., c.d.f., and the LT of p.d.f. for the non-negative random variable ξ_m. The LT of ξ_m is given as:

$$f^*_{\xi_m}(s) = \mathbf{E}[e^{-s\xi_m}] = \prod_{i=1}^{m} \mathbf{E}[e^{-st_{g,i}}] = \left(\frac{\mu_g}{s + \mu_g}\right)^m \tag{5.15}$$

where $\mathbf{E}(X)$ represents the expected value of random variable X. From the LT of ξ_m, the p.d.f. and the c.d.f. are thus given by:

$$f_{\xi_m}(t) = \frac{\mu_g(\mu_g t)^{m-1}}{(m-1)!} e^{-\mu_g t}; \tag{5.16}$$

$$F_{\xi_m}(t) = \int_0^t f_{\xi_m}(x)dx = 1 - \sum_{k=0}^{m-1} \frac{(\mu_g t)^k}{k!} e^{-\mu_g t} \tag{5.17}$$

It is known that the distribution of ξ_m is the Erlang distribution with parameter μ_g and m [40]. Hence, the probability $Pr(M = m)$ can be written as:

$$
\begin{aligned}
Pr(M = m) &= Pr(\xi_{m-1} < t_c \leq \xi_m) \\
&= \int_0^\infty [F_{\xi_{m-1}}(t) - F_{\xi_m}(t)] f_{t_c}(t) dt \\
&= \frac{\mu_g^{m-1}}{(m-1)!} w_{m-1}(\mu_g)
\end{aligned}
\tag{5.18}
$$

where the item $w_k(x)$ is defined as:

$$
w_k(x) = \int_0^\infty t^k e^{-xt} f_{t_c}(t) dt = (-1)^k \left(\frac{d^k f_{t_c}^*(s)}{ds^k} \right) \bigg|_{s=x}
\tag{5.19}
$$

with the initial condition $w_0(x) = f_{t_c}^*(x)$. Checking the normalization condition, we found that:

$$
\sum_{i=1}^\infty \frac{\mu_g^{i-1}}{(i-1)!} w_{i-1}(\mu_g) = 1
\tag{5.20}
$$

Due to the unreliability of a wireless channel, a call may be dropped without completing the required M cycles. Let $\widehat{M}$ denote the actual number of cycles a user experienced. If the physical link is successfully reestablished during all the *bad* channel states the user experiences, then $\widehat{M}$ is equal to M and hence the call is normally completed. As a consequence, we have the conditional probability:

$$
Pr(\widehat{M} = m | M = m) = P_s^{m-1}
\tag{5.21}
$$

The call completion probability, P_{cc}, is defined as the probability that the call connection can be completed successfully. Thus, P_{cc} is expressed as:

$$
\begin{aligned}
P_{cc} &= \sum_{m=1}^\infty Pr(\widehat{M} = m, M = m) \\
&= \sum_{m=1}^\infty Pr(\widehat{M} = m | M = m) Pr(M = m) \\
&= \sum_{m=1}^\infty \frac{\mu_g^{m-1}}{(m-1)!} w_{m-1}(\mu_g) \cdot P_s^{m-1}
\end{aligned}
\tag{5.22}
$$

In particular, we have the following specific cases for commonly used call holding time distributions. For exponential call holding time with p.d.f. $f_{t_c}(t) = \mu_c e^{-\mu_c t}$, the call completion probability simplifies to:

$$P_{cc} = \frac{\mu_c}{\mu_c + \mu_g - \mu_g P_s}$$

An *n*-order hyper-exponential distributed call holding time with p.d.f. function

$$f_{t_c}(t) = \sum_{i=1}^{n} \alpha_i \eta_i e^{-\eta_i t}; \quad \text{with} \quad \sum_{i=1}^{n} \alpha_i = 1, \quad (n \in \mathcal{N}, 0 \le \alpha_i \le 1, \eta_i > 0)$$

$$(5.23)$$

yields the call completion probability

$$P_{cc} = \frac{1}{n} \sum_{i=1}^{n} \alpha_i \frac{\eta_i}{\eta_i + \mu_g - \mu_g P_s} \tag{5.24}$$

For an *n*-stage Erlang distributed call holding time with mean $1/\mu_c$, variance $V_c = 1/(n\mu_c^2)$, and the p.d.f. $f_{t_c}(t)$ and the c.d.f. $F_{t_c}(t)$,

$$f_{t_c}(t) = \frac{(n\mu_c)^n t^{n-1}}{(n-1)!} e^{-n\mu_c t}, \quad (n \in \mathcal{N}, t > 0) \tag{5.25}$$

$$F_{t_c}(t) = 1 - \sum_{i=0}^{n-1} \frac{(n\mu_c t)^i}{i!} e^{-n\mu_c t}, \quad (n \in \mathcal{N}, t > 0) \tag{5.26}$$

The LT of the Erlang call holding time is given by:

$$f_{t_c}^*(s) = \left(\frac{n\mu_c}{s + n\mu_c} \right)^n \tag{5.27}$$

In this case, the call completion probability is given by:

$$P_{cc} = \sum_{i=1}^{\infty} \binom{n+i-2}{i-1} \left(\frac{\mu_g P_s}{n\mu_c + \mu_g} \right)^{i-1} \tag{5.28}$$

where $\binom{x}{y} = \frac{x!}{y!(x-y)!}$ for non-negative integers x and y.

5.4 Call Completion Probability with Generalized Channel Model

5.4.1 Wireless Channel Model

In Section 5.3, we employed the Gilbert-Elliott model [6,7] to develop the call completion probability for the sake of analysis simplicity. The Gilbert-Elliott model is one of the earliest and most commonly used channel models. As explained in Section 5.3.1, there are two states, *good* and *bad*, in this simple Markov chain model. It is assumed that the channel alternately stays in the *good* or *bad* state for an exponentially distributed duration. However, the simple Gilbert-Elliott model may not be able to capture the sharp change in signal; various wireless channel models have been proposed to reflect the periods of signal degradation in wireless networks. Fritchman [8] generalized the two states of Gilbert's wireless channel model as a Markov chain with $N_F (\geq 2)$ states. Two classes are grouped in these N_F finite states. The first class is comprised of N_g error-free states, while the second one consists of the remaining $N_b = N_F - N_g$ states with each state representing a different error state. A corresponding model was developed with the period of the error-free state expressed as the summation of N_g exponentials and the duration of the error state given by the summation of N_b exponentials. Furthermore, Wang and Moayeri [9] proposed and studied a Finite State Markov Channel (FSMC).

As a consequence, owing to the diversity and the suitability of different channel models, we assume that the duration in the *good* (or error-free) or *bad* (or erroneous) state follows a general distribution. The time duration of the *good* state during the $i^{th}(i \in \mathcal{N} \equiv \{1, 2 \ldots\})$ cycle is denoted as $t_{g,i}$ and $t_{b,i}$ for the interval of the *bad* state during cycle i. We note that $t_{g,i}(i \in \mathcal{N})$ are generally distributed i.i.d. random variables with the p.d.f. $g(t)$, the c.d.f. $G(t)$, and the average $1/\mu_g$, and that $t_{b,i}(i \in \mathcal{N})$ are the generally i.i.d. distributed random variables with the p.d.f. $b(t)$, the c.d.f. $B(t)$, and the average $1/\mu_b$. The LT of $g(t)$ and $b(t)$ are denoted as $g^*(s)$ and $b^*(s)$, respectively.

As explained in Section 5.3.2, link reestablishment is the procedure for a call to attempt to rebuild the broken link between the MS and BS [15,16]. Similarly, we define the link reestablishment successful probability as the probability that the physical link is successfully reestablished during the channel *bad* state. Due to the i.i.d. property of random variables $t_{g,i}(i \in \mathcal{N})$ and $t_{b,i}(i \in \mathcal{N})$, the link reestablishment successful probabilities are identical during the channel *bad* state in any cycle; we similarly focus on the first cycle to derive the link reestablishment successful probability P_s. A physical link is said to be failed in reestablishing in case all the reestablishment attempts are failed, which implies that the wireless channel remains in the *bad* state after the maximum number of reestablishing attempts. Hence, the probability P_s is given by:

$$P_s = 1 - Pr(T_{mc} + N_{LER}T_{LER} < t_{b,1})$$

$$= 1 - \int_\Delta^\infty b(t)dt$$

$$= B(\Delta) \tag{5.29}$$

It is seen that the probability P_s is independent of the length of the wireless channel *good* state.

With the Gilbert-Elliott wireless channel model [6,7], the duration of the *good* state and the *bad* state are exponentially distributed; that is:

$$k(t) = \mu_k e^{-\mu_k t}; \quad k \in \{g, b\} \tag{5.30}$$

Then, we have the probability P_s under such a simple channel model:

$$P_s^{GE} = 1 - e^{-\mu_b \Delta} \tag{5.31}$$

This expression is the same as the result Equation 5.4. For the Fritchman channel model [8], we denote the duration of the *good* state and the *bad* state with the following distribution functions:

$$k(t) = \sum_{i=1}^{N_k} \alpha_{k,i}\eta_{k,i}e^{-\eta_{k,i}t}; \quad k \in \{g, b\} \tag{5.32}$$

where $\sum_{i=1}^{N_k} \alpha_{k,i} = 1$, $(N_k \in \mathcal{N}, 0 \leq \alpha_{k,i} \leq 1, \eta_{k,i} > 0)$. Then, we obtain the probability as:

$$P_s^{Fritchman} = \sum_{i=1}^{N_b} \alpha_{b,i}(1 - e^{-\eta_{b,i}\Delta}) \tag{5.33}$$

5.4.2 Call Completion Probability

Similarly following the system model in Figure 5.1, we develop the call completion probability under generalized wireless channel models and the general call holding time distributions. Define ξ_m as the summation of the number of *m* wireless channel *good* state time durations; that is,

$$\xi_m = (t_{g,1} + \cdots + t_{g,(m-1)} + t_{g,m}) \cdot 1_{m \geq 1} \tag{5.34}$$

Let $f_{\xi_m}(t)$, $F_{\xi_m}(t)$, and $f_{\xi_m}^*(s)$ denote the p.d.f., c.d.f., and LT of p.d.f. for the non-negative random variable ξ_m. The LT of ξ_m is given as:

$$f_{\xi_m}^*(s) = \mathbf{E}[e^{-s\xi_m}] = \prod_{i=1}^{m} \mathbf{E}[e^{-st_{g,i}}] = [g^*(s)]^m \tag{5.35}$$

A call connection is normally completed when the link reestablishment is successful during each wireless channel *bad* state, if available. Because no messages are exchanged between the MS and BS during the channel *bad* state represented by the dotted line in Figure 5.1, the length of the *bad* state will not contribute effectively to the requested call connection time. As a consequence, we develop the equation for the call completion probability as:

$$\begin{aligned}
P_{cc} &= \sum_{m=0}^{\infty} Pr(\xi_m < t_c \leq \xi_{m+1}) P_s^m \\
&= \sum_{m=0}^{\infty} \int_0^{\infty} Pr(\xi_m < x \leq \xi_{m+1}) f_{t_c}(x) dx P_s^m \\
&= \int_0^{\infty} \left(\sum_{m=0}^{\infty} Pr(\xi_m < x \leq \xi_{m+1}) P_s^m \right) f_{t_c}(x) dx
\end{aligned} \tag{5.36}$$

Based on Equation 5.36, we define:

$$a(m, x) = Pr(\xi_m < x \leq \xi_{m+1}) \tag{5.37}$$

$$a^*(m, s) = \mathcal{L}(a(m, x)) = \int_0^{\infty} a(m, x) e^{-sx} dx \tag{5.38}$$

$$a^{**}(z, s) = \sum_{m=0}^{\infty} a^*(m, s) z^m = \sum_{m=0}^{\infty} \int_0^{\infty} a(m, x) e^{-sx} dx z^m \tag{5.39}$$

$$A(x) = \sum_{m=0}^{\infty} Pr(\xi_m < x \leq \xi_{m+1}) P_s^m \tag{5.40}$$

$$A^*(s) = \mathcal{L}(A(x)) = \int_0^{\infty} A(x) e^{-sx} dx \tag{5.41}$$

where the operator $\mathcal{L}(\cdot)$ represents the LT. The item $a(m, x)$ can be physically interpreted as the probability that a call connection experiences m *good* channel state for a fixed interval x. Proceeding with the development of $a(m, x)$, we have:

$$a(m, x) = \int_0^x f_{\xi_m}(u) \int_{x-u}^\infty g(t)dt\,du$$

$$= \int_0^x f_{\xi_m}(u)[1 - G(x - u)]du$$

$$= \int_0^x f_{\xi_m}(u)du - \int_0^x f_{\xi_m}(u)G(x - u)du$$

$$= F_{\xi_m}(x) - f_{\xi_m}(x) \circledast G(x) \tag{5.42}$$

where the operator $\circledast$ represents the convolution operation. Then, from Equation 5.42, we obtain the LT of $a(m, x)$ as:

$$\mathcal{L}(a(m, x)) = \mathcal{L}(F_{\xi_m}(x) - f_{\xi_m}(x) \circledast G(x))$$

$$= \mathcal{L}(F_{\xi_m}(x)) - \mathcal{L}(f_{\xi_m}(x)) \cdot \mathcal{L}(G(x))$$

$$= \frac{[g^*(s)]^m}{s} - [g^*(s)]^m \cdot \frac{g^*(s)}{s}$$

$$= \frac{[g^*(s)]^m[1 - g^*(s)]}{s} \tag{5.43}$$

For $|z| \leq 1$, from Equations 5.39 and 5.43, we have:

$$a^{**}(z, s) = \sum_{m=0}^\infty \mathcal{L}(a(m, x))z^m$$

$$= \sum_{m=0}^\infty \frac{[g^*(s)]^m[1 - g^*(s)]}{s}z^m$$

$$= \frac{1 - g^*(s)}{s} \sum_{m=0}^\infty [zg^*(s)]^m$$

$$= \frac{1 - g^*(s)}{s[1 - zg^*(s)]} \tag{5.44}$$

As a consequence, the LT for $A(x)$ is given by:

$$A^*(s) = \sum_{m=0}^\infty \mathcal{L}(a(m, x)) P_s^m$$

$$= a^{**}(P_s, s)$$

$$= \frac{1 - g^*(s)}{s[1 - P_s g^*(s)]} \tag{5.45}$$

Following a similar technique in [27] and applying the Residue theorem [41], the call completion probability (Equation 5.65) becomes:

$$P_{cc} = \int_0^\infty A(x) f_{t_c}(x) dx$$

$$= \int_0^\infty f_{t_c}(x) \left[\frac{1}{2\pi j} \int_{\sigma-j\infty}^{\sigma+j\infty} A^*(s) e^{sx} ds \right] dx$$

$$= \frac{1}{2\pi j} \int_{\sigma-j\infty}^{\sigma+j\infty} A^*(s) \left(\int_0^\infty f_{t_c}(x) e^{sx} dx \right) ds$$

$$= \frac{1}{2\pi j} \int_{\sigma-j\infty}^{\sigma+j\infty} A^*(s) f_{t_c}^*(-s) ds$$

$$= -\sum_{s_0 \in \sigma_c} \operatorname*{Res}_{s=s_0} [A^*(s) f_{t_c}^*(-s)]$$

$$= -\sum_{s_0 \in \sigma_c} \operatorname*{Res}_{s=s_0} \left[\frac{[1 - g^*(s)] f_{t_c}^*(-s)}{s[1 - P_s g^*(s)]} \right] \tag{5.46}$$

where j is the imaginary unit ($j^2 = -1$) and σ is sufficiently small. σ_c denotes the set of the poles $f_{t_c}^*(-s)$ in the right half complex plane. $Res_{s=s_0}$ represents the residue at $s = s_0$.

Theorem 5.1 *The call completion probability in wireless network in the presence of unreliable wireless channel is given by:*

$$P_{cc} = -\sum_{s_0 \in \sigma_c} \operatorname*{Res}_{s=s_0} \left[\frac{[1 - g^*(s)] f_{t_c}^*(-s)}{s[1 - P_s g^*(s)]} \right] \tag{5.47}$$

where $g^*(s)$ is the LT of channel good state distribution function; $f_{t_c}^*(s)$ is the LT of call holding time and P_s is the link reestablishment successful probability in Equation 5.29.

In particular, we have the following specific cases for commonly used call holding time distributions and wireless channel models. For the Gilbert-Elliot channel model, the durations of the channel *good* state and *bad* state are exponentially distributed with the p.d.f. shown in Equation 5.30. For the exponential call holding time, we have the following corollary:

Corollary 5.1 *The call completion probability in wireless networks in the presence of the Gilbert-Elliot wireless channel with exponential call holding time is* *given by:*

$$P_{cc} = \frac{\mu_c}{\mu_c + \mu_g - P_s\mu_g} \qquad (5.48)$$

The result (Equation 5.48) is obtained in Section 5.3.4 using an alternative approach.

For the hyper-exponential call holding time with the p.d.f. in Equation 5.23, we have the following corollary:

Corollary 5.2 *The call completion probability in wireless networks in the presence of the Gilbert-Elliot wireless channel with hyper-exponential call holding time is given by:*

$$P_{cc} = \sum_{i=1}^{n} \alpha_i \frac{\eta_i}{\eta_i + \mu_g - P_s\mu_g} \qquad (5.49)$$

For an *n*-stage Erlang call holding time [24,39] with p.d.f. in Equation 5.25, we have the following corollary:

Corollary 5.3 *The call completion probability in wireless networks in the presence of the Gilbert-Elliot wireless channel with Erlang call holding time is given by:*

$$P_{cc} = \left(\frac{n\mu_c}{n\mu_c + \mu_g - P_s\mu_g} \right)^n \qquad (5.50)$$

where we have applied the result below:

$$\underset{s = n\mu_c}{Res} \left[b^*(s) f_{t_c}^*(-s) \right] = \underset{s = n\mu_c}{Res} \left[b^*(s) \left(\frac{n\mu_c}{-s + n\mu_c} \right)^n \right] \qquad (5.51)$$

$$= \frac{(-n\mu_c)^n}{(n-1)!} b^{(n-1)}(n\mu_c) \qquad (5.52)$$

with $b^{(n-1)}(s)$ representing the $(n-1)$-order derivative of $b^*(s)$ with respect to s.

Note that the Erlang distribution can be easily extended into a hyper-Erlang distribution, which has been proven to be able to arbitrarily closely approximate the distribution of any positive random variable as well as measured data [43]. In particular, its general approximation property has been applied in different scenarios. Yeo and Jun [29] employed the hyper-Erlang distribution to model the call holding time. Fang [28], Ashtiani et al. [30] proposed the hyper-Erlang model for the cell residence time. Recently, Lin and Chen [25] modeled the serving GPRS support node (SGSN) residence time as a hyper-Erlang distribution to evaluate authentication signaling traffic in the UMTS network.

For the hyper-Erlang call holding time with p.d.f. [28,42,43],

$$f_{t_c}(t) = \sum_{i=1}^{H} q_i \frac{(n_i \theta_i)^{n_i} t^{n_i-1}}{(n_i - 1)!} e^{-n_i \theta_i t};$$

$$\sum_{i=1}^{H} q_i = 1, H \in \mathcal{N}, n_i \in \mathcal{N}, 0 \le q_i \le 1, \theta_i \ge 0 \qquad (5.53)$$

and its LT

$$f_{t_c}^*(s) = \sum_{i=1}^{H} q_i \left(\frac{n_i \theta_i}{s + n_i \theta_i} \right)^{n_i} \qquad (5.54)$$

we have the following corollary:

Corollary 5.4 *The call completion probability in wireless networks in the presence of the Gilbert-Elliot wireless channel with hyper-Erlang call holding time is given by:*

$$P_{cc} = \sum_{i=1}^{H} q_i \left(\frac{n_i \theta_i}{n_i \theta_i + \mu_g - P_s \mu_g} \right)^{n_i} \qquad (5.55)$$

Note that the link reestablishment successful probability P_s in Corollary 1 to Corollary 4 is given by Equation 5.31.

For the Fritchman channel model, the wireless link *good* state and *bad* state durations follow the hyper-exponential distribution shown in Equation 5.32. For the exponential call holding time, we have the following corollary:

Corollary 5.5 *The call completion probability in wireless networks in the presence of the Fritchman wireless channel with exponential call holding time is given by:*

$$P_{cc} = \frac{1 - \sum_{i=1}^{N_g} \frac{\alpha_{g,i} \eta_{g,i}}{\mu_c + \eta_{g,i}}}{1 - P_s \sum_{i=1}^{N_g} \frac{\alpha_{g,i} \eta_{g,i}}{\mu_c + \eta_{g,i}}} \qquad (5.56)$$

For the hyper-exponential call holding time (Equation 5.23), we have the following corollary:

Corollary 5.6 *The call completion probability in wireless networks in the presence of the Fritchman wireless channel with hyper-exponential call holding time is given by:*

$$P_{cc} = \sum_{i=1}^{H} q_i \left(\frac{1 - \sum_{j=1}^{N_g} \frac{\alpha_{g,j}\eta_{g,j}}{\theta_i+\eta_{g,j}}}{1 - P_s \sum_{j=1}^{N_g} \frac{\alpha_{g,j}\eta_{g,j}}{\theta_i+\eta_{g,j}}} \right) \tag{5.57}$$

For an *n*-stage Erlang call holding time (Equation 5.25), we have the following corollary:

Corollary 5.7 *The call completion probability in wireless networks in the presence of the Fritchman wireless channel with Erlang call holding time is given by:*

$$P_{cc} = \frac{(-1)^{n-1}(n\mu_c)^n}{(n-1)!} A^{(n-1)}(n\mu_c) \tag{5.58}$$

where $A^*(s)$ is expressed as:

$$A^*(s) = \frac{1 - \sum_{i=1}^{N_g} \frac{\alpha_{g,i}\eta_{g,i}}{s+\eta_{g,i}}}{s\left[1 - P_s \sum_{i=1}^{N_g} \frac{\alpha_{g,i}\eta_{g,i}}{s+\eta_{g,i}}\right]} \tag{5.59}$$

For hyper-Erlang call holding time (Equation 5.53), we have:

Corollary 5.8 *The call completion probability in wireless networks in the presence of the Fritchman wireless channel with hyper-Erlang call holding time is given by:*

$$P_{cc} = \sum_{i=1}^{H} q_i \frac{(-1)^{n_i-1}(n_i\theta_i)^{n_i}}{(n_i-1)!} A^{(n_i-1)}(n_i\theta_i) \tag{5.60}$$

where $A^*(s)$ is defined by Equation 5.59.

Note that the link reestablishment successful probability P_s in Corollary 5 to Corollary 8 is given by Equation 5.33 in case of the Fritchman wireless channel.

5.5 Call Completion Probability with Unreliable Link and Insufficient Resource

In Section 5.3 and Section 5.4, we concentrated on the problem when a call connection might be dropped due to the impaired channel and consequently neglect the blocking due to insufficient resource or bandwidth. Generally speaking, the resource insufficiency and the wireless channel unreliability process at the same time, and they depend on each other due to the common constraint of the call holding time. However, conditioning on a given specific call connection, the two factors independently affect the call behavior. Based on this point, the proposed analytical model in Figure 5.1 will be extended to derive the call complete probability by focusing on a typical call trajectory while taking into account the two independent processes for the specific call, instead of the single wireless channel impairment.

Figure 5.2 shows the model of the typical trajectory of an MS roaming in a wireless mobile network with the unreliable wireless channel, block (1), and resource insufficiency, block (2). In terms of the link unreliability effect, the call connection alternately experiences channel good state and bad state. This is similar to the call behavior in the two previous sections. With respect to resource insufficiency, the call may be blocked during a new call connection request or terminated during a handoff operation when moving out of the current serving cell and entering a new one. Denote $t_{crt,k}(k \in \mathcal{N})$ as the cell residence time an MS stays in cell k with the generic form t_{crt} and the average value $1/\eta_{crt}$. The p.d.f. of t_{crt} and its corresponding LT are denoted by $f_{t_{crt}}(t)$ and $f_{t_{crt}}^*(s)$, respectively. Let t_{crt}^r be the residual cell residence time in the first cell that the call initializes the

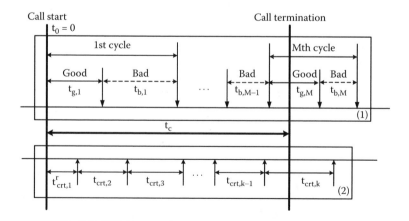

Figure 5.2 The time diagram for a typical call in mobile network with unreliable wireless channel and insufficient bandwidth.

connection request. Then, referring to the residual life theorem [42, p. 172], the p.d.f. of t_{crt}^r is given by:

$$f_{t_{crt}^r}(t) = \eta_{crt} \int_t^\infty f_{t_{crt}}(t) dt \qquad (5.61)$$

with the LT of t_{crt} p.d.f. given by:

$$f_{t_{ctr}^r}^*(s) = \frac{\eta_{crt}[1 - f_{t_{crt}}^*(s)]}{s} \qquad (5.62)$$

In addition, we define χ_k as the time summation of k cell residence time, that is:

$$\chi_k = t_{crt,1}^r \cdot \mathbf{1}_{k=1} + \left(t_{crt,1}^r + \sum_{i=2}^k t_{crt,i} \right) \cdot \mathbf{1}_{k \geq 2} \qquad (5.63)$$

Then, the LT of χ_k is given by:

$$f_{\chi_k}^*(s) = f_{t_{crt}^r}^*(s)[f_{t_{crt}}^*(s)]^{k-1}$$

$$= \frac{\eta_{crt}[1 - f_{t_{crt}}^*(s)][f_{t_{crt}}^*(s)]^{k-1}}{s}; \quad k \in \mathcal{N} \qquad (5.64)$$

5.5.1 Call Completion Probability

Let P_n and P_b denote the new call blocking probability and the handoff call blocking probability, respectively. A call connection is normally completed when the following three events are valid:

1. Event S1: the originating call is accepted by the target BS.
2. Event S2: at each handoff operation during the active call connection, there are sufficient resources and the call connection is not blocked.
3. Event S3: during each wireless channel bad state, the wireless link is successfully reestablished.

Note that the latter two events, S2 and S3, are not generally independent due to the common constraint of the call holding time. However, S2 and S3 are conditionally independent due to the independent impacts of the wireless link and the resource on a given specific call connection. In addition, no messages are exchanged between the MS and BS during channel bad state, which is represented by the dashed line in Figure 5.2. As a consequence, we develop the equation for the call completion probability as:

$$
P_{cc} = (1 - P_n) \sum_{m=0}^{\infty} \sum_{k=0}^{\infty} Pr(\xi_m < t_c \le \xi_{m+1}; \chi_k < t_c \le \chi_{k+1})
$$
$$
\times P_s^m (1 - P_h)^k
$$

$$
= (1 - P_n) \sum_{m=0}^{\infty} \sum_{k=0}^{\infty} \int_0^{\infty} Pr(\xi_m < x \le \xi_{m+1})
$$
$$
Pr(\chi_k < x \le \chi_{k+1}) f_{t_c}(x) dx P_s^m (1 - P_h)^k
$$

$$
= (1 - P_n) \int_0^{\infty} \sum_{m=0}^{\infty} Pr(\xi_m < x \le \xi_{m+1}) P_s^m
$$
$$
\sum_{k=0}^{\infty} Pr(\chi_k < x \le \chi_{k+1})(1 - P_h)^k f_{t_c}(x) dx \quad (5.65)
$$

Based on Equation 5.65, we define:

$$
a(m, x) = Pr(\xi_m < x \le \xi_{m+1}) \tag{5.66}
$$

$$
a^{**}(z, s) = \sum_{m=0}^{\infty} \int_0^{\infty} a(m, x) e^{-sx} dx z^m \tag{5.67}
$$

$$
d(k, x) = Pr(\chi_k < x \le \chi_{k+1}) \tag{5.68}
$$

$$
d^{**}(z, s) = \sum_{k=0}^{\infty} \int_0^{\infty} d(k, x) e^{-sx} dx z^k \tag{5.69}
$$

$$
A(x) = \sum_{m=0}^{\infty} Pr(\xi_m < x \le \xi_{m+1}) P_s^m \tag{5.70}
$$

$$
A^*(s) = \mathcal{L}(A(x)) \tag{5.71}
$$

$$
D(x) = \sum_{k=0}^{\infty} Pr(\chi_k < x \le \chi_{k+1})(1 - P_h)^k \tag{5.72}
$$

$$
D^*(s) = \mathcal{L}(D(x)) \tag{5.73}
$$

The item $a(m, x)$ can be physically interpreted as the probability that a call connection experiences m good channel state for a fixed interval x. The item $d(k, x)$ can be physically interpreted as the probability that a call connection traverses k cells in a fixed interval x. Following similar techniques as in the previous section, we obtain Equation 5.45 for $A^*(s)$. Proceeding with the development for $d(k, x)$, we have:

$$d(0, x) = \int_x^\infty f_{\chi_1}(u)\,du = 1 - F_{\chi_1}(x) \tag{5.74}$$

and

$$
\begin{aligned}
d(k, x) &= \int_0^x f_{\chi_k}(u)[1 - F_{t_{crt}}(x - u)]\,du \\
&= \int_0^x f_{\chi_k}(u)\,du - \int_0^x f_{\chi_k}(u) F_{t_{crt}}(x - u)\,du \\
&= F_{\chi_k}(x) - f_{\chi_k}(x) \circledast F_{t_{crt}}(x); \quad k \geq 1
\end{aligned}
\tag{5.75}
$$

Based on Equation 5.74, the LT of $d(0, x)$ is given by:

$$
\begin{aligned}
\mathcal{L}(d(0, x)) &= \frac{1}{s} - \frac{f_{\chi_1}^*(s)}{s} \\
&= \frac{s - \eta_{crt}[1 - f_{t_{crt}}^*(s)]}{s^2}
\end{aligned}
\tag{5.76}
$$

Based on Equation 5.75, the LT of $d(k, x)$ is given by:

$$
\begin{aligned}
\mathcal{L}(d(k, x)) &= \frac{f_{\chi_k}^*(s)}{s} - f_{\chi_k}^*(s)\frac{f_{t_{crt}}}{s} \\
&= \frac{f_{\chi_k}^*(s)[1 - f_{t_{crt}}^*(s)]}{s} \\
&= \frac{\eta_{crt}[f_{t_{crt}}^*(s)]^{k-1}[1 - f_{t_{crt}}^*(s)]^2}{s^2}; \quad k \geq 1
\end{aligned}
\tag{5.77}
$$

For $|z| \leq 1$, from Equations 5.68, 5.76, and 5.77, we can express:

$$d^{**}(z, s) = \frac{s - \eta_{crt}[1 - f_{t_{crt}}^*(s)]}{s^2} + \frac{\eta_{crt}[1 - f_{t_{crt}}^*(s)]^2 z}{s^2[1 - zf_{t_{crt}}^*(s)]} \tag{5.78}$$

As a consequence, the LT for $D(x)$ is given by:

$$
\begin{aligned}
D^*(s) &= d^{**}(1 - P_b, s) \\
&= \frac{s - \eta_{crt}[1 - f_{t_{crt}}^*(s)]}{s^2} + \frac{\eta_{crt}(1 - P_b)[1 - f_{t_{crt}}^*(s)]^2}{s^2[1 - (1 - P_b)f_{t_{crt}}^*(s)]}
\end{aligned}
\tag{5.79}
$$

On the basis of the above reasoning, the call completion probability (Equation 5.65) becomes:

$$P_{cc} = (1 - P_n)\int_0^\infty A(x) D(x) f_{t_c}(x)\,dx \tag{5.80}$$

For the special case when the channels are reliable as assumed in the previous studies, this result is the same as [27, Theorem 1]. Furthermore, when t_c follows an exponential distribution and the channels are reliable, this result is same as those in references [24] and [31].

5.6 Numerical Results

In this section, we focus on analytical model validation and the effect of wireless link error characteristics on call behavior with respect to call completion probability. We choose $N_{LER} = 3$ and $T_{LER} = 1000$ milliseconds, [16]. Both T_{mc} and T_u are regarded as exponentially distributed random variables with an average of 6.0 seconds and 6.0 seconds, respectively. The average duration of the channel good state is fixed at 24.0 seconds. The cell residence time follows the Gamma distribution [37] with an average value $1/\eta_{crt} = 120.0$ seconds (if not specified) and the variance $V_{crt} = 0.5\eta_{crt}^{-2}$. The new call blocking probability is $P_n = 0.05$ and the handoff call blocking probability is $P_b = 0.02$.

To validate our analytical model, we simulate the MS behavior in a two-dimensional plane using Monte Carlo simulation to obtain the simulation result that follows [38]. A subprogram is written to simulate an MS's behavior. We present a comparison between the simulation result under the Fritchman channel model. In this case, we suppose that the channel good state or bad state duration follows a second-order hyper-exponential distribution with p.d.f.

$$k(t) = 0.2 \times 0.4\mu_k e^{-0.4\mu_k t} + 0.8 \times 1.6\mu_k e^{-1.6\mu_k t}; \quad k \in \{g, b\} \quad (5.81)$$

We will show the performance estimation discrepancy due to the ignorance of the wireless link unreliability. In this example, the call holding time follows the second-order hyper-exponential (H2) distribution $0.2 \times 0.4\mu_c e^{-0.4\mu_c t} + 0.8 \times 1.6\mu_c e^{-1.6\mu_c t}$ with mean 180.0 seconds. The Fritchman wireless channel model in Equation 5.81 is employed.

Figure 5.3 shows the call completion probability in terms of the ratio η_{crt}/μ_c in the presence or absence of the wireless link unreliability impact. In the figure, "w/o link" denotes the traditional result without considering the unreliable wireless link, while "$\mu_b = 1/0.5$" and "$\mu_b = 1/2.0$" represent the results in the presence of the unreliable physical link with an average bad state duration of 0.5 seconds and 2.0 seconds, respectively. It is clear that P_{cc} is substantially overestimated without taking into account the wireless link effect. The discrepancy reduces with the larger η_{crt}/μ_c, which corresponds to faster mobility. This is because, in such cases, the resource insufficiency dominates, while the wireless link unreliability becomes insignificant as a consequence of the call completion.

Figure 5.4 shows the call completion probability in terms of the average channel bad state duration with the Fritchman wireless channel model

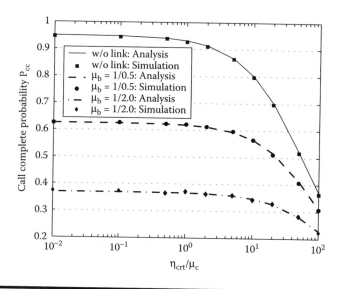

Figure 5.3   The call completion probability P_{cc} in terms of the ratio η_{crt}/μ_c with or without the presence of the wireless link unreliability effect.

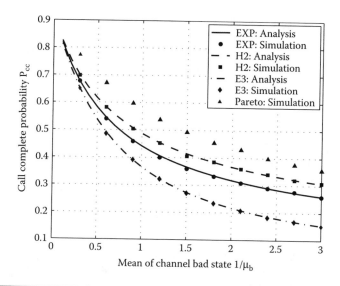

Figure 5.4   The call completion probability P_{cc} in terms of the average channel bad state duration μ_b^{-1} with the Fritchman wireless channel model under the different call holding times; EXP: exponential; H_2: hyper-exponential; E_3: three-stage Erlang.

under the different call holding times. The comparison indicates that the performance metrics show a significant discrepancy when using different distribution functions for the call holding time.

5.7 Conclusions and Future Work

In this chapter, we studied wireless network tele-traffic modeling and analysis over the lossy link. We derived the closed-form formula for the performance metrics call completion probability under the general call holding time and the generalized wireless channel model. It was shown that network performance will be substantially overestimated without considering the unreliable wireless link impact. In addition, the comparison demonstrates that there is a considerable performance gap in applying different call holding time distributions. The results presented are helpful in designing more practical and efficient call admission control schemes integrating the impaired wireless channel effect and in analyzing the reliability or availablity characteristics of the wireless system. An interesting future direction is to design more efficient call admission control algorithms considering the link lossy characteristics.

References

[1] http://grouper.ieee.org/groups/802/11/
[2] http://grouper.ieee.org/groups/802/16/
[3] http://www.gsmworld.com/
[4] http://www.umts-forum.org/
[5] http://www.bluetooth.com/
[6] E.N. Gilbert, "Capacity of a burst-noise channel," *Bell System Technical Journal*, Vol. 39, pp. 1253–1265, September 1960.
[7] E.O. Elliott, "Estimates of error rates for codes on burst-noise channels," *Bell System Technical Journal*, Vol. 42, pp. 1977–1997, September 1963.
[8] B.D. Fritchman, "A binary channel characterization using partitioned Markov chains," *IEEE Trans. Inform. Theory*, Vol. 13, No. 2, pp. 221–227, April 1966.
[9] H.S. Wang and N. Moayeri, "Finite-state Markov channel — a useful model for radio communication channels," *IEEE Trans. Veh. Technol.*, Vol. 44, No. 1, pp. 163–171, February 1995.
[10] Y. Zhang and B. Soong, "The effect of unreliable wireless channel on the call performance in mobile network," *IEEE Trans. Wireless Commun.*, Vol. 4, No. 2, pp. 653–661, March 2005.
[11] Y. Zhang and B. Soong, "Performance of mobile network with wireless channel unreliability and resource insufficiency," to appear in *IEEE Trans. Wireless Commun.*, Vol. 5, No. 5, pp. 990–995, May 2006.

[12] M. Krunz and J. Kim, "Fluid analysis of delay and packet discard performance for QoS support in wireless networks," *IEEE J. Sel. Areas Commun.*, Vol. 19, No. 2, pp. 384–395, February 2001.

[13] R. Fantacci and S. Nannicini, "Performance evaluation of a reservation TDMA protocol for voice/data transmission in personal communication networks with nonindependent channel errors," *IEEE J. Sel. Areas Commun.*, Vol. 19, No. 9, pp. 1636–1646, September 2000.

[14] Y. Kim and S. Li, "Capturing important statistics of a fading/shadowing channel for network performance analysis," *IEEE J. Sel. Areas Commun.*, Vol. 17, No. 5, pp. 888–901, May 1999.

[15] ETSI, Radio Link Protocol for Data and Telematic Services on the Mobile Station–Base Station System (MS-BSS) Interface and the Base Station System–Mobile Switching Centre (BSS-MSC) Interface, GSM Specification 04.22, Version 5.0.0, December 1995.

[16] 3rd Generation Partnership Project; Technical Specification Group Radio Access Network; Radio Resource Control (RRC); Protocol Specification (Release 1999); 3GPP TS 25.331 V3.9.0 (2001-12).

[17] D. Hong and S.S. Rappaport, "Traffic model and performance analysis for cellular mobile radio telephone systems with prioritized and nonprioritized handoff procedure," *IEEE Trans. Veh. Technol.*, Vol. 35, No. 3, pp. 77–92, 1986.

[18] Y.B. Lin, S. Mohan, and A. Noerpel, "Queueing priority channel assignment strategies for PCS handoff and initial access," *IEEE Trans. Veh. Technol.*, Vol. 43, No. 3, pp. 704–712, August 1994.

[19] Y. Zhang and B. Soong, "Handoff dwell time distribution effect on mobile network performance," *IEEE Trans. Veh. Technol.*, Vol. 54, No. 4, pp. 1500–1508, July 2005.

[20] Y.B. Lin, R. Anthony, and D.J. Harasty, "The subrate channel assignment strategies for PCS handoffs," IEEE *Trans. Veh. Technol.*, Vol. 45, pp. 122–130, February 1996.

[21] I.S. Yoon and B.G. Lee, "DDR: a distributed dynamic reservation scheme that supports mobility in wireless multimedia communications," *IEEE J. Select. Areas Commun.*, Vol. 19, No. 11, pp. 2243–2253, November 2001.

[22] T.-S.P. Yum and K.L. Yeung, "Blocking and handoff performance analysis of directed retry in cellular mobile systems," *IEEE Trans. Veh. Technol.*, Vol. 44, No. 3, pp. 645–650, August 1995.

[23] S. Tekinary and B. Jabbari, "A measurement based prioritization scheme for handovers in cellular and microcellular networks," *IEEE J. Select. Areas Commun.*, Vol. 10, No. 8, pp. 1343–1350, October 1992.

[24] Y.B. Lin and I. Chlamtac, "Effects of Erlang call holding times on PCS call completion," *IEEE Trans. Veh. Technol.*, Vol. 48, No. 3, pp. 815–823, May 1999.

[25] Y.B. Lin and Y.K. Chen, "Reducing authentication signaling traffic in third-generation mobile networks," *IEEE Trans. Wireless Communications*, Vol. 2, No. 3, pp. 493–501, May 2003.

[26] Y. Fang and Y. Zhang, "Call admission control schemes and performance analysis in wireless mobile networks," *IEEE Trans. Veh. Technol.*, Vol. 51, No. 2, pp. 371–384, March 2002.

[27] Y. Fang, I. Chlamtac, and Y.B. Lin, "Call performance for a PCS network," *IEEE J. Select. Areas Commun.*, Vol. 15, No. 8, pp. 1568–1581, October 1997.

[28] Y. Fang, "Hyper-Erlang distribution model and its application in wireless mobile networks," *ACM Wireless Networks*, Vol. 7, No. 3, pp. 211–219, May 2001.

[29] K. Yeo and C.-H. Jun, "Modeling and analysis of hierarchical cellular networks with general distributions of call and cell residence times," *IEEE Trans. Veh. Technol.*, Vol. 51, No. 6, pp. 1361–1374, November 2002.

[30] F. Ashtiani, J.A. Salehi, and M.R. Aref, "Mobility modeling and analytical solution for spatial traffic distribution in wireless multimedia networks," *IEEE J. Select. Areas Commun.*, Vol. 21, No. 10, pp. 1699–1709, December 2003.

[31] W. Li and A.S. Alfa, "Channel reservation for handoff calls in a PCS network," *IEEE Trans. Veh. Technol.*, Vol. 49, No. 1, pp. 95–104, January 2000.

[32] W. Li, K. Makki, and N. Pissinou, "Performance analysis of a PCS network with state dependent calls arrival processes and impatient calls," *Elsevier Computer Communications*, Vol. 25, No. 5, pp. 507–515, 2002.

[33] V.-A. Bolotin, "Modeling call holding time distributions for CCS network design and performance analysis," *IEEE J. Select. Areas Commun.*, Vol. 14, No. 4, pp. 433–438, April 1994.

[34] C. Chang, C.-J. Chang, and K.-R. Lo, "Analysis of a hierarchical cellular system with reneging and dropping for waiting new and handoff call," *IEEE Trans. Veh. Technol.*, Vol. 48, No. 4, pp. 1080–1090, July 1999.

[35] P.V. Orlik and S.S. Rappaport, "A model for teletraffic performance and channel holding time characterization in wireless cellular communication with general session and dwell time distributions," *IEEE J. Select. Area Commun.*, Vol. 16, No. 5, pp. 788–803, June 1998.

[36] M.A. Marsan, G. Ginella, R. Maglione, and M. Meo, "Performance analysis of hierarchical cellular networks with generally distributed call holding times and dwell times," *IEEE Trans. Wireless Commun.*, Vol. 3, No. 1, pp. 248–257, January 2004.

[37] M.M. Zonoozi and P. Dassanayake, "User mobility modeling and characterization of mobility patterns," *IEEE J. Select. Area Commun.*, Vol. 15, No. 3, pp. 1239–1252, September 1997.

[38] G.S. Fishman, *Monte Carlo: Concepts, Algorithms and Applications*, Springer Verlag, 1997.

[39] N.A.J. Hastings and J.B. Peacock, *Statistical Distributions*, New York: Wiley, 1975.

[40] N.L. Johnson, *Continuous Univariate Distributions-1*, New York: Wiley, 1970.

[41] W.R. LePage, *Complex Variables and the Laplace Transform for Engineers*, New York: Dover Publications, 1980.

[42] L. Kleinrock, *Queueing Systems*, New York: John Wiley & Sons, 1975.

[43] F.P. Kelly, *Reversibility and Stochastic Networks*, New York: Wiley, 1979.

[44] T.S. Rappaport, *Wireless Communications: Principles and Practice*, Prentice Hall, 1996.

[45] W.C. Jakes, *Microwave Mobile Communications*, New York: IEEE, 1993.

Chapter 6

Heterogeneous Wireless Networks: Optimal Resource Management and QoS Provisioning

Nidal Nasser

Contents

6.1 Overview

Future wireless and mobile networks will rely on more than one type of radio access technology, including cellular networks, wireless local area networks (WLANs), and multi-hop/ad hoc variable topology networks. Seamless intersystem mobility across heterogeneous networks will be one of the main features in future-generation wireless networks. This change in the technological landscape further challenges today's radio systems by the increasing amount of capacity-demanding services. The services span from traditional conversational audio to conversational video, voice messaging, streamed audio and voice, fax, telnet, interactive games, Web browsing, file transfer, paging, and e-mailing. No single radio system can effectively cover all these services from a multi-service point of view if Quality-of-Service (QoS) requirements are to be met. Consequently, the development moves toward interworking the different but complementary radio systems that together can provide an unparalleled level of service delivery. Specifically, future wireless networks, called fourth-generation (4G) wireless, will be able to offer personalized service delivery over the most efficient/preferred network, depending on the user profile and the type of data to transmit. 4G wireless networks will be generally characterized by heterogeneity in architectures, protocols, and air interfaces. Interoperability between heterogeneous systems will be provided through different inter-technology coupling techniques and terminal support to several radio interfaces. However, the notion of seamless mobility in heterogeneous wireless networks raises some challenges from the point of view of radio resource management (RRM) and QoS provisioning; for example:

■ How does one manage resources and control end-to-end QoS for mobile users accessing different services over heterogeneous networks, given that each network possesses its own resource management schemes and policies?

■ Will the wireless environment in the context of heterogeneity be able to provide strict QoS guarantees to real-time applications? Or, can the networks provide at least a gradual change in resource availability in order to let the applications adapt?

Therefore, the need for new RRM schemes to answer these questions and to satisfy the diverse QoS requirements of heterogeneous wireless networks becomes more important. The objective of this chapter is to provide a breadth and depth of RRM techniques for provisioning QoS in heterogeneous wireless networks. The breadth part covers the basic concepts related to resource management. The depth part, on the other hand, presents a technical case study. This includes developing an admission control policy based on the stochastic control technique Markov decision process (MDP).

6.2 Basic Radio Resource Management (RRM) and Challenges

RRM is a set of algorithms that control the usage of radio resources. RRM functionality aims to guarantee QoS and to maximize the overall system capacity. A common definition for capacity is the maximum traffic load that the system can accommodate under some predefined service quality requirements. The basic RRM can be classified into the following components: hand off and mobility management, call admission control, load control, channel allocation and reservation, and packet scheduling. In this section we present an overview of RRM components; in particular, emphasis is given to admission control and hand off and mobility management between different access networks, both of which are critical in a heterogeneous network environment as they balance the load across networks while maintaining sufficient QoS for mobile users.

Giving access to users they roam between networks or within a single network is governed by an admission control scheme. Admission control enables a wireless network to carry the largest amount of traffic for a given amount of spectrum. It ensures that the QoS perceived by each user is above the minimum guaranteed. Admission control and mobility management schemes facilitate load balancing between heterogeneous wireless networks — users can be forced to hand over to another network to make way for users with more demanding bandwidth requirements and can thus prioritize users. It may be possible using an admission control algorithm to admit a user to multiple networks simultaneously and use multiple connections to deliver services to the user and thus achieve a higher QoS than that offered from a single network. Much work has been done on call admission control in homogeneous networks. The Universal Mobile Telecommunications System (UMTS)-based wireless cellular network, known also as a third-generation (3G) network, is an example of such a network. Before admitting a new user, admission control needs to check that the admittance will not sacrifice the planned coverage area or the quality of the existing connections. In UMTS, admission control accepts or rejects a request to establish a radio access bearer in the radio access network based on the interference levels, as the capacity in UMTS is interference limited. The introduction of a new user increases the interference level to existing users as all connections share the same radio channel. Signal-to-interference ratio (SIR) values are periodically measured to determine the level of interference at each base station. The SIR is the ratio of the received signal strength to the total interference from all interfering mobiles. The admission control algorithm is executed when a bearer is set up or modified. The admission control functionality can have a decentralized approach where information regarding neighbour cell loads can be obtained from an entity controlling those cells. The admission control algorithm estimates the load increase that the establishment of the bearer would cause in the radio network. This must be estimated for both the uplink and downlink for asymmetrical traffic. The requesting bearer gains access only if both the uplink and downlink admission control admit it; otherwise it is rejected because of the excessive interference that it would produce in the network. The admission control procedure across heterogeneous wireless networks is much more complicated than in a homogeneous wireless network such as that described above. If multiple networks are available to a user at any one time, then choosing the most optimal network for a particular service delivery and choosing the correct time to execute a vertical hand off[1] to improve the QoS for all users are important factors. A mobility management system can be used to control the migration of users from one system to another. The user, network, or both can govern the mobility management and admission control procedures. Giving total control to the user can result in network instability as users compete for network resources, while a network-controlled system will ignore user preferences and QoS requirements. The most optimal mobility management and admission control scheme should encompass both aspects — mobile assisted call admission control and mobility management.

6.3 Case Study: Optimal Call Admission Control Policy in Heterogeneous Wireless Networks

In this section we present a case study for controlling the traffic load in heterogeneous wireless networks; it consists of two different wireless access technologies: UMTS and WLAN. We use the Markov decision process technique to develop an optimal call admission control policy for heterogeneous wireless networks. The aim of this policy is to provide a balance between the two opposing objectives of the network operator (or service provider) and mobile users. The former wants to achieve high system utilization so that more users can be accommodated by the system and more revenue can thus be obtained, while the latter wants to receive better QoS.

---

[1] Transfer an active connection between two heterogeneous network interfaces is known as a vertical hand off.

The remainder of this section is organized as follows. Section 6.3.1 presents an overview of the integration between UMTS and WLAN. Section 6.3.2 discusses the hand offs in heterogeneous wireless networks followed by Section 6.3.3, which presents the recent related research work in the literature. Section 6.3.4 gives the contributions of this chapter. The system model is described in Section 6.3.5. The Markov decision-based admission model is identified and formulated by five components in Section 6.3.6. Simulation results are shown in Section 6.3.7.

6.3.1 Background

Currently, the wireless networking scene is dominated by two distinct networking platforms: (1) cellular networks, which passed through multiple generations (1G, 2G, and 3G); and (2) WLANs championed by the IEEE 802.11 networks. Recent trends indicate that 3G networks and WLANs will coexist to offer public wireless broadband services to end users [1]. The two platforms offer characteristics that complement each other perfectly.

The 3G cellular systems such as UMTS and Code Division Multiple Access (CDMA2000) will support real-time and non real-time multimedia services with data rates from 144 kbp/s to 2 Mbp/s with wide coverage and nearly universal roaming. However, the costs of acquiring the necessary radio spectrum and the required network equipment upgrades are very high. This is in contrast to WLAN systems such as IEEE 802.11a/b/g, which provide affordable services and bit rates surpassing those of 3G systems, up to 11 Mbp/s with IEEE 802.11b and 54 Mbp/s with IEEE 802.11a/g. However, the coverage offered by WLANs is quite limited and lacks roaming support.

The complementary characteristics of 3G cellular systems (slow, wide coverage) and WLAN (fast, limited coverage) make it attractive to integrate these two technologies to provide ubiquitous wireless access. The purpose of integrating 3G systems and WLANs is to make it possible to use the best parts of both systems. High bandwidth WLANs are used for data transfer where available and 3G systems can be used where WLAN coverage is lacking. Development and standardization efforts are currently underway for defining suitable architectures for 3G/WLAN integration. However, designing a network architecture that efficiently integrates 3G systems and WLANs is a challenging task that needs a lot of research efforts.

Integrating two very different technologies, such as 3G and WLAN, introduces a number of technical and logistical issues that must be resolved to maximize the benefits reaped from such integration. Transferring an active call between access points (APs) or base stations (BSs) is called horizontal hand off. The horizontal hand off has long been an issue within the wireless telecommunication field. However, a higher level of hand off complexity, and thus issues, is introduced to the differences between internetworked heterogeneous wireless networks. This transfer between different types of wireless networks is known as a vertical hand off [2]. An example of a vertical hand off is when a mobile user moves back and forth between 3G and WLAN networks. Seamless intersystem mobility across such access heterogeneity will be the capital feature in next-generation wireless networks. In such networks, it will be necessary to support seamless hand offs of mobile users without causing disruption to their ongoing connections. As a result, the need for seamless hand offs across the different wireless networks is becoming increasingly important.

One of the chief issues that aids in providing a seamless hand off is the ability to correctly decide whether to carry out a vertical hand off at any given time. This could be accomplished by taking into consideration two key issues: (1) network conditions for vertical hand off decisions and (2) connection maintenance [3]. These two issues must be tightly coupled to move seamlessly across different network interfaces. To attain a positive vertical hand off, the network state ought to be constantly obtainable by means of a suitable hand off metric. In multi-network environments, this is very challenging and difficult to achieve as there does not exist a single factor than can provide a clear idea of when to hand off. Signal strength, which is the chief hand off metric measured in horizontal hand offs, cannot be utilized for vertical hand off decisions due to the overlay nature of heterogeneous networks and the different physical techniques used by each network. This criterion of a vertical hand off is one of the chief challenges for seamless mobility. The solution to this challenge is the proper management of scarce resources to provide better QoS for mobile users and to achieve high system utilization so that more users can be accommodated by the system. Sometimes, these two goals are conflicting and trade-offs must be made. Providing a proper balance between system utilization and users' QoS satisfaction is the focus of this chapter.

6.3.2 Hand offs in Heterogeneous Wireless Networks

As discussed above, hand offs can be defined as the transition of signal transmission between different cells. A hand off scheme is required to preserve connectivity as devices move about, and at the same time curtail any disturbance to ongoing transfers. Therefore, hand offs must exhibit low dropping rates, sustain minimal amounts of data loss, as well as scale to large networks. Hand off schemes have been thoroughly studied and deployed in cellular systems, and are escalating in importance in other networks, such as WLANs, as research in 4G wireless communications increases in popularity. Hand offs can be classified as either horizontal or vertical, as depicted in Figure 6.1.

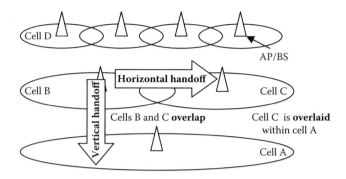

Figure 6.1   Horizontal versus vertical hand offs.

Horizontal hand off is the changeover of signal communication from one AP/BS to a geographically neighboring AP/BS supporting the same technology, as the user roams about. Horizontal hand off is also referred to as intra-technology hand off. Every time a mobile cellular host crosses from one cell into a neighbouring cell (supporting the same technology), the network routinely and automatically exchanges the coverage responsibility from one AP/BS to another. In a properly operating network, the hand off takes place smoothly and efficiently, without gaps in communications and without uncertainty as to which BS should be dealing with the mobile user. Mobile users need not get involved for horizontal hand off to take place, nor do they have to sense the hand off process or identify which BS is managing the signals at any certain time.

Horizontal hand off is the most widespread definition of hand off due to the extensive research that has taken place in this field in the past several years. Vertical hand off, on the other hand, is a more recent and exciting scheme that promises to transfigure the way we communicate. While horizontal hand off is a handover among BSs in service by the same wireless network interface, vertical hand off takes place between different network interfaces that usually represent different technologies. Vertical hand off architectures and schemes will play a major role in the Multimedia Independent Handover standard (IEEE 802.21) [1] and will pave the road for the emergence of 4G overlay, multi-network environments.

There are two types of vertical hand offs: (1) upward and (2) downward [4]. An upward vertical hand off is roaming to an overlay with a larger cell size and lower bandwidth such as wide area networks (cellular networks), and a downward vertical hand off is roaming to an overlay with a smaller cell size and larger bandwidth such as WLANs. Downward vertical hand offs are less time critical because a mobile user can always remain connected to the upper overlay and not hand off at all.

6.3.3 Related Work

In this section we present various hand off decisions and policy schemes that were proposed by researchers and deeply analyze these ideas in terms of their ability to provide QoS guarantees while meeting user and network requirements and constraints. Related works for integrating 3G and WLAN have been recently discussed in the literature. However, more work needs to be done.

In a recent study [5], we define a vertical hand off decision function that provides handover decisions when roaming across heterogeneous wireless networks. New optimizations for vertical hand off decision algorithms were developed by Zhu and McNair [6] to maximize the benefit of the hand off for both the user and the network. Ylianttila et al. [7] studied the performance of horizontal hand off and vertical hand off between WLAN and General Packet Radio Services (GPRS) networks and presented an optimization scheme for performing vertical hand off.

McNair et al. [8] introduced a new intersystem hand off protocol that uses boundary cells that allow the mobile user to roam into a different network. The mobile user receives and compares signals from the detected networks and hand off is made to a new network if it has stronger signal strength.

Makela et al. [10] proposed a hand off algorithm between WLANs and cellular networks. They argued that the vertical hand off policy is different, depending on the direction of hand off. When handing over from cellular network to WLAN, the delay in the hand off transmission region must be short, so the preferred hand off point is the first time the signal strength degrades. On the other hand, when handing over from WLAN to cellular networks, because a cellular network covers a wide area and the hand off time is not critical, the preferred hand off point from cellular to WLAN is the first time the signal strength in the WLAN reaches an acceptable level. Park et al. [9] also proposed a similar scheme that presents a seamless vertical hand off procedure between IEEE 802.11 WLANs and CDMA2000 cellular networks. In their algorithm, traffic is classified into real-time (from WLAN to CDMA) and nonreal-time (from CDMA to WLAN) services and a hand off decision policy similar to the one used by Makela et al. [10] is applied. These solutions are very interesting, but they provide specific solutions for specific networks.

Similarly, Zhang et al. [11] proposed a novel mobility management system for vertical hand off between UMTS and WLAN. The system contains a connection manager that is responsible for detecting wireless network changes when making hand off decisions. They consider two hand off scenarios. In the first scenario when moving from UMTS to WLAN, the objective of the hand off is to improve the quality of service (because the hand off to the WLAN is optional). They assume that the WLAN would provide higher bandwidth and lower cost and therefore recommend hand off to WLAN in this case. In the second scenario, when users step out of a WLAN, their connection manager's goal is to switch to UMTS before the WLAN line breaks, but at the same time, it tries to stay in the WLAN as long as possible because, again, they assume that the WLAN provides lower cost and better QoS. They measure signal strength to determine WLAN unavailability.

Chen et al. [12] presented a smart decision model that tries to smartly perform hand off to the highest-quality network at the most suitable time. Their proposed model is able to make vertical hand off decisions based on the properties of available network interfaces such as link capacity, power consumption, and link cost. In another project [13], they proposed a Universal Seamless hand off Architecture (USHA) to deal with both horizontal and vertical hand off scenarios. USHA achieves seamless hand off by following the middleware design philosophy [14], and by integrating the middleware with existing Internet services and applications.

Ylianttila et al. [15] investigated the hand off characteristics in a compound vertical setting and established the relation between the hand off delay and throughput diminution perceived by the user throughout the handover period. An assessment of hand off decision effectiveness is suggested, but without making an allowance for the QoS associated features. In their conclusion they assert that it is better to remain connected via a WLAN as long as the data rate in the WLAN is higher than in cellular UMTS, and this is true for any two systems having different rates of data transfer. Their assumption is simply invalidated in the presence of high background traffic on the WLAN.

In future mobile generations, service providers and operators may allow their customers to define their personal set of mobility parameters; these vertical hand off parameters are chosen according to user preferences and are tailored specifically for personal needs [15]. Wang et al. [16] described a policy-enabled hand off scheme that enables users to articulate policies, depending on what is the finest wireless system at any instance, and make trade-offs amid network characteristics such as monetary cost, power utilization, and QoS.

In view of the fact that diverse layers of the network architecture will sustain different data rates, hand off can be achieved with virtually no delay whatsoever if applications detect the change in service level and adjust their QoS demands accordingly. Experiments conducted on hand off performance from networks with a low bit rate to others with higher bit rates have shown that reasonably rapid hand offs can be attained with only small amounts of buffering and retransmission necessary for loss-free handovers [17]. On the contrary, significant amounts of buffering and retransmissions are required in the case of hand off from fast to slower networks; otherwise, data loss will be experienced [17].

Ylianttila et al. [18] explained how neural networks and fuzzy logic concepts can be used to control vertical hand off procedures. They use hand off between WLAN and GPRS as an example. To enhance features such as power saving, by powering down unused interface cards, the proposed scheme allocates several levels of alerts (stable, unstable, or poor WLAN), which would aid in the preparation for an upcoming hand off. Ylianttila et al. used several hand off factors in their proposed solution and these included the signal strength, beacon packets, signal-to-noise ratio (SNR), bit-error-rate (BER), packet error rates, and hysteresis margin of signal strength.

Alsenmyr et al. [19] at Ericsson have proposed their own mechanism for vertical hand off between WCDMA and GSM. They focus on the handover mechanism but do not give any details regarding hand off decisions. Heickero et al. [20], nonetheless, discussed and recommended routing traffic among Global System for Mobile Communications (GSM) and WCDMA based on sharing.

Myers [21] explained how an understanding of human perception and the simulation of a mobile IP network can be used to tackle the relevant issues of providing acceptable QoS for mobile multimedia. Although this scheme does not explicitly discuss vertical hand off issues, it clearly discusses many factors, such as packet delay, dropped packets, and latency, that must be analyzed for enhanced service during hand off.

Wang [22] proposed a dynamic resource allocation model that is mainly suitable for the multimedia transmission on heterogeneous wireless networks. This model has two goals. First, it induces the resource map of the current cell using the beacon codes received from the neighboring cells. It then extracts the mobility pattern of the mobile users from past hand off records using the back-propagation neural network. With the mobility pattern, the system then infers which cells would be possible targets of the mobile user when it moves. Second, it uses rules of application layers (e.g., the communication cost, maximal moving velocity, maximal data rate, etc.) to select candidate cells from the target ones chosen above. It then executes dynamic resource allocations on the candidate cells for supporting the mobile user with guaranteed QoS in advance. In this way, it decreases the hand off call dropping probability and new call blocking probability, and further, increases the resource utilization on the heterogeneous wireless networks. [22]. We remark that none of the above studies consider the stochastic optimization analysis of admitting mobile users to the integrated networks. This actually motivates us to present an optimization stochastic tool that the service provider can use to optimize the performance of the integrated networks.

6.3.4 Contributions

In this study we consider a wireless network that consists of a WLAN cell overlaid within a 3G cellular network. Thus, a mobile user moving within the wireless network can vertically hand off between the 3G cellular

Figure 6.2   Service coverage of the 3G and WLAN networks.

network and the WLAN. Accepting or rejecting new arrivals (new calls or vertical hand off calls) is a critical decision that the service provider should choose based on the network load conditions. The service provider objective is to minimize the blocking rate of new arrivals (new calls and vertical hand off calls) while maximizing system utilization. We argue that this problem is best formulated in terms of decision theory. In this chapter we use the stochastic control technique MDP to study and examine the relationship between optimal decision, which the service provider should apply, and traffic parameters in the network.

6.3.5   System Model

Consider an integrated 3G–WLAN network consisting of a set of 3G cells and a number of WLANs located inside each cell. Assume that all the 3G cells and underlying WLANs are stochastically identical, so we can focus our analysis on one 3G cell and the underlying WLAN, as shown in Figure 6.2. In this work, we are concerned with improving the performance of WLAN cell. Therefore, the main objective of our work is to minimize the blocking rate of new calls and vertical hand off calls and to maximize the system utilization of the WLAN cell.

Consider a WLAN cell that has a total capacity of C bandwidth units (bu). Three types of calls share the bandwidth of the cell: (1) new calls; (2) vertical hand off calls from 3G to WLAN, which we called downward hand offs; and (3) vertical hand off calls from WLAN to 3G, which we called upward hand offs. The bandwidth for each call is given by c, which represents the number of bandwidth units that is adequate for guaranteeing desired QoS for this call with certain traffic characteristics.

As for traffic characterization, we consider a simple model for a WLAN cell perspective. New call arrivals into a cell are assumed to be Poisson with rate λ_n. Vertical hand off calls, downward and upward, are also assumed to be Poisson with rates λ_{b_1}, λ_{b_2}, respectively. The call holding time (CHT) is assumed to follow an exponential distribution with mean $1/\mu$. Based on a detailed study of mobility in wireless networks [23], we assume that cell residence time (CRT) is exponentially distributed. Thus, we assume that the mean CRT for 3G users in a WLAN is $1/b_1$, and the mean CRT for WLAN users in 3G is $1/b_2$. The channel occupancy time is the minimum of the CHT and the CRT. As the minimum of exponentially distributed random variables is also exponentially distributed, then the channel occupancy time is therefore assumed to be exponentially distributed with means $1/(\mu + b_1 + b_2)$.

6.3.6 Markov Decision Process Model

In this section we represent the state of the WLAN cell inside the 3G-WLAN integrated network as an MDP. Then the MDP problem is formulated to obtain the optimal decision.

MDP is a stochastic dynamic process that can model an optimization problem in which the time intervals between consecutive decision epochs are not identical but follow a probability distribution. In our proposed model, the consecutive decision epochs are assumed to follow the exponential distribution. Our system is observed at discrete time points t = 0, 1, 2, $\cdots$. At each observation, the system is classified to be one of the possible states, which are finite and countable. The set of possible states forms the state spaces. For each state that belongs to the state space, a set of actions is given, which is also finite and countable. Whenever there is an arrival call (new or vertical hand off), the system must make some decisions based on the current system state. These decisions include whether the system will accept the call. The system selects its decisions from a finite decision (action) space. The possible decision for all traffic types is accept or reject. If the system is in state x and action a is chosen, then:

1. The next state, y, of the system is chosen according to the transition probability p_{xy}^a.
2. The time until the transition from x to y occurs is $\tau(x, a)$.

After the transition occurs, an action is again chosen and 1 and 2 above are repeated. Markovian properties are satisfied if at a decision epoch the action a is chosen in the current state x, and the state at the next decision epoch depends only on the current state x and the chosen action a. They are thus independent of the past history of the system.

The procedure for getting the optimal solution is as follows. We first model the system in terms of MDP properties, which we discuss in detail below. This will include first defining the system state, the state space, and the action space. Second, we need to formulate the expected sojourn time analytically. Third, we define the transition probability between the system states. Fourth, we define a reward function that the service provider wants to maximize. Finally, we formulate the MDP analysis as a linear programming problem. We then solve the linear programming formulas. The solution was proved mathematically to be the optimal solution [24].

The proposed MDP model can be uniquely identified by the following five formularization components.

6.3.6.1 State Space

We denote the current state of the system (WLAN cell) by a vector x as follows:

$$x = (x_n, x_{b_1}, x_{b_2}) \tag{6.1}$$

where x_n is the number of new calls, x_{b_1} is the number of downward hand off calls, and x_{b_2} is the number of upward hand off calls. The state space Ω is a set of all possible combinations of occupied bandwidth units in the system and given by:

$$\Omega = \{x | x = (x_n, x_{b_1}, x_{b_2}), x_n, x_{b_1}, x_{b_2} \geq 0, (x_n + x_{b_1} + x_{b_2}) \leq C\} \tag{6.2}$$

6.3.6.2 Action Space

When the system is in state x, an accept/reject decision must be made for each type of possible arrival. Call arrival events include new call, downward hand off call, and upward hand off call arrivals. The decision is made only at the occurrence of a call arrival. At events of call completion or hand off to other WLAN cells, decisions will not be made. The action space A is a set of vectors consisting of three binary elements, that is:

$$A = \{a | a = (a_n, a_{b_1}, a_{b_2}); a_n, a_{b_1}, a_{b_2} \in \{0(reject), 1(accept)\}\} \tag{6.3}$$

where a_n, a_{b_1}, a_{b_2} are actions for new calls, downward hand off calls, and upward hand off calls, respectively. They take the value of 0 for rejecting and 1 for accepting that type of call. The action space A_x for state $x \in \Omega$ can be written as:

$$A_x = \begin{cases} \{a = (1, 1, 1)\} & \text{if } x = (1, 1, 1) \\ \{a = (0, 0, 0)\} & \text{if } x_n, x_{b_1}, x_{b_2} \geq C \\ \{a | a = (a_n, a_{b_1}, a_{b_2}); a_n, a_{b_1}, a_{b_2} \in \{0, 1\}\} & \text{Otherwise} \end{cases} \tag{6.4}$$

For example, when $(a_n, a_{b_1}, a_{b_2}) = (0, 1, 1)$, this indicates that only new calls will be rejected.

6.3.6.3 Sojourn Time

Let $\tau(x, a)$ be the sojourn time, the expected time until a new state is entered, when the system is in the present state $x \in \Omega$ and when action $a \in A_x$ is chosen. The value of sojourn time can be expressed by:

$$\tau(x, a) = [\lambda_n a_n + \lambda_{b_1} a_{b_1} + \lambda_{b_2} a_{b_2} + x_n \mu + x_{b_1} b_1 + x_{b_2} b_2]^{-1} \tag{6.5}$$

where a_n, a_{b_1}, a_{b_2} take a binary value, that is, with 1 for accepting a call and 0 for rejecting a call.

6.3.6.4 Transition Probability

Let us denote a call arrival event as a vector e of three binary values as follows:

$$e = (e_n, e_{b_1}, e_{b_2}) = \begin{cases} (1, 0, 0) & \text{if new call arrival} \\ (0, 1, 0) & \text{if downward handoff call arrival} \\ (0, 0, 1) & \text{if upward handoff call arrival} \end{cases} \quad (6.6)$$

Then, the transition probability from state $x \in \Omega$ with action $a \in A_x$ to state y can be written as:

$$P(y|x, a) = \begin{cases} a_n \lambda_n . \tau(x, a) & \text{if } y = x + e_n \\ a_{b_1} \lambda_{b_1} . \tau(x, a) & \text{if } y = x + e_{b_1} \\ a_{b_2} \lambda_{b_2} . \tau(x, a) & \text{if } y = x + e_{b_2} \\ x_n \mu . \tau(x, a) & \text{if } y = x - e_n \\ x_{b_1} b_1 . \tau(x, a) & \text{if } y = x - e_{b_1} \\ x_{b_2} b_2 . \tau(x, a) & \text{if } y = x - e_{b_2} \\ 0 & \text{if } y = x \end{cases} \quad (6.7)$$

6.3.6.5 Reward Function

The reward $r(x, a)$ of state x when action a is taken is expressed as:

$$r(x, a) = w_n(x_n + e_n . a_n) + w_{b_1}(x_{b_1} + e_{b_1} . a_{b_1}) + w_{b_2}(x_{b_2} + e_{b_2} . a_{b_2}) \quad (6.8)$$

where w_n, w_{b_1}, w_{b_2} are the weighting factors for each call type, respectively. When the weighting factors are equal to 1, the objective reward function is to maximize system utilization.

The linear programming algorithm is a well-known technique to find out Markov decision policies. It has several advantages. First, it is convenient to add more constraints without modifying the structure significantly. Second, it allows us to analyze the sensitivity of the obtained solution. The simplex method is commonly adopted to find the optimal solution. Instead of evaluating the objective function for all candidates satisfying the constraints, this method examines only better candidates, which are known in advance that the objective function will have a large value. Using the uniformization transformation [24], we can transfer a continuous-time Markov chain with nonidentical decision into an equivalent continuous-time Markov process in which decision epochs are generated by a Poisson process at a uniform rate. After the use of uniformization, the transition process from one state to another can be descried by a discrete-time Markov chain. We can therefore solve the following linear programming for a continuous-time MDP problem with the standard linear programming formulation for the discrete-time MDP problem after the uniformization transformation.

The linear programming formulas associated with the MDP for maximizing reward are given below with decision variable π_{xa}, $x \in \Omega$, $a \in A_x$:

$$\text{Maximize} \quad \sum_{x \in \Omega} \sum_{a \in A_x} r(x, a)\tau(x, a)\pi_{xa} \qquad (6.9)$$

$$\text{Subject to:} \quad \sum_{x \in \Omega} \sum_{a \in A_x} \tau(x, a)\pi_{xa} = 1 \qquad (6.10)$$

$$\sum_{a \in A_x} \pi_{ya} = \sum_{x \in \Omega} \sum_{a \in A_x} p_{xy}^{a}\pi_{xa}, \quad y \in \Omega \qquad (6.11)$$

$$\pi_{xa} \geq 0, \, x \in \Omega, \quad a \in A_x \qquad (6.12)$$

The variables π_{xa} satisfying Equations 6.9 to 6.12 can be viewed as the steady-state probabilities of being in state x and choosing action a. The optimal action in each state can be chosen among all actions in each state because the value of π_{xa} will be zero for all but one action is each state. This implies that the optimal decision is a deterministic function of x.

6.3.7 Numerical Results

The main aim here is to indicate that the MDP technique, described above, can be used as a tool in a 3G-WLAN integrated network to find out optimal decisions that the service provider should apply to minimize the downward hand off and upward hand off dropping probabilities and to maximize system utilization. The experimental results here are based on the simulation of a system that was adopted from the literature [25].

Figure 6.3 demonstrates the simulation model used in this chapter. The network topology is set up in NS-2, such that the mobile user and stationary server are both wirelessly connected via two heterogeneous network interfaces. The system consists of two networks: (1) network A represents UMTS and has a low bandwidth of 384 kbp/s, and (2) network B, on the other hand, has a higher bandwidth than network A, 1 Mbp/s, and represents a WLAN. Nonetheless, each connection is independent of the other.

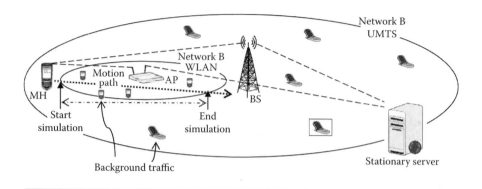

Figure 6.3 The simulated network topology.

Initially, the mobile user is connected to the server through network A's interface (i.e., the primary path is from the mobile user to the interface on network A). The secondary path is from the mobile user to network B's interface. The mobile user shares the network's bandwidth with various other background traffic sources. As the mobile user moves about, the amount of background traffic can vary and consequently the performance of both networks will also vary.

The assumptions for our simulation study are as follows:

■ The total bandwidth capacity of the simulated cell is C bandwidth units (bu).

■ The mobile user is connected to an FTP server. The bandwidth for each call is c bandwidth units.

■ New call requests are generated in the cell according to a Poisson process with rate λ_n (calls/second), respectively. We assume that the call holding time of a connection is exponentially distributed with mean μ^{-1} (seconds).

■ For the mobility model, we consider three parameters: the initial position of a mobile, its direction, and its speed. A newly generated call can appear anywhere in the cell with equal probability. When a new call is initiated, a mobile is assigned a random initial position derived from a uniform probability distribution function over the cell area. As for hand off calls, the initial position is determined when the downward vertical hand off event is scheduled as described below. A mobile is assigned a random direction upon entering a WLAN cell. The distribution of the direction reflects the correlation between the different cells. A constant, randomly selected speed is assigned to a mobile when it enters a WLAN cell, either at call initiation or after hand off. The speed is obtained from a uniform probability distribution function ranging between V_{min} and V_{max}.

From the three parameters described above, along with the radius of the cell R_{WLAN}, and the network topology, the simulation tool calculates the mobile residence time in the cell. It also determines whether an upward vertical hand off is taking place, as well as the initial position of the mobile in the UMTS cell. In [23], an analytical model is developed to find the average residence time for a new call, t_{nc}, given by:

$$t_{nc} = \frac{8 R_{WLAN} E[1/V]}{3\pi} \qquad (6.13)$$

While the average residence time of a vertical hand off calls, t_h, is given by:

$$t_h = \frac{\pi R_{WLAN}}{2 E[V]} \qquad (6.14)$$

where R_{WLAN} is the radius of the WLAN cell and V is the average speed of a mobile in the cell. Therefore, the hand off rate of new calls b_{nc} equals $1/t_{nc}$, and the hand off rate of vertical hand off calls b equals $1/t_h$.

The simulation model is very flexible, in which all the above parameters are provided as input to the simulation program. Thus, this allows us to test the system with different scenarios. In this chapter we limit our experimental tests to the simulation parameter values that are shown in Table 6.1. However, we believe that the higher the bandwidth capacity, the more efficiency our policy can achieve. This is because solving the linear programming problem to find the optimal call admission decisions is an offline procedure. That is, the decisions are obtained before operating the CAC. In addition, techniques for solving large-scale linear programming problems such as [26] can be applied to the cases of large wireless systems and wireless systems with larger capacity. It should be also noted that although there may be some discrepancy between the real bandwidth

Table 6.1   Simulation Parameters

Parameter	Value	Unit
C	50	bu
c	2	bu
λ_n	2	calls/sec
$1/\mu$	500	sec
V_{min}	2	km/hr
V_{max}	4	km/hr
UMTS cell radius	500	meter
WLAN cell radius	100	meter

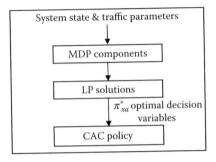

Figure 6.4   Admission controller.

values of the connection and the values in Table 6.1, we believe that our experiments can reflect the real systems behavior.

Part of the access point is the Admission Controller (AC) that operates the CAC policy. The AC components are shown in Figure 6.4 and operate as follows. The system and traffic parameters are modeled first. Then, these parameters are used to formulate the MDP problem mathematically, where the MDP components are calculated. The MDP model is converted to associate linear programming equations. This set of equations is solved by a professional optimization tool, such as LINGO, to derive the decision solutions. The decision solutions are mapped into optimal CAC policy. The optimal CAC policy is embedded in the network simulation model. Solving the linear programming problem to find the optimal call admission decisions is an offline procedure. That is, the decisions are obtained before invoking the CAC mechanism.

We evaluate our MDP-based CAC policy in terms of the following performance metrics: system utilization (U), downward hand off dropping probability (DDP), and upward hand off dropping probability (UDP).

Figure 6.5 provides performance results for three different cases under light to heavy traffic load in terms of call arrival rate, λ (calls/sec). In each case we use different values for the weights. The values of U, DDP, and UDP are compared in each case. We reveal from these results that the proposed MDP-based admission model achieves optimal utilization and satisfies user QoS requirements where low hand off dropping probabilities for both downward and upward users are achieved. We observe also that as the weights for hand off types increase, the utilization decreases. Thus, our MDP-based admission model significantly reduces both the downward hand off probability and the upward hand off probability without decreasing much the value for system utilization.

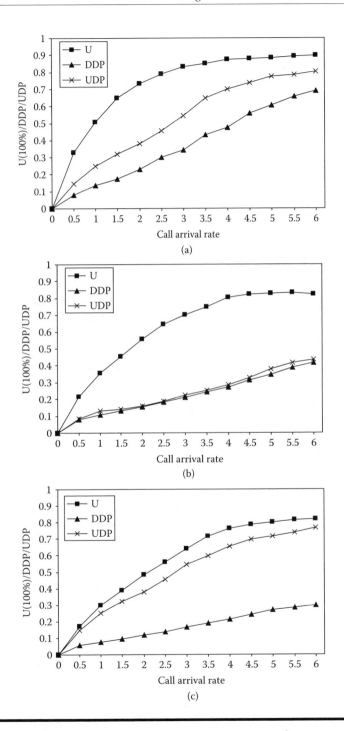

Figure 6.5   Simulation results: (a) $w_n = 1, w_{b1} = 1, w_{b2} = 1$; **(b)** $w_n = 1, w_{b1} = 5,$ $w_{b2} = 5$; **(c)** $w_n = 1, w_{b1} = 5, w_{b2} = 10$.

6.4 Summary

Providing broadband services with QoS guarantees in heterogeneous wireless networks poses great challenges due to scarce radio bandwidth. Effective radio resource management techniques are important for the efficient utilization of the limited bandwidth. In this chapter, a Markov decision-based call admission control model is developed for heterogeneous wireless networks. The model can be implemented as an optimization stochastic tool that the service provider can use to optimize the performance of integrated networks. Our simulation results achieve optimal system utilization and satisfy user QoS requirements. The requirements of mobile users are hence satisfied in periods with different loads, including overload. Moreover, the implemented policy ensures efficient utilization of resources. This latter facet is highly desirable by service providers.

We remark that although the complexity and solution space for the linear programming Equations 6.9 to 6.12 increase exponentially as C increases, there are techniques for solving large-scale linear programming problems such as the Interior-Point method [26] that can be applied to cases of large wireless systems or wireless systems with larger capacity. In reality, the values of C are typically not unreasonably high.

The disadvantage of our approach in real systems is that the action space is quite large and requires additional storage space. However, it can be manageable for reasonable values of C.

References

[1] http://www.ieee802.org/21/

[2] Chuanxiong Guo, Zihua Guo, Qian Zhang, and Wenwu Zhu, "A seamless and proactive end-to-end mobility solution for roaming across heterogeneous wireless networks," *IEEE Journal on Selected Areas in Communications*, Vol. 22, No. 5, June 2004, pp. 834–848.

[3] Xu Yang, J. Bigham, and L. Cuthber, "Resource management for service providers in heterogeneous wireless networks," *IEEE Wireless Communications and Networking Conference*, Vol. 3, March 2005, pp. 1305–1310.

[4] R. Chakravorty, P. Vidales, L. Patanapongpibul, K. Subramanian, I. Pratt, and J. Crowcroft, "On Inter-network Handover Performance using Mobile IPv6." University of Cambridge Computer Laboratory Technical Report, June 2003.

[5] A. Hasswa, N. Nasser, and H. Hassanein, "Generic Vertical Hand off Decision Function for Heterogeneous Wireless Networks," *IEEE and IFIP International Conference on Wireless and Optical Communications Networks (WOCN)*, Dubai, UAE, March 2005, pp. 239–243.

[6] F. Zhu and J. McNair, "Optimizations for Vertical Hand off Decision Algorithms," *IEEE Wireless Communications and Networking Conference (WCNC)*, March 2004, pp. 867–872.

[7]   M. Ylianttila, J. Makela, and P. Mahonen, "Optimization scheme for mobile users performing vertical hand offs between IEEE 802.11 and GPRS/EDGE networks," *IEEE PIMRC*, Vol. 1, Sept. 2002, pp. 15–18.

[8]   J. McNair, I.F. Akyildiz, and M. Bender, "An Inter-System Hand off Technique for the IMT-2000 System," in *Proc. of IEEE INFOCOM 2000*, March 2000.

[9]   Hyosoon Park, Sunghoon Yoon, Taehyoun Kim, Jungshin Park, Misun Do, and Jaiyong Lee, "Vertical Hand off Procedure and Algorithm between IEEE802.11 WLAN and CDMA Cellular Network," *Lecture Notes in Computer Science (LNCS)*, No. 2524, 2003, pp. 103–112.

[10] J. Makela, M. Ylianttila, and K. Pahlavan, "Hand off Decision in Multi-Service Networks," *The 11th IEEE International Symposium on Personal, Indoor and Mobile Radio Communications (PIMRC)*, September 2000.

[11] Qian Zhang, Chuanxiong Guo, Zihua Guo, and Wenwu Zhu, "Efficient mobility management for vertical hand off between WWAN and WLAN," *IEEE Communications Magazine*, Vol. 41, No. 11, November 2003, pp. 102–108.

[12] Ling-Jyh Chen, Tony Sun, Benny Chen, Venkatesh Rajendran, and Mario Gerla, "A Smart Decision Model for Vertical Hand off," *The 4th ANWIRE International Workshop on Wireless Internet and Reconfigurability (ANWIRE 2004)*, Athens, Greece, 2004.

[13] L.-J. Chen, T. Sun, B. Cheung, D. Nguyen, and M. Gerla, "Universal Seamless Hand off Architecture in Wireless Overlay Networks," Technical Report TR040012, UCLA CSD, 2004.

[14] Armando Fox, Steven Gribble, Yatin Chawathe, Eric Brewer, and Paul Gauthier, "Cluster-Based Scalable Network Services," *Proceedings of the Symposium on Operating Systems Principles*, October 1997.

[15] M. Ylianttila, J. Mkel, and P. Mhnen, "Supporting Resource Allocation with Vertical Hand offs in Multiple Radio Network Environment," *IEEE International Symposium on Personal Indoor and Mobile Radio Communications*, 2002.

[16] H. Wang, R. Katz, and J. Giese, "Policy-Enabled Hand offs across Heterogeneous Wireless Networks," *WMCSA*, 1999.

[17] L. Taylor, R. Titmus, and C. Lebre, "The challenges of seamless handover in future mobile multimedia networks," *IEEE Personal Communications, Special Issue on Advanced Mobile Communication Systems Managing Complexity in a Competitive and Seamless Environment*, 1999.

[18] M. Ylianttila, R. Pichna, J. Vallstrm, and J. Mkel, "Hand off Procedure for Heterogeneous Wireless Networks," *Global Telecommunications Conference*, 1999.

[19] Gertie Alsenmyr, Joakim Bergstrm, Mattias Hagberg, Anders Miln, Walter Mller, Hkan Palm, Himke van der Velde, Pontus Wallentin, and Fredrik Wallgren, "Handover between WCDMA and GSM," Ericsson Review No. 01, 2003.

[20] Roland Heicker, Stefan Jelvin, and Bodil Josefsson, "Ericsson Seamless Network," Ericsson Review No. 02, 2002.

[21] M.B. Myers, "Predicting and measuring quality of service for mobile multimedia," *3G Mobile Communication Technologies*, Conference Publication No. 471, IEEE 2000.

[22] Ching-Hsiang Wang, A Dynamic Resource Allocation for Vertical Hand off on Heterogeneous Wireless Networks, Department and Graduate Institute of Information Management, Master's thesis, June 2003.

[23] K. Yeung and S. Nanda, "Channel management in microcell/macrocell cellular radio systems," *IEEE Transactions Vehicular Technology*, Vol. 45, No. 4, November 1996, pp. 601–612.

[24] H. Tijms, "Stochastic Modeling and Analysis: A Computational Approach," Wiley, New York, 1986.

[25] A. Hasswa, N. Nasser, and H. Hassanein, "Performance Evaluation of a Transport Layer Solution for Seamless Vertical Mobility," *IEEE WIRELESS-COM*, Maui, Hawaii, USA, June 2005, pp. 779–784.

[26] G.Y. Zhao, "Interior-point methods with decomposition for solving large-scale linear programs," *Journal of Optimization Theory and Applications*, Vol. 102, No. 1, 1999, pp. 169–192.

Chapter 7

Medium Access Control in Wireless Ad Hoc Networks

Jialing Zheng, Maode Ma, Yan Zhang,
and Masayuki Fujise

Contents

7.1   Overview

Ad hoc networks have recently attracted increasing interest because of their great application potential. The absence of predesigned infrastructures and the intrinsic characteristics of wireless communication render great challenges to the control and management of ad hoc networks; thus, multiple factors must be considered comprehensively to design Medium Access Control (MAC) layer protocols. In this chapter, recent MAC layer protocols for ad hoc networks are investigated. With the classification and trade-off analysis, readers can gain a more thorough understanding of ad hoc MAC layer protocols and the latest trend in this area.

7.2   Introduction

Ad hoc networks can be rapidly deployed and maintained without manipulating a predesigned infrastructure. Nodes can communicate with each other over wireless channels either directly (if they are close enough) or through intermediate nodes that relay their packets. The fact that each node must handle control and management information and relay packets brings much more traffic to ad hoc networks than infrastructure wireless ones. The problems of hidden terminal and exposed terminal make the situation even worse, as shown in Figure 7.1 and Figure 7.2. The function of the MAC layer is to allocate the multi-access channel so that each node can transmit its packets successfully even in the presence of undue interference from other nodes. The main objective of MAC layer protocol design is to implement the function of the MAC layer, improve overall and per-node throughput, reduce delay, and guarantee fairness and QoS (Quality-of-Service).

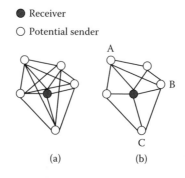

Figure 7.1   The hidden terminal problem: (a) is a fully connected topology; in (b), both nodes A and C are within the range of the receiver, but they are out of range of each other. If both nodes sense the channel is idle and try to send, collision will happen. This is the hidden terminal problem.

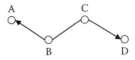

Figure 7.2 **The exposed terminal problem: node B is sending to A. When C tries to transmit to D, it detects the carrier and backs off. But this is unreasonable because the transmission from C to D will not interfere with the transmission from B to A. C is "exposed" to B, and this problem is called the exposed terminal problem.**

From the perspective of hierarchy, ad hoc networks can be grouped into two main categories: (1) flat structure and (2) clustered structure. Flat structure considers a fully distributed approach, where all nodes are both nodes and routers, and the notion of centralized control is absent. Clustered structure designates one node in each group of nodes to emulate the function of a base station and handle localized control of the group. Gupta and Kumar [1] set up a simplified model of a flat ad hoc network in which all nodes interfere in an omni-directional fashion with a power decay law and derived that when the number of nodes n increases, the throughput per node will decrease at $\frac{1}{\sqrt{n}}$. This result further indicates that the throughput per node diminishes to zero as the number of nodes increases. IEEE 802.11x are currently the most popular standards for WLAN and single-hop ad hoc networks due to robustness, simple design, and ease of implementation. Although the MAC protocols of IEEE 802.11 work well in single-hop ad hoc networks, the overall and per-node throughput will exacerbate when they are applied to multi-hop ad hoc networks. Consequently, flat topology is impractical for large-scale ad hoc networks. A clustered or hierarchical structure may be an effective and efficient solution, but each node must be able to function as a cluster head, which will generate extra cost to nodes. It is difficult to divide the nodes into clusters in these networks because there must be some trade-off. So far, there is no agreed-upon dividing criterion.

From the perspective of number of channels, MAC schemes can be categorized into single-channel schemes and multi-channel ones. In single-channel schemes, all nodes in a network share the same single channel, and all the control packets (if any) and data packets are transmitted in this channel. When the number of nodes is small and the traffic load is light, single-channel schemes can achieve a high utility and low probability of collision. Multi-channel schemes apply to scenarios where single-channel schemes cannot cope with the high probability of collision caused by dense nodes and heavy traffic load. In multi-channel networks, control packets can use a dedicated channel to avoid collision with data packets. Control packets can be used to dynamically allocate data channels for data packets. In multi-channel schemes, the transceiver of a node must be capable of operating on multiple channels and switch between them quickly, or a node must have multiple transceivers. There are two extremes among the MAC layer approaches in multi-access communications [2]. One is the "free-of-all" approach in which nodes normally send new packets immediately, hoping for no interference from other nodes. The other extreme is the "perfect scheduled" approach in which there is some order (time divison multiple access or TDMA, for example) in which nodes receive reserved intervals for channel use. That is, one extreme is to "fully contend," and the other is "contention-free." In ad hoc networks, "contention-free" is impossible due to the lack of predetermined coordinating nodes. We can view the contention-based category as a collection of "random access" and "reservation/collision resolution."

With rapid technology progress in physical layer facility, new features have been introduced to MAC schemes. The exploitation of MAC schemes has experienced from fixed transmission power to tunable power and from omni-directional antennas to smart antennas. Use of these techniques helps alleviate the problems of hidden terminal and exposed terminal and improve resource utility by more concurrent transmissions. Both directional antenna and power control can improve spatial reuse and they are highly correlated. Some recent MAC protocols combined these two features and achieved good results. Distributed Coordination Function (DCF) of IEEE 802.11 suffers from the fairness problem and realizes a best-effort service (i.e., does not guarantee any QoS level to nodes and packet flows). How to guarantee fairness and provide QoS in ad hoc networks is a great challenge. These two aspects have been a keen focus recently.

In this chapter, MAC protocols are first classified in terms of topology and then further classified in terms of channel separation. In both categories of flat topology — that is, single-hop flat topology and multi-hop flat topology — protocols with special physical layer characteristics are classified into protocols with consideration of power and protocols using directional antennas. Then, protocols with the objective of fairness and QoS are investigated. Because the categories are not totally independent of each other, a given protocol may fall into more than one category. Finally, open issues and future research directions are discussed. The classification is organized as follows:

- Flat structure:
    - Single-hop flat topology:
        - Single-channel schemes
        - Multi-channel schemes
        - MAC protocols with consideration of power
        - MAC protocols using directional antennas
        - MAC protocols to support fairness and QoS

- Multi-hop flat topology:
    - Single-channel schemes
    - Multi-channel schemes
    - MAC protocols with consideration of power
    - MAC protocols using directional antennas
    - MAC protocols to support fairness and QoS
- Clustered structure

7.3   MAC Protocols for Network of Flat Structures

7.3.1   Single-Hop Flat Topology

Flat topology adopts a fully distributed approach, and the notion of centralized control does not exist. In a flat single-hop network, a sender can directly communicate with a destination, without the need for intermediate nodes to relay packets. Single-hop networks are much simpler than multi-hop networks in that the hidden terminal and exposed terminal problems occur only in multi-hop environment. Research on MAC protocols for single hop ad hoc network can provide reference for design of MAC protocol for multi-hop ad hoc networks.

7.3.1.1   Single-Channel Schemes

Single-channel schemes consider the medium as a single channel, and all the nodes use the common channel to transmit both control packets and data packets.

7.3.1.1.1   Random Access Schemes

Carrier Sense Multiple Access (CSMA) [3] was the MAC layer protocol used in the first generation of packet radio networks, which was the predecessor of ad hoc networks. A CSMA protocol works as follows. A station willing to transmit senses the medium. If the medium is busy (i.e., some other station is transmitting), the station will defer its transmission to a later time; if the medium is sensed free, then the station is allowed to transmit. CSMA has large advantages over random Aloha access mode. Lal et al. [4] evaluated the one-hop performance of MAC for multiple-beam antenna nodes and concluded that under heavy offered load conditions, CSMA is a good choice with nodes that have multiple-beam smart antennas, despite the performance loss due to the beam synchronization, providing a stable throughput that approaches unity and is invariant to fluctuations in the offered load. Slotted Aloha, on the other hand, is capable of providing higher peak throughput in a narrow range of offered loads as more switched beams are employed, but performance drastically reduces beyond optimum offered loads.

7.3.1.1.2 Reservation/Collision Resolution Schemes

Multiple Access Collision Avoidance (MACA) [5] introduces a handshake between a sender and a receiver. The sender initiates the handshake by sending an RTS (Request to Send) message to the receiver. Upon receiving the RTS, if the receiver is ready, it replies with a CTS (Clear to Send) message. Through the CTS, other potential senders in the vicinity will defer from transmitting. Once the handshake is completed successfully, the transmission will be guaranteed without risk of collision. MACA outperforms CSMA. MACA Wireless (MACAW) [6] introduces ACK and DS to MACA. Before sending a DATA packet, a station sends a short Data-Sending packet (DS), the function of which is similar to a "carrier" of CSMA. An acknowledgment packet, ACK, is returned from a receiver to a sender immediately upon completion of data reception. ACK allows faster recovery from the errors caused by the unreliability of the wireless medium at the data link layer. The RTS/CTS/DATA/ACK sequence is adopted by many MAC layer protocols afterward. In all the RTS/CTS-based protocols, nodes involved in a collision retransmit according to some algorithms. The simplest is Binary Exponential Back-off: if a node is collided for the second time, it backs off for twice the time it does at the previous attempt. A small back-off factor may reduce the waste of resources and is adopted by MACAW.

Fullmer and Garcia-Luna-Aceves [7] introduced the concept of a channel access discipline called floor acquisition multiple access (FAMA), which consists of carrier sensing and a collision-avoidance dialogue between a source and the intended receiver of a packet. The objective of a FAMA protocol is for a station that has data to send to acquire control of the channel (which is called the floor) before sending any data packets, and to ensure that no data packet collisions occur. The MACA protocol and its derivatives (e.g., MACAW) become a variant of FAMA protocols when RTS and CTS transmissions last long enough. FAMA-NTR (Nonpersistent Transmit Request) [7] combines CSMA with the RTS-CTS exchange of MACA.

The IEEE 802.11 [8] standard defines two possible network configurations: (1) the infrastructure and (2) ad hoc configurations. An ad hoc WLAN is composed solely of stations within communication range of each other and they are able to communicate with each other directly. The basic access mechanism, distributed coordination function (DCF) is basically a carrier sense multiple access with collision avoidance (CSMA/CA) mechanism. DCF is an RTS-CTS-DATA-ACK sequence and is almost identical to MACAW, except that the CSMA mechanism is combined with MACAW to lower the probability of RTS collision. CSMA is performed both through physical and virtual mechanisms. The RTS and CTS frames contain a Duration/ID field that defines the period of time that the medium is to be reserved to transmit the actual data frame and the returning ACK frames. All stations receiving either the RTS or the CTS will set their Virtual Carrier Sense indicator

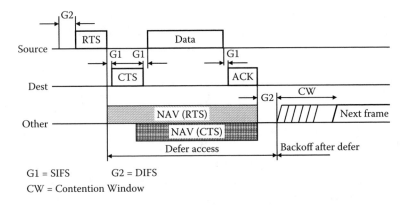

Figure 7.3   IEEE 802.11 DCF.

(called NAV, for Network Allocation Vector) for the given duration, and will use this information together with the Physical Carrier Sense when sensing the medium. As shown in Figure 7.3, the CSMA/CA algorithm mandates that a gap of minimum specified duration exist between contiguous frame sequences. For a station to transmit, it will sense the medium to determine if another station is transmitting. If the medium is determined to be busy, the station will defer until the end of the current transmission. Just after the medium becomes idle following a busy medium is when the highest probability of a collision exists. This is because multiple stations could have been waiting for the medium to become available again. This situation necessitates a random backoff procedure to resolve medium contention conflicts. After deferral, or prior to attempting to transmit again immediately after a successful transmission, the station will select a random backoff interval and will decrement the backoff interval counter while the medium is idle. The time interval between frames is called the inter-frame space (IFS). Four different IFS (namely SIFS, PIFS, DIFS, and EIFS) are defined to provide priority levels for access to the wireless media. The RTS/CTS mechanism does not need to be used for every data frame transmission. Because the additional RTS and CTS frames add overhead inefficiency, the mechanism is not always justified, especially for short data frames. IEEE 802.11 is now a popular solution for WLAN; deployment of IEEE 802.11 in ad hoc networks, however, does not always yield good performance [9,10]. Therefore, research on the performance improvement of mobile ad hoc networks using IEEE 802.11 deserves special attention. Many protocols based on IEEE 802.11 have been proposed. Some of them improve the performance of IEEE 802.11 for single-hop networks, while others extend IEEE 802.11 from a single-hop scenario to the multi-hop environment, where there is, thus far, no perfect solution.

In the Load Awareness Medium Access Control Protocol (LA) [11], IEEE 802.11 DCF is combined with the token passing scheme. Because an active list should be maintained for the token passing scheme, LA is under the assumption that nodes operate in a fully connected environment. LA defines two traffic load thresholds: *Threshold_A* and *Threshold_B* ($A > B$). In a light-loaded environment, nodes contend to access the channel. As the traffic load goes higher than A, the access scheme is switched to the token passing scheme until the traffic load is down to B. LA switches between these two modes according to the real-time traffic load of the network. Every single node uses its waiting time as the measurement of A and B. If the traffic load is not evenly distributed among the nodes, because any station that seizes the channel and finds it has waited longer than Threshold A will initiate the token passing scheme, those nodes with much less traffic load may encounter larger delays than in the DCF mode, causing unfairness.

7.3.1.1.3   Summary

Single-channel MAC schemes are comparatively simple and require relatively simple hardware. When the traffic load is light, random access schemes perform well. When the traffic load is heavy, the performance will be penalized for excessive collision. Using a CSMA/CA and RTS/CTS sequence has been a common way to contend and dynamically reserve the channel. The RTS/CTS sequence trades off bandwidth against DATA collisions. ACK is often used to improve MAC layer reliability. But RTS/CTS themselves cannot avoid hidden terminal and exposed terminal problems and control packets may collide with data packets.

7.3.1.2   *Multi-Channel Schemes*

A major problem of single shared channel schemes is that the probability of collision increases with the number of nodes. It is possible to alleviate this problem by transmitting concurrently on multiple channels at the cost of bandwidth resources frequency division multiple access (FDMA) or by increasing delay (TDMA). A physical channel can be divided into multiple logical channels based on TDMA or code division multiple access (CDMA). The use of CDMA increases the capacity by allowing multiple successful transmissions simultaneously within the limits of multiple access interference (MAI). Many protocols aimed at how to fully utilize the CDMA techniques have emerged. Most of them focus on how to assign codes or spreading factors. In the literature, there are several basic types of code assignment [12,13]. The simplest one is that all nodes monitor the same common code for any packet arrival and send data on the common code. In *receiver-based code assignment schemes*, a transmitter must look up a code assignment table to find out the code of its receiver and then sends packets on the *receiver-based code*. The receiver must monitor its code all the time for any packet arrival. In *transmitter-based code assignment schemes*, a transmitter sends its data on its own code. The receiver must monitor all transmitter-based codes of its neighbors because the receiver does not know in advance incoming transmission will be on which node. In *pairwise-based code assignment schemes*, each pair of nodes is assigned a unique code. A transmitter will look up a code assignment table to find out the code to communicate with a specific receiver. The receiver must monitor a set of codes simultaneously.

Some multi-channel schemes use one or more dedicated control channels and one or more data channels. Among those schemes, control channels are used to transmit control packets or send busy tones, which are usually simple sine waves. Control packets are usually much shorter than data packets, and limiting collision of control packets to the control channels can improve the utility of data channels. The advantage of using busy tones is that consumption of small bandwidth can inform the potential collision nodes of the ongoing transmission. Other multi-channel schemes use multiple peer channels.

7.3.1.2.1   Use of Separate Control and Data Channels

In [13–15] control packets and data packets are sent on different spreading codes with small cross-correlation values. Spreading factors can be varied according to the network congestion of the network. In Fantacci et al. [13], RTS and CTS control packets are transmitted on a common code, while data packets are transmitted on a receiver-based code. This code is notified to the sender within CTS packets. The proposed protocol foresees the possibility to use a set of different spreading codes during handshaking. Spreading factors are chosen in an adaptive way, by taking into account the network congestion state in terms of number of retransmission of RTS/CTS packets necessary to start a new communication. In particular, failure of the handshaking phase is due to the collision of RTS/CTS packets from different nodes or to multiple access interference (MAI) due to the active nodes that are transmitting with other codes. The spreading factor (SF) of the common code varies in an interval [SF_{min}, SF_{max}]. At the first transmission attempt, the value of spreading factor for the RTS/CTS packets is SF_{min}. Whenever the handshaking phase fails, the spreading factor value is doubled, and a new transmission attempt of the control packet is carried out after a backoff phase. Before SF_{max} is reached, the new instant of transmission is selected with a uniform probability in an interval equal to the control packet transmission time. An exponential backoff is adopted after the highest spreading factor is reached. This protocol demands a complex transceiver that can handle modulation/demodulation of variant spreading factors.

Joint detection allows many concurrent, asynchronous packet transmissions to occur, thereby enhancing the capacity of a system. Kota and

Schlegel [16] proposed a method based on CDMA and joint detection, in which the header part of a packet and the data part of the packet use different access schemes. The spreading code of the header is identical to all users. The header part consists of the access preamble and code identifier (ID). The data portion of each packet is spread by a unique random sequence identified by the code identifier field (CID), which allows joint detection of the asynchronous, overlapping packets. The headers are spread with a large enough processing gain to allow detection even in severe interference. Header access is similar to a pure ALOHA system. Receivers for the proposed system operate in two stages that deal with header and data detection, respectively. After successfully detecting the header, exact timing information, frame synchronization, and code ID information are passed on to stage II, which is a software process dedicated to the detection of the packet. A multi-user detector is used in stage II. Stage I then becomes free again to search for new packet arrivals. Using CSMA instead of ALOHA for the header of packets further modifies the protocol.

7.3.1.2.2 Use of Multiple Peer Channels

In Multi-channel MAC (MMAC) [17], N non-overlap channels are used. All channels have the same bandwidth and all nodes are synchronized. Beacons transmitted periodically divide time into beacon intervals. A small *announcement traffic indication message (ATIM)* window is placed at the start of each beacon interval. The nodes that have packets to transmit negotiate channels with the destination nodes during this window on the predefined default channel. The default channel is one of the multiple channels and is used for sending data outside the ATIM window. When the negotiation is successful, the sender and the receiver switch to the selected channel and start to exchange RTS/CTS before sending DATA packets.

Realp and Perez-Neira [18] proposed a hybrid CDMA-TDMA MAC protocol. It assumes that time is slotted and each time slot is assigned to one node. A node can use multiple codes assigned to it to transmit its packets simultaneously in the time slot assigned to it, while other nodes contend for the remaining codes. It provides a very good analysis model, but the number of nodes is fixed and how the time slots and codes are allocated in a distributed fashion is not detailed.

Stine proposed Synchronous Collision Resolution (SCR) [19] to solve the near-far effect of CDMA. The system time is divided into slots and the header part of a slot named CR signaling is further divided into signaling slots organized into groups of slots called phases. Nodes contend in the CR signaling, and the winning nodes will transmit during the rest of the time slot simultaneously.

7.3.1.2.3 Summary

Multi-channel MAC protocols using multiple peer channels or multiple data channels enable pairs of nodes to communicate on different channels simultaneously to increase network throughput. Multi-channel schemes demand more complex hardware than single-channel schemes. A multi-channel MAC scheme typically needs to address two issues: (1) channel assignment and (2) collision avoidance. Busy tones can guard the transmission of control packets or data packets against collision. If a dedicated channel is used for control packets, collisions between control packets and data packets can be avoided, and channel assignment can be implemented by exchange of control packets. However, when traffic is heavy, because all the data transmissions must start with control packets on the control channel, the single control channel may be a bottleneck and the data channels may be under-utilized. Some protocols combine time division and multiple channels, but synchronization is difficult in ad hoc networks. Some protocols assume that each node knows all of its neighbor's code or each node can listen to each code; then there is no need to assign code (or channel).

7.3.1.3 MAC Protocols with Consideration of Power

In any type of mobile network, power consumption is a key consideration because the mobile terminals are power-constrained. The basic idea of power-saving mechanisms is to let the wireless interface of a node be turned off when the station is not going to communicate with other stations during a certain period. The major objective of power control is to reduce the power consumption rationally, prolong the system lasting time but without decreasing the throughput of individual nodes and the overall system. Most of the power of a mobile node is consumed by signal transmission, and thus it is of great interest to control the signal transmission power to increase the lifetime of mobile nodes [20].

In IEEE 802.11 DCF, wireless nodes have two power modes: (1) active and (2) power saving. The power management scheme divides time into beacon intervals. At the beginning of each beacon interval, power-saving nodes wake up for a short time period, called the *announcement traffic indication message (ATIM)* window. In the ATIM window, nodes exchange control frames to inform their power-saving counterparts to remain awake until the end of the beacon interval for receiving their data frames. After an ATIM window, all nodes follow the DCF protocol to transmit their data frames. The size of the ATIM window affects the power consumption and throughput of a network considerably, and a fixed ATIM window size is not favorable under variable traffic conditions. However, the IEEE 802.11 power scheme does not specify how to tune the size of the ATIM window. In Wu and Tseng [21], the ATIM window is adjusted dynamically to adapt to the traffic status. The buffered data frame's duration is piggybacked in an ATIM frame. After an ATIM, all nodes can maintain the same transmission table recording the duration of ongoing transmissions. The transmissions are scheduled according to the transmission duration. Short packets are transmitted earlier.

There are protocols aimed at maximizing the residual system lifetime, which is defined as the minimal residual lifetime of nodes. To save energy of nodes with less energy left, nodes with high energy are transmitted more frequently. The Battery Level Aware MAC Protocol (BLAM) [22] modified the random access nature of the IEEE 802.11 DCF to a prioritized access protocol, where the priority of a node to access the medium is determined by its remaining energy. Low-energy nodes are transmitted less frequently to the same energy.

7.3.1.3.1   Summary

In power-saving schemes, nodes sleep when they are not involved in transmission or receiving. Use of the sleep mode cannot only save energy, but also reduce collision. However, at least nodes should be awake periodically in case other nodes have data for them. How to schedule and implement the waking time is an important issue. Too frequent waking will discount the effect of power saving, while too much sleep will defer the transmission of data.

7.3.1.4   MAC Protocols Using Directional Antennas

Directional antennas are usually implemented as an array of (omni-directional) antenna elements connected with a combining network, to produce a radiation pattern that allows transmission in a particular direction or, analogously, reception from a specific direction [23,24]. Directional antennas can improve the resource utility by spatial multiplexing. However, the use of directional antennas on mobile terminals introduced the complex issue of how to find the desired direction for transmission or reception. This depends on a GPS-like device or direction-of-arrival methods (DOA) and time difference of arrival (TDOA) position location technique.

DOA-ALOHA [25] modified the slotted Aloha using directional antennas and a busy tone. Nodes are aware of the angular location of each neighbor by GPS. As shown in Figure 7.4, all transmitters transmit a simple tone (i.e., a sine wave) during the DOA minislot toward their intended receivers. Once a receiver determines the DOA of all transmitters it can hear by a DOA algorithm, it forms its directed reception beam toward the one that has the maximum power and forms nulls in all the other identified directions. After receiving a packet during the second minislot, the receiver looks at the header and rejects the packet if it was not the intended destination. If a receiver beamforms incorrectly in a given timeslot, it records that direction

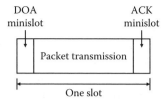

Figure 7.4 DOA-ALOHA.

in a single-entry cache. In the next slot, if the maximum signal strength is again in the direction recorded in the single-entry cache, the node ignores that direction and beamforms toward the second strongest signal. If the node receives a packet correctly (i.e., it was the intended recipient), it does not change the cache. If it receives a packet incorrectly, it updates the cache with this new direction. If there is no packet in a slot from the direction recorded in the cache, the cache is reset. However, if the traffic load is very heavy and the sender is far from the destination, the probability of receiving the wrong packet will increase.

Some protocols using directional antenna are based on the IEEE 802.11 and adopt different combinations of omni-directional/directional RTS/CTS/DATA/ACK. In [24], the positions of nodes are supposed to be known by GPS and each node knows all its neighbors' positions. As shown in Figure 7.5a, RTS and data packets are transmitted by directional antenna, while CTS is transmitted by omni-directional antenna. In [26–28], it is assumed that all nodes have no information of other nodes at first. RTS and CTS are transmitted by omni-directional antenna to enable source and destination nodes to identify each other. In [27, 28], schemes of power control are also proposed. In E-MAC [29], each node has a *rotational-sector-receive-mode.*

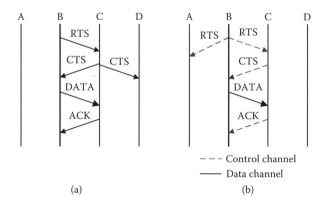

Figure 7.5   Directional antenna and RTS/CTS handshake.

In this mode, a node rotates its directional antenna sequentially in all directions at 30-degree intervals, covering the entire 360-degree space in the form of sequential directional receiving in each direction and senses the received signal at each direction. Each node periodically collects its neighborhood directional information and forms an Angle-Signal Table (AST). In addition, each control packet is transmitted with a preceding tone with a duration such that the time to rotate a receiver's rotational receive beam through 360 degrees is a little less than the duration of the tone. Each node periodically transmits an omni-directional beacon and its neighbors on receiving the beacon decode it and make its entry in the AST. Because RTS and CTS are sent with a preceding tone and contain the source address, they also serve as beacons, which minimizes the overhead of transmitting beacons at high traffic.

NULLHOC [30] employs a multipath, multiple-input, multiple-output (MIMO) model. NULLHOC uses the knowledge of MIMO channels between antenna arrays at different nodes to design, transmit, and receive beamformers that achieve a specified gain to the desired node while nulling existing communications between other nodes.

7.3.1.4.1 Summary

All the above protocols use directional antennas to improve spatial reuse and propose various schemes. However, the protocols using GPS fail to depict exact mechanism of GPS information exchange and the consequent overhead is not clear. The power of omni-directional transmission and directional transmission are usually assumed to be equal. Beamforming is often assumed perfect, and how to consider the effect of side and back lobes in the evaluation of network performance remains insufficiently studied [29]. When the nodes are in a line, which is the usual case of vehicle communication, the advantage of directional antennas will discount greatly. In this scenario, power control must be introduced to combine with directional antennas.

7.3.1.5 MAC Protocols to Support Fairness and QoS

7.3.1.5.1 Fairness

DCF of IEEE 802.11 may lead to "capture effects," which means that some nodes grab the shared channel and other nodes suffer from starvation. This is also known as the *fairness* problem. DCF faces this problem because the binary backoff algorithm favors the last succeeding node [31]. Several protocols [32–36] designed to alleviate this problem propose various backoff algorithms.

Distributed Fair Scheduling (DFS) [32] is based on IEEE 802.11 and Start-Time Fair Queuing (SFQ) [37]. DFS assumes that all packets at a node belong to a single queue and the algorithm can be easily extended when multiple queues are maintained at each node. The distributed mechanism for determining the smallest start time is based on the idea of a *backoff interval* utilized in IEEE 802.11 MAC. The proposed MAC protocol borrows SFQ's idea of transmitting the packet whose start time is smallest, as well as SFQ's mechanism for updating the virtual time. Meanwhile, implementation of the virtual time and the process of determining the smallest start time are distributed.

7.3.1.5.2 QoS

With the growing popularity of ad hoc networks and the diversity of traffic data types, it is reasonable to expect that users will demand some level of QoS. Some of the QoS-related parameters are available bandwidth, end-to-end delay, jitter, probability of packet loss/errors, etc. Different traffic types require different network services. Much research has been done on the QoS on the network layer, but the MAC layer also plays a critical role in providing QoS.

The IEEE 802.11e draft specification [34] introduces Enhanced DCF (EDCF), which defines mechanisms to enable QoS. EDCF supports eight priorities of Traffic Class (TC). Within a node, the eight TCs have independent transmission queues. Each TC starts a backoff after detecting the channel being idle for an Arbitration Inter-Frame Space (AIFS). The AIFS is at least as large as the DIFS. Both the AIFS and the CW can be chosen individually for each TC. The CW_{Max} value sets the maximum possible value for the CW and is intended to be the same for all TCs as in DCF. If the backoff counters of two or more parallel TCs reach zero at the same time within a node, an inner scheduler will favor the TC with the highest priority — while other TCs will backoff as if a collision has occurred in the medium. Gannoune et al. [38,39] proposed to extend EDCF with dynamic adaptation algorithms of the maximum contention window (CW_{Max}) and the minimum contention window (CW_{Min}) that enable each station to tune the size of the CW_{Max} used in its backoff algorithm at runtime. Xiao and Li [40] proposed two local data-control schemes and an admission-control scheme for ad hoc networks with the IEEE 802.11e. In the schemes, each node maps the measured traffic-load condition into backoff parameters locally and dynamically. In the proposed distributed admission-control scheme, based on measurements, each node makes decisions on the acceptances/rejections of flows by themselves, without the presence of access points.

Xiao and Pan [41] proposed a priority scheme by differentiation services between the delay-sensitive real-time (RT) traffic and the best-effort (BE) traffic [41]. For the RT traffic, retransmission is not used, regardless of whether the previous transmission is successful. The BE class follows the original DCF. The RT class has a smaller backoff window size than the BE class.

The Directional Transmission and Reception Algorithm (DTRA) [42] is a TDMA-based MAC algorithm for load-dependent negotiation of slot reservations. To support QoS, DTMA set three priority queues for each neighbor. When making a reservation, if there are not enough available slots for higher-priority traffic, slots allocated for a lower priority can be asked to be deallocated according to its releasing metric or simply preempted by higher-priority traffic. To ensure fairness, a threshold on the maximum number of slots for each priority class should be set.

7.3.1.5.3 Summary

Protocols designed to alleviate the fairness problem usually have different backoff algorithms, expecting to give equal opportunities to nodes that did not capture the channel. Guarantee of QoS depends on several layers, such as the network layer, application layer, and MAC layer. At the contention-based MAC layer, we can only alleviate the problem of QoS.

7.3.2 Multi-Hop Flat Topology

In multi-hop ad hoc networks, not all the nodes are within transmission range of each other. A data packet from a source node may need to traverse intermediate nodes before it reaches the destination node. A node not only needs to transmit data packets produced by it, but also forwards data packets from nodes within its receiving range. This feature of multi-hop determines that communication between different pairs is closely related to each other. The high throughput of single-hop links between immediate neighbor nodes does not necessarily mean high end-to-end throughput from source nodes to their destinations. Problems of hidden terminal and exposed terminal make the design of the MAC layer protocol very difficult. The achievable capacity depends on network size, traffic pattern, and detailed local radio interactions. Li et al. [43] examined these factors alone and in combination. They argue that the key factor deciding whether large ad hoc networks are feasible is the locality of traffic and present specific criteria to distinguish traffic patterns that allow scalable capacity from those that do not. The dynamics of the MAC layer are tightly connected to the dynamics of the physical layer. Carvalho et al. [44] introduced a modeling framework for multi-hop ad hoc networks that explicitly takes into account the impact of the physical layer on the operation and performance of the MAC layer. In this section we use the same classification convention as in the single-hop flat topology.

7.3.2.1 Single-Channel Schemes

Deepanshu et al. [45] proposed an algorithm to mitigate the exposed terminal problem in IEEE 802.11 ad hoc networks. In this algorithm, if a node hears an RTS packet followed by the start of a DATA packet, it identifies itself as an exposed node. Whenever there is data transmission in progress (called the primary transmission), the exposed node tries to squeeze in a parallel or secondary transmission for better overall throughput. Once the DATA transfer from B to A begins, node C commences DATA transmission to node D directly, without making use of the RTS/CTS exchange, as shown in Figure 7.2. A secondary transmission can begin only after a node has identified itself as exposed, and it must end at the same time as the primary transmission, so the data packets of the secondary transmission must be smaller than those of the primary transmission. The increase in throughput is due to a large number of small packets getting through in parallel transmission. So the advantage of this protocol greatly depends on the percentage of small data packets.

Medium Access via Collision Avoidance with Parallelism (MACA-P) [46] seeks to increase the feasible set of concurrent transmissions by introducing a control gap T_{data} between the RTS/CTS and DATA/ACK phases. This scheme is similar to the one proposed by Zhang [42]; the difference is that a secondary transmission uses RTS-CTS rather than transmitting directly. As shown in Figure 7.6, the two pairs of transmissions must synchronize the start of DATA transmission (reception) and ACK reception (transmission) by adjusting the control gap. Like the scheme of Zhang [42], the advantage of MACA-P also depends on the percentage of small data packets.

In most protocols based on RTS/CTS, nodes transmit only one packet after capturing the channel and the overhead of capturing the channel will be imposed on each packet. Ozugur et al. [47] proposed a P_{ij} persistent CSMA-based algorithm that also uses RTS/CTS. However, if a node captures the channel, it transmits multiple data packets. Each data packet is acknowledged by ACK. After the transmission of data packets and their ACK packets, the source node transmits an end-of-burst (EOB) packet and starts waiting for the end-of-burst confirmation (EOBC) packets.

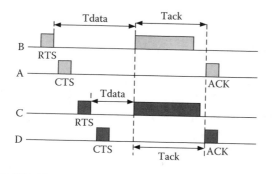

Figure 7.6   MACA-P.

DPC/ALP [48] also relies on the basic RTS/CTS handshake. A node sends its RTS by progressively increasing its transmission power until the receiver hears it. If the receiver replies, a connection is established; otherwise, the sender will back off. During data transmission, the sender transmits at the minimum power needed. This scheme can save energy to some extent in static networks; but if the receiver and the sender are moving away during transmission, the connection will be down.

In MACA by invitation (MACA-BI) [49], the RTS part of the RTS/CTS handshake is suppressed, leaving only the CTS message that can be viewed as an "invitation" by a receiver to transmit. The CTS is renamed as RTR (Ready to Receive). The receiver decides how many packets to invite from previous history and the backlog notification carried by data packets from the transmitter. The efficiency of the invitation scheme relies on the stationarity of the traffic pattern. In nonstationary traffic situations, a node can transmit an explicit RTS if the queue length or delay has exceeded a given threshold before an RTR is received from the intended destination. In such a scenario, MACA-BI reduces to MACA if traffic burstiness prevents timely invitations.

In Garcia-Luna-Aceves and Tzamaloukas [50], it is pointed out that MACA-BI does not prevent data packets sent to a given receiver from colliding with other data packets sent concurrently in the neighborhood of the receiver. The authors modified the MACA-BI and proposed three kinds of receiver-initiated protocols. To make the RTR-data handshake in MACA-BI collision-free, Receiver-Initiated Medium Access Single polling (RIMA-SP) makes two modifications. A polled node should transmit data packets only if they are addressed to the polling node. A new control signal No-Transmission Request (NTR) is introduced. Before answering an RTR, a polled node must wait for a collision-avoidance waiting period. During that period, if any channel activity is heard, the receiver (polling node) that originated an RTR sends an NTR to cancel the invitation. Otherwise, if nothing happens during the waiting period, the polled sender transmits its data, if it has any to send to the polling node. In RIMA-DP (Dual-use Polling), an RTR entry is both a request for data from a polled node and a transmission request for a polling node to send. Both polling and polled nodes can send data in a round of collision avoidance. This is because the RTR makes all the neighbors of the polling node back off, and the data from the polled node makes all its neighbors back off, which can then be used by the polling node to send its data. In RIMA-BP (Broadcast Polling), an RTR can be sent to multiple neighbors. An additional control packet RTS is introduced to ensure that transmissions that collide last a short period and do not carry user data.

Multiple Access with ReduCed Handshake (MARCH) [51] attempts to reduce control packets. The RTS/CTS handshake is used only by the first hop of a route to forward data packets while for the rest it utilizes a CTS-only

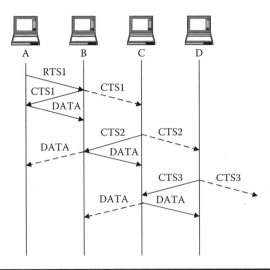

Figure 7.7   Handshake mechanism of MARCH protocol.

handshake. As shown in Figure 7.7, when a node overhears CTS from its upstream, it will transmit CTS to invite data from its upstream node. The data will be relayed from upstream to downstream hop by hop. A node must know which nodes is its upstream node, or else CTS will collide when multiple nodes send CTS simultaneously.

7.3.2.1.1 Summary

In multi-hop networks, channels can be reused spatially. However, because not all the nodes can hear each other, the exposed terminal problems and hidden terminal problems are very severe. To improve throughput, protocols must overcome exposed terminal problems and hidden terminal problems and improve the spatial reuse of channels.

7.3.2.2   Multi-Channel Schemes

Kyasanur et al. [52] derived the lower and upper bounds on the capacity of static multi-channel wireless networks, where the number of interfaces, m, is smaller than the number of channels, c. It is shown that the capacity of multi-channel networks exhibits different bounds that depend on the ratio between c and m.

7.3.2.2.1   Use of Separate Control and Data Channels

Busy Tone Multiple Access (BTMA) [53] was proposed to eliminate the hidden terminal problem of CSMA. In BTMA, the total available bandwidth is divided into two channels: (1) a message channel and (2) a busy-tone (BT) channel. During the period of receiving, the receiver transmits a BT (usually a sine wave) to indicate that the message channel is busy. When other nodes detect the BT, they will wait for some time and try later.

Dual Busy Tone Multiple Access (DBTMA) [54] divides the channel into two narrow-band BT channels and one common data channel. Senders use the RTS packets to initiate channel request. Two BTs are then used to protect the RTS packets and the data packets, respectively. One of the BTs, the transmitting BT, which is set up by the RTS transmitter, is used to protect the RTS packets. Another BT, the receiving BT, issued by the receiver, acknowledges the RTS packet and provides continuous protection for the incoming data packets. Nodes sensing any BT defer from sending their RTS packets on the channel. DBTMA almost completely solves both the hidden- and exposed-terminal problems.

In Fang et al. [55], a control channel is adopted in which nodes periodically send hello messages in order to be detected by neighbors. Each node in the network has a unique transmission code and knows all of its neighbors' transmission codes in advance. The node with the smallest transmission code among its neighbors is granted access to the channel. The transmission codes are dynamically changed (rotated among the nodes). This scheme can be viewed as a form of TDMA. Because all nodes must know all of their neighbors' transmission codes, this scheme cannot cope with large-scale networks where nodes are roaming in and out. In Directional Medium Access Protocol (DMAP) [56], RTS/CTS/ACK are transmitted in a dedicated control channel, while DATA packets are transmitted in a data channel.

Xu et al. [57] propose a multi-transceiver multiple access (MTMA) protocol for ad hoc networks in which each node has multiple sub-nodes equipped with independent wireless transceivers. The whole wireless channel is divided into multiple sub-data channels and a common control channel. Each sub-node dynamically reserves an idle traffic channel by RTS/CTS dialogue on the common channel that enables a node to perform parallel communication with other nodes.

7.3.2.2.2 Use of Multiple Peer Channels

7.3.2.2.2.1 TDMA Extended Group TDMA [58] uses two-layered time-division operation for resource allocation and link scheduling. The outer layer allocates disjoint fractions of time (depending on residual energy) to activate distinct receiver groups (predetermined using network topology). For each receiver group, the inner layer creates time orthogonality (using throughput properties) among interfering transmitter groups.

7.3.2.2.2.2 Reservation/Collision Resolution Schemes Nasipuri and others [59–61] proposed several variants of IEEE 802.11 that divide the available bandwidth into several channels. Packet transmission channels can be selected randomly [59], based on receiver [60], or based on power [61].

Preference is given to the channel that was used for the last successful transmission. In [59] and [61], the total bandwidth is divided into N physical channels. In [60], a dedicated control channel is used and the total bandwidth is divided into N+1 non-overlapping frequency bands, one control channel and the rest data channels. Because control packets rely on CSMA alone and are prone to hidden terminal problems, a separate control channel will eliminate the probability of interference between control and data packets. In Jain et al. [61], a sender senses carrier on all the channels and selects the channel with the lowest sensed power. The use of multiple channels may provide some performance advantages in reducing collisions and enabling more concurrent transmissions, but nodes must be able to simultaneously sense carrier on all the channels for incoming transmissions.

The Simple Tone Sense (STS) [62] protocol is designed for multi-hop ad hoc networks with multiple directional antenna stations. Each node records the busy/idle status of its transmission sectors, and each node is assigned a tone that is unique to its neighbors. When a station detects a packet addressed to it, it broadcasts its assigned tone for T_1 seconds to its neighbors in different directions. When detecting a tone of duration T_1, all stations will change the status of the sector corresponding to the received tone to busy. After a node has received a packet correctly, it acknowledges the source station by broadcasting its assigned tone for T_2 seconds. When detecting a tone for T_2 seconds, all nodes will change the status of the corresponding sector to idle. A variation of the STS protocol, called the Variable Power Tone Sense (VPTS) protocol, is also designed to further reduce interference. In both protocols, all nodes are static.

In Garcia-Luna-Aceves and Raju [63], interference is described as either being *direct* or *secondary*. Direct interference occurs when two nodes simultaneously initiate transmission to each other. Secondary interference occurs at a receiver due to simultaneous transmissions by two transmitters that cannot hear each other. In their protocol, a code allocated to a node is different from the codes allocated to its two-hop neighbors. This can eliminate collisions to some extent. Yeh [64] proposed Multiple Access with Spread Spectrum (MASS) and various techniques for MASS, including spread spectrum scheduling, spread spectrum data and code assignment techniques. Announcement-based Conflict Avoidance (ACA) is a proactive code assignment algorithm. All nodes periodically announce the codes they are using or will use when the channel is idle or (relatively) lightly loaded. A node records in its code table the codes that have been announced by other nearby nodes. When a new code is needed, the node checks its code table and selects a code that is not used or will not cause high cross-correlations with other codes. In the ROC Code Verification (ROCCV) scheme, a node that initiates communication first randomly selects a code or a set of codes that will not conflict or cause high cross-correlations and sends RTS (transmitter initiated) or CTS (receiver initiated) with the code.

Then the receiver/transmitter or neighbors will test the code and negotiate with the node that initiates communication. In Randomly Initiated Code Hopping (RICH), a node can decide the codes to be used by itself without negotiation with nearby nodes because the sequences of codes are extremely long. MASS provides several possible spread spectrum scheduling and code assignment techniques, but does not provide sufficient theoretical analysis and comparison.

Receiver-Initiated Channel-Hopping with Dual Polling (RICH-DP) [65] takes advantage of the characteristics of frequency hopping spread spectrum (FHSS) and does not require carrier sensing or the assignment of unique codes to nodes. The basic collision-avoidance handshake sequence is similar to that of RIMA-DP [50]. In RICH-DP, all the nodes assume a common frequency-hopping sequence, so that nodes can listen on the same channel at the same time. Nodes then carry out a receiver-initiated collision-avoidance handshake to determine which sender-receiver pair should remain in the present hop in order to exchange data, while all other nodes continue hopping on the common hopping sequence.

7.3.2.2.3 Summary

In multi-hop ad hoc networks, channels can be spatially reused. But the hidden terminal and exposed terminal problems hinder further improvement in throughput. Because CDMA uses the medium all the time, it may be an optimum choice of channel separation. Multiple-channel protocols generally accommodate more users than single-channel protocols.

7.3.2.3 MAC Protocols with Consideration of Power

In single-hop ad hoc networks, protocols with consideration of power saving usually assume that a network is fully connected or there is global clock synchronization. Hence, node wake-up/sleep patterns can depend on the clock. But in multi-hop ad hoc networks, global clock synchronization is very difficult, and a node must predict when another node will wake up to receive packets. A node may not be aware of the existence of another node that is in sleep mode. This will have a detrimental effect on protocols that depend on neighborhood information. Adjusting the transmission power of nodes will also impede neighbor discovery. Following are some MAC protocols with power management.

7.3.2.3.1 Power Saving

In Power Aware Multi-Access protocol with Signaling (PAMAS) [66], two conditions under which a node should turn itself off are identified: (1) if a node has no packets to transmit, it ought to power itself off if a neighbor begins transmitting; and (2) if at least one neighbor of a node is transmitting and another is receiving, the node ought to power off because it cannot transmit or receive a packet (even if its transmit queue is nonempty).

The aim of PAMAS is to reduce power waste caused by not only needless transmission, but also needless overhearing.

Group TDMA [58] at any time instance only activates the receiver group for which the transmitters have the highest residual cumulative energy. Although protocols with consideration of energy-level awareness can last the lifetime of the entire network, the fairness problem for individual nodes is inevitable.

Tseng et al. [67] proposed three sleep schemes to improve the PS mode in IEEE 802.11 for its operation in multi-hop networks, and each with a different wake-up pattern — namely, *domination-awake-interval*, *periodically-fully-awake-interval*, and *quorum-based protocols*. For each PS node, it divides its time axis into a number of fixed-length intervals called beacon intervals. As shown in Figure 7.8, in each beacon interval there are three windows called *active window*, *beacon window*, and *MTIM window*, respectively. During the active window, the PS node should turn on its receiver to listen to any packet and take proper actions as usual. The beacon window is used for the PS node to send its beacon, which will allow others to be aware of its existence. The MTIM window is used for other nodes to send their MTIM frames to the PS node. The MTIM frames serve the similar purpose as ATIM frames in IEEE 802.11. Excluding these three windows, a PS node with no packet to send or receive may go to sleep. The structure of the beacon interval varies with the three protocols. The domination-awake-interval protocol keeps a PS node awake sufficiently long to ensure that neighboring nodes know each other. The periodically-fully-awake-interval has two types of beacon intervals: namely, low-power intervals and fully awake intervals. These two types of beacon intervals are interleaved to save power. In the quorum-based protocol, a PS node only needs to send beacons in $O(1/n)$ of the all beacon intervals. Each of these protocols can guarantee an upper bound on packet delay if there is no collision in the beacon window.

Sensor MAC (S-MAC) [68] is designed for sensor networks where energy conservation and self-configuration are more critical than throughput and delay. S-MAC is close to the periodically-fully-awake-interval except that S-MAC synchronizes the sleep schedules of neighboring nodes. Nodes

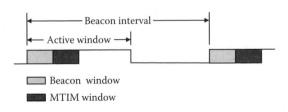

Figure 7.8   Structure of a beacon interval.

exchange their schedules by periodically broadcasting a SYNC packet to their immediate neighbors. A node talks to its neighbors at their scheduled listen time, thus ensuring that all neighboring nodes can communicate even if they have different schedules.

7.3.2.3.2 Power Control

Wattenhofer et al. [69] proposed a scheme according to which a transmitter gradually increases its transmission power until it finds at least one neighbor in every cone of angle $\alpha = 2\pi/3$ centered at the transmitter (a $5\pi/6$ angle was later proven to guarantee network connectivity). Recall that in DPC/ALP [48], a node sends its RTS by progressively increasing its transmission power until the receiver can hear it. DCAPC [70] uses a contrary way to adjust the transmission power. In DCAPC, a node is initially unaware of the appropriate power levels, so it transmits with maximum power. After it reaches the receiver, both the sender and the receiver will learn the appropriate transmission power.

In [71–73], nodes transmit RTS/CTS at maximum power but send DATA/ ACK at minimum necessary power. The minimum necessary power varies for traffic pairs with different distances, or different interference levels at the receiver side. In Jung and Vaidya [74], the power level of the DATA/ACK will increase in a short time to the maximum to keep neighbors within the carrier sensing zone aware of the ongoing transmission. Jia et al. [75] point out that the limitation of throughput improvement of [71–74,76] lies in RTS and CTS when transmitted at the maximum power level, which prevents potential concurrent sessions from proceeding in the vicinity of the transmitter and the receiver. In Jia et al. [75] it is proposed that nodes choose a transmission power level based on its traffic distance d and an estimate of the interference level it experiences. The power function, $\delta - PCS$, is the following: $P(t) = P_{max}(d/d_{max})^{\delta}$, where d_{max} is the transmission range corresponding to P_{max}, δ is a constant between 0 and α, and α is the power attenuation factor. Each δ value corresponds to a different power control scheme; and in one scheme, the value δ is uniform to each node in the network. Different δ values have different preferences in terms of traffic pair distance. The larger δ, the longer traffic distance is preferred. In practice, nodes do not know the distance to another node d and cannot use the above-mentioned power function directly. A method is proposed to solve this issue: communication nodes can estimate the channel gain during an RTS/CTS handshake. The distance variable of the power function can be converted into a gain variable. In the Power Controlled Multiple-Access Protocol (PCMA) [77], a "variable bounded power" collision suppression model is proposed. The packet handshake sequence on the data channel is RPTS-APTS-DATA-ACK, while the request-power-to-send (RPTS)/acceptable-power-to-send (APTS) handshake is similar to the RTS/CTS handshake. Each active receiver advertises the maximum additional noise power it can tolerate, given its current received signal and noise power levels. The noise tolerance advertisement or BT is pulsed periodically in a BT channel. The signal strength of the pulse indicates the tolerance to additional noise. A potential transmitter first senses the carrier by listening to the BT for a minimum time period to detect the upper bound of its transmitter power for all control (RPTS, APTS, ACK) and data packets. When the destination receives the RPTS, it measures the received power and computes the channel gain. The destination sends the APTS at a power level computed by receiving threshold, channel gain, and signal-to-reference ratio. The suggested power level at which data packets should be sent is contained in the APTS packet. It is demonstrated that PCMA allows for a greater number of simultaneous transmissions than IEEE 802.11 by adapting the transmission ranges to be the minimum value required to satisfy successful reception at the intended destination.

In POWMAC [78], multiple pairs of neighboring nodes exchange their control packets in an access window before transmitting data packets. The access window consists of an adjustable number of fixed-duration access slots. Collision avoidance information (CAI) is inserted in the control packets, and the received signal strengths of control packets are used to dynamically bound the transmission power of potentially interfering terminals in the vicinity of a receiving node. Control packets are transmitted at an adjustable power level so that they reach all and only potentially interfering terminals, while data packets are transmitted at a reduced power level to reach only the intended receiver.

PEM [79] enhances IEEE 802.11 by adjusting transmission power in an effort to increase throughput and to reduce power consumption. Stations that can transmit simultaneously without interfering with each other are scheduled to transmit simultaneously. Each node uses the Maximum Independent Set algorithm to find the maximum number of simultaneous transmission pairs without interference.

7.3.2.3.3 Summary

Due to the lack of clock synchronization in multi-hop ad hoc networks, power saving using the sleep mode is challenging. Each node may have its own sleep/wake circle. To transmit and receive packets, waking intervals of different nodes must overlap. The protocols in Tseng et al. [67] ensure the overlap by elaborately selecting the structure of intervals, while S-MAC allows neighbors to exchange their schedules. Sleep mode may affect the connectivity and hence the routing function. To keep correct neighborhood information, periodic beacon information can be used to declare a node's existence. Adjusting the transmission power can also affect routing, so power management must be combined with the routing protocol.

7.3.2.4 MAC Protocols Using Directional Antennas

The use of directional antennas can avoid collisions by confining transmission or reception into a sector. The benefit of spatial reuse achieved by a MAC protocol that uses directional mode in all transmissions can outweigh the benefit of a conservative collision avoidance MAC protocol that sends some omni-directional control packets to silent potential interfering nodes [80]. In crowded multi-hop networks, there will be many nodes in a sector of a transceiver, even if the angle is small. The angle of sectors must be deliberately selected. In Nasipuri and Das [61], algorithms for determining the orientation and broadcasting angles of the directional antennas are also designed to make the number of stations in each sector as evenly distributed as possible.

In DMAP [56], nodes use out-of-band signaling for control messages, as shown in Figure 7.5(b). On receiving RTS, the destination node calculates the power control factor β, and includes β in the DCTS (directional CTS). The source node calculates the transmission power of data packet based on β.

ADAPT [81] maintains a DOA table and a Smart Virtual Carrier Sensing (SVCS) table. The former table records the node IDs and angles of the most recently encountered nodes. The latter table records the angles and lasting time of upcoming communications. Its virtual carrier sensing is similar to that of the IEEE 802.11 and is used to defer transmission that is possible to collide. Each packet transmitted (i.e., RTS, CTS, DATA, ACK) carries a value indicating the remaining communication duration. ADAPT also takes into account the mobility of nodes and the deafness problem. It is assumed that each node includes its coordinates in every packet it transmits, and adaptive antennas have the ability to automatically estimate the angle-of-arrival of an incoming signal using appropriate signal processing algorithms. However, it is not explained in detail how to acquire the DOA information in the first place when nothing has been transmitted. Angle of Arrival and Directional Virtual Carrier Sensing (DVCS) are also used in Takai et al. [82].

Smart antennas based Wider-range Access MAC Protocol (SWAMP) proposed by Takata et al. [83] consists of two access modes: (1) Omni-directional transmission range Communication mode (OC-mode) and (2) Extend omni-directional transmission range Communication mode (EC-mode). OC-mode is selected when the receiver is near the transmitter or when the transmitter does not know the location of the receiver. OC-mode is similar to the scheme depicted in Nasipuri et al. [26] except that all its control packets RTS/CTS/SOF are transmitted by omni-directional antennas. EC-mode is selected when a receiver is out of the range of a transmitter's omni-directional beam. EC-mode can extend the transmission range. Takata et al. [84] further investigated the optimization of parameters associated with location information staleness of SWAMP.

In Selective CSMA with Cooperative Nulling (SCSMA/CN) [85], all packets are transmitted omni-directionally, but a receiving beamform is used to receive data packets and ACKs. Whenever a listening node detects an RTS/CTS interaction between two other nodes, the source node and all the nodes that receive the CTS packet simultaneously transmit a short cooperative nulling (CN) packet on the channel. During this period, the destination node beamforms to the desired transmitted CN sent by the source node. Because the neighbor nodes are also transmitting at this point (as interferers), the beamforming at the destination node will attempt to maximize the desired signal and null these interfering transmissions. The beamforming weights obtained by the destination node are recorded. Depending on the type of beamforming, the CN packet transmitted by the source node may contain a known pilot signal (containing a known bit sequence, for example) so that the destination node can beamform to this desired signal. Following this, the destination node and all the neighbors of the source node send CN packets in the same fashion. The source node can beamform, and the beamforming weights are recorded.

In directional mode, a node can point its beam toward a specified direction with a gain greater than in omni-directional mode. Multi-hop RTS MAC (MMAC) [86] attempts to exploit the extended transmission range of directional antennas. Two kinds of neighbors are defined: (1) Direction-Omni (DO) neighbor and (2) Direction-Direction (DD) neighbor. A node sends an RTS along the DO neighbor route to the DD neighbor (destination), and requests the destination node to point its receiving beam toward the RTS sender at a specific point in time in the future. The RTS can be forwarded by several nodes before it reaches the destination. On receiving the RTS, the destination transmits the CTS in the direction of the RTS sender and waits for the arrival of the DATA packet. CTS/DATA/ACK are transmitted and received in directional mode.

7.3.2.4.1 Summary

STS was designed on the assumption that all nodes are static. However, in mobile ad hoc networks, a tone cannot be ensured to be unique in its neighborhood. Hence, STS cannot be applied to mobile ad hoc networks directly. To use directional antennas, nodes must know and update the location information of other nodes. If nodes are not evenly distributed, a large number of nodes may concentrate in certain transmission sectors, causing severe congestions in these sectors. How to determine the angles of directional antenna is not discussed sufficiently in the literature.

7.3.2.5 Fairness and QoS

In mobile ad hoc networks, supporting real-time traffic is very difficult because of the lack of infrastructure, the shared radio medium, and the potentially rapid change in topology. Multi-hop ad hoc networks can achieve QoS to some extent through the combination of MAC protocols, routing protocols, and other resource allocation related protocols. Some MAC layer solutions are proposed to alleviate the QoS problem.

7.3.2.5.1 Consider Fairness and QoS at the MAC Layer

In Ozugur and Naghshineh [47], link access probabilities P_{ij} are calculated at the source node in two ways using connection-based and time-based media access methods. Each active user broadcasts information on either the number of logical connections or the average contention time to the stations within the communication reach. Different links from the same source node can have different priorities.

The Reservation CSMA/CA protocol proposed by Joe [87] is based on a hierarchical approach consisting of two sublayers. The lower sublayer of the MAC protocol provides a fundamental access method such as IEEE 802.11 to support asynchronous data traffic over mobile ad-hoc networks. The upper sublayer is designed to support real-time periodic traffic with QoS requirements. Time is divided into frames, and a sender that wants to transmit real-time traffic must reserve a slot by a three-way handshake with the receiver prior to actual data transmission.

Sobrinho et al. [88] proposed a distributed MAC scheme that provides QoS real-time access to ad hoc CSMA wireless networks. Real-time packets have a higher priority over data packets. Real-time nodes contend for access to the channel by jamming the channel with pulses of energy, the durations of which is a function of the delay incurred by the nodes until the channel becomes idle. As shown in Figure 7.9, Distributed Link Scheduling Multiple Access (D_LSMA) [89] segregates the MAC layer into upper MAC and lower MAC layers. The lower MAC layer is similar to that in the IEEE 802.11 standard. The upper layer serves as an intelligent scheduler that makes

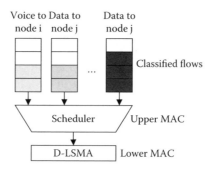

Figure 7.9   D_LSMA.

decisions depending on traffic volume and flow QoS requirements at that node. The upper MAC layer can be designed to support a mix of traffic types with separate packet queues and different scheduling policies. This type of segregation by traffic type is an important feature in multi-hop ad hoc networks where nodes have to handle both local traffic and cross-traffic with different service and bandwidth requirements. By far, research on how to discriminately treat local traffic and cross-traffic on MAC layer is a relatively new field and is far from sufficient.

In Nandagopal et al. [90], a general analytical framework is proposed in which fairness is modeled through a four-step process. By the derivation of a flow contention graph and a resource constraint graph, the design of fairness mechanisms is well defined. It is shown that the fairness model of the system is determined by the utility function that must be maximized. A general mechanism is also presented to translate a given fairness model into a corresponding contention resolution algorithm.

7.3.2.5.2   Cross-Layer Design to Support QoS Service

The issue of QoS is often considered separately at an individual network layer such as the network layer or the MAC layer. Although improvement can be achieved at a particular layer, side effects may be incurred at other layers because layers are closely coupled in real networks. Cross-layer design is necessary for improving network performance comprehensively.

Martinez et al. [91] proposed a cross-layer design with smart antenna and QoS support. Spatial Reuse Time Division Multiple Access (STDMA) is used as a medium access control mechanism and smart antennas are adopted to improve system performance. STDMA is a variant of TDMA. In STDMA, multiple nodes can use the same transmission slot if the interference between them is below a certain threshold. The objective of the cross-layer structure is to collect information in three different levels. Application layer metrics (ALM), MAC layer metrics (MLM), and Network layer metrics (NLM) are collected to create a new schedule and provide different services for traffic with different priorities. The number and position of each assigned slot in the frame are related to the QoS offered to each service class. A queue loaded with high-priority traffic will be assigned with more slots than another queue loaded with low-priority traffic. The time between assigned slots will influence the jitter. Two different queueing management mechanisms, FPQ and WFQ, were analyzed.

Yang and Sankar [92] discussed the performance optimization challenges of an ad hoc network and how cross-layer approaches can improve its performance. A protocol stack is proposed that involves cross-layer processing. Transmission rate adaptation in the MAC layer is based on the channel signal strength information from the physical layer. Congestion information from MAC layer is used in the routing protocol Dynamic Source Routing (DSR).

The less congested routes rather than the shortest routes are more likely used in this scheme. It increases the fairness of the overall ad hoc network and also reduces the total latency. Due to the unique characteristics of ad hoc networks, cross-layer design may be a good solution. Application layer sends hard QoS requests to lower layer, which should schedule transmission according to the dynamic requests. In case of heavy congestion, it is impossible to fulfill the hard QoS, the cross-design should allow negotiation between the upper layers and the lower layers.

7.3.2.5.3 Summary

Unlike best-effort traffic, real-time flows and aggregated traffic between wireless routers need relatively deterministic bandwidths and bounded delays, usually requiring some combination of reservation, priority control, scheduling, and dynamic resource management. Consideration of the QoS problem only in MAC layer is not sufficient. QoS is difficult to evaluate and deploy because it is relevant to different traffic types, the number of nodes, the number of traffic flows, etc. A cross-layer design combines the MAC layer and the upper layers, and is a possible approach.

7.4 MAC Protocols for Clustered Networks

Because a large-scale ad hoc network with a flat structure cannot guarantee performance, many cluster and hierarchy algorithms have been proposed to solve the scalability issue of routing problems. A typical cluster structure is illustrated in Figure 7.10. One node among each group is elected to act as the clusterhead according to some rules. A clusterhead normally serves as a local coordinator for its cluster members, performing intra-cluster transmission arrangement, data forwarding, etc. The role of a clusterhead is

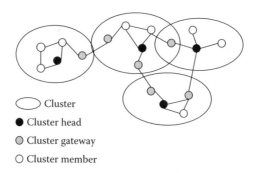

⬭ Cluster
● Cluster head
◐ Cluster gateway
○ Cluster member

Figure 7.10   Clustering.

very similar to that of a base station in an infrastructure network. A cluster gateway can access neighboring clusters and forward information between clusters. A class member is usually called an ordinary node, which is a non-clusterhead node without any inter-cluster links [93]. A complete and thorough classification and analysis of clustering schemes is given in Yu and Chong [93]. A cluster structure enables local central control and makes management much easier. However, due to the mobility of nodes, the density of nodes changes from time to time, and clusters need to be merged or divided dynamically. A member node needs to know which cluster it belongs to, and a clusterhead should know which nodes belong to it. Many dynamic clustering schemes have been proposed to adapt to the change of networks. In Cai et al. [94], a channel access-based clustering scheme is proposed, where member nodes contend to be the cluster leader in a control channel. Two busy tones are transmitted in the control channel to protect data transmission. Each node periodically broadcasts beacon signals to identify its existence. Beacon signals transmitted by the leaders (leader beacon signal) are different from those transmitted by the cluster members (normal beacon signal). A clusterhead uses beacon signals from its member nodes to maintain its member list. Cluster members use beacon signals from the clusterhead to decide if it is necessary to determine a new clusterhead.

Clusterheads are often selected according to factors such as node ID, connectivity range, energy level, number of neighbors, mobility, etc. One of the challenges in clustering is how to maintain the stability of clusters. From the perspective of routing, clustering can reduce the routing information traversing the entire network, and makes the network more stable in the view of each mobile terminal. From the perspective of medium access, a clustered structure facilitates the spatial reuse of resources to increase the system capacity. Also, with the help of its clusterhead, a cluster can better schedule its transmission events. This can save resources used for retransmission resulting from reduced transmission collision [93]. The following are some protocols designed to make use of the facilities of a clustered structure.

IEEE 802.15.4 [95] is now considered the MAC protocol with the most potential for wireless sensor networks. Depending on the application requirements, a low-rate WPAN can operate in either of two topologies: (1) the star topology or (2) the peer-to-peer topology. An example of the use of the peer-to-peer communications topology is the cluster-tree. Any of the full function devices (FFDs) can act as a coordinator and provide synchronization services to other devices or other coordinators. When a device wishes to transfer data to a coordinator in a beacon-enabled network, it first listens for the network beacon. When the beacon is found, the device can be synchronized. At the appropriate point, the device transmits its data frame, using slotted CSMA-CA, to the coordinator. When a device wishes to transfer data in a non-beacon-enabled network, it simply transmits

its data frame, using unslotted CSMA-CA, to the coordinator. In both scenarios, the coordinator acknowledges the successful reception of the data by transmitting an optional acknowledgment frame. When a coordinator wishes to transfer data to a device in a beacon-enabled network, it indicates in the network beacon that the data message is pending. The device periodically listens to the network beacon and, if a message is pending, transmits a MAC command requesting the data, using slotted CSMA-CA. The pending data frame is then sent using slotted CSMA-CA. The device acknowledges successful reception of the data by transmitting an acknowledgment frame. When a coordinator wishes to transfer data to a device in a non-beacon-enabled network, it waits for a MAC command transmitted from the device requesting for data.

Clique-based Randomized Multiple Access (CRMA) [96] forms collections of nodes, or cliques, separated by one hop. In multicast cliques (cliques with more than two members), a master-slave arrangement is used, whereby a clique has only one master that controls access to the channel — slaves speak only when spoken to.

Two-Phase Coding Multi-channel MAC Protocol (TPCMMP) [97] is a CDMA-based multi-channel MAC protocol. A network is divided into cells with a hexagonal cellular geographic structure and all cells are of the same radius, r, which equals the transmission range of the nodes. The first-phase code is employed for differentiating cells and the second-phase code is used for distinguishing nodes in one specific cell. A cell leader is responsible for maintaining and assigning the second-phase codes to its cell members. A node relies on its location information acquired from a position device such as GPS to know its first code.

Differentiated DCF (DDCF) [98] provides node-based distributed priority and access differentiation. It defines four roles for nodes within a cluster with decreasing priority: (1) clusterhead, (2) first-level spine nodes, (3) second-level spine nodes, and (4) leaf nodes according to distances from the nodes to the clusterhead. DDCF is based on EDCF of IEEE 802.11e, but the contending parameters such as IFS, CW_{Min} and CW_{Max} are differentiated for nodes of different roles, not for different packet flows.

7.4.1 Summary

MAC schemes of clustered structure often emulate those of cellular network with the clusterheads serving as virtual base stations. But in cellular networks, base stations are fixed and the QoS issue is much simpler than that of ad hoc networks. In clustered ad hoc networks, clusterheads are selected from the common nodes and all nodes must have the capability of being a clusterhead. Due to mobility, clusterheads must be reselected from time to time. How to schedule medium access when a cluster is switching its clusterhead is another difficult problem yet to be solved.

7.5 Discussion

In single-hop flat networks, IEEE 802.11 is popular and can perform well except that problems of fairness and QoS need further research. Performance studies and improvements on IEEE 802.11 have been the subject of many contributions. When the basic schemes of RTS/CTS/DATA/ACK handshake and carrier sensing are used in multi-hop networks to overcome hidden or exposed problems, control packets cannot be prevented from colliding with other control packets or data packets. If the data packets are small relative to the control packets, system efficiency will be penalized for the overhead of control packets because the successful transmission of each data packet needs successful transmission of three control packets. To protect transmission through busy tones, each node needs two transceivers, which will increase hardware costs.

Splitting the medium into one control channel and one or more data channels may avoid collisions of control packets and data packets. But because all the transmission must start from negotiating in the control channel, the only control channel becomes a bottleneck in some environments. Using multiple channels may reduce collisions and improve system throughput, but the overall channel utility against throughput is not clear. In some protocols, nodes try to reserve the medium in a contention window, but these schemes need precise synchronization and the structure of frames is not flexible, and thus cannot adapt to the scale of networks. How to alleviate hidden or exposed problems as much as possible remains a challenging issue.

How to use directional antennas to avoid collisions and to increase resource utility is now a hot topic. Due to the fundamental principle of directional antennas, they do not apply to handheld devices, and antenna-equipped intelligent transportation systems will be a potential application. Link power control is essential in exploiting the benefits of beamforming antennas to their fullest potential [99]. Usually, directional antennas are considered with power control and the combination of these two aspects enables us to control the transmission range flexibly and eliminate the hidden terminal problems and exposed terminal problems as much as possible. There are two major categories of smart antennas: (1) switched beam systems and (2) adaptive array systems. Most protocols using directional antennas in the literature focus on switched beam systems. Because cost barriers have prevented their use in commercial systems until recent years, relatively little work has been done on adaptive array systems. The advent of powerful low-cost digital signal processors (DSPs), general-purpose processors (and ASICs), and innovative software-based signal-processing techniques (algorithms) have made intelligent antennas practical. Smart antenna systems will be a promising area of study in the future. As far as QoS is concerned, protocols that consider how to treat local traffic and cross-traffic with different service and bandwidth requirements are expected.

A flat structure is not practical for large-scale networks. Many clustering schemes have been proposed, but most of them focus on how to select the cluster leader and how to maintain cluster stability, which is often realized through a control channel. Few of them provide details on how clusterheads manage their clusters through resource assignment, synchronization, and scheduling.

In our survey of ad hoc network MAC protocols, it is evident that the majority of these protocols were derived heuristically and were aimed at improving some measurement under a specific assumption. In simulation-based studies, the physical layer is often oversimplified. A direction for future work is to provide a comprehensive model and design the network with a cross-layer approach. However, cross-layer design will enhance the coupling between layers, and the model is less flexible and will not meet the requirements of various settings. A cross-layer model may be better used for some specific scenario.

References

[1] P. Gupta and P.R. Kumar, "The Capacity of Wireless Networks," *IEEE Transactions on Information Theory*, Vol. 46, pp. 388–404, March 2000.

[2] D. Bertsekas and R. Gallager, *Data Networks (2nd ed.)*, Chapter 4. Prentice Hall, 1992.

[3] L. Kleinrock and F.A. Tobagi, "Packet Switching in Radio Channels. Part I. Carrier Sense Multiple-Access Models and Their Throughput-Delay Characteristics," *IEEE Transactions on Communications*, Vol. 23, pp. 1400–1416, December 1975.

[4] D. Lal, V. Jain, Q.-A. Zeng, and D.P. Agrawal, "Performance Evaluation of Medium Access Control for Multiple-Beam Antenna Nodes in a Wireless LAN," *IEEE Transactions on Parallel and Distributed Systems*, Vol. 15, pp. 1117–1129, December 2004.

[5] P. Karn, "MACA: A New Channel Access Protocol for Packet Radio," *Proceedings of ARRL/CRRL Amateur Radio 9th Computer Networking Conference*, 1990, pp. 134–140.

[6] V. Bharghavan, A. Demers, S. Shenker, and L. Zhang, "MACAW: A Media Access Protocol for Wireless LANs," *Proceedings of ACM SIGCOMM'94*, 1994, pp. 212–225.

[7] C.L. Fullmer and J.J. Garcia-Luna-Aceves, "Floor Acquisition Multiple Access (FAMA) for Packet-Radio Networks," *Proceedings of ACM SIGCOMM'95*, 1995, pp. 262–273.

[8] ANSI/IEEE Std. 802.11, "Part 11: Wireless LAN Medium Access Control (MAC) and Physical Layer (PHY) Specifications," 1999.

[9] S. Xu and T. Saadawi, "Does the IEEE 802.11 MAC Protocol Work Well in Multihop Wireless Ad Hoc Networks?," *IEEE Communications Magazine*, Vol. 39, pp. 130–137, June 2001.

[10] C. Chaudet, D. Dhoutaut, and I.G. Lassous, "Performance Issues with IEEE 802.11 in Ad Hoc Networking," *IEEE Communications Magazine*, Vol. 43, pp. 110–116, July 2005.

[11] C.-M. Chao, J.-P. Sheu, and I-C. Chou "A Load Awareness Medium Access Control Protocol for Wireless Ad Hoc Network," *Proceedings of IEEE ICC '03*, 2003, Vol. 1, pp. 438–442.

[12] W.-C. Hung, K.L.E. Law, and A. Leon-Garcia, "A Dynamic Multi-channel MAC for Ad-hoc LAN," *Proceedings of the 21st Biennial Symposium on Communications*, 2002, pp. 31–35.

[13] R. Fantacci, A. Ferri, and D. Tarchi, "A MAC Technique for CDMA Based Ad-hoc Networks," *Proceedings of IEEE Wireless Communications and Networking Conference (WCNC) 2005*, Vol. 1, pp. 645–650.

[14] R. Fantacci, A. Ferri, and D. Tarchi, "Medium Access Control Protocol for CDMA Ad Hoc Networks," *IEE Electronics Letters*, Vol. 40, pp. 1131–1133, 2004.

[15] E.S. Sousa and J.A. Silvester, "Spreading Code Protocols for Distributed Spread-Spectrum Packet Radio Networks," *IEEE Transactions on Communications*, Vol. 36, pp. 272–281, March 1998.

[16] P. Kota and C. Schlegel, "A Wireless Packet Multiple Access Method Exploiting Joint Detection," *Proceedings of IEEE ICC'03*, Vol. 4, pp. 2985–2989, 2003.

[17] J. So and N. Vaidya, "Multi-Channel MAC for Ad Hoc Networks: Handling Multi-Channel Hidden Terminals Using A Single Transceiver," *Proceedings of the 5th ACM International Symposium on Mobile Ad Hoc Networking and Computing (MobiHoc'04)*, May 2004, pp. 222–233.

[18] M. Realp and A.I. Perez-Neira, "Decentralized Multiaccess MAC Protocol for Ad-Hoc Networks," *Proceedings of the 14th IEEE Personal, Indoor and Mobile Radio Communications (PIMRC'03)*, September 2003, Vol. 2, pp. 1634–1638.

[19] J.A. Stine, "Exploiting Processing Gain in Wireless Ad Hoc Networks Using Synchronous Collision Resolution Medium Access Control Schemes," *Proceedings of IEEE Wireless Communications and Networking Conference (WCNC)*, March 2005, Vol. 1, pp. 612–618.

[20] M. Krunz, A. Muqattash, and S.-J. Lee, "Transmission Power Control in Wireless Ad Hoc Networks: Challenges, Solutions and Open Issues," *IEEE Network*, Vol. 18, pp. 8–14, 2004.

[21] S.-L. Wu and P.-C. Tseng "An Energy Efficient MAC Protocol for IEEE 802.11 WLANs," *Proceedings of the Second Annual Conference on Communication Networks and Services Research (CNSR'04)*, May 2004, pp. 137–145.

[22] S. Gobriel, R. Melhem, and D. Moss, "BLAM: An Energy-Aware MAC Layer Enhancement for Wireless Ad Hoc Networks," *Proceedings of IEEE Wireless Communications and Networking Conference (WCNC)*, March 2005, Vol. 3, pp. 1557–1563.

[23] Y.-B. Ko, V. Shankarkumar, and N.H. Vaidya, "Medium Access Control Protocols Using Directional Antennas in Ad Hoc Networks," *Proceedings of IEEE INFOCOM'00*, March 2000, Vol. 1, pp. 13–21.

[24] D. Lal, R. Gupta, and D.P. Agrawal, "Throughput Enhancement in Wireless Ad Hoc Networks with Spatial Channels C A MAC Layer

Perspective," *Proceedings of Computers and Communications 2002 (ISCC 2002)*, pp. 421–428.

[25] H. Singh and S. Singh, "DOA-ALOHA: Slotted ALOHA for Ad Hoc Networking Using Smart Antennas," *Proceedings of IEEE 58th Vehicular Technology Conference (VTC 2003-Fall)*, Vol. 5, pp. 2804–2808.

[26] A. Nasipuri, S. Ye, J. You, and R.E. Hiromoto, "A MAC Protocol for Mobile Ad Hoc Networks Using Directional Antennas," *Proceedings of IEEE Wireless Communications and Networking Conference (WCNC'00)*, September 2000, Vol. 3, pp. 1214–1219.

[27] N.S. Fahmy, T.D. Todd, and V. Kezys, "Ad Hoc Networks with Smart Antennas Using IEEE 802.11-Based Protocols," *Proceedings of IEEE ICC'02*, Vol. 5, pp. 3144–3148.

[28] N.S. Fahmy, T.D. Todd, and V. Kezys, "Distributed Power Control for Ad Hoc Networks with Smart Antennas," *Proceedings of IEEE 56th Vehicular Technology Conference (VTC'02)*, 2002, Vol. 4, pp. 2141–2144.

[29] T. Ueda, S. Tanaka, D. Saha, S. Roy, and S. Bandyopadhyay, "An Efficient MAC Protocol with Direction Finding Scheme in Wireless Ad Hoc Network using Directional Antenna," *Proceedings of Radio and Wireless Conference 2003 (RAWCON)*, pp. 233–236.

[30] J.C. Mundarath, P. Ramanathan, and B.D. Van Veen, "NULLHOC: A MAC Protocol for Adaptive Antenna Array Based Wireless Ad Hoc Networks in Multipath Environments," *Proceedings of IEEE GLOBECOM '04*, Vol. 5, pp. 2765–2769.

[31] B. Bensaou, Y. Wang, and C.C. Ko, "Fair Medium Access in 802.11 based Wireless Ad-Hoc Networks," *Proceedings of ACM International Symposium on Mobile Ad Hoc Networking and Computing (MobiHoc)*, 2000, pp. 99–106.

[32] N.H. Vaidya and P. Bahl, "Fair Scheduling in Broadcast Environments," Microsoft Research, Technical Report MSR-TR-99-61, December 1999.

[33] C. Wu, J. Feng, and P. Fan, "On a New Queue Backoff Fair Algorithm for Ad Hoc Networks," *Proceedings of Parallel and Distributed Computing, Applications and Technologies, 2003 (PDCAT'2003)*, pp. 335–339.

[34] IEEE 802.11WG, "Draft Supplement to Part II: Wireless Medium Access Control (MAC) and Physical Layer (PHY) Specifications: Medium Access Control (MAC) Enhancements for Quality of Service (QoS)," IEEE Standard 802.11e/D3.3.2, November 2002.

[35] J. Deng and R.-S. Chang, "A Priority Scheme for IEEE 802. 11 DCF Access Method," *IEICE Transactions on Communications*, Vol. E82-B, No. 1, pp. 96–102, January 1999.

[36] X. Pallot and L.E. Miller, "Implementing Message Priority Policies over an 802.11 Based Mobile Ad Hoc Network," *Proceedings of IEEE Military Communications Conference (MILCOM)*, 2001, pp. 860–864.

[37] P. Goyal, H.M. Vin, and H. Cheng, "Start-Time Fair Queueing: A Scheduling Algorithm for Integrated Services Packet Switching Networks," *IEEE/ACM Transactions on Networking*, Vol. 5, No. 5, pp. 690–704, October 1997.

[38] L. Gannoune, S. Robert, N. Tomar, and T. Agarwal, "Dynamic Tuning of the Maximum Contention Window (CWmax) for Enhanced Service

Differentiation in IEEE 802.11 Wireless Ad-Hoc Networks," *Proceedings of IEEE Vehicular Technology Conference (VTC)*, September 2004, Vol. 4, pp. 2956–2961.

[39] L. Gannoune and S. Robert, "Dynamic Tuning of the Contention Window Minimum (CWmin) for Enhanced Service Differentiation in IEEE 802.11 Wireless Ad-Hoc Networks," *Proceedings of 15th IEEE International Symposium on Personal, Indoor and Mobile Radio Communications (PIMRC 2004)*, September 2004, Vol. 1, pp. 311–317.

[40] Y. Xiao and H. Li, "Local Data Control and Admission Control for QoS Support in Wireless Ad Hoc Networks," *IEEE Transactions on Vehicular Technology*, Vol. 53, Issue 5, pp. 1558–1572, September 2004.

[41] Y. Xiao and Y. Pan, "Differentiation, QoS Guarantee, and Optimization for Real-Time Traffic over One-Hop Ad Hoc Networks," *IEEE Transactions on Parallel and Distributed Systems*, Vol. 16, Issue 6, pp. 538–549, 2005.

[42] Z. Zhang, "DTRA: Directional Transmission and Reception Algorithms in WLANs with Directional Antennas for QoS Support," *IEEE Network*, Vol. 19, Issue 3, pp. 27–32, 2005.

[43] J. Li, C. Blake, D.S.J. De Couto, H.I. Lee, and R. Morris, "Capacity of Ad Hoc Wireless Networks," *Proceedings of ACM Mobicom*, July 2001, pp. 61–69.

[44] M.M. Carvalho and J.J. Garcia-Luna-Aceves, "A Scalable Model for Channel Access Protocols in Multihop Ad Hoc Networks," *Proceedings of ACM Mobicom*, 2004, pp. 330–334.

[45] S. Deepanshu, C.-W. Leena, and I. Sridhar, "Mitigating the Exposed Node Problem in IEEE 802.11 Ad Hoc Networks," *Proceedings of The 12th International Conference on Computer Communications and Networks (ICCCN 2003)*, pp. 157–162.

[46] A. Acharya, A. Misra, and S. Bansal, "MACA-P: a MAC for Concurrent Transmissions in Multi-hop Wireless Networks," *Proceedings of the First IEEE International Conference on Pervasive Computing and Communications (PerCom)*, 2003, pp. 505–508.

[47] T. Ozugur, M. Naghshineh, P. Kermani, and J.A. Copeland, "Fair Media Access for Wireless LANs," *Proceedings of GLOBALCOM'99*, 1999, pp. 570–579.

[48] J.-W. Kim and N. Bambos, "Power-Efficient MAC Scheme Using Channel Probing in Multi-rate Wireless Ad Hoc Networks," *Proceedings of IEEE Vehicular Technology Conference (VTC)*, 2002-Fall, Vol. 4, pp. 2380–2384.

[49] F. Talucci, M. Gerla, and L. Fratta, "MACA-BI (MACA By Invitation): A Receiver-Oriented Access Protocol for Wireless Multihop Networks," *Proceedings of IEEE International Symposium on Personal, Indoor and Mobile Radio Communications (PIMRC)*, 1997, Vol. 2, pp. 435–439.

[50] J.J. Garcia-Luna-Aceves and A. Tzamaloukas. "Reversing the CollisionAvoidance Handshake in Wireless Networks," *Proceedings of ACM/IEEE Mobicom*, August 1999, pp. 120–131.

[51] C.-K. Toh, V. Vassiliou, G. Guichal, and C.-H. Shih, "MARCH: A Medium Access Control Protocol for Multihop Wireless Ad Hoc Networks," *Proceedings of IEEE Military Communications Conference (MILCOM)*, 2000, Vol. 1, pp. 512–516.

[52] P. Kyasanur and N.H. Vaidya, "Capacity of Multi-Channel Wireless Networks: Impact of Number of Channels and Interfaces," *Proceedings of Mobicom 2005*, pp. 43–57.

[53] F.A. Tobagi and L. Kleinrock, "Packet Switching in Radio Channels. Part II C. The Hidden Terminal Problem in Carrier Sense Multiple-Access and the Busy-Tone Solution," *IEEE Transactions on Communications*, Vol. 23, pp. 1417–1433, 1975.

[54] Z.J. Hass and J. Deng, "Dual Busy Tone Multiple Access (DBTMA), A Multiple Access Control Scheme for Ad Hoc Networks," *IEEE Transactions of Communications*, Vol. 50, No. 6, pp. 975–985, June 2002.

[55] Z. Fang, B. Bensaou, and J. Yuan, "Collision-Free MAC Scheduling Algorithms for Wireless Ad Hoc Networks," *Proceedings of IEEE GLOBECOM '04*, Vol. 5, pp. 2770–2774.

[56] A. Arora, M. Krunz, A. Muqattash, "Directional Medium Access Protocol (DMAP) with Power Control for Wireless Ad Hoc Networks," *Proceedings of IEEE GLOBECOM '04*, Vol. 5, pp. 2797–2801.

[57] C. Xu, G. Liu, W. Cheng, and Z. Yang, "Multi-Transceiver Multiple Access (MTMA) for Mobile Wireless Ad Hoc Networks," *Proceedings of IEEE ICC'05*, May 2005, Vol. 5, pp. 2932–2936.

[58] Y.E. Sagduyu and A. Ephremides, "Energy and Throughput Efficiency in Wireless Ad Hoc Networks through Group TDMA," *Proceedings of IEEE Conference on Decision and Control (CDC)*, December 2004, Vol. 3, pp. 2836–2841.

[59] A. Nasipuri, J. Zhuang, and S.R. Das, "A Multichannel CSMA MAC Protocol for Multihop Wireless Networks," *Proceedings of IEEE Wireless Communications and Networking Conference (WCNC)*, 1999, Vol. 3, pp. 1402–1406.

[60] N. Jain, S.R. Das, and A. Nasipuri, "A Multichannel CSMA MAC Protocol with Receiver-Based Channel Selection for Multihop Wireless Networks," *Proceedings of International Conference on Computer Communications and Networks (ICCCN)*, 2001, pp. 432–439.

[61] A. Nasipuri and S.R. Das, "Multichannel CSMA with Signal Power-Based Channel Selection for Multihop Wireless Networks," *Proceedings of IEEE Vehicular Technology Conference (VTC)*, 2000, Vol. 1, pp. 211–218.

[62] T.-S. Yum and K.-W. Hung, "Design Algorithms for Multihop Packet Radio Networks with Multiple Directional Antennas Stations," *IEEE Transactions on Communications*, Vol. 40, Issue 11, pp. 1716–1724, November 1992.

[63] J.J. Garcia-Luna-Aceves and J. Raju, "Distributed Assignment of Codes for Multihop Packet-Radio Networks," *Proceedings of IEEE Military Communications Conference (MILCOM)*, November 1997, Vol. 450, pp. 450–454.

[64] C.-H. Yeh, "Spread Spectrum Techniques for Solving MAC-layer Interference Issues in Mobile Ad Hoc Networks," *Proceedings of IEEE Vehicular Technology Conference (VTC)*, 2004, Vol. 3, pp. 1339–1344.

[65] A. Tzamaloukas and J.J. Garcia-Luna-Aceves, "A Receiver-Initiated Collision-Avoidance Protocol for Multi-Channel Networks," *Proceedings of IEEE INFOCOM'01*, 2001, Vol. 1, pp. 188–198.

[66] S. Singh and C.S. Raghavendra, "PAMAS C Power Aware Multi-Access Protocol with Signaling for Ad Hoc Networks," *ACM Computer Communications Review*, Vol. 28, No. 3, pp. 5–26, 1998.

[67] Y.-C. Tseng, C.-S. Hsu, and T.-Y. Hsieh, "Power-Saving Protocols for IEEE 802.11-based Multi-Hop Ad Hoc Networks," *Proceedings of IEEE INFO-COM'02*, June 2002, Vol. 1, pp. 200–209.

[68] W. Ye, J. Heidemann, and D. Estrin, "Medium Access Control with Coordinated Adaptive Sleeping for Wireless Sensor Networks," *IEEE/ACM Transactions on Networking*, Vol. 12, Issue 3, pp. 493–506, June 2004.

[69] R. Wattenhofer, L. Li, P. Bahl, and Y.-M. Wang, "Distributed Topology Control for Power Efficient Operation in Multihop Wireless Ad Hoc Networks," *Proceedings of IEEE INFOCOM'01*, 2001, pp. 1388–1397.

[70] Y.-C. Tseng, S.-L. Wu, C.-Y. Lin, and J.-P. Sheu, "A Multi-Channel MAC Protocol with Power Control for Multi-Hop Mobile Ad Hoc Networks," *International Conference on Distributed Computing Systems Workshop*, 2001, pp. 419–424.

[71] J. Gomez, A. T. Campbell, M. Naghshineh, and C. Bisdikian, "Conserving Transmission Power in Wireless Ad Hoc Networks," *Proceedings of International Conference on Network Protocols (ICNP)*, November 2001, pp. 24–34.

[72] S. Agarwal, R.H. Katz, S.V. Krishnamurthy, and S.K. Dao, "Distributed Power Control in Ad Hoc Wireless Networks," *Proceedings of IEEE International Symposium on Personal, Indoor and Mobile Radio Communications (PIMRC)*, 2001, Vol. 2, pp. F-59–F-66.

[73] J. Gomez, A.T. Campbell, M. Naghshineh, and C. Bisdikian. "PARO: Supporting Dynamic Power Controlled Routing in Wireless Ad Hoc Networks," *ACM/Kluwer Journal on Wireless Networks*, Vol. 9, No. 5, pp. 443–460, 2003.

[74] E.-S. Jung and N. Vaidya, "A Power Control MAC Protocol for Ad Hoc Networks," *Proceedings of ACM MobiCom*, 2002, pp. 36–47.

[75] L. Jia, X. Liu, G. Noubir, and R. Rajaraman, "Transmission Power Control for Ad Hoc Wireless Networks: Throughput, Energy and Fairness," *Proceedings of IEEE Wireless Communications and Networking Conference (WCNC)*, 2005, Vol. 1, pp. 619–625.

[76] M.B. Pursley, H.B. Russell, and J.S. Wysocarski, "Energy-Efficient Transmission and Routing Protocols for Wireless Multi-Hop Networks and Spread-Spectrum Radios," *Proceedings of the EUROCOMM 2000*, pp. 1–5.

[77] J.P. Monks, V. Bharghavan, and W.-M.W. Hwu, "A Power Controlled Multiple Access Protocol for Wireless Packet Networks," *Proceedings of IEEE INFOCOM'01*, April 2001, Vol. 1, pp. 219–228.

[78] A. Muqattash and M. Krunz, "POWMAC: A Single-channel Power-Control Protocol for Throughput Enhancement in Wireless Ad Hoc Networks," *IEEE Journal on Selected Areas in Communications*, Vol. 23, No. 5, pp. 1067–1084, May 2005.

[79] K.-P. Shih, C.-Y. Chang, C.-M. Chou, and S.-M. Chen, "A Power Saving MAC Protocol by Increasing Spatial Reuse for IEEE 802.11 Ad Hoc WLANs," *Proceedings of the 19th IEEE International Conference on Advanced Information Networking and Applications (AINA'05)*, March 2005, Vol. 1, pp. 420–425.

[80] Y. Wang and J.J. Garcia-Luna-Aceves, "Spatial Reuse and Collision Avoidance in Ad Hoc Networks with Directional Antennas," *Proceedings of IEEE GLOBECOM'02*, Vol. 1, pp. 112–116.

[81] T. Spyropoulos and C.S. Raghavendra, "ADAPT: A Media Access Control Protocol for Mobile Ad Hoc Networks Using Adaptive Array Antennas," *Proceedings of IEEE International Symposium on Personal, Indoor and Mobile Radio Communications (PIMRC)*, 2004, Vol. 1, pp. 370–374.

[82] M. Takai, J. Martin, A. Ren, and R. Bagrodia, "Directional Virtual Carrier Sensing for Directional Antennas in Mobile Ad Hoc Networks," *Proceedings of ACM International Symposium on Mobile Ad Hoc Networking and Computing (MobiHoc)*, 2002, pp. 183–193.

[83] M. Takata, K. Nagashima, and T. Watanabe, "A Dual Access Mode MAC Protocol for Ad Hoc Networks using Smart Antennas," *Proceedings of IEEE ICC'04*, June 2004, Vol. 7, pp. 4182–4186.

[84] M. Takata, M. Bandai, and T. Watanabe, "An Extended Directional MAC for Location Information Staleness in Ad Hoc Networks," *Proceedings of the 25th IEEE International Conference on Distributed Computing Systems Workshops*, 2005, pp. 899–905.

[85] N.S. Fahmy and T.D. Todd, "A Selective CSMA Protocol with Cooperative Nulling for Ad Hoc Networks with Smart Antennas," *Proceedings of IEEE Wireless Communications and Networking Conference (WCNC)*, 2004, Vol. 1, pp. 387–392.

[86] R.R. Choudhury, X. Yang, R. Ramanathan, and N.H. Vaidya, "Using Directional Antennas for Medium Access Control in Ad Hoc Networks," *Proceedings of ACM Mobicom'02*, pp. 59–70.

[87] I. Joe, "QoS-aware MAC with Reservation for Mobile Ad-Hoc Networks," *Proceedings of IEEE Vehicular Technology Conference (VTC)*, September 2004, Vol. 2, pp. 1108–1112.

[88] J. L. Sobrinho and A. S. Krishnakumar, "Quality-of-Service in Ad Hoc Carrier Sense Multiple Access Wireless Networks," *IEEE Journal on Selected Areas in Communications*, Vol. 17, Issue 8, pp. 1353–1368, 1999.

[89] Z. Wu and D. Raychaudhuri, "D-LSMA: Distributed Link Scheduling Multiple Access Protocol for QoS in Ad-Hoc Networks," *Proceedings of IEEE GLOBECOM'04*, Vol. 3, pp. 1670–1675.

[90] T. Nandagopal, T.-E. Kim, X. Gao, and V. Bharghavan, "Achieving MAC Layer Fairness in Wireless Packet Networks," *Proceedings of ACM Mobicom'00*, August 2000, pp. 87–98.

[91] I. Martinez and J. Altuna, "A Cross-Layer Design for Ad Hoc Wireless Networks with Smart Antennas and QoS Support," *Proceedings of the 15th IEEE International Symposium on Personal, Indoor and Mobile Radio Communications (PIMRC 2004)*, September 2004, Vol. 1, pp. 589–593.

[92] N. Yang and R. Sankar, "Effects of Cross-Layer Processing on Wireless Ad Hoc Network Performance," *Proceedings of IEEE International Conference on Wireless and Mobile Computing, Networking And Communications (WiMob'2005)*, August 2005, Vol. 3, pp. 284–290.

[93] J.Y. Yu and P.H.J. Chong, "A Survey of Clustering Schemes for Mobile Ad Hoc Networks," *IEEE Communications Surveys & Tutorials*, Vol. 7, Issue 1, pp. 32–48, First Qtr. 2005.

[94] Z. Cai, M. Lu, and X. Wang, "Channel Access-Based Self-Organized Clustering in Ad Hoc Networks," *IEEE Transactions on Mobile Computing*, Vol. 2, Issue 2, pp. 102–113, April–June 2003.

[95] IEEE 802.15.4 WPAN-LR Task Group, "Wireless Medium Access Control (MAC) and Physical Layer (PHY) Specifications for Low-Rate Wireless Personal Area Networks (LR-WPANs)," 2003, standard.

[96] P. G. Flikkema and B. West, "Clique-Based Randomized Multiple Access for Energy-Efficient Wireless Ad Hoc Networks," *Proceedings of IEEE Wireless Communications and Networking Conference (WCNC'03)*, March 2003, Vol. 2, pp. 977–981.

[97] L. Zhang, B.-H. Soong, and W. Xiao, "Two-Phase Coding Multichannel MAC Protocol with MAI Mitigation for Mobile Ad Hoc Networks," *IEEE Communications Letters*, Vol. 8, Issue 9, pp. 597–599, September 2004.

[98] L. Bononi et al., "A Differentiated Distributed Coordination Function MAC Protocol for Cluster-based Wireless Ad Hoc Networks," *Proceedings of the 1st ACM International Workshop on Performance Evaluation of Wireless Ad Hoc, Sensor, and Ubiquitous Networks*, October 2004, pp. 77–86.

[99] R. Ramanathan, "On the Performance of Ad Hoc Networks with Beamforming Antennas," *Proceedings of ACM International Symposium on Mobile Ad Hoc Networking and Computing (MobiHoc)*, 2001, pp. 95–105.

Chapter 8

Optimal Scheduling and Resource Allocation in the CDMA2000 1xEV-DV: A Balanced Approach

*Shirley Mayadewi, Qian Wang,
and Attahiru Sule Alfa*

Contents

8.1 Overview 236
8.2 Introduction 237
8.3 Performance Measures of Cellular System 239
    8.3.1 Constraints on Network Resources.......................... 239
    8.3.2 Performance Measures of Forward Link 240
    8.3.3 Performance Measures of Reverse Link 243
8.4 System Model.. 245
    8.4.1 Forward Link.. 245
        8.4.1.1 Scheduling Problem Discussion 246
        8.4.1.2 Resource Allocation Problem Discussion......... 247
        8.4.1.3 Formulation on the Forward Link 249
    8.4.2 Enhancements on Reverse Link 1xEV-DV 250
        8.4.2.1 Transmission Characteristics 251

8.1   Overview

Third-generation (3G) standards are designed to meet the increasing demand for high-speed packet data transmission. CDMA2000 1xEV-DV is one
of the 3G standards that can support the integration of multimedia services
on a single 1.25-MHz carrier. In this chapter we develop joint scheduling and resource allocation schemes for both forward and reverse link of
CDMA2000 1xEV-DV system. The existing resource management schemes
mostly consider the benefits of either service providers or subscribers. On
the other hand, our resource management schemes not only benefit the
subscribers by optimizing the data rate or (Quality-of-Service) QoS supported, but also satisfy the service providers by maximizing the revenue
generated. The schemes also provide flexible trade-offs between the maximization of data rate, QoS, and revenue. The combined scheduling and
resource allocation problems are formulated into binary integer linear programming (LP) with multiple constraints. First, we investigated solving the
binary integer LP using exact algorithms, which are Branch and Bound
(B&B) and Complete Enumeration (CE). The results show that these algorithms are inefficient for this problem, as they require high computational
time. We then created heuristic algorithms that consume much shorter computational times although they sometimes compromise the objective function values. These heuristics are more appropriate for our scheduling and
resource allocation schemes, which require short computation time for implementation. In the end, we are able to make recommendations to service providers on how to choose the algorithms to properly schedule the
transmissions.

8.2   Introduction

The increasing subscriber demand for reliable wireless service in the area of voice, Internet, and data communication has motivated service providers to upgrade their networks to facilitate higher capacity with improved capabilities. The service providers' goal is to integrate a wide variety of services onto a common infrastructure. In the telecommunications industry, service providers have spent billions of dollars to buy the spectrum [1] and consequently they want to efficiently utilize it to generate as high a revenue as possible. Nevertheless, in the competitive world of cellular systems, providers should also be able to support satisfactory service, both fast and reliable, for the subscribers, at a low cost. Providing such service is not trivial because they have to use limited spectrum and network resources to serve the growing number of subscribers. Evidently, as inefficient spectrum and resource utilization decreases the quality of service provided, the revenue gained by the service providers may not be optimal. Hence, service providers need to find a method to efficiently utilize their resources while balancing their cost-effectiveness level and the satisfaction level of subscribers. The problems above are encountered by many cellular service providers, including those who are implementing Code Division Multiple Access2000 (CDMA2000) 1x standard for their network. First, we discuss the development of cellular-based communication systems.

The first generation (1G) of mobile communication systems is analog and only carries voice traffic. It utilizes Frequency Division Multiple Access (FDMA) to provide circuit-switched, voice-only service. Examples of 1G systems are Advanced Mobile Telephone System (AMPS) and Total Access Communications System (TACS). The second generation (2G) was introduced in the early 1990s and already implemented digital encoding. It uses Time Division Multiple Access (TDMA) to modulate information on the same carrier and supports both voice and limited speed data transfer. The most popular 2G system in Europe is Global System for Mobile communications (GSM). On the other hand, TDMA, which is also known as IS (Interim Standard)-136, is the most common 2G system in North America. In this chapter, we limit our discussion to that of the TDMA-based system when referring to 2G systems.

To accomplish higher data rates, add-ons were developed for GSM and TDMA. TDMA evolved to cdmaOne. CdmaOne is comprised of two standards: IS-95a and IS-95b. IS-95 networks use one or more 1.25-MHz carriers and operate within the 800-MHz and 1900-MHz frequency bands. Based on its data rate, IS-95b is categorized as a 2.5G standard. IS-95b was the first cellular system standard that implemented CDMA technique. As CDMA offers many benefits, such as high throughput, high capacity, transmit diversity, and allows the service provider to allocate resources with high flexibility, it has become the basis of the new generation of cellular systems, the so-called third generation (3G).

In 1999, the International Telecommunications Union (ITU) approved five radio interfaces for 3G systems. One of them is the improved version of IS-95, called CDMA2000 1x; 1x refers to the single 1.25-MHz carrier that is utilized to transmit both voice and data packets. Using the CDMA2000 1x standard, service providers should be able to support both high-speed data and carrier-quality voice on a single 1.25-MHz carrier. Transmitting voice and data on the same carrier is not trivial as they have different characteristics. For example, voice is more sensitive to delay than data and, conversely, data is more sensitive to error than voice. In the first release of CDMA2000 1x (Revision 0), voice and data are transmitted using the same channel, limiting the peak data rate supported to only 153 Kbps. For that reason, the ITU approved CDMA2000 1xEvolution (EV)-Data Only (DO) in 2001.

The first release of 1xEV-DO was designed to enable high data rate transmission of up to 2.4 Mbps on its forward link (FL) and 153 Kbps on its reverse link (RL) by dedicating a separate 1.25-MHz carrier for data traffic. If service providers would like to provide voice services on the 1xEV-DO platform, they can either use Voice-over-IP (VoIP) to integrate voice traffic to the same carrier as data or transmit the voice traffic on a separate 1.25-MHz carrier. However, VoIP has long latency and digital packetizing issues [2], which can be considered major problems in high-speed wireless communication. Transmitting voice and data on separate carriers is not desirable by service providers as it is not spectrally efficient. Hence, ITU has continuously pushed the evolution of the CDMA2000 1x standard since its first release. Enhancements are added on CDMA2000 1x layers to increase the spectrum capacity (i.e., higher data rates) and efficiency. The first release of CDMA2000 1x, Revision 0, was developed to Revision A then to Revision B. On Revision A and B, the forward fundamental channel (F-FCH) and the reverse fundamental channel (R-FCH) are used to transmit both voice and data at variable rates of up to 14.4 Kbps. If a higher rate is required, data packets can be transmitted over a supplemental channel (F-SCH/R-SCH) with a peak rate of 307.2 Kbps. The ITU approved CDMA2000 1x, Revision C, in 2002 and Revision D in 2003. Revision C and D are also known as CDMA2000 1xEV-Data Voice (DV). Revision C specifies a higher data rate on its FL (up to 3.1 Mbps), which is supported by the addition of a new channel called Forward Packet Data Channel (F-PDCH). On the other hand, a new Reverse Packet Data Channel (R-PDCH) is introduced in Revision D to increase the data rate on the RL (up to 1.8 Mbps). As part of the CDMA2000 family, 1xEV-DV is also designed to support multimedia services (video, data, and voice) with significantly higher rates on the same carrier. In response to the fast-growing numbers of wireless packet data service users, a new multiplexing technique, the time and code division multiplexing (TDM/CDM), is implemented on both FL and RL of

Table 8.1 Evolution of CDMA2000

Year	Standard			Peak Rate on FL	Peak Rate on RL
1995	CDMAOne		IS-95A	14.4 Kbps	14.4 Kbps
1998			IS-95B	64 Kbps	64 Kbps
1999	CDMA2000	1x	Release 0	153 Kbps	153 Kbps
2000		1x	Release A	307.2 Kbps	307.2 Kbps
2001		1xEV-DO		2.4 Mbps	153 Kbps
2002		1x	Release B	307.2 Kbps	307.2 Kbps
2002		1xEV-DV	Release C	3.1 Mbps	307.2 Kbps
2003			Release D	3.1 Mbps	1.8 Mbps

1xEV-DV. With TDM/CDM, the base station (BS) can serve more than one mobile station (MS) within one or more time slots. Therefore, it is clear that 1xEV-DV is known to greatly improve system capacity and enable wireless service providers to utilize their spectrum more efficiently. Note that the terms "MS," "user," and "subscriber" have the same meaning and are used interchangeably in this chapter.

The new features added to 1xEV-DV allow service providers to flexibly allocate the resources for voice and data traffic based on various considerations such as traffic load, revenue maximization, or user satisfaction. Further details regarding the enhancements are discussed in a later section in this chapter. Table 8.1 shows the development of cellular systems from 2G to 3G.

8.3   Performance Measures of Cellular System

8.3.1   Constraints on Network Resources

In one cell, BS is the main controller of the transmission process on both FL and RL. The scheduling decision made by the BS on the RL is highly dependent on the initial information obtained from MSs, such as the data rate request, reverse channel condition, and MS buffer status. In contrast, on the FL, MSs are less involved in the resource allocation process as the main information needed by BS from MSs is the forward channel condition. Whenever BS performs scheduling and resource allocation, it must consider the channel condition of each MS and the availability of network resources. The main resource constraints on FL are the limited amount of transmission power and the number of Walsh codes that can be allocated by BS to MSs. Conversely, on RL, due to the near/far problem, the received power at the BS from all MSs ideally should be uniform, and their sum should be below a certain threshold value. Hence, on FL and RL, there are different factors that must be taken into consideration while designing the scheduling and resource allocation schemes.

8.3.2 Performance Measures of Forward Link

Various scheduling and resource allocation methods have been proposed for the FL of a CDMA-based cellular system. Most of them focus on the earlier revisions of CDMA2000 1x, WCDMA, and CDMA2000 1xEV-DO. However, few works have been published on the FL of 1xEV-DV as EV-DV must be able to integrate high-speed data and carrier-quality voice transmission on a single channel, which is not trivial. As voice has been a major application in wireless communication for such a long time, there have been many well-founded results on voice transmission on CDMA-based systems. Nevertheless, as high-speed data transmission on integrated voice-data carriers is an area that has not been as well developed as voice transmission, we narrow our research to optimizing data transmission on the FL of 1xEV-DV.

Due to mobility, each MS has a time-varying and location-dependent channel condition. In 1xEV-DV, MSs use the pilot power, which is sent by several different BSs, to calculate the FL's signal-to-interference plus noise ratio (SINR). Afterward, the MSs send this information back to the BS with the best SINR value through Reverse Channel Quality Indicator Channel (R-CQICH) every 1.25 ms. In this chapter we consider that the SINR value, which represents the forward channel condition, is accurately determined. Opportunistic scheduling can be implemented to exploit the good channel condition in this multi-user system. In this chapter it is done by allowing the BS to transmit a larger packet size to users with a lower interference level. However, users that continuously have poor channel conditions will be at a disadvantage with this scheduling scheme. In [3–6], various methods were proposed to solve the fairness problems that arise in the cellular system implementing opportunistic scheduling. Kulkani and Rosenberg [3] and Panigrahi and Khaleghi [4] propose a throughput maximization scheduling technique for TDM-based systems. The short-term fairness is enhanced by giving a strict guarantee on the maximum starvation period experienced by each class of subscribers and each packet type. Kim [5] covers dynamic power and rate allocation for slotted CDMA systems based on the users' SINR feedback. The fairness among users is supported by controlling the rate allocation at the BS such that the measured latency (waiting time, until a packet of fixed slot length is received at the MS) is maintained between any prespecified maximum and minimum latencies [5]. The method proposed only allows the BS to serve one user in every slot with different transmission rates. In this current work, we aim to enhance short-term fairness in a multi-user, multi-rate slotted CDMA system by implementing different maximum delay limits for different subscription classes and packet types based on the service level agreements (SLAs) between subscribers and service providers. The reliability of channel quality feedback depends on the scheduling latency even if the feedback is error-free [6]. In our work, the scheduling delay is kept to a minimum by allowing the BS to update the MSs' information and resource availability status every 1.25 ms.

Various works have focused on determining the optimal power and data rate allocation on the FL of CDMA-based systems, with the goal of maximizing throughput [7–10] or the degree of utilization of network resources [11]. Vannithamby and Sousa [8] consider a slotted structure on a Wideband CDMA (WCDMA) system with homogeneous packet and user class. They allow variable transmission lengths and frames for packet transmissions. The number of slots is equal to the number of users requesting service [8]. In real-world applications, due to mobility, the number of MSs in the coverage area of a BS changes constantly. Moreover, the proposed method may introduce prolonged delay if scheduling is performed every M time slots, where M is the number of users at a scheduling instant. A static CDM-based resource allocation scheme with power and code constraints is proposed by Kwan and Leung [7]. They only consider a single class of user, and do not specify how the users are chosen. No fairness control is considered in their scheme. However, they show that throughput can be improved if traffic load is accounted for in the resource allocation scheme. The model includes the technique called Adaptive Modulation and Coding (AMC), which allows the use of different modulation schemes for supporting various data rates. According to Revision C standard, the FL of CDMA2000 1xEV-DV should be enhanced with AMC on its new F-PDCH. On F-PDCH, AMC is jointly used with the mixed TDM/CDM mode transmission, which allows the BS to support simultaneous transmissions to multiple users with variable transmission lengths and rates of up to 3.1 Mbps.

Agrawal et al. [9] propose a gradient-based scheduling scheme for a slotted structure system with utility used as the fairness constraint. Their utility is a function of each user's throughput, and their goal is to find the maximum weighted sum throughput for an FL channel where weights are determined by the gradient of utility. Joint resource allocation and base station assignment with the goal of maximizing throughput under maximum power and data rate constraints are analyzed in Lee et al. [10] who cover various important aspects affecting the performance of resource allocation on FL, such as retransmission, variable data rates, maximum data rate, transmission schemes, BS assignment, and MS selection strategy. However, they consider neither the history of previous transmission nor fairness control. Shabany and Navaie [11] propose downlink joint resource allocation and BS assignment scheme for a slotted multi-cell, heterogeneous user CDMA system with the objective of maximizing the degree of network resources utilization. They formulate the resource allocation scheme under a power constraint based on a dynamic pricing platform and predetermined utility function. The decisions regarding BS coverage area, BS assignment, and rate assignment are based on the traffic load and carried out on various time scales.

Although the technique proposed in Xu et al. [12] is dedicated to wireless systems in general, it can be implemented in CDMA-based cellular networks. They studied a dynamic power and rate allocation scheme for a system with differentiated services and user classes. They utilized online measurements to achieve the required Quality-of-Service (QoS) under a given pricing structure. At the BS, each packet type is stored in separate buffers and the allocation decision is made based on the available resources (effective bandwidth/rate) and queue length. The article takes into account the packet arrival history for predicting future traffic load and resources. The scheduling and resource allocation is performed using loss probability, average queue delay, and utility as the optimization metrics, regardless of the users' channel condition.

A dynamic resource scheduling scheme for the forward link of a 1xEV-DV system with variable packet types (voice, video, and data) has been proposed by Ci and Guizani [13]. They consider throughput, delay, and delay violation probability as their QoS measures. Although they also use the parameters specified in the 1xEV-DV standard, their scheduling method is designed only for forward transmission on a single F-PDCH, which can support just one MS. Hence, it is unknown how well the method will respond to the presence of multiple heterogeneous MSs in a cell. In this chapter we consider that two F-PDCHs will be used so that more than one MS can be supported at a time.

Resource allocation for data or voice service in the FL of a CDMA system with heterogeneous users has been studied in Zhou et al. [14,15]. The objective of the methods proposed in those articles is to set the price and power allocated, based on either revenue maximization or utility maximization.

To stay competitive in the telecommunications industry, while choosing the scheduling and allocation method for their network, service providers must consider how the method will benefit both themselves and their subscribers. The benefit gained by service providers is commonly represented by the revenue generated. On the other hand, the subscribers' satisfaction level can be measured by the data rate and the QoS they receive. After analyzing the presented schemes, we conclude that the resource management methods proposed in the past mostly put the benefits offered to one of the factions (either subscribers or providers) as the main objective and those provided to the other as the derivative. In contrast, our proposed scheme will benefit both service providers and subscribers, as the revenue generated by service providers will be considered as a part of the objective function along with the data rate. Our scheme also offers higher flexibility to the service providers. Simply by varying the values of two variables, service providers are given the options of prioritizing either revenue maximization or data rate maximization, or balancing both. Boariu [17] has also proposed a power and code allocation method that allows the trade-off reward and the throughput capacity in a flexible manner. The variable used to support the trade-off is the number of users that can be served simultaneously. Nevertheless, the Revision C standard [18] has ruled that the maximum number of users that can be served simultaneously is fixed at two. As a matter of fact, the results of Boariu's experiments also show that system performance is optimal when the BS is only allowed to serve a maximum of two users at a time.

Although not compulsory, industry ultimately must comply with the ITU-approved standard to ensure that their services and equipment are working well all around the world. Many of the methods presented above are designed based on the individual authors' perspectives. On the contrary, the parameters and techniques used in this chapter follow the general guidelines of the 1xEV-DV Revision C [18,19] and Revision D [20,21] standards. Therefore, our schemes could be easily implemented in industry.

8.3.3 Performance Measures of Reverse Link

The RL of 1xEV-DV is interference limited. In each time slot, because the transmission on the RL can be done in TDM/CDM mode, several MSs are allowed to transmit simultaneously. Thus, the transmission power from the other MSs is regarded as interference for one particular MS. The accumulated power of the received signals represents the level of interference at the BS. The RL design is to ensure that the rise-over-thermal (RoT), which is associated with the total received power at the BS, does not exceed the threshold level as long as there is RL data to be transmitted [22]. However, the transmission power is associated with a certain data rate. A higher data rate requires higher transmission power, which subsequently causes more interference for other users. The total interference in a sector is composed of the interference caused by the target user, the interference caused by the other users, and thermal noise. Because multiple MSs can be served simultaneously, it is crucial to schedule the transmission and control the power of each MS so as to maximize the resource utilization while keeping the RoT at the BS under the maximum threshold.

Several studies on the RL of CDMA2000 Revision D are reported in [23–26]. They analyze the throughput and QoS, such as delay, supported in Revision D. Derryberry and Pi [23] give an overview of the main enhancements on Revision D. Wu et al. [25] and Derryberry et al. [26] present system-level performance analysis according to the new RL enhancements. Wu et al. [25] investigate the RL QoS performance of CRC (Common Rate Control) and DRC (Dedicated Rate Control) modes. They use throughput and delay of various user classes as the performance measures. The experimental results show that the CRC and DRC algorithms will support different

QoS requirements, respectively. The RL scheduling and radio resource allocation for data traffic were studied as well by Pi et al. [23]. They analyze and compare the system performance of Revision D, in terms of throughput and delay, to those of the earlier revisions. Kwon et al. [24] introduce three basic radio resource management mechanisms: (1) rate-control (RC) scheme, (2) time-scheduled (TS) scheme, and (3) rate control with quick start (RCQS) scheme (a hybrid of RC and TS). They also evaluate and compare the performance of three radio resource management techniques in terms of average system throughput, packet delay, and the size of the overheads. We further discuss related contributions that are not specific for 1xEV-DV Revision D in the following paragraphs.

The works reported in [22,27–30] focus on improving system throughput with QoS consideration based on the Revision C standard. Zhang et al. [28] propose an optimal power allocation method based on greedy and fair policies so as to maximize system utility subject to peak transmit power, total received power from all data users, and a minimum SINR for each user. In [22,27,29], new scheduling and rate allocation schemes are proposed. Chung et al. [27] propose a scheme that enables dedicated rate control and group rate control with the one on the MSs' location. It can support a high average data rate, total amount of transmitted data, and short delay compared to other existing methods for Revision C. Pi and Derryberry [29] propose a scheduling and rate allocation scheme, taking advantage of multi-user diversity considering rate request, the allocated data rate, and the received power from mobiles while providing certain fairness among users. Chung and Cho [22] propose a transmission procedure considering users' location, speed, and the amount of data to be transmitted with the objective of improving average throughput and goodput. Teerapabkajorndet and Krishnamurthy [30] introduce a power control scheme based on the game theoretic framework for multi-cell wireless data networks.

Several analyses on the system performance in terms of throughput, capacity, and QoS-based metrics are presented in [31–33]. Derryberry et al. [31] use Revision C as the basis of their study. The system-level performance, such as throughput and outage, is observed under mixed voice and data scenarios. Yeo and Cho [32] evaluate the performance of IS-856 (also known as 1xEV-DO) RL rate control by modeling it as a Markov process. The throughput and the outage probability are employed as two performance metrics. Sarkar [33] concentrates on voice traffic and studies the CDMA2000 system capacity under various channel conditions.

The approaches in [34,35] are developed to improve RL capacity for general CDMA2000 systems. Damnjanovic et al. [34] simulates the scheduling on the RL of CDMA2000 1x and evaluates the performance in terms of throughput, interference, and fairness. A new RL bandwidth allocation scheme with the goal of reducing interference and increasing capacity for real-time applications is studied by Heyaime and Prabhu [35].

A traffic model for RL, taking into account the interference limitation and soft handoff, is constructed by Ashtiani et al. [36]. Due to its flexibility, the model is suitable for analyzing traffic and managing the handoff in a dynamic environment.

Based on our observations of the existing literature, we conclude that most of the work focuses on increasing the throughput, capacity, and fairness and on decreasing the delay, outage probability, and interference. Those parameters are important to support the QoS provided to subscribers. However, as mentioned in the previous section, it is also important to consider both service providers and subscribers in designing the scheduling and resource allocation scheme on the RL. Therefore, this chapter also develops a scheduling scheme for data transmissions on the RL that can balance the benefits gained by both subscribers and service providers.

8.4   System Model

8.4.1   Forward Link

Prior to CDMA2000 1xEV-DV Revision C, both data and voice packets were transmitted over F-FCH and F-SCH. As both channels adopt the CDM technique, one or more Walsh Code(s) is dedicated to each MS until the packet transmission is completed. F-FCH is used to transmit voice and data packets at variable rates of up to 14.4 Kbps. If the packet transmitted exceeds the rate limitation of F-FCH, the Supplemental Channel (F-SCH) can be used to accommodate the extra load at a range of data rates up to 307.2 Kbps. Those two channels are the major traffic channels that carry non-control packets. A new channel, called F-PDCH, is introduced in the FL of 1xEV-DV Revision C. F-PDCH is designed to support data packet transmissions with rates up to 3.1 Mbps. CDMA2000 1xEV-DV is backward compatible with CDMA2000 1x. This feature gives providers high flexibility as it allows the 1xEV-DV operator to provide voice services only, data services only, or mixed voice and data services to the subscribers. Voice traffic can be transmitted over the conventional F-FCH channel in a circuit switching manner. On the other hand, when only data services are required, F-PDCH is used on the traffic channel. Although F-PDCH is designed for high-speed data transmissions, it can also be used to transmit voice traffic whenever necessary. F-PDCH can be implemented in conjunction with the other channels of CDMA2000 1x. The requirements for the older traffic channels (e.g., F-FCH and F-SCH) do not change in 1xEV-DV. Hence, in this chapter, we focus attention on developing the resource management schemes for F-PDCH and integrating the F-PDCH with the existing channels.

   In a 1xEV-DV based system, a single forward traffic channel can support a maximum of two F-PDCHs. There are 1 to 28 Walsh codes, each with chip size of 32, available to be shared by two F-PDCHs. Each F-PDCH transmits information to one specific MS at a time [21]. One feature of 1xEV-DV that is not available in the previous versions of CDMA2000 systems is the mixed TDM/CDM technique. TDM/CDM can only be implemented on F-PDCH. With this technique, the BS can use both F-PDCHs to serve two MSs simultaneously with maximum total data rate of 3.1 Mbps. F-PDCH also supports three different transmission lengths: 1.25, 2.5, or 5 ms. Nevertheless, if the BS chooses to transmit to two MSs simultaneously, the same transmission length must be used for both transmissions.

In 1xEV-DV, each packet is turbo-encoded, and then divided into four sub-packets. Each sub-packet is differentiated by sub-packet identification (SPID). Because the first sub-packet (SPID = 0) has the most important information, it is commonly transmitted first. Subsequently, any other sub-packets can be sent if a retransmission is needed [37]. For experimental purposes, in the later numerical example in this chapter, only the first sub-packet (SPID = 0) is transmitted for every encoded packet, and all transmissions are considered successful.

8.4.1.1 Scheduling Problem Discussion

On the system level, service providers can further categorize subscribers into different classes based on the subscription fee paid. This categorization is then utilized to give each subscriber a priority level. Hence, upon entering a cell, MS i is assigned a benefit value (B_i), which represents the revenue gained by the service provider from MS i. To maintain short-term fairness, a compensation credit (μ_i) is assigned to MS i. If MS i does not get served at time slot t, its priority level will be increased by incrementing the value of a priority variable $\lambda_i(t)$ by one. Thus, the priority level ($L_i(t)$) of MS i can be represented by:

$$L_i(t) = B_i + \lambda_i(t) * \mu_i \qquad (8.1)$$

The subscribers categorization scheme specified above is illustrated in Table 8.2. (TS is transmission size, TLS is the transmission length, C refers

Table 8.2   Various Combinations Available for FL of 1xEV-DV

TS	TLS (Slot)	C	M	Data Rate (Kbps)
	1	3	16-QAM	326.4
	2	3	8-PSK	163.2
408 bits		1	16-QAM	81.6
		2	8-PSK	81.6
	4	4	QPSK	81.6
		2	16-QAM	163.2

Table 8.3 Subscription Class Categorization

Subscription Class	Packet Type	B_i	μ_i	Max_λ_i (Slots)
	RT	100	0.5	220
Class 1	NRT	80	0.09	224
	Voice	100	1	180
	RT	95	0.5	215
Class 2	NRT	70	0.09	214
	Voice	90	1	170
	RT	90	0.5	210
Class 3	NRT	60	0.09	204
	Voice	80	1	160

to the number of codes, and M is the modulation mode). Delay bound (γ_i) is introduced to limit the transit delay that can be tolerated by the packet requested by MS i. In Table 8.3, this delay limit is represented by Max_λ_i, and the relationship between γ_i and Max_λ_i is:

$$Max\_\lambda_i = B_i + (\gamma_i) * \mu_i \qquad (8.2)$$

If at t, $L_i(t)$ is equal to Max_λ_i, MS i must be served with all the available resources. We realize that in using the compensation scheme presented above, there is a possibility that many users will reach the delay limit at the same time. If the BS does not have enough resources to serve these MSs, they will be dropped. This is very undesirable, as it may cause bottleneck problems at the BS and high dropping rates in a system with a large number of users. Therefore, the B_i and μ_i values must be chosen carefully, depending on the network environment. Further research could be done in determining these values.

The scheduling problem in the FL can be summarized as follows:

■ **Objective:** To choose the MSs based on their priority level. This, in turn, will maximize the revenue obtained by the service provider.
■ **Constraint:** The maximum delay bound for each MS must be obeyed so that short-term fairness and high QoS level are enhanced.

8.4.1.2 Resource Allocation Problem Discussion

The FL of CDMA2000 1xEV-DV supports data packet transmission with various data rates. As a matter of fact, Revisions C and D have specified the combinations of physical layer parameters (such as number of Walsh codes, modulation order, and transmission length) that identify a specific data rate. Table 8.2 provides examples of the various combinations for a transmission size of 408 bits. Nonetheless, the standards do not specify how the combination should be chosen to support upper-layer QoS, such as delay, packet loss, and Bit-Error-Rate (BER). Because our scheduling scheme should be able to determine these parameters, we must first identify the correlation between these parameters and the QoS supported.

The physical layer standard of 1xEV-DV Revision C [18] specifies six fixed encoder packet sizes: 408 bits, 792 bits, 1560 bits, 2328 bits, 3096 bits, and 3864 bits. The FL of 1xEV-DV also supports various transmission rates, ranging from 81.6 Kbps to 3.1 Mbps, with the use of the AMC scheme. The coding scheme refers to the error correction coding scheme used to combat the presence of data corrupting interferences. Only one coding scheme can be implemented on F-PDCH: that is, Turbo Coding with the rate of 1/5. The modulation schemes available, which are QPSK, 8-PSK, and 16-QAM, can be used interchangeably, depending on the channel condition. If an MS has good channel quality, the BS can use higher modulation order to serve that MS with a high data rate and low amount of power and codes used. Each modulation mode has a minimum SINR requirement to guarantee a certain BER value. For the FL numerical examples, we consider the minimum SINR for QPSK, 8-PSK, and 16-QAM at 4, 7, and 10 dB, respectively. Note that these values are just approximations based on the values required by other coding techniques [38,39], as no exact values have been established yet for Turbo Coding with the rate of 1/5.

Let t represent the slot instant when the scheduling and allocation is performed. The BS uses the result of the scheduling algorithm calculation at t to allocate the resources at $t + 1$. $h_i(t)$ is the path gain of transmission from the BS to MS i. $P_i(t)$ is the transmission power from the target BS to MS i, $P_o(t)$ is the transmission power from other interfering BSs, and $h_{oi}(t)$ is the path gain of transmission from other BSs to MS i. $I_B(t)$ is the background thermal noise and $s_i(t)$ characterizes the number of Walsh codes assigned to MS i. The SINR per code model can be represented by:

$$SINR_i(t) = \frac{1}{\eta_i(t)} * \frac{h_i(t) * P_i(t)}{\left(\sum_{o \neq i} P_o(t) * h_{oi}(t)\right) + I_B(t)} \quad (8.3)$$

The resources (transmission power and number of Walsh codes) available at the BS are limited. As mentioned previously, the maximum total code that can be used at the BS is 28. The maximum power available depends on the physical structure of the BS. The power required by each transmission can be obtained from Equation 8.3. Thus, Equation 8.3 can be rewritten as:

$$SINR_i(t) = \frac{1}{\eta_i(t)} * \frac{P_i(t)}{\frac{\left[\sum_{o \neq i} P_o(t) * h_{oi}(t)\right] + I_B(t)}{h_i(t)}}, \quad (8.4)$$

The interference can be represented by the following equation.

$$\rho_i(t) = \frac{\left[\sum_{o \neq i} P_o(t) * h_{oi}(t)\right] + I_B(t)}{h_i(t)}, \tag{8.5}$$

The transmission power required for MS i can be expressed by the following equation.

$$P_i(t) = \eta_i(t) * SINR_m * \rho_i(t), \tag{8.6}$$

where $SINR_m$ is the minimum SINR requirement for modulation order m to guarantee a certain BER.

Thus, the resource allocation problem can be summarized as follows:

- **Objective:** To maximize the total data rates generated.
- **Constraints:**
  - Among many MSs requesting service simultaneously, the BS can only serve, at most, two of them simultaneously.
  - For each MS, the maximum number of combinations that can be chosen is one.
  - The maximum power and number of Walsh codes available at the BS are limited.
  - If the BS chooses to transmit to two users, it must use the same transmission length for both transmissions.

8.4.1.3 Formulation on the Forward Link

The scheduling and resource allocation problems on the FL can be combined and formulated into the following equations:

- **Objective:**

$$Max \sum_{i=1}^{N(t)} \sum_{v=1}^{K_i(t)} [(\theta_1(t) * L_i(t)) + (\theta_2(t) * Rate_{i,v}(t))] * y_{i,v}(t) \tag{8.7}$$

where i refers to the MS and v represents the combination number. $K_i(t)$ is the total number of combinations allocated to MS i and $N(t)$ is the number of MSs requesting service. $\theta_1(t)$ and $\theta_2(t)$ are the unit converters (weights) and their values depend on whichever is more important for the service providers, the revenue generated or data rate provided. The data rate generated by choosing (i, v) is symbolized by $Rate_{i,v}$. The value of decision variable $y_{i,v}(t)$ can either be zero or one. If $y_{i,v}(t)$ is equal to one, MS i with combination v is chosen. In summary, the goal of this scheduling and allocation scheme is to pick a set of $\{i,v\}$ that can maximize the objective function specified in Equation 8.7.

■ **Subject to the following constraints:**

1. The maximum number of MS can be served simultaneously is two:

$$\sum_{i=1}^{N(t)} \sum_{v=1}^{K_i(t)} y_{i,v}(t) \leq 2 \qquad (8.8)$$

2. For each MS, only one combination of power, code, and slot can be chosen:

$$\sum_{v=1}^{K_i(t)} y_{i,v}(t) \leq 1 \qquad (8.9)$$

where $i = \{1, 2, \ldots, N(t)\}$

3. Total power and code used should be less than the maximum power and code available at the BS:

$$\sum_{i=1}^{N(t)} \sum_{v=1}^{K_i(t)} Num_{i,v,z}(t) * y_{i,v}(t) \leq MaxNum_z(t) \qquad (8.10)$$

where z is the constraint number and $z = \{1, 2, \ldots, q\}$. Consequently, q = maximum number of constraints used. As in this case, the constraints used are power ($P_{i,v}(t)$) and code ($C_{i,v}(t)$); thus, $q = 2$. $Num_{i,v,z}$ is the weight of MS i, combination v, constraint z. If $z = 1$, then $Num_{i,v,1}(t) = P_{i,v}(t)$; and if $z = 2$, then $Num_{i,v,2}(t) = C_{i,v}(t)$. $MaxNum_{i,v,z}(t)$ is the maximum number of constraint z. Thus, if $z = 1$, then $MaxNum_1(t)$ represents maximum power available ($MaxP(t)$). Similarly, if $z = 2$, then $MaxNum_2(t)$ represents maximum code available ($MaxC(t)$).

4. Transmission length used for two simultaneous transmissions must be the same.

8.4.2 Enhancements on Reverse Link 1xEV-DV

In this section we briefly discuss the major enhancements on the RL of 1xEV-DV Revision D. The goals of RL design in 1xEV-DV are to increase the RL peak rate (up to 1.8 Mbps) and sector throughput (up to 600 Kbps), and maintain backward compatibility as well. Revision D optimizes utilization of the RL resource which is represented by sector load or RoT. To meet the new requirements of 1xEV-DV, three major RL channels introduced in Revision D [20] for enhancing the high data rate can be explained as follows:

1. Reverse Packet Data Channel (R-PDCH): a reverse traffic channel transmits higher-level data and control information from an MS to a BS. The data rates can be used on R-PDCH with a 10-ms frame size are 19.2, 38.4, 79.2, 156, 309.6, 463.2, 616.8, 924, 1231, 1538.4, and 1845.6 Kbps.
2. Reverse Packet Data Control Channel (R-PDCCH): a control channel to transmit control information for the sub-packet that is being transmitted on the R-PDCH and the mobile status indicator bit.
3. Reverse Request Channel (R-REQCH): a control channel for the MS to report available maximum power that an MS can support and MS's buffer status.

Several FL enhancements introduced in Revision D to support the high-speed transmission on R-PDCH include:

1. Forward Grant Channel (F-GCH): a control channel used by the BS to transmit transmission control information on the R-PDCH to the MSs.
2. Forward Indicator Control Channel (F-ICCH): a control channel used by the BS to transmit rate control messages to the MSs.

8.4.2.1 Transmission Characteristics

According to CDMA2000 standard Revision D, the modulation format of R-PDCH can be BPSK, QPSK, or 8-PSK. The modulation depends on the encoder packet size (EPsize), which is specified as 192, 408, 792, 1560, 3096, 4632, 6168, 9240, 12312, 15384, or 18456 bits. BPSK modulation is used by the MS for encoder packet sizes of 192, 408, and 792 bits. QPSK modulation is used for encoder packet sizes of 1560, 3096, 4632, 6168, 9240, 12312, and 15384 bits. 8-PSK modulation is applied for an encoder packet size of 18456 bits [20]. A time slot is equal to the duration of a sub-packet, which is the duration that each MS receives rate control command from the BS and updates its transmission rate. 1xEV-DV allows an encoder packet to be transmitted using from one to three 10-msec sub-packets. An R-PDCH frame starts only when the system time is an integral multiple of 10 msec.

The allowable RL transmission rates on R-PDCH can be denoted by a finite rate set $R = \{9.2, 19.2, 40.8, \ldots, 1845.6\}$ Kbps. Consider a cell that has N mobiles connected to it. Let the transmission rate of mobile i be j and $j \in R$. The reverse link transmission rate is associated with a nominal transmission power to pilot power ratio (TPR). TPR expresses the total power transmitted relative to the pilot transmit power, a quantity referred to as RPDCH_TPR_Table in the standard [20]. The RoT value that represents the utilization level on the RL can be calculated using the following equation:

$$Z(t) = 10 * log_{10} \left(1 + \sum_{i=1}^{N} \frac{TPR_i(t) * P_{pilot}(t)}{N_0 * W} \right) \tag{8.11}$$

where $TPR_i(t)$ denotes the ratio of the total transmission power to the pilot power of mobile i in the time slot [t, t+1], and $P_{pilot}(t)$ denotes the current transmit pilot power of mobile i in the time slot [t, t+1]. W is the chip bandwidth, N_0 is the thermal noise density, and $N_0 * W$ represents the background noise power (including the intracell interference) in Watts.

Each MS maintains an authorized TPR table, and updates its transmission power based on the rate control command (up/hold/down) from the BS through F-ICCH. Table 8.4 shows the default R-PDCH Nominal Attribute Gain (TPR) [20].

8.4.2.2 Rate Control Schemes

The forward rate control channel can be used by the BS to transmit rate control information for controlling the data rates on R-PDCH from one or more MSs. The MS operates in DRC mode when a single R-PDCH is controlled by a rate control subchannel, while it works in CRC mode when multiple R-PDCHs are controlled by one rate control subchannel [20]. The rate control commands are sent from BS every time slot. The two rate control modes discussed above operate as follows:

1. CRC mode: MSs in a sector or a group of MSs are controlled by a common, one-symbol command that represents up/down/hold.
2. DRC mode: individual MS is controlled by a dedicated one-symbol command that represents up/down/hold.

Table 8.4 Default R-PDCH Nominal Attribute Gain

Rate Index	R-PDCH Rate (Kbps)	Nominal TPR (1/8 dB)
1	19.2	6
2	40.8	30
3	79.2	54
4	156.0	77
5	309.6	95
6	463.2	109
7	616.8	119
8	924.0	133
9	1231.2	144
10	1538.4	153
11	1845.6	170

8.4.2.3 System Model Operation

System design depends on the physical layer structure. Based on new en-hancements of the physical layer architecture in Revision D, the RL trans-mission procedure and scheduling operation can be described as follows. In each time slot, each MS sends transmission rate request information, which is based on authorized transmission power, along with the infor-mation about the amount of data to transmit and the buffer size to the BS through R-REQCH. The BS receives the request signal from all the MSs. The total RoT contribution to the BS is computed. Our scheduler determines scheduling and rate allocation for each MS using the proposed algorithms. The grant message is sent back to each MS through F-ICCH. The authorized MSs then transmit high-speed data traffic with allocated data rate through R-PDCH. The modulation, coding, and HARQ (Hybrid Automatic Repeat re-Quest) information are conveyed through R-PDCCH and at the same time used to help the receiver decode the data delivered on R-PDCH.

8.4.2.4 Formulation on the Reverse Link

The resource allocation problem is to select which users to be scheduled for transmission in order to maximize revenue and minimize loss subject to the RoT limitation. Therefore, we expect to select users who can generate more revenue as well as have more packets transmitted in the buffer. We formulate this into a knapsack problem with multiple constraints. The model can be expressed as:

■ **Objective:**

$$Max \sum_{i=1}^{N(t)} \sum_{j=R_{min}^i(t)}^{R_{max}^i(t)} \left(r_{i,j}(t) + k_i(t) * \frac{\alpha_i(t)}{\beta_i(t)} \right) * x_{i,j}(t) \qquad (8.12)$$

■ **Subject to two constraints:**

$$\sum_{i=1}^{N(t)} \sum_{j=R_{min}^i(t)}^{R_{max}^i(t)} (w_{i,j}(t) * x_{i,j}(t)) \leq RoT\ Limitation \qquad (8.13)$$

$$\sum_{j} x_{i,j}(t) \leq 1 \quad \forall i \qquad (8.14)$$

At any time slot [t, t + 1], $r_{i,j}(t)$ is the obtained revenue from mobile i transmitting at data rate j; $k_i(t)$ is objective function balancing parameter for user i, which can be defined by service providers based on their concerns (revenue or loss); $\alpha_i(t)$ denotes the amount of data of MS i to be transmitted; $\beta_i(t)$ denotes buffer size of MS i; $w_{i,j}(t)$ is the weight (contribution to the RoT) of MS i transmitting at data rate j; $N(t)$ is the total number of MSs; $R^i_{min}(t)$ is the minimum available data rate to transmit for MS i; $R^i_{max}(t)$ is the maximum available data rate to transmit for MS i; $x_{i,j}(t)$ is the decision variable, which can be represented as a binary number 0 or 1, $x_{i,j}(t) = 1$ if MS i transmitting at rate j is selected, otherwise $x_{i,j}(t) = 0$.

Definition: Weight $w_{i,j}(t)$ is defined as the power contribution of mobile i to the RoT value. We assume the BS has perfect power control. The pilot power from each MS received at the BS is exactly $P_r(t)$. Thus, the total transmissions power of MS i with transmit rate j can be represented as:

$$w_{i,j}(t) = P_{i,j(total)}(t)$$

$$= TPR_{i,j}(t) * P_r(t) \tag{8.15}$$

where $TPR_{i,j}(t)$ denotes the TPR value of mobile i with transmit rate j.

In the objective function (Equation 8.12), $k_i(t)$ is determined based on QoS requirements. Maximizing $k_i(t)(\alpha_i(t)/\beta_i(t))$ implies minimizing loss. $k_i(t)$ is given based on the service provider's expectation. Additionally, parameter $k_i(t)$ can be dynamically adjusted according to an individual user's situation. For example, it can be increased when the buffer is almost full, that is, $\alpha_i(t)/\beta_i(t)$ close to 1. Conversely, it can be decreased when the buffer is nearly empty; that is, $\alpha_i(t)/\beta_i(t)$ close to 0. The greater the value of $k_i(t)$, the more QoS concerned and the less lost opportunities obtained. Hence, larger $k_i(t)$ can improve the QoS, but may cause a certain revenue loss. Note that $k_i(t) \geq 0$.

8.5 Proposed Solutions

8.5.1 Solutions for the Forward Link

The combined scheduling and resource allocation problem can be categorized as a binary integer linear programming (LP) problem with multiple constraints. There are various methods that can be used to obtain the solution for the binary integer LP problem. Branch and Bound (B&B) and Complete Enumeration (CE) are the two exact algorithms designed to generate the optimal solution. A heuristic algorithm, which commonly generates a suboptimal solution, could be created if the results of exact algorithms are not satisfactory. In our case, we create a heuristic (HEU) algorithm because the processing times required by exact algorithms are significantly high. The results of the experiments using all three algorithms are presented in the next section. First, let us take a quick look at the description of each algorithm.

The B&B algorithm operates by relaxing the initial candidate solution sets and subsequently constructing a tree of possible sets. The candidate solution sets are the $\{i, v\}$ that have the same transmission length and satisfy Equation 8.8, Equation 8.9, and Equation 8.10. The optimal binary integer solution is found by keeping the best solution discovered as the algorithm performs an exhaustive search over all of the branches created. The B&B discards the sets that do not improve the best solution. The B&B algorithm would search all 2^δ nodes, where δ in our case is the total number of combinations assigned to all MSs in that scheduling instant.

The CE algorithm solves the optimization problem by trying all possible combinations of discrete variable values. It initially creates a list of all candidate solution sets ($\{i, v\}$s). Following that, the CE algorithm calculates the objective value generated by each $\{i, v\}$ as it constructs the list. Afterward, the algorithm sorts the list in decreasing order based on the objective value. The optimal solution is essentially the one located on the top of the list. As it is possible for several candidate solutions to have identical objective values, a new rule can be introduced to define the best solution. In the current example, the solution is considered optimal if it consumes the least amount of transmission power.

The HEU algorithm is a simple ratio-based algorithm, created with the goal of minimizing the computational time required. Initially, HEU rules out the (i, v) pairs that do not satisfy the power and code constraints as candidate solutions. It then calculates the ratio of the objective value over the amount of resources required by the remaining pairs of (i, v). The group of (i, v) pairs is then sorted in decreasing order based on the ratio. The (i, v) on the top of the list is chosen as the first solution. Afterward, HEU searches down the list to find the second (see Equation 8.8) pair of (i, v) that also satisfies the other constraints. The process is repeated until we obtain several sets of (i, v) pairs. The $\{i, v\}$ with the best total objective value is then chosen as the final solution.

8.5.1.1 Performance Analysis of B&B, CE, and HEU Algorithms

All experiments in this chapter are implemented and run on SunFire 480 Server with four 900 MHz, UltraSparc-III processors and 16 GB RAM.

In this section, all three algorithms are utilized to solve the problem for one scheduling instant (static). Hence, $t = 1$. The traffic from a target BS in an unsectored cell is analyzed. For this static scheduling example, assume that the BS can only transmit with transmission length of one time slot (1.25 msec). Note that this assumption reduces the number of possible combinations that can be assigned to each MS, thus decreasing the number of candidate solutions. For the static example, only data and video packets are considered present. As in this example, the scheduling is only performed for one time slot, the assignment of encoder packet size

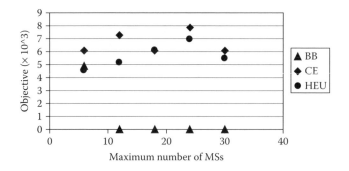

Figure 8.1 Objective value versus number of MSs.

depends only on the MSs' SINR value and packet type. The encoder packet size options available for video users are 1560, 2328, 3096, and 3864 bits. Data users are assigned smaller encoder sizes: 408, 792, 1560, and 2328 bits. Each MS is assigned random packet types and subscription classes. The maximum number of MSs that exist in the system ($N(t)$) is varied from 6 to 36.

Recall that the scheme is initially designed to perform dynamic scheduling every 5 msec, at most. Hence, the processing time of each algorithm must be kept to a minimum at all times. In this example, we consider the worst-case scenario and set the upper bound for computational time to 10 seconds. If the algorithm is unable to finish the processing task in 10 seconds, it must stop the search and return whatever solution it can find. *bintprog*, which is a built-in B&B MATLAB function, is used in all examples with the B&B algorithm. The results are shown in Figure 8.1 and Figure 8.2. The processing times for B&B, CE, and HEU recorded in Figure 8.2 include the additional program setup time.

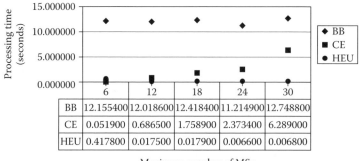

Figure 8.2 Processing time versus number of MSs.

Figure 8.1 shows that the objective values generated by B&B are not optimal for all $N(t)$s. Apparently, *bintprog* has to terminate the program as the maximum processing time allowed is exceeded and return the best solution that it can find at that point in time. When the number of MSs increases, B&B is unable to find any solution within 10 seconds and return the objective values of zeros. On the other hand, CE always gives the optimal objective values at all time. Nonetheless, the computational time needed by CE to solve the problem increases exponentially with number of MSs. As expected, the objective values generated by HEU are sub-optimal in most cases compared to CE. However, as shown in Figure 8.2, the computational time of HEU algorithm is independent on the number of MSs.

8.5.1.2 Dynamic Scheduling

In the next example, CE and HEU are implemented to perform dynamic scheduling and resource allocation in the presence of variation in network environment. The goal of the experiment is to analyze if the proposed scheme is capable of adjusting to the changes while maintaining desirable results. For this dynamic experiment, we consider three packet types: data, video, and voice. If admitted, each voice user will be assigned one Walsh code. We assume that voice is prioritized over data and video. Data and video traffic can only be assigned the Walsh codes that are not used by voice. Nevertheless, to maintain the high rate for data and video, the number of Walsh codes on F-PDCH that can be used by voice users must be limited. In this dynamic experiment, the maximum number of Walsh codes for voice users is set to 4.

There can be a maximum of 15 MSs ($N(t) = 15$) present in the cell, and each MS is assigned a random SINR value, subscription class, and packet type. The number of voice users varies from one to five. We consider that there are always packets waiting at the BS for all MSs (infinite buffers).

The encoder packet sizes that can be assigned to video and data are the same with static scheduling, which are 1560, 2328, 3096, 3864 bits and 408, 792, 1560, 2328 bits, respectively. In this experiment, we set the upper delay bound (in-transit delay) for voice to be 100 msec. On the other hand, the upper delay bound for video and data are relaxed to 200 msec and 2 seconds. The values assigned to each subscription class are specified in Table 8.3. When MS i exceeds the delay bound, its packet will be dropped. The encoder size of the packet dropped is considered 1560 for video and 408 for data. The encoder packet size assignment depends on the MSs' SINR, packet type, and delay status. Taking delay status into consideration is expected to reduce the packet loss probability.

The detailed parameters for this dynamic example can be found in Table 8.5. The results are shown in Figure 8.3, Figure 8.4, Figure 8.5, Figure 8.6, Figure 8.7, and Figure 8.8.

Table 8.5   Parameters for Dynamic Example

Bandwidth	1.25 MHz
Modulation	QPSK ($SINR_m = 4$ dB) 8-PSK ($SINR_m = 7$ dB) 16-QAM ($SINR_m = 10$ dB)
Chip rate	1.2288 Mcps
Maximum power	20 Watts
Maximum codes for data and video	28
Maximum codes for voice	4
Transmission length options	1.25, 2.5, or 5 msec

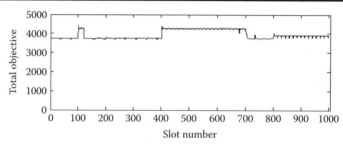

Figure 8.3   Objective, HEU with $\theta_1 = 12$ and $\theta_2 = 1$.

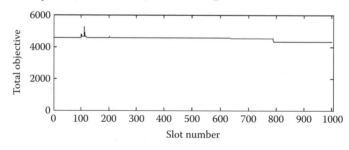

Figure 8.4   Objective, CE with $\theta_1 = 12$ and $\theta_2 = 1$.

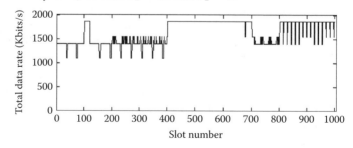

Figure 8.5   Data rate, HEU with $\theta_1 = 12$ and $\theta_2 = 1$.

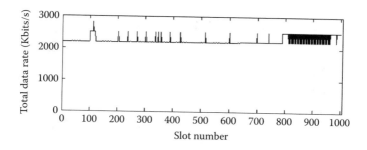

Figure 8.6 Data rate, CE with $\theta_1 = 12$ and $\theta_2 = 1$.

Both HEU and CE generate competitive objective values, with very little fluctuation over the entire 1,000 time slots. The revenues generated by HEU in Figure 8.7 are comparable to those generated by CE in Figure 8.8. Near the end of the experiment, the total priority level of both CE and HEU are notably lower than the earlier ones. We predict that the decline is due to the short-term fairness scheme that we implement. At the beginning of simulation, the BS would serve the MSs that have high priority levels. After a certain period of time, when the MSs do not get served, their priority value would increase. On top of that, their delay values would also be accumulated. Recall that the encoder packet assignment also depends on the MSs' delay status. Hence, those MSs with higher delay values would be assigned the combinations with higher encoder packet sizes. Consequently, this would increase the data rates supported and thus raise their chance of being chosen at that scheduling period.

From Figure 8.5 and Figure 8.6 we can see that near the completion of the experiment, the data rates generated are slightly higher than the previous ones. This shows that the scheme probably would allocate more resources to serve the MSs with high delay.

Table 8.6 and Table 8.7 show the numerical results of dynamic experiments with various θ_1 and θ_2 values. When θ_1 is set to 4 in Table 8.6, the priority given to service providers is lower than that of Table 8.7. Although

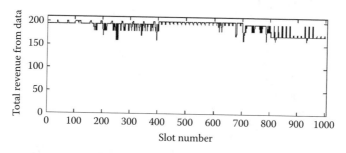

Figure 8.7 Revenue, HEU with $\theta_1 = 12$ and $\theta_2 = 1$.

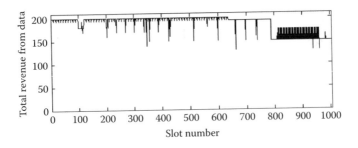

Figure 8.8 Revenue, CE with $\theta_1 = 12$ and $\theta_2 = 1$.

the average rate generated by HEU is less than that of CE, HEU still manages to maintain the data rate that is over the average sector throughput recommended by the ITU, that is, 1 Mbps. However, the results might change when we consider the presence of packet error and retransmission.

Although in most cases the delay experienced by video and data packets is slightly higher when the scheduling is performed using HEU, it is still much lower than the upper bound assigned. There is no packet dropped in all three experiments. Note that the delay for data traffic for both CE and HEU increases as θ_1 increases. This is rather unexceptional as video users offer higher revenue than data users and would further be prioritized as the weight of revenue increases.

The results in Table 8.6 and Table 8.7 show that both algorithms are responding well to the change in θ_1 and θ_2 values. Hence, it confirms that both algorithms are capable of supporting the goal of our proposed scheme, that is, to offer high flexibility to the service providers. With our scheme, service providers are allowed to prioritize either the service provider or

Table 8.6 Numerical Results for Simulation over 1,000 Slots (1.25 msec)

Performance Measure	HEU	CE
Average objective values	2479.9	3063.8
Average data rate	1.8 Mbps	2.4 Mbps
Total packets transmitted	2.3 Mbits	3 Mbits
Video transmitted	2 Mbits	1.5 Mbits
Data transmitted	0.3 Mbits	1.5 Mbits
Total voice, video, data packets lost	0	0
Total revenue from data and video	163435	172350
Total revenue from voice	82009	81784.5
Average video delay (Max = 200 msec)	4.6 msec	4.6 msec
Average data delay	32.7 msec	5.12 msec
Average processing time per slot	0.029 sec	0.665 sec

Note: $\theta_1 = 4$ and $\theta_2 = 1$.

Table 8.7   Numerical Results for Simulation over 1,000 Slots (1.25 msec)

Performance Measure	HEU	CE
Average objective values	3941.5	4510.2
Average data rate	1.63 Mbps	2.24 Mbps
Total packets transmitted	2.04 Mbits	2.8 Mbits
Video transmitted	1.86 Mbits	2.11 Mbits
Data transmitted	0.177 Mbits	0.683 Mbits
Total voice, video, data packets lost	0	0
Total revenue from data and video	190190	188335
Total revenue from voice	81848.5	82070.5
Average video delay (Max = 200 msec)	4.04 msec	2.88 msec
Average data delay	56.75 msec	19.13 msec
Average processing time per slot	0.037 sec	0.691 sec

Note: $\theta_1 = 12$ and $\theta_2 = 1$.

the subscriber, or balance both. Nevertheless, service providers should be careful in choosing the values of θ_1 and θ_2 so as not to lose the major advantage offered by this scheme, that is, to satisfy themselves and their subscribers. HEU also manages to keep the average processing time per slot on the order of milliseconds at any time. On average, HEU is 20 times faster than CE. Therefore, HEU would work very well in a situation where the processing time is highly crucial.

8.5.2   Solutions for the Reverse Link

8.5.2.1   Algorithm Description

1. **B&B Solution:** The B&B method is one of the existing algorithms to solve the binary integer linear programming problem. Due to the node searching strategy, the B&B algorithm presents longer computational time when more users send requests. Thus, it is not efficient for the model and obviously cannot meet the requirement of scheduling time. To cope with this problem, two heuristic algorithms are proposed.

   We let F_i be a binary integer feasible solution of objective function at an intermediate rate of mobile i, and C_i be the first node objective relaxation value at an intermediate rate of mobile i. Then the difference between F_i and C_i is defined as the objective value gap G_i using following equation:

   $$G_i = \frac{(C_i - F_i)}{C_i} \tag{8.16}$$

   G_i is used to compare the performance over three solutions.

2. **HAI (Heuristic Algorithm I):** The user with high unit price plan, low weight, and large amount of data in the buffer will have a greater chance of being scheduled. For each user i with transmission rate j, HAI calculates the expected achieved revenue $r_{i,j}$, $w_{i,j}$, (α_i/β_i), and ratio $R_{i,j}$. The ratio function is defined as:

$$R_{i,j} = \frac{r_{i,j} + k_i * \left(\frac{\alpha_i}{\beta_i}\right)}{w_{i,j}} \qquad (8.17)$$

Consider that the RL system works in DRC mode. In DRC mode, the BS can allocate the rate that is, at most, one level higher or, at most, two levels lower than the initial rate requested by the MS. Hence, for each MS, the BS has four rate options. If N is the total number of users in the cell, the possible number of combinations (i, j) available to be chosen by the BS is 4*N. HAI starts by listing all possible combinations of (i, j). It then sorts the list in decreasing order based on the ratio value $(R_{i,j})$. Afterward, HAI starts to choose the (i, j) from the top of the sorted list to the end and only keeps the (i, j) combinations that satisfy Equations 8.13 and 8.14.

3. **HAII (Heuristic Algorithm II):** HAII is an improved version of HAI. After running several experiments using HAI, we notice that the objective values generated by HAI are sometimes lower than those of B&B. This is due to the fact that the selection process in HAI is purely based on the ratio values. However, the ratio value is not always proportional to the objective values generated for (i, j). Note that the (i, j) combinations in the sorted list can be further classified based on their ratio values. For example, there are X ratio groups in the list. In HAI, the algorithm always starts the selection process from the first ratio group (i.e., the group with the highest ratio value). HAII uses the same concept as HAI. However, it performs a more thorough search over the combinations using starting points from different ratio groups to run HAI. Thus, HAI is run X times and in the end, HAII generates X groups of (i, j) combinations that have different objective values. The group of (i, j) combinations with the best objective value is chosen as the solution of HAII.

8.5.2.2 Illustrative Example Results

The performance is evaluated in terms of average objective value gap, average RoT consumption, and average processing time for scheduling. As a balancing parameter, k_i is introduced in the objective function, the behavior of parameter k_i is also discussed through experiments.

The assumptions used in this section include:

1. MSs always have enough power and enough data to transmit at their highest transmission rates.
2. The pilot power from each MS received at the BS is $P_r = -80$ dBm.
3. The threshold of RoT $= 7$ dB, chip bandwidth $W = 1.23$ MHz, $N_0 = -60$ dBm$/1.23$ MHz.
4. 0 to 20 data users in a single cell.
5. Users' requests follow a uniform distribution.
6. Time slot $= 10$ msec.
7. k_i is randomly selected between 1 and 50; (α_i/β_i) is randomly selected between 0 and 1.

We ran five experiments with 5, 7, 10, 15, and 20 users sending requests in one cell in one time slot. In each scenario, 20 sets of experimental data were collected for performance evaluation.

Figure 8.9 shows the average processing time required by all three algorithms. We observe that the processing time of the B&B solution increases significantly as there are more users in a cell. On the contrary, the processing time of HAI and HAII appear to be insensitive to the number of users. The processing time of HAII is about 2.5 times higher than that of HAI. Based on the 10-msec constraint of the scheduling period, the B&B solution becomes unrealistic when more mobiles send requests at the same time. This is expected. However, the proposed heuristic algorithms HAI and HAII meet the scheduling time requirement and HAI works better.

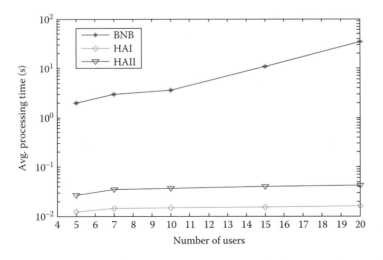

Figure 8.9 **Average processing time versus number of users sending requests.**

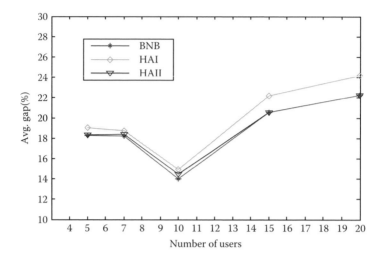

Figure 8.10 Average gap versus number of users sending requests.

The average objective function value gap versus the number of users for three algorithms is shown in Figure 8.10. The curves for B&B and HAII are closer in most of the scenarios and work better than HAI.

Figure 8.11 shows the RoT utilization of three solutions. When more than five users request to transmit high-speed data simultaneously, the RoT resource is highly utilized for all three algorithms. In this case, the three algorithms are approximately equivalent.

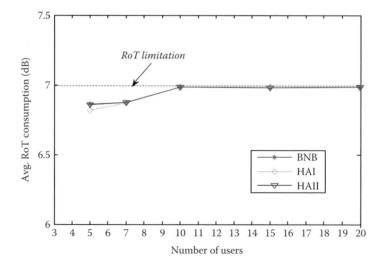

Figure 8.11 Average RoT consumption versus number of users sending requests.

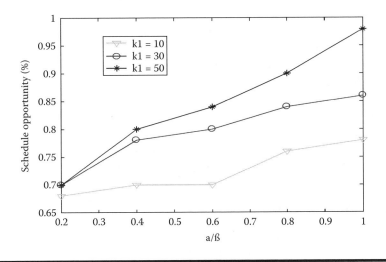

Figure 8.12   Schedule opportunity versus buffer status (α_i / β_i).

Because the three solutions achieve similar performance in terms of objective value gap, any one of them can be employed to study the behavior of objective function balance parameter k_i. Here we choose HAI due to its fast processing. Figure 8.12 presents the schedule opportunity in the scenario of given various values of k_i and α_i/β_i. The scheduling opportunity of MS i is the number of times that MS i is allowed to transmit over the total number of experiments. In this trial, we have five users sending requests.

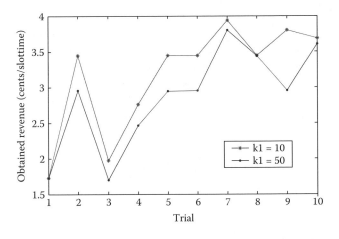

Figure 8.13   Balance between obtained revenue and packet loss. Obtained Revenue versus trials for user with full buffer (5 users send requests).

Assume $k_2 = k_3 = k_4 = k_5 = 10$, $\alpha_2/\beta_2 = 0.3$, $\alpha_3/\beta_3 = 0.7$, $\alpha_4/\beta_4 = 0.05$, and $\alpha_5/\beta_5 = 0.5$. We concentrate on evaluating the scheduling opportunity of user 1 by granting different values of k_1 and α_1/β_1. As expected, given a fixed ratio value of α_i/β_i, larger k_i will create more scheduling possibilities. A larger k_i combined with a greater ratio value α_i/β_i could create more than 90 percent of scheduling opportunities. However, this might be achieved by losing some revenue, as shown in Figure 8.13. Figure 8.13 presents the trade-off between the revenue maximization and loss minimization of user 1. The packet loss can be reduced by increasing the value of parameter k_1 for user 1. However, it decreases the obtained revenue.

8.6 Conclusion

This section summarizes the results of our work, the contributions of this chapter, and the recommendations for future work. In this chapter we proposed a novel scheduling and resource allocation approach for 3G networks that presents a viewpoint different from other existing solutions. Our resource management schemes can also offer service providers high flexibility by allowing them to maximize either the revenue gained or the data rate/QoS supported, or balance both.

Our schemes can be easily implemented in real industry as we adopt the parameters and specifications given in the ITU-approved MAC and physical layer standards of CDMA2000 1xEV-DV. We use different approaches in solving the scheduling problems in both the FL and RL, as the two have different constraints. On the RL, the RoT level at the BS is the major transmission constraint. On the FL, the limitations are the amount of transmission power and the number of Walsh codes available at the BS.

The problems were formulated into binary integer LP with multiple constraints. Several exact algorithms and heuristic algorithms were used to solve the problems. The performance of the algorithms were evaluated through illustrative experiments. In conclusion, the proposed heuristic algorithms are more suitable to be implemented for time-constrained scheduling and resource allocation in both FL and RL than the exact algorithms. However, if the network consists of a relatively small number of users, service providers can use the exact algorithms to obtain better solutions.

8.6.1 Recommendations for Future Work

The following is a list of research ideas that should be considered in the future:

- Compare our schemes with existing methods.
- Consider the presence of packet error, retransmission, multiple cells, handoff, and maximum/minimum data rate constraints.

- Run dynamic experiments over a longer period of time with more intense variation in the network environment.
- Carry out further research on the integration of short-term fairness for our schemes.
- Use real data from industry to understand how the scheme will perform under practical conditions.

Acknowledgment

The research of Attahiru Sule Alfa is partially supported by a grant from the Natural Sciences and Engineering Research Council of Canada.

References

[1] D. Owens, (2001, Aug.) Squeezing Spectrum. [Online]. Available: http://horizontest.bvdep.com/telecom/default.asp?journalid=3&func=articles&page=0108t05&year=2001&month=8.

[2] K. Fitchard, (2005, Mar.) Is EV-DV dead? [Online]. Available: http://telephonyonline.com/mag/telecom_evdv_dead/.

[3] S.S. Kulkani and C. Rosenberg, "Opportunistic scheduling policies for wireless systems with short term fairness constraints," in *IEEE Global Telecommunications Conference*, December 2003.

[4] D. Panigrahi and F. Khalegi, "Enabling trade-offs between system throughput and fairness in wireless data scheduling techniques," in *Proc. of World Wide Congress (3G Wireless)*, October 2003.

[5] D.I. Kim, "Optimum packet data transmission in cellular multirate CDMA systems with rate-based slot allocation," in *IEEE Transactions on Wireless Communication*, Vol. 3, No. 1, January 2004.

[6] S. Rao and S. Vasudevan, "Resource allocation and fairness for downlink shared data channels," in *IEEE Wireless Communications and Networking Conference*, March 2003.

[7] R. Kwan and C. Leung, "Downlink scheduling optimization in CDMA networks," in *IEEE Communications Letters*, Vol. 8, No. 10, October 2004.

[8] R. Vannithamby and E. Sousa, "Resource allocation and scheduling schemes for WCDMA downlinks," in *IEEE International Conference on Communications*, June 2001.

[9] R. Agrawal, V. Subramanian, and R. Berry, "Joint scheduling and resource allocation in CDMA systems," in *Proceedings of WiOpt04*, March 2004.

[10] J.W. Lee, R.R. Maxumdar, and N.B. Shroff, "Joint resource allocation and base-station assignment for the downlink in CDMA networks," to appear at *IEEE/ACM Transactions on Networking*, 2006.

[11] M. Shabany and K. Navaie, "Joint pilot power adjustment and base station assignment for data traffic in cellular CDMA networks," in *IEEE/Sarnoff Symposium on Advances in Wired and Wireless Communication*, April 2004.

[12] P. Xu, G. Michailidis, and M. Devetsikiotis, "Online scheduling for resource allocation of differentiated services: optimal settings and sensitivity analysis," Statistical and Applied Mathematical Sciences Institute, Tech. Rep., March 2004. [Online]. Available: www.samsi.info/TR/tr2004-13.pdf.

[13] S. Ci and M. Guizani, "A dynamic resource scheduling scheme for CDMA2000 systems," in *The 3rd ACS/IEEE International Conference on Computer Systems and Applications, Invited Paper*, 2005.

[14] C. Zhou, M.L. Honig, S. Jordan, and R. Berry, "Forward-link resource allocation for a two-cell voice network with multiple service classes," in *IEEE Wireless Communications and Networking Conference*, March 2003.

[15] C. Zhou, M.L. Honig, and S. Jordan, "Two-cell power allocation for wireless data based on pricing," in *Allerton Conference on Communication, Control and Computing*, September 2001.

[16] P. Liu, P. Zhang, S. Jordan, and M.L. Honig, "Single-cell forward link power allocation using pricing in wireless networks," in *IEEE Transactions on Wireless Communications*, October 2004.

[17] A. Boariu, "A resource allocation algorithm for downlink CDMA system," in *15th IEEE Symposium on Personal, Indoor and Mobile Radio Communications*, 2004.

[18] *TIA/EIA/IS-2000-3-C Physical Layer Protocol*, Telecommunications Industry Association Std., 2003.

[19] *TIA/EIA/IS-2000-3-C Medium Access Protocol*, Telecommunications Industry Association Std., 2003.

[20] *TIA/EIA/IS-2000-3-D Physical Layer*, Telecommunications Industry Association Std., 2004.

[21] *TIA/EIA/IS-2000-3-D Medium Access Protocol*, Telecommunications Industry Association Std., 2004.

[22] Y. Chung and D. Cho, "A novel transmission procedure for throughput maximization in 1xEV-DV reverse link," in *IEEE 55th Vehicular Technology Conference*, Vol. 1, May 2002.

[23] R. Derryberry and Z. Pi, "Overview and performance of the CDMA2000 1xEV-DV enhanced reverse link," in *IEEE Eighth International Symposium on Spread Spectrum Techniques and Applications*, 2004.

[24] H. Kwon, Y. Kim, J. Han, and D. Kim, "An efficient radio resource management technique for the reverse link in CDMA2000 1xEV-DV," in *IEEE on Wireless Communications and Networking Conference*, Vol. 1, March 2005.

[25] T. Wu, P. Hosein, R. Vannithamby, Y. Yoon, and A. Soong, "Common and dedicated rate control with variable QoS support for reverse-link high-speed packet data service in 1xEV-DV," in *IEE Fifth International Conference on 3G Mobile Communication Technologies*, October 2004.

[26] R. Derryberry, A. Hsu, and W. Tamminen, "Overview of CDMA2000 revision D," Nokia, 2004.

[27] Y. Chung, C. Koo, B. Bae, H. Lee, and D. Cho, "An efficient reverse link data rate control scheme for 1xEV-DV system," in *54th IEEE Vehicular Technology Conference*, Vol. 2, October 2001.

[28] D. Zhang, S. Oh, and N. Sindhushayana, "Optimal resource allocation for data service in CDMA reverse link," in *IEEE Wireless Communications and Networking Conference*, Vol. 3, March 2004.

[29] R. T. Derryberry and Z. Pi, "Overview and performance of the CDMA2000 1xEV-DV enhanced reverse link," in *IEEE Eighth International Symposium on Spread Spectrum Techniques and Applications*, 2004.

[30] T. Teerapabkajorndet and P. Krishnamurthy, "A reverse link power control algorithm based on game theory for multi-cell wireless data networks," in *IEEE 29th on Global Telecommunications Conference Workshops*, 2004.

[31] R. Derryberry, L. Ma, and Z. Rong, "Voice and data performance of the CDMA2000 1xEV-DV system," Nokia White Paper, 2002.

[32] W. Yeo and D. Cho, "A Markovian approach for modeling IS-856 reverse link rate control," in *IEEE International Conference on Communications*, Vol. 6, June 2004.

[33] S. Sarkar, "Reverse link capacity for CDMA2000," in *IEEE 53rd Vehicular Technology Conference*, Vol. 4, May 2001.

[34] J. Damnjanovic, A. Jain, T. Chen, and S. Sarkar, "Scheduling the CDMA2000 reverse link," in *IEEE 56th Vehicular Technology Conference*, Vol. 1, September 2001.

[35] C. Heyaime and V. Prabhu, "TraPS: Traffic-based packet scheduling for the CDMA2000 reverse link," in *IEEE Personal, Indoor and Mobile Radio Communications*, Vol. 3, September 2003.

[36] F. Ashtiani, J. Salehi, and M. Aref, "A flexible dynamic traffic model for reverse link CDMA cellular networks," in *IEEE Transactions on Wireless Communications*, Vol. 3, January 2004.

[37] L. Hsu, M. W. Cheng, and I. Niva, "Evolution towards simultaneous high-speed packet data and voice services: an overview of CDMA2000 1xEV-DV," in *10th International Conference on Telecommunications*, 2003.

[38] Comtech EF Data,"Higher order modulation and turbo coding options for the CDM-600 satellite modem," Tech. Rep., November 2001.

[39] Q. Liu, S. Zhou, and G.B. Giannakis, "Queueing with adaptive modulation and coding over wireless links: cross-layer analysis and design," in *IEEE Transactions on Wireless Communications*, Vol. 4, No. 3, May 2005.

Chapter 9

A Cost-Controlled Bandwidth Adaptation Algorithm for Multimedia Wireless Networks

*Abd-Elhamid M. Taha, Hossam S. Hassanein,
and Hussein T. Mouftah*

Contents

9.1   Overview

Bandwidth Adaptation Algorithms (BAAs) assume an important role in multimedia wireless networks. They exploit the nature of adaptive multimedia applications to the benefit of both the user and the network. Their basic operation entails varying the user allocations depending on the demand intensity and the network conditions. Several proposals have been made for BAAs, each with a different objective and approach. However, a common drawback in previous algorithms is the persistent engagement of bandwidth adaptation whenever it is requested by an admission control module. This leads to high operational cost and results in user dissatisfaction.

In future multimedia networks, such persistence can be more costly, especially when considering the highly dynamic nature of traffic patterns. In such networks, terminals will be equipped with multiple radio interfaces and will be able to access multimedia services through different networks, and even maintain simultaneous connections through the different interfaces. In addition to the already troublesome intra-access technology handoffs, inter-access technology handoffs will also be viable. Hence there is the need for simplified radio resource management modules that maintain low operational cost while maximizing the revenues of the service provider.

This chapter introduces the Stochastically Triggered Bandwidth Adaptation Algorithm (STBAA). Through a probabilistic trigger, we are able to control the operational cost and provide means for controlling an enhanced trade-off between admission and operational guarantees. The core of this algorithm is a simplified, structured, and optimized BAA with no assumptions on the underlying traffic model. We also provide measures that ensure stable operation and maintainable user satisfaction.

9.2   Introduction

It is difficult to ignore the recent surge in using multimedia applications. Equally inexcusable is to underestimate the increasing demand for their usage. Webcasting, Voice-over-IP (VoIP), video streaming, and many other functionalities are becoming indispensable elements of lifestyle.

The demand for multimedia applications is naturally extending to the venue of wireless and mobile computing. Readily, service providers are toying with delivering full-length movies to the mobile end user [1]. With many advances underway, not only will real-time video streaming to mobile end user be proliferated, but also interactive video and audio exchange will become attainable for the masses.

In arriving to this current state, wireless and mobile networks have evolved from relying on a connection-oriented infrastructure to become peripherals of a universal, data-centric network. Specifically, the reliance on IP and IP-compatible platforms enabled an efficient extension of the Internet framework to the wireless network. Not only is this extension enabling the accommodation of the increasing demand for wireless multimedia applications, but it is also realizing means for embracing different wireless platforms, both existing and to come. In tying together the various wireless technologies, the Internet will also realize inter-technology handoffs. This is becoming possible as mobile terminals are being equipped with more than one radio interface, and designed to connect to more than one access technology, even establish and maintain simultaneous connections for different sessions. Accordingly, network selection based on preferences and requirements, in addition to other elements such as a user's mobility profile, becomes persistently possible and not just at session initiation.

From the network perspective, however, these advancements require revising the designs of traditional frameworks for radio resource management (RRM). User demand patterns, the viability of accessing multimedia services through different networks, in addition to the elements of both intra- *and* inter-technology mobility, all call for RRM modules that are robust and efficient in both time and cost. Such modules, in addition to their sustenance to the general characteristics of radio interfaces, should also sustain the coexistence of different access technologies.

The initial introduction of multimedia applications in wireless networks motivated the conception of a certain RRM functionality, namely bandwidth adaptation, which exploits their adaptive nature. The term "adaptive nature" refers to the fact that the quality of delivery can be varied without affecting the delivery of the content. This characteristic enables a BAA to respond to both network conditions and demand patterns. The extent to which a BAA can vary the quality of delivery, however, is controlled by the guarantees the RRM framework attempts to uphold. Such guarantees, detailed in the service level agreements (SLAs) set between the service provider and the users, include admission guarantees (e.g., blocking and dropping probabilities) and operational guarantees describing a Quality-of-Service (QoS) a connection receives while it is active.

For example, BAAs respond to severe network conditions (e.g., rising interference) or added demand by reducing the allocations of the active users. In the case of added demand, a BAA is engaged in seeking to maintain certain admission guarantees. When substantial bandwidth becomes abundant, either due to users departing the network or improving medium conditions, the BAA attempts to distribute the newly available bandwidth between connections with reduced allocations. Here, BAA is engaged to maintain operational guarantees. In either operation, however, the BAA handles a delicate trade-off between satisfying the admission guarantees and fulfilling the operational requirements.

Despite the advantages of BAA, however, its operation can be associated with different costs. To begin with, there is the basic operational cost relative to computing the adjustable bandwidth and selecting users to have their allocations adjusted. The persistent operation of a BAA (e.g., triggering the module for every arriving request when the network is overloaded) might result in excessive or frequent adjustments, both of which can be undesirable from a user perspective, and can have a devastating effect on the service provider's economics. Furthermore, adjusting a user's allocation requires an exchange between the user and the network. Thus, there is an overhead cost associated with bandwidth adaptation.

As will be discussed in Section 9.4, previous proposals for BAA do address some of the above concerns. However, a common drawback in previous proposals is that BAA is engaged whenever requested, without considering the operational cost or evaluating the economic worth of this engagement. That is, there is a general lack of cost control. Moreover, certain proposals are designed based on certain assumptions regarding demand characteristics. Against the demand heterogeneity to be found in future networks, such models stand at a disadvantage.

We believe that such control is achievable by understanding the trade-off between admission and operational guarantees. In this chapter we propose means to control the costs relative to the bandwidth adaptation procedure. This control is achieved by exchanging the common deterministic trigger of BAA with a stochastic one. Moreover, by designing a structured and optimized BAA core, we enhance the aforementioned trade-off. The BAA core is also designed based on generalized arguments, and is not based on specific assumptions with respect to demand or traffic patterns.

This chapter is organized as follows. In Section 9.3 the general elements and considerations of BAAs are discussed to make the overview of previous proposals, provided in Section 9.4, more accessible. The literature detailed in Section 9.4 is discussed with an emphasis on the objectives, metrics, and approaches of each proposal when selecting connections to be adapted. The motivation and rationale for our work are presented in Section 9.5. Our proposal, the Stochastically Triggered Bandwidth Adaptation Algorithm, is described in detail in Section 9.6. The results of evaluating the general goodness of our algorithm are shown in Section 9.7. Finally, in Section 9.8, we conclude and hint at possible future work.

9.3 Elements and Considerations of BAAs

The objective of this chapter is to detail the general characteristics of BAAs prior to exploring the different previous proposals and introducing the facets of our work. In what follows, we refer to the general elements and considerations that have been taken into account in other BAA proposals.

9.3.1 Definitions

For the consideration of a BAA, a service or a class of service is commonly characterized by a set of allocations, Ω, where $\Omega = \{\omega_1, \omega_2, \ldots \omega_i \ldots, \omega_N\}$ with N being the number of possible allocations for the service. The allocations are strictly increasing in the order of the subscript i, that is $\omega_i < \omega_{i+1}$, for all i. Of the possible allocations, a certain allocation, denoted ω_{ref}, is commonly referred to as the target or reference allocation. It is this allocation that a BAA, possibly with other modules in an RRM framework, attempts to provide for the user as much as possible. In Ω, ω_{ref} can hold the value of either the minimum or maximum allocation, or a value in between. During its lifetime, if a connection is given an allocation larger than ω_{ref}, the connection is said to be upgraded.[1] Similarly, if a connection is given an allocation less than ω_{ref}, the connection is said to be downgraded. Furthermore, in this work we refer to the operation of reducing user allocations *downgradation* and the operation of increasing user allocations as *upgradation*. It is hence that we will sometimes interchange the use of *adaptation* and *gradation*, in addition to using the latter in reference to the status of a certain class.

It is important here to note that the definition of Ω for a certain class or grade of service is the first element in defining the extent to which a

---

[1] The reader should note that part of this work is an attempt to present generalized arguments for Bandwidth Adaptation Algorithms (BAAs). In doing this, we also devise and encourage the use of certain terms, such as *gradation* and *gradeability*, the objective of which is to establish a concise and nonambiguous terminology for the different aspects of BAAs.

class can be up- or downgraded. For example, if Ω for a certain class is defined with only one possible allocation, then this single allocation becomes the only possible allocation that a user can receive. If Ω is defined with two or more allocations, then the extent of upgradation, or *upgradeability*, is defined by the difference between the target allocation and the maximum allocation. Similarly, for such a class, the extent of downgradation, or *downgradeability*, is defined by the difference between the target allocation and the minimum allocation in Ω. In general reference to either downgradeability or upgradeability, we use the term "*gradeability.*"

9.3.2 The Role of a BAA

Figure 9.1 illustrates exemplar actions that can be performed by a BAA. When a network is unable to admit further users with all its active users at their target allocation, the system can initiate the selection of certain users to be downgraded so that sufficient bandwidth can be released to accommodate the incoming request. When responding to devastating network conditions, the objective of the BAA becomes not so much "making room" as downgrading the nominal allocations to the practical capabilities of the network. The objectives of upgradation are equally varied. Whenever a call departs the network, BAA checks whether there are connections with downgraded allocations. If so, the objective of the BAA is to bring as many connections as possible to their target allocations. In the instance of network conditions improving, the objective of the upgradation module is to upgrade the nominal allocations of the active calls to the actual capabilities of the network.

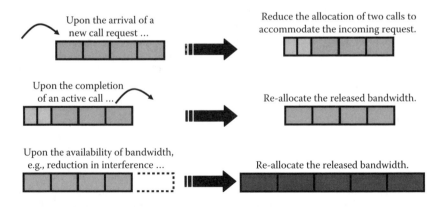

Figure 9.1   Different scenarios for the operation of a BAA.

9.3.3   Trigger Frequency

It is certainly possible to consider different types of triggers for the operation of a BAA. A BAA can be set to react only to changes in the state of the system. In such a reactive setting, the cost of engaging a BAA becomes highly dependent on the dynamicity of both the users and the medium. However, a BAA can also be set to operate in a proactive setting. For example, adaptation can be set to be periodically engaged at the end of fixed time intervals. During these intervals, the system conditions and user demand would be measured in ways that suffice making a sound adaptation decision at the end of each interval. While the costs relative to a BAA in such a setting become more or less fixed, this approach bears the following complexities. First, the time interval would need to be either appropriately fixed or continually adjusted relative to, say, the average of the call holding or residence time. Second, the adaptation decision at the end of each interval to either downgrade or upgrade would respectively need a prediction of the bandwidth to be required or available during the following interval. Such a prediction mechanism calls for elaborate considerations in both measurement (sampling) and computation. Notwithstanding, such a solution may indeed be possible, provided the existence of a prediction module within the collective RRM framework.

9.3.4   Required Measurements

At the most basic level, a BAA needs to know the extent to which it can downgrade or upgrade a user or a certain class of users. As aforementioned, this is primarily defined by Ω. Hence, whenever bandwidth adaptation is to be engaged, the system needs to compute how much of the allocated bandwidth is either downgradeable or upgradeable, depending on the type of the trigger.

However, the gradeability of a certain class is not a sufficient measure to judge whether a certain class is viable for gradation. For example, if a class has been persistently downgraded over a certain period of time, the system, where possible, should seek to downgrade other classes more often. This temporal monitoring of gradation can materialize using different measures.

For example, a system can calculate the number of downgraded or upgraded calls in a network, and favor between the different classes based only on this number, independent of the allocations associated with the graded users. Another possible measure would be the percentage of users graded in each class. A measure relative to the actual allocations made to the different connections is the average offset between the allocation and the reference allocation. Note that, in addition to utilizing such measures in favoring between the different classes, these measures can be used in deciding whether a certain class should be downgraded. For example, in the case of using the percentage of downgraded users or their average offset from their target allocation, threshold values can be used that, if violated for a certain class, the class would not be considered for downgradation.

In using any of the above measures, or other measures that might seem appropriate in systems with different settings, the consideration of a temporal average, or median, serves two purposes: (1) to protect users of a certain class from persistent gradation, and (2) to avoid the effect of high fluctuations in such measurements.

9.3.5   Conclusiveness of a BAA

A BAA can be either conclusive or inconclusive. A *conclusive BAA* would not be engaged unless it is known that it will provide the bandwidth required to, say, accommodate an incoming call, or that there are downgraded calls that need a newly available bandwidth. An *inconclusive BAA* is akin to a search program and is engaged for the dual purposes of checking whether there is sufficient bandwidth to be released (distributed) and releasing (distributing) this bandwidth if it is available (required).

There is usually a relationship between the conclusiveness of a BAA and the type of measurements employed. There is also an inherent trade-off between the cost of the measurement and the cost of the algorithm. However, it is possible to argue that utilizing a conclusive algorithm would, in general, reduce the operational cost of a BAA. This is because the measurements required by a BAA (i.e., the gradeable bandwidth and the gradeability of a certain class) can be readily and persistently made by the collective RRM framework within which the BAA is employed.

However, it is possible that the nature of the wireless network hinders the utilization of a conclusive algorithm. This is especially true for networks that are affected by the increased number of users such as contention-based networks (e.g., a WLAN) or interference-limited systems. In such systems, it becomes difficult to ascertain different measures required by the BAA at the instant it is engaged. Nevertheless, the nature of these systems can still be circumvented by controlling access into the network and maintaining the access level in a range where such measures can be easily quantified. This, of course, does come at a cost of network underutilization.

9.4   Overview of Previous Proposals

Previous proposals for bandwidth adaptation can be broadly classified based on their assumptions regarding the nature of the underlying traffic model. Under the assumption of continuous or fluid allocations, bandwidth adaptation becomes easier to handle and implement. This is especially true when attempting to realize a form of fairness either at the level of one class of service or between different classes. A model of digitized or discrete allocations, however, stands closer to the nature of practical allocations, which are predominantly made at specific rates. Nevertheless, the design and implementation of bandwidth adaptation for discrete allocations are more challenging.

In what follows we restrain ourselves to the discussion of BAA proposals made for traffic models with discrete allocations. Exemplar proposals for bandwidth adaptation based on fluid models are the works of El-Kadi, Olariu, and Abdel-Wahab [2] and of Seth and Fapojuwo [3].

Talukdar et al. [4] note that the decision to downgrade or upgrade the bandwidth allocated to a certain call is coupled with an exchange overhead that relays the decision to the associated terminal. Given the scarcity of media resources in wireless networks, it might very well be an objective of the service provider to reduce the overhead resulting from the adaptation process. Nevertheless, the authors also note that the objectives of realizing fairness and minimizing adaptation overhead are conflicting objectives.

Talukdar et al. [4] propose three algorithms: (1) Minimum Adaptation, (2) Fair Adaptation, and (3) Average-Fair Adaptation. The objective of the first algorithm is to strictly minimize the number of users downgraded to satisfy a certain request. The second algorithm attempts to consistently achieve a max-min fairness in allocating and reallocating the resources to the active calls. The Average-Fair algorithm attempts to strike a balance between the objectives of minimizing overhead and achieving fairness by considering the temporal averages of bandwidth allocations.

In Talukdar et al.'s setting, the general operational aim for a class is to maximize its allocation as much as possible. While the authors detail algorithms for the single-class scenario, they do show that their work is extendible to the multi-class setting. For Minimum Adaptation, calls are ordered in terms of the bandwidth they can release before reaching their minimum possible allocation in `availlist` and the bandwidth they can acquire to reach their maximum possible allocations in `downgradedlist`. The algorithm selects the minimum number of users to satisfy an incoming request or reallocate a newly released bandwidth by accessing these two lists. The operation of Minimum Adaptation results in an unfair distribution of allocations. Fair Adaptation works through a single module that persistently divides the allocatable bandwidth between all the users. While fairness in this manner is certainly achieved, it comes at a high cost of overhead. The Average-Fair algorithm strikes a balance between the two objectives. It utilizes the same modules used by Minimum Adaptation, but adds the *average bandwidth allocation* to the tuple by which the calls are sorted. For example, users with both the highest releasable bandwidth *and* the highest average bandwidth allocation are the ones selected for downgradation.

Kwon et al. [5–7], detail a bandwidth adaptation, or reallocation, algorithm that provides a guaranteed state of gradation. In Kwon et al. [5], the authors detail a reallocation algorithm to operate with a specific admission control module proposed by Naghshineh and Schwarz [8]. The objective of the proposed algorithm is to reduce the *Degraded Period Ratio* (DPR) per call, which represents the portion of time a call spends in the system while being allocated less than its target allocation. In the adaptation algorithm, the DPR is only used in the upgradation process. In the downgradation process, the algorithm bounds the degree of downgradation by necessity. Specifically, the system first downgrades as few calls as necessary to make the target bandwidth available for an incoming call. If the released bandwidth is insufficient, the system then tries to release sufficient bandwidth by downgrading the calls to their minimum bandwidth.

The general objective of the algorithm is to allocate as much as possible to the arriving call request. Accordingly, the algorithm first attempts to allocate more than the target allocation using only the available bandwidth. If the available bandwidth is insufficient, and sufficient bandwidth can be released without reducing any call below the target bandwidth, only the remaining bandwidth is released. If the target bandwidth (or more) cannot be allocated, the algorithm attempts to allocate the minimum bandwidth first through only the available bandwidth. If this is infeasible, then active calls are downgraded, but not below their target bandwidth. If the bandwidth released is still insufficient, the active calls are further downgraded to the minimum allocation possible. If after exhausting the above options the system cannot provide even the minimum allocation, the incoming call is blocked. The upgradation algorithm works by trying to *correct* the DPR per ratio. Implicitly, a sense of fairness can be said to be realized. Upgradation works as follows. Upon a call departure, the remaining calls are first ordered decreasingly according to their DPR. Taking one call at a time, if the available bandwidth can suffice allocating the target bandwidth or more, the allocation is made. Otherwise, other calls are reduced to make such an allocation possible without making their allocations below their target bandwidth. The procedure is repeated until all active calls are processed.

Kwon et al. [6] adjust the gradation algorithms to work with a measurement-based admission control. Instead of maintaining the measurement state of DPR for each user, the authors propose a per-class indicator, called the degradation probability, P_D, defined as the temporal average of the ratio between the number of downgraded calls to the total number of ongoing calls in the system. Moreover, to prevent fluctuation of the measure, the authors propose means to isolate outliers. Finally, when considering a call for admission, the proposed algorithm takes into account the average measurement of the neighboring cells, not just the cell to which the call request is made. Relative to Kwon et al. [5], the core of the BAA in Kwon et al. [6] is the same. The two algorithms only differ in the measurement and triggers. Also, because P_D is used instead of the per-call DPR, no sorting takes place in the upgradation procedure and calls are processed in random order.

In [7], Kwon et al. provide an expanded exposition of their work. A minor adjustment is made to the upgradation algorithm by sorting the calls decreasingly by their current allocations prior to processing. However, the authors state that other measures can be used for ordering the calls, "such as the amount of degraded time, if available."

A different approach is made by Kwon et al. [9], Kwon et al. [10]; and Kwon et al. [11]. The common emphasis in these three works is overcoming the possible complexities of optimal adaptation algorithms through near-optimal bandwidth adaptation. Kwon et al. [9] detail the operational requirements from a BAA, in addition to its possible objectives. They note that the requirement of maximizing revenue and the objective of maximizing a call's quality are operationally synonymous. This is because a call's revenue can be tied to the resources allocated; thus, the higher a call's allocation, the higher the provider's revenue. The authors also note that a BAA should have low operational complexity. Similar to Talukdar et al. [4], Kwon et al. [9] also state that there is a trade-off between minimizing the number of adaptations and maximizing the fairness among users. However, the motivation of Kwon et al. [9] is different from that of Talukdar et al. [4] — while Talukdar et al. [4] are motivated by reducing the overhead associated with adaptation, Kwon et al. [9] are concerned with reducing the number of variations or disruptions experienced by a call during its lifetime.

The authors describe three algorithms: (1) BAA for Revenue Only (BAA-R), which aims strictly at maximizing the revenue of active calls; (2) BAA for Revenue and Anti-Adaptation (BAA-RA), which attempts to maximize the revenue of active calls while minimizing the number of adaptations; and (3) BAA for Revenue and Fairness (BAA-RF), which attempts to maximize the revenue while establishing a specific sense of fairness within each class and between the different classes. For BAA-R and BAA-RA, the core of the algorithm is based on a graph formulation. In simple terms, the state space of possible downgradations is set up as a tree graph, with the root being the state of no downgradations, and each column branching from the root indicating the possible states of downgradations for a call, increasingly toward the outer leaf. Associated with each edge in the graph is a tuple that indicates the cost incurred and the bandwidth released passing through the edge (i.e., utilizing the associated downgradations). In the case of BAA-R, the cost of an edge is the difference between the revenues of the allocations divided by the gained bandwidth. For BAA-RA, an additional cost is incorporated into the calculation that represents the user's dislike for undergoing adaptation. In either case, the algorithm becomes a search algorithm that moves from the root of the graph outward, establishing a solution space that minimizes the cost but makes available the bandwidth requested.

The third algorithm, BAA-RF, operates in a different manner. The algorithm first tries to distribute the bandwidth allocated to each class evenly between the different users. However, due to the discrete nature of allocations, an even distribution may not always be feasible. The algorithm is hence set to distribute either the next lower or the larger allocation of the average allocation to each user. If such is possible, then intra-class fairness is said to be achieved. For inter-class fairness (i.e., fairness between the different classes), the algorithm attempts to distribute total allocations in the system between the classes, depending on the number of active users in each class and their associated bandwidth. In doing so, allocations within certain classes can be toggled between the two aforementioned allocations. In exploring the possible allocation states that satisfy both inter-class and intra-class fairness, the objective becomes selecting the allocations that maximize the revenue while observing the maximum capacity of the system.

While the three algorithms are described in [9], the BAA-R is analyzed in further details in [10], and the three algorithms are further elaborated upon in [11].

Xiao et al. [12,13,16] and Xiao and Chen [14,15] address bandwidth adaptation from a different perspective and utilize different metrics. Xiao et al. [12] introduce the notions of *Degradation Ratio* (*DR*) and *Degradation Degree* (*DD*). They also introduce the notion of *Degraded Area Size* (*DAS*). The temporal average of the number of users downgraded in the system is *DR*. The temporal average of the summation of the offset between the assigned bandwidth and target bandwidth of the downgraded users is *DD*. The *DAS* is the temporal average of the product of the number of users downgraded each by the offset of the user's allocation from the target allocation. The objective of the algorithm in [12] is to maintain a *DAS* less than or equal to a target *DAS* (i.e., DAS_{qos}). This stated, the authors provide an admission control that always accepts handoffs, while denying access to new calls when the *DAS* is below DAS_{qos}. Note that the *DAS* considered in the latter condition may be that of the cell to which the call request was made, or the average *DAS* for both the cell and its neighbors. For upgradation, Xiao et al.'s BAA orders calls according to their *DD*, and upgrades as many calls as possible to the target allocation. For downgradation, the extent to which active calls are downgraded depends on the type of the call. If the arriving call is a handoff call, active calls can be reduced to the minimum allocation possible. If the arriving call is a new one, active calls can only be downgraded to the target allocation.

Xiao et al. [13] propose another BAA with the objective of achieving fairness at both the inter- and intra-class levels. This work is similar in spirit to Kwon et al.'s [9] BAA-RF, but attempts to achieve another sense of fairness. Different from their work in [12], Xiao et al. utilize a staged admission control where calls are momentarily admitted until sufficient bandwidth is verified to be available. Instead of using the *DAS* metric, the *DD* and the *DR* for each class are verified to be below their respective thresholds. A BAA called Fair Bandwidth Allocation (FBA) is then initiated. Inter-class fairness is achieved by dividing the allocations based on the arrival rates of the different classes. The manner in which the arrival rates are estimated is not discussed by the authors but it is assumed that the arrival rates for each class is exponential. For intra-class fairness, FBA accommodates the discrete nature of allocation by the manner in which it performs adaptation; that is, when upgrading, calls that are most downgraded are processed first, and when downgrading, calls that are most upgraded are processed first.

Xiao and Chen [14] further introduce two different notions. The first is that of utilizing a weighted average in the measurement of *DD* and *DR*, controlling the sensitivity of the temporal average to the value of the most recent measurement. It is these adjusted averages that are used in the admission decision. Similar to the admission control in [13], the admission control here is also staged. The second notion introduced in this work is that of a leveled adaptation. In the Two-Level BAA (TL-BAA), for example, downgradation is first performed by downgrading all calls to their target allocations. If this does not provide sufficient bandwidth, then calls are downgraded to the minimum allocation possible. The upgradation of the TL-BAA operates in a similar manner. In the generalized K-Level BAA (KL-BAA), where K is the number of possible allocations in a class, adaptation is performed by reducing or increasing the adaptation of all calls by only one allocation at a time. This work is further elaborated upon by Xiao et al. [16].

The notion of multi-class fairness is revisited by Xiao and Chen in [15]. Unlike FBA, the authors propose maintaining predefined ratios between the *DD* and *DR* of the different classes. They also propose a KL-BAA that is somewhat different from the one proposed in [14], as it operates with the objective of maintaining the predefined ratios.

Nasser and Hassanein [17,18] investigate tying the class to be adapted with the class of the incoming call. In their collective RRM framework, a cell's capacity is dynamically partitioned between the different classes. When a call from a certain class makes a request, the admission control goes into the following decision sequence. If there is sufficient unallocated bandwidth available, the call is accepted. If there is insufficient bandwidth in the call's class partition while bandwidth is available for other classes, the call's class is downgraded. If there is sufficient bandwidth in the call's class, but insufficient bandwidth in the cell, all classes are considered for downgradations in order of their target bandwidth. If there is insufficient bandwidth in both the call's class and the system, the call's class and the system are downgraded in sequence.

Other approaches have been made in the literature. For example, a more complex BAA proposal was made by Xiao et al. [19,20] where, based on strict Markovian assumptions, the authors utilize a Semi-Markov Decision Process (SMDP) to generate an optimized state-action profile for the network. With a similar objective, Yu, Wong, and Leung [21,22] combined SMDP and MDP with reinforcement learning. While such proposals provide optimal solutions for both admission and adaptation, the complexity and scalability of SMDP-based solutions should not be overlooked.

9.5 Motivations and Objectives

Reviewing the work discussed above, one notices that different proposals were made for BAAs with different objectives. It can also be understood that fulfilling certain objectives comes at the cost of dissatisfying other measures. For example, attempting to achieve any sense of fairness, which represents a form of satisfying operational guarantees, comes at the cost of dissatisfying admission guarantees. Meanwhile, attempting to maximize the number of admitted users comes at the cost of dissatisfying already active users. It is possible, however, to strike a balance between the different objectives, which is precisely the objective of TBA's Average-Fair BAA.

There are other points that can be equally observed. Some algorithms require holding a measurement state (e.g., *DPR*) for each connection. While this results in an ultimate control of user status, it bears costly processing and memory space. Given the expected dynamicity of future wireless networks, a solution depending on such per-user granularity in terms of measurement would not scale.

Certain BAAs also make strong assumptions regarding the underlying traffic model, or are restricted to a certain type of admission control. This is a drawback that should be addressed. A BAA in future wireless networks should be able to cope with unpredictable traffic, and should not rely heavily on any traffic assumptions.

The notion of staged admission is plausible as long as low dynamicity is assumed. That is, to momentarily accept calls while verifying that there is sufficient downgradeable bandwidth can be acceptable as long as the minimum inter-arrival time allows for processing the calls independently. The verification procedure can naturally be multi-threaded, but this leads to more complexities. A preferred setting would be one where such delays and complexities can be avoided without the dependence on nonscaleable measurement for user or class status. Utilizing a carefully designed conclusive BAA would achieve such an objective.

The fact that adaptation is persistently triggered whenever possible does not serve the economic and operational objectives of service providers.

Persistent adaptation results in intensive processing, high overhead, and user dissatisfaction, each of which bear short- and long-term costs. From another perspective, the above algorithms do not allow means to judge the worth of performing the adaptation at any given instant.

The work presented in this chapter attempts to address the above issues. Specifically, we aim to design a BAA based on generalized arguments. The BAA should also assume sufficient flexibility to be tailored to different operational objectives. In doing this, we design a BAA with loose assumptions on the underlying traffic model, and simplify, structure, and optimize the procedure for user selection in both the downgradation and the upgradation modules. We distinguish between the network level and the class level, and control the selection of users across different classes to reduce the operational cost. To that end, the nature of our proposed algorithm is conclusive, avoiding measurements with per-user granularity and relying mostly on measurements readily accessible in any wireless system.

A more important feature of our BAA, however, is that it provides means for enhancing and controlling the trade-off between the admission and operational guarantees. The same means also provide for controlling the operational cost and selecting the proper time to initiate the adaptation procedure.

9.6 Stochastically Triggered Bandwidth Adaptation Algorithm

The essence of the Stochastically Triggered Bandwidth Adaptation Algorithm (STBAA) is to replace the deterministic manner in which adaptation is initiated with a probabilistic trigger. That is, when sufficient bandwidth is verified to be downgradeable or upgradeable, the algorithm first consults a certain probabilistic threshold. In case of downgradation, the probabilistic trigger may serve as protection against users undergoing further reduction in their allocations. The role of the trigger in the upgradation process, especially in a network with highly dynamic traffic, is to reduce the number of adaptations that users can undergo in a very short period. Hence, the role of the trigger and other design features in the algorithm detailed below is to provide stable, controllable service delivery to wireless end users, while at the same time reducing the operational cost of the BAA.

In what follows we detail the different aspects of STBAA. The general architectural assumptions include that the BAA is operating at either the level of a cell or at the level of a higher entity (e.g., a radio network controller (RNC)) that overlooks a cluster of cells. As the operation of a BAA is tied to the operation of an admission control module, it is assumed that both algorithms are implemented at the same level.

9.6.1 Operational Overview

Upon receiving a call request, an admission control would first verify that there is sufficient unallocated bandwidth that can satisfy the incoming request. If so, the call is accepted.

If the call request is received when the network is at its full capacity, or when the unallocated bandwidth is insufficient to grant the request, the admission control considers adaptation. The manner in which adaptation is invoked is made in stages. At the first stage, the system would verify from the different measurements that the allocations of the active users can be downgraded to satisfy the incoming request. Once this is verified, the system generates a random number from a (0,1) uniform distribution and compares this number against a persistently valuated probabilistic threshold. Only if the randomly generated number is less than the probabilistic threshold would downgradation be engaged. If neither the downgradeable bandwidth is sufficient nor the probabilistic threshold constraint satisfied, the call is rejected.

When a call departs the network, the system contemplates upgradation. The system first verifies whether there are any downgraded calls. If there are, the system resorts to a probabilistic trigger. If there are no downgraded calls, or if the probabilistic trigger fails, upgradation is not performed.

9.6.2 Architectural Overview

The above detailed operations require the coordination of three separate modules. The first module is concerned with collecting and computing the various measures required for the operation of the gradation or adaptation mechanisms. The second module handles the valuation of the probability threshold and can interact with the measurement module to consider the status of the system in this valuation. The third module, called the gradation module, comprises two submodules: one for downgradation, the other for upgradation. The gradation modules interact with the measurement module for their operation and require the valuation of the probability threshold to be engaged.

We detail the three modules below. Prior to doing so, however, some basic notations and definitions are in order.

9.6.3 Notations and Definitions

Define a number of classes, N, where class N is the highest class. A set of allocatable bandwidths, Ω_i, is associated with each class i such that $\Omega_i = \{\omega_{i,1}, \omega_{i,2}, \cdots\}$, where i and j of bandwidth ω refer to its class and rank, respectively. The higher j is the higher the allocation of the rank, that is, $\omega_{i,j} < \omega_{i,j+1}$. For each class, define $\omega_{i,ref}$ as the value that the system guarantees the user, such that $\omega_{1,ref} \leq \cdots \leq \omega_{N,ref}$.

During its lifetime, if a call is allocated a bandwidth that is greater than $\omega_{i,ref}$, the call is said to be upgraded. Similarly, if a call is allocated a bandwidth less than $\omega_{i,ref}$, the call is said to be downgraded. If $|\Omega_i| = 1$, then the class is defined to have one allocatable bandwidth that can neither be upgraded nor downgraded. If $|\Omega_i| \geq 1$, then the system allows for the allocations in this class to be upgraded and downgraded, depending on the class definition. Hence, the purpose of Ω_i is to define the set of bandwidths upon which the provider and the user agree for class i calls to undergo during different system conditions. Note that that the value of $\omega_{i,ref}$ need not be a single value but can be a range of values — in which case the definitions of upgraded and downgraded refer to a call being allocated more or less than, respectively, $\max\{\omega_{i,ref}\}$ and $\min\{\omega_{i,ref}\}$. In this work, however, we only consider classes with a single reference allocation.

9.6.4 The Measurement Module

The objective of the measurement module is to compute the current satisfaction of each class, in addition to providing the gradation modules with the gradeable bandwidth for the system.

For each class, the measurement module computes the average gradation. Denote $b_{i,u}$ as the bandwidth allocated to call u of class i. Define $d_{i,u}$ as the difference between $b_{i,u}$ and $\omega_{i,ref}$, that is, $d_{i,u} = b_{i,u} - \omega_{i,ref}$. Note that, depending on position of $\omega_{i,ref}$ within the class definition, $d_{i,u}$ can be either positive, zero, or negative.

The total gradation of class i, D_i, can be defined as $D_i = \sum_u d_{i,u}$ where $u \in U_i$ and U_i is the set of users belonging to class i and currently in the system.

The average gradation for class i, which is the measure of its satisfaction, is $d_i = \frac{D_i}{|U_i|}$.

The total gradeable bandwidth for each class (i.e., each class's downgradeable and upgradeable bandwidth) is computed as follows. Denote by $D_{i,max}$ the total gradation of class i when all its active users are allocated the maximum allowable bandwidth, that is, when $b_{i,u} = \max\{\Omega_i\}$, $\forall u \in U_i$. Similarly, denote by $D_{i,min}$ the total gradation of class i when all its active users are allocated the minimum allowable bandwidth, that is, when $b_{i,u} = \min\{\Omega_i\}$, $\forall u \in U_i$.

The total upgradeable bandwidth of class i (i.e., how far can class i be upgraded) is $D_{i,up} = D_{i,max} - D_i$. Similarly, the total downgradeable bandwidth of class i is $D_{i,down} = D_i - D_{i,min}$.

For the stable operation of STBAA, the gradation metric should not be used in its instantaneous value. Also, to protect the different classes from being over-downgraded, the classes should not report their total downgradeable bandwidth in full. Similarly, to accommodate high dynamicity, a class's total upgradeable bandwidth should not relay its upgradeable bandwidth in full.

The measurement module computes a temporal average of the gradation, denoted by $d_i(\tau)$, where τ indicates the instant at which the temporal average is computed. There are two possible ways to compute $d_i(\tau)$: (1) through periodic sampling, and (2) through sampling whenever adaptation is engaged.

The gradeable bandwidth relayed to the gradation modules is referred to as the allowed gradeable bandwidth. Denote the allowed downgradeable bandwidth of class i by BD_i. Given α_i, then $BD_i = \lfloor \alpha_i \cdot D_{i,down} \rfloor$, where α_i is the *allowance ratio* and holds a value between 0 and 1. The allowed downgradeable bandwidth for the system as a whole, BD, can then be calculated by $BD = \sum_i BD_i$. The allowed upgradeable bandwidth for class i, denoted by BU_i, and for the system BU can be computed in a similar manner.

Note that the allowance ratio for either the downgradeable or upgradeable bandwidth need not to be fixed throughout system operation. For example, α_i could be made a function of the current average gradation of the class, that is, $\alpha_i = f_{a_i}(d_i(\tau))$. The function $f_{a_i}(d_i(\tau))$ would allow more bandwidth to be downgradeable when $d_i(\tau)$ indicates high satisfaction, and less bandwidth to be downgraded when otherwise.

9.6.5 Valuation of the Probability Threshold

The valuation of the probability threshold, denoted p_t, can be done in different ways, depending on the objective of the valuation and also on whether the valuation is made for upgradation or downgradation.

In its simplest form, p_t can take on a constant value, for example, $p_t = p$ where $p \in [0, 1]$. The ramifications of utilizing this simple form, however, are nontrivial as they directly represent the control of the trade-off between the admission guarantees and the operational guarantees. The higher the probability, the more the BAA core will be engaged and the more the users will be downgraded. Reducing the threshold probability will result in a lower number of admissions, and hence a lower number of adaptations, but will result in higher operational satisfaction.

Maintaining the independence from user status, the valuation can still be programmable. For example, the threshold probability can be made to increase whenever a downgradation is requested and denied, and reset to an initial value whenever the downgradation request is granted. The increase can be made linear, exponential, etc. An exponential increase can take on the form $p_t = r^{x-s}$ where $x \in [0, \cdots, s]$ and $x \in \mathbb{R}^+$ is the number of requests made until the calculation instant. Here, the values of r and s dictate the responsiveness of the threshold. Figure 9.2 shows several examples for $s = 5$. The probability of adaptation gradually increases with every

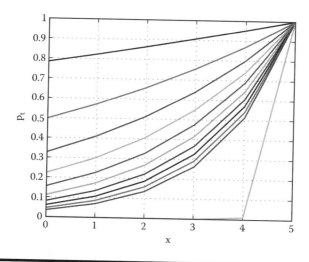

Figure 9.2 Plots for $p_t = r^{(xs)}$ with $s = 5$ and different values of r.

request until adaptation is performed. By the fifth request, the probability threshold takes on the value of 1. The upper ten lines in the figure show p_t with r differing between 1.05 and 2 in steps of 0.1. The lowest line shows p_t for $r = 100$, a value that makes p_t behave in a manner as if engaging adaptation every fifth request.

The valuation of the probability threshold, however, need not be the same for all call requests. For example, a probability threshold can be different for each class. More importantly, the valuation can be made to depend on the status of each class. For example, the valuation of probability threshold for class i, denoted $p_{t,i}$, can be directly related to $d_i(\tau)$; that is, $p_{t,i} = f_{p_{t,i}}(d_i(\tau))$. In tying $f_{t,i}$ to the class status, powerful protection is offered to the users of the respective class. It is here that the versatility of stochastic triggers can be observed. Through $f_{p_{t,i}}$, different considerations can be instilled into the valuation of the probability threshold. For example, $f_{p_{t,i}}$ can be used to make appropriate considerations regarding the worth of performing adaptation at the instant a request is made, in light of the status and other attributes of the system. An example of other attributes would be the demand intensity.

A point should be made here regarding the roles and objectives of the allowance ratio and a status-based valuation for the probability threshold. While these two elements appear to have contradictory roles, it should be observed that each serves a complementary purpose. The objective of the probability threshold is to control when a gradation should be made. However, the objective of the allowance ratio is to control the extent of any gradation action once a decision is made to engage either downgradation or upgradation.

9.6.6 The Gradation Modules

There are two gradation modules. The downgradation module is concerned with selecting the users whose allocations will be reduced. The upgradation module is concerned with selecting users whose allocations will be increased. As shown below, the general considerations of both modules are somewhat similar.

For gradation, we employ a hierarchical design that enables the control of fairness at both the system level and the class level. Furthermore, the utilization of a hierarchical design reduces the operational complexity. By first selecting the classes to be downgraded and then selecting the users under each chosen class, the modules avoid needless consideration of all the classes every time a request for adaptation is made.

9.6.6.1 Considerations of the Downgradation Module

Generally speaking, there is a cost incurred by the system when downgrading a user from a certain allocation to a lower one. In more specific terms, this cost can represent either a short-term cost, (e.g., the difference in the revenue rate) or a long-term cost (e.g., customer loss to a competitor as a result of frequent downgradations or adaptation). An added advantage of a hierarchical design is that different cost elements can be induced into the design at different levels. For example, the favoring between the different classes can be based on long-term cost elements, while user selection at the class level can be based on short-term cost elements.

As aforementioned, downgradation is performed in two steps, each at a different level. The first step is performed at the system level and returns the required downgradation for each class i, denoted AD_i. The second step is performed at the class level and returns the number of calls to be downgraded from rank k to rank l of class i, for all $k < l$.

The downgradation module can be set to seek any given objective that is a function of the downgradation for each class. Denote such an objective function by Φ_a. As an example, we assume that each unit of degradation at the system level is associated with a cost c_i, and make the objective of the downgradation at the system level to minimize the total degradation cost at the system level; that is:

$$\text{Minimize} \qquad \Phi_a = \sum_i c_i \cdot AD_i \qquad (9.1)$$

The extent of downgradation for each class, however, is limited by the class's downgradeability (i.e., BD_i):

$$AD_i \leq BD_i \qquad i \in \{1, \dots, N\} \qquad (9.2)$$

Also, the summation of the downgraded bandwidth from all classes should satisfy the bandwidth requested from the downgradation module, denoted by B_{req}:

$$\sum_i AD_i \geq B_{req} \tag{9.3}$$

Once the downgradation for each class has been computed at the system level, the downgradation module now performs the selection of users. Naturally, only classes with positive AD_i are considered for downgradation at this level. We maintain the exemplar objective expressed at the system level. However, it should be noted that, as in the case of the system level, the downgradation module at the class level can be made to seek any objective.

Denote by m_{kl} the number of calls to be downgraded from rank k to rank l, for all $k > l$. Assume that there is a cost associated with downgrading a call from rank k to rank l and denote this cost by c_{kl}. The objective at the class level hence becomes:

$$\text{Minimize} \qquad \Phi_i = \sum_{k,l} c_{kl} \cdot m_{kl} \qquad \forall k > l \tag{9.4}$$

The first constraint to apply in engaging the class-level downgradation procedure is that the bandwidth released by user selection should satisfy the value of AD_i set by the downgradation procedure at the system level:

$$\sum_{k,l} m_{kl} \cdot (\Omega_{i,k} - \Omega_{i,l}) = AD_i \qquad \forall k > l \tag{9.5}$$

The second constraint, or set of constraints, to apply is a balancing one. Specifically, it should be exercised that the total number of calls to be downgraded from each rank k cannot exceed the number of active users in the same rank:

$$\sum_l m_{kl} \leq |U_{i,k}| \tag{9.6}$$

The constraints displayed above are the most basic constraints for any downgradation module. Naturally, more constraints can be added. For example, at the class level, it is possible to add constraints that prohibit downgradation by more than one allocation.

9.6.6.2 Considerations of the Upgradation Module

Similar to the downgradation module, the upgradation module operates at both the system and class levels. The objective of the upgradation module is to distribute a certain available bandwidth, denoted by B_{avail}. At the system level, the model computes the upgradation for each class i, denoted by AU_i. In computing the various AU_i's, the system considers both revenues and costs. The possible costs incurred by the system during the upgradation process are those resulting from user experiencing variations in service quality. The considerations for the short-term and long-term revenues for the upgradation module are akin to the costs considered for the downgradation module.

Denote the revenue per unit upgrade for class i by e_i. The objective is hence to maximize the total upgradation revenue. An objective of the upgradation module can hence be to consider users of classes whose selection can maximize the overall revenue:

$$\text{Maximize} \qquad \sum_i e_i \cdot AU_i \qquad\qquad (9.7)$$

At the system level, the upgradation considers constraints similar in nature to the system-level constraints imposed on downgradation. Namely, constraints regarding the allowable upgradeable bandwidth and another regarding accommodating B_{avail}:

$$AU_i \leq BU_i \qquad i \in \{1, \ldots, N\} \qquad\qquad (9.8)$$

$$\sum_i A_i \leq B_{avail} \qquad\qquad (9.9)$$

At the class level, the upgradation module continues to seek maximizing the revenue of the upgradation process. Denote by e_{kl} the revenue associated with upgrading a call from rank k to rank l. Also denote by n_{kl} the number of calls to be downgraded from rank k to rank l, for all $k < l$. The upgradation objective at the class level hence becomes:

$$\text{Maximize} \qquad \sum_{k,l} e_{kl} \cdot n_{kl} \qquad \forall k < l \qquad\qquad (9.10)$$

The first constraint to apply is the class absorbing AU_i:

$$\sum_{k,l} n_{kl} \cdot (\Omega_{i,l} - \Omega_{i,k}) = AU_i \qquad \forall k < l \qquad\qquad (9.11)$$

The second constraint maintains that the calls upgraded from any rank k do not exceed the number of active users in the same rank:

$$\sum_k n_{kl} \leq |U_{i,k}| \qquad\qquad (9.12)$$

9.6.7   The Algorithm

We now detail how the different elements described above come together in the operation of STBAA.

Table 9.1 shows the algorithm for bandwidth downgradation which, for reference, is given the name **tryDowngradation**. When an admission control module calls **tryDowngradation**, it passes on the bandwidth required to be released, that is, B_{req}. The algorithm first computes the values required for verifying that there is sufficient bandwidth in the system to satisfy B_{req}. These values include the temporal average of gradation for each class i, $d_i(\tau)$; the allowed downgradeable bandwidth for class i, BD_i; and the allowed downgradeable bandwidth in the system, BD. The algorithm then verifies that BD is sufficient to satisfy B_{req}. If sufficient downgradeable bandwidth exists, **tryDowngradation** proceeds to valuate the probability threshold, p_t, in the manner outlined in Section 9.6.5. A random number, $p_{randgen}$ is then generated from a uniform distribution between 0 and 1.

Table 9.1   Algorithm for bandwidth downgradation

---

tryDowngrade(B_{req})

Compute $d_i(\tau)$ and BD_i for each class, and BD, as in Section 9.6.4;

if $BD \geq B_{req}$

    Valuate p_t, as in Section 9.6.5;

    $p_{randgen} = uniform(0, 1)$;

    if $p_{randgen} > p_t$

        return reject;

    else

        Valuate AD_i to satisfy B_{req} using equations 9.1 to 9.3;

        for all $AD_i > 0$

            Valuate $m_{k,l}$ to satisfy AD_i using equations 9.4 to 9.6;

        rof

        Perform downgradation;

        return accept;

    fi

else

    return reject;

fi

Table 9.2   Algorithm for bandwidth upgradation

---

tryUpgrade(B_{avail})

Compute $d_i(\tau)$ and BU_i for each class, and BU, as in Section 9.6.4;

if $BU > 0$

    Valuate p_t, as in Section 9.6.5;

    $p_{randgen} = uniform\,(0, 1)$;

    if $p_{randgen} > p_t$

        return `reject`;

    else

        Valuate AU_i to satisfy B_{avail} using equations 9.7 to 9.9;

        for all $AU_i > 0$

            Valuate $n_{k,l}$ to satisfy AU_i using equations 9.10 to 9.12;

        rof

        Perform upgradation;

        return `accept`;

    fi

else

    return `reject`;

fi

If $p_{randgen}$ holds a value greater than p_t, the downgradation is denied. If not, **tryDowngradation** proceeds with selecting classes and users for downgradation according to the constraints described in Section 9.6.6. Once users are selected, downgradation is performed and the admission control module is notified that the downgradation was successful.

The upgradation algorithm, shown in Table 9.2, operates in a similar manner. Upon noticing the availability of a substantial bandwidth, B_{avail}, the admission control module passes on this value to the **tryUpgradation** algorithm. In turn, **tryUpgradation** verifies whether there are any downgraded calls in the system and valuates the probability threshold. If there are downgraded calls, and the probability threshold is satisfied, **tryUpgradation** proceeds with the upgradation. In not, the admission control is notified that upgradation was not performed.

It should be noted here that p_t need not be the same for both downgradation and upgradation. For example, the service provider may opt for a low p_t for upgradation to protect active users from undergoing frequent adaptations, especially in the case of high demand intensity, where an upgradation may prove to be pointless. However, a different policy can be set by utilizing a high-valued p t, which would represent a greedy approach to upgradation and an attempt to increase user allocations whenever possible.

9.7 Evaluating the Performance of STBAA

In what follows, we detail the setup and sample results for simulation experiments that were carried out to evaluate the general goodness of STBAA. Prior to describing our implementation, however, certain remarks are in order.

The algorithm was implemented within a generic RRM framework with an admission control that only admits users at their target allocation. We made no measures for admission prioritization such as those made for minimizing the handoff dropping probability, nor did we employ resource preemption. We believe that such specific objectives stand beyond the precise requirements for a BAA and are more appropriately achieved by other, carefully sought RRM modules for admission and congestion control. Nevertheless, we maintain that the proposed BAA algorithm can be incorporated in any RRM framework.

Further note should also be taken regarding comparisons with other proposals. As discussed in Section 9.4, different proposals were previously made, each with a different objective. It should hence be noted that providing a comparison between the different proposals presented thus far in the literature poses a cumbersome task. More often than not, the adaptation algorithm overlaps with the general admission control of the proposed framework, making it difficult to evaluate the proposed adaptation procedure and to distinguish the benefits reaped specifically from the BAA.

In essence, the motivation behind STBAA goes far beyond adapting an arbitrary BAA that is augmented with a stochastic trigger. It is naturally of interest to compare STBAA with such an augmented module. Nevertheless, such a comparison stands beyond the aim of the following evaluation.

9.7.1 Simulation Setup

The simulation environment consists of a single cell with a maximum capacity of 100 units. Two classes are defined. Class 1 has $\Omega_1 = \{6, 8\}$ and an exponentially distributed holding time with a mean of 180 time units. Class 2 has $\Omega_2 = \{2, 4, 6\}$ and a fixed holding time of 60 time units. In both classes, the reference allocation is the highest allocation possible. The request inter-arrival time for both classes is exponentially distributed with a mean varying from 5 to 10 time units. The simulation experiments ran for 3600 time units. Each result shown represents the average outcome of ten experiments.

The simulations were made using release 13 of MATLAB. The linear programs were solved using GLPK of the GNU project [23]. For the interface between MATLAB and GLPK, Giorgettie's MEX interface was used [24].

9.7.2 Preliminary Evaluation

A preliminary setting is used to verify the basic operation of the BAA, that is, without employing a probabilistic trigger. For both classes, a fixed α_i of 0.6 was used when the temporal mean of d_i was below the half-mark, that is, −1 and −2 for class 1 and 2, respectively. Otherwise, $\alpha_i = 0$. No allowance ratio was set for upgradation.

The costs and profits for the adaptation module were set as follows. Both the costs and profits for both classes were set to unity. The cost of downgrading a call from rank 2 to rank 1 in class 1 was set to unity. The costs of downgrading a call from ranks 3 to 2, 3 to 1, and 2 to 1 were respectively set to 5, 3, and 2. The profit of upgrading a call from rank 1 to rank 2 in class 1 was set to unity. The profits of upgrading a call from ranks 1 to 2, 1 to 3, and 2 to 3 were respectively set to 2, 5, and 3.

The results for this setting are shown in Figures 9.3 to 9.5. In Figures 9.3 and 9.4, the blocking probability for classes 1 and 2 are respectively shown, with and without adaptation. In Figure 9.5, the average gradations for classes 1 and 2 with adaptation are shown. Note that the gradation without adaptation for both classes is zero. The basic trade-off between

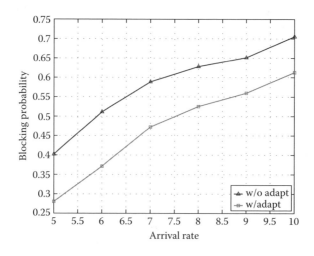

Figure 9.3 Blocking probability for class 1 without and with adaptation.

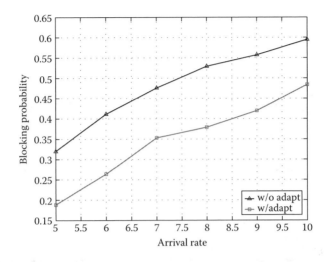

Figure 9.4   Blocking probability for class 2 without and with adaptation.

admission ratios and satisfaction is recognized, and the general advantage of the adaptation module is hence verified.

9.7.3   The Effect of Constant Probability Thresholds

Using the same settings as above, the decision to trigger downgradation was considered against a fixed probability threshold that was varied

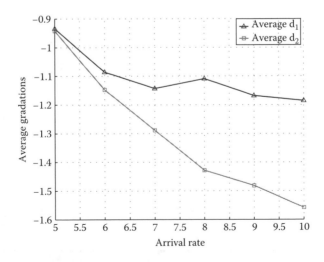

Figure 9.5   Gradations for classes 1 and 2 with adaptation.

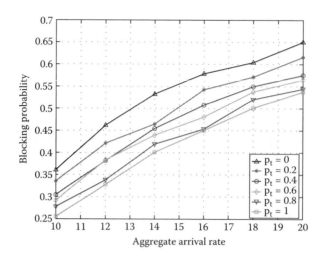

Figure 9.6 System blocking probability threshold, p_t, equal to 0, 0.2, 0.4, 0.6, 0.8, and 1.

from 0 (no adaptation) to 1 (full adaptation) in steps of 0.2. For upgradation, the probability threshold was set to 1. The results are shown in Figures 9.6 to 9.8.

It can be observed in Figure 9.6 that considerable gain is achieved by engaging adaptation only 20 percent of the time, that is, with $p_t = 0.2$.

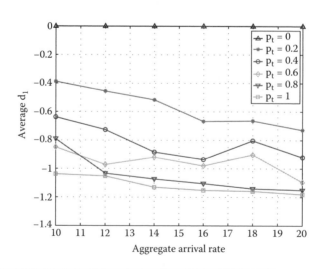

Figure 9.7 Gradation of class 1 with probability threshold, p_t, equal to 0, 0.2, 0.4, 0.6, 0.8, and 1.

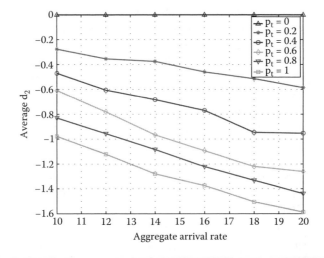

Figure 9.8 **Gradation of class 2 with probability threshold, p_t, equal to 0, 0.2, 0.4, 0.6, 0.8, and 1.**

Furthermore, and noting the proximity of the blocking probabilities for values of $p_t = 0.6$ and 0.8 to the full adaptation line, it is apparent what can be achieved by minor reductions in the operational cost. The trade-off between the adaptation and blocking probability is visible. The trade-off between raising the threshold probability and d_i can be observed in Figure 9.7 and Figure 9.8. These are the very trade-offs that can be exploited during system operation in balancing between admission and user satisfaction.

9.8 Conclusions and Future Work

In the Stochastically Triggered Bandwidth Adaptation Algorithm (STBAA), the deterministic manner in which BAAs are usually initiated is replaced by a probabilistic trigger. Through this trigger, the service provider is able to tailor the operation of the STBAA to different objectives. If the objective was to control the trade-off between admission and operational guarantees, the service provider can vary the value of the probability threshold. More detailed probabilistic triggers can also be envisaged. For example, different probability thresholds can be utilized for different classes. In this case, the valuation of the probability threshold can depend on the status of the class (e.g., its current or average level of satisfaction) or the class's demand intensity. In a more complex setting, the probabilistic trigger can be used when evaluating the monetary worth of engaging the adaptation process.

While this work does show that considerable gains can be attained without the persistent engagement of bandwidth adaptation, further investigation and analysis are required to realize the full potential of probabilistic triggers. For example, we need to identify the proper design guidelines to be used in implementing per-class probabilistic thresholds. This includes enumerating the considerations involved in the valuation of each threshold. Other analyses are required to examine the transient performance of the STBAA, and how different STBAA elements interact with instantaneous considerations.

As aforementioned, the characteristics of future networks demand rethinking the popular notions and implementations used for designing RRM frameworks. This work is aimed at realizing a time- and cost-efficient RRM module. We believe that certain design principles employed in the STBAA can be employed in other modules, such as admission control, congestion control, and mobility management. For example, it is possible for mobility management to probabilistically consider the gradual downgradation or upgradation of certain calls if they are known or predicted to follow a certain mobility profile. Such would be especially desirable in the setting of wireless networks with heterogeneous access. A network recognizing a user's mobility toward a more capable network can initiate gradual upgradation for other calls in the system.

Another interesting notion motivated by the heterogeneous setting is that of joint RRM functionalities. For example, instead of making admission decisions in each network independent of the status of other networks, allocations can be made in a joint fashion for the different networks. This leads to reduced operational costs and higher revenues. In Taha et al. [25,26], we investigate joint admission control modules for heterogeneous wireless networks. A further instance of joint functionalities expands on the adaptation example described above. When a terminal is recognized to be moving toward a certain network, the target network can be notified of this mobility so that, if need be, it can initiate gradual downgradation to accomodate the incoming user.

However, even the notion of bandwidth adaptation takes on a different role in future networks. The traditional notion considers the reconfiguration or reallocation of user allocations based on the possible set of allocations in a single network. In future settings, the coverage of different networks may overlap, resulting in what is commonly referred to as wireless overlay networks. With mobile terminals equipped with more than one interface (i.e., the capability to connect to more than one type of access technology), bandwidth adaptation assumes a new definition. Specifically, service providers will be able to consider the reconfiguration of user allocations across different networks within the coverage overlay. In Taha et al. [27] we explore this redefinition through identifying means for exploiting inter-technology handoffs to the benefit of both the user and the service provider.

References

[1] http://www.gameshout.com/news/122005/article1942.htm.

[2] M. El-Kadi, S. Olariu, and H. Abdel-Wahab. "A Rate-Based Borrowing Scheme for QoS Provisioning in Multimedia Wireless Networks," *IEEE Transactions on Parallel and Distributed Systems*, Vol. 13, No. 2, pp. 156–166, February 2002.

[3] M. Seth and A.O. Fapojuwo, "Adaptive Resource Management for Multimedia Wireless Networks," in *Proceedings of the IEEE 58th Vehicular Technology Conference — Fall*, Vol. 3, pp. 1668–1672, October 2003.

[4] A.K. Talukdar, B.R. Badrinath, and A. Acharya, "Rate Adaptation Schemes in Networks with Mobile Hosts," *Proceedings of the ACM/IEEE International Conference on Mobile Computing and Networking*, pp. 169–180, October 1998.

[5] T. Kwon, Y. Choi, C. Bisdikian, and M. Naghshineh, "Call Admission Control for Adaptive Multimedia in Wireless/Mobile Networks," *Proceedings of the ACM International Workshop on Wireless Mobile Multimedia*, pp. 111–116, October 1998.

[6] T. Kwon, Y. Choi, C. Bisdikian, and M. Naghshineh, "Measurement-Based Call Admission Control for Adaptive Multimedia in Wireless/Mobile Networks," *Proceedings of the IEEE Wireless Communications and Networks Conference*, Vol. 2, pp. 540–544, September 1999.

[7] T. Kwon, Y. Choi, C. Bisdikian, and M. Naghshineh, "QoS Provisioning in Wireless/Mobile Multimedia Networks Using an Adaptive Framework," *Wireless Networks*, Vol. 9, No. 1, pp. 51–59, January 2003.

[8] M. Naghshineh and M. Schwarz, "Distributed Call Admission Control in Mobile/Wireless Networks," in *Proceedings of the International Symposium on Personal, Indoor and Mobile Radio Communications*, Vol. 1, pp. 289–293, September 1995.

[9] T. Kwon, I. Park, Y. Choi, and S. Das, "Bandwidth Adaptation Algorithms with Multi-Objectives for Adaptive Multimedia Services in Wireless/Mobile Networks," *Proceedings of the ACM International Workshop on Wireless Mobile Multimedia*, pp. 51–59, August 1999.

[10] T. Kwon, J. Choi, Y. Choi, and S. Das, "Near Optimal Bandwidth Adaptation Algorithm for Adaptive Multimedia Services in Wireless/Mobile Networks," in *Proceedings of IEEE Vehicular Technology Conference — Fall*, Vol. 2, pp. 874–878, September 1999.

[11] T. Kwon, Y. Choi, and S. Das, "Bandwidth Adaptation Algorithms for Adaptive Multimedia Services in Mobile Cellular Networks," *Wireless Personal Communications*, Vol. 22, No. 3, pp. 337–357, September 2002.

[12] Y. Xiao, C.L.P. Chen, and Y. Wang, "Quality of Service and Call Admission Control for Adaptive Multimedia in Wireless/Mobile Networks," *Proceedings of the IEEE National Aerospace and Electronics Conference*, pp. 214–220, October 2000.

[13] Y. Xiao, C.L.P. Chen, and Y. Wang, "Fair Bandwidth Allocation for Multi-Class of Adaptive Multimedia Services in Wireless/Mobile Networks," *Proceedings of the IEEE Vehicular Technology Conference*, Vol. 3, pp. 2081–2085, May 2001.

[14] Y. Xiao and C.L.P. Chen, "QoS for Adaptive Multimedia in Wireless/Mobile Networks," *Proceedings of the IEEE International Symposium on Modeling, Analysis and Simulation of Computer and Telecommunication Systems*, pp. 81–88, August 2001.

[15] Y. Xiao and C.L.P Chen, "Improving Degradation and Fairness for Mobile Adaptive Multimedia Wireless Networks," *Proceedings of the IEEE International Conference on Computer Communications and Networks*, pp. 598–601, October 2001.

[16] Y. Xiao, C.L.P. Chen, and B. Wang, "Bandwidth Degradation QoS Provisioning for Adaptive Multimedia in Wireless/Mobile Networks," *Computer Communications*, Vol. 25, No. 13, pp. 1153–1161, August 2002.

[17] N. Nasser and H. Hassanein, "Connection-Level Performance Analysis for Adaptive Bandwidth Allocation in Multimedia Wireless Networks," *Proceedings of the IEEE International Conference on Performance, Computing and Communications*, pp. 61–68, April 2004.

[18] N. Nasser and H. Hassanein, "Prioritized Multi-Class Adaptive Framework for Multimedia Wireless Networks," *Proceedings of the IEEE International Conference on Communications*, Vol. 7, pp. 4295–4300, June 2004.

[19] Y. Xiao, C.L.P. Chen, and Y. Wang, "An Optimal Distributed Call Admission Control for Adaptive Multimedia in Wireless/Mobile Networks," *Proceedings of the International Symposium on Modeling, Analysis and Simulation of Computer and Telecommunication Systems*, pp. 477–482, August/September 2000.

[20] Y. Xiao, C.L.P. Chen, and Y. Wang, "Quality of Service Provisioning Framework for Multimedia Traffic in Wireless/Mobile Networks," in *Proceedings of the IEEE International Conference on Computer Communications and Networks*, pp. 644–648, October 2000.

[21] F. Yu, V.W.S. Wong, and V.C.M. Leung, "A New QoS Provisioning Method for Adaptive Multimedia in Cellular Wireless Networks,"*Proceedings of the IEEE Annual Joint Conference of the Computer and Communications Societies*, Vol. 3, pp. 2130–2141, March 2004.

[22] F. Yu, V.W.S. Wong, and V.C.M. Leung, "Efficient QoS Provisioning for Adaptive Multimedia in Mobile Communication Networks by Reinforcement Learning," *Proceedings of the International Conference on Broadband Networks*, pp. 579–588, October 2004.

[23] http://www.gnu.org/software/glpk/glpk.html.

[24] http://www-dii.ing.unisi.it/ giorgetti/downloads.html.

[25] A.-E.M. Taha, H.S. Hassanein, and H.T. Mouftah, "On Robust Allocation Policies in Wireless Heterogeneous Networks," *Proceedings of the 1st International Conference on QoS in Heterogeneous Wired/Wireless Networks*, pp. 198–205, October 2004.

[26] A.-E.M. Taha, H.S. Hassanein, and H.T. Mouftah, "The Effect of Joint Allocation Policies on Preference-Triggered Vertical Handoffs," *Proceedings of the IEEE International Conference on Wireless and Mobile Computing, Networking and Communications*, pp. 57–63, August 2005.

[27] A.-E.M. Taha, H.S. Hassanein, and H.T. Mouftah, "Exploiting Vertical Handoffs in Next Generation Radio Resource Management," *Proceedings of the IEEE International Conference on Communications*, pp. 12–14, June 2005.

Chapter 10

Advanced Radio Resource Management for Future Mobile Networks

Jijun Luo and Honglin Hu

Contents

10.1   Introduction

Along with the rapid growth of wireless communication during the past few decades, the mobile communication system has been widely accepted as a vehicle to convey information between people. However, the future mobile radio network faces challenges of high *Quality-of-Service* (QoS) requirements by supporting high mobility and throughput for multimedia services with the encountered scarcity of spectrum resources. The frequency spectrum physically limits the capacity of a radio network; thus, effective solutions to increase spectrum efficiency are required.

Terminal, user, service, and *network* are four primary participants in a mobile communication system. The effective solutions setting the communication configurations between the *terminal* and *network* are managed by us to meet the requirements of the *services* demanded by the *users*. That is, we need to find the proper communication mechanisms between the communication apparatuses based on the characteristics of the *users* and the *services*. Digital communication systems are replacing the old analog ones (First-Generation or 1G) due to the improved spectrum efficiency through digitalization, implemented source coding, channel coding and modulation schemes, and other higher-layer protocols.

Future radio systems will require high connectivity and multiple functions. One trend involved in the development of antenna technology concerns the network being able to be supported by multiple antenna resources. The antenna elements can be distributed, so that a *connection everywhere* and *optimizable* network can be organized. Distributed antenna systems called *distributed wireless communication systems* (DWCS) are currently proposed in some literature; they aim to offer the optimal network constellation in terms of the location of *Access Points* (APs) and antenna beamforming bestowed by smart antenna techniques [37]. Such systems will result in better connectivity, in the sense of low signal attenuation resulting from smaller cell size and high diversity gain. Facing the era with new experiences, advanced *radio resource management* (RRM) mechanisms should be developed accordingly.

On the other hand, we face nowadays *heterogeneous networks* distinguished by different *radio access techniques* (RATs). In addition to the RATs in the Second-Generation (2G) systems (e.g., GSM [Global System for Mobile communications], IS-95 [Interim Standard 95]), Third-Generation (3G) systems such as UMTS (Universal Mobile Telecommunications System) and various WLAN (wireless local area network) systems have been standardized and already operate in some regions. In addition, new wireless standards such as WiMAX (Worldwide Interoperability for Microwave Access) are emerging in an effort to extend the limited coverage of WLANs.

Consequently, along with the heterogeneity of different radio networks, we also have to face the situation of coexisting networks. Therefore, further questions arise as to how to manage the traffic in heterogeneous networks in an efficient way.

Trends for terminals are mainly toward high levels of integration in terms of multi-antenna, multi-mode, and multi-band, where the multi-mode capability enables terminal access to different RATs, and the multi-band capability enables terminal access to different carrier frequencies simultaneously. As the 3GPP documentation [3] describes, four possible types of multi-mode 3G terminals are defined according to their capabilities of concurrent reception and monitoring more than one radio link.

10.2   Radio Resource Management in General

Typical problems are encountered when we design a radio network, including signal attenuation, terminal noise, fast fading due to multi-path phenomena, shadowing, multiple access interference (MAI), and other system-specific problems. These typical problems challenge us in using radio resources efficiently. The radio resource is not only, by definition, the radio spectrum, but also the access rights for individual mobile users, the time period a mobile user is active, channelization codes, transmission power, connection mode, etc. The radio resource requires that management functions be designed in different time scales.

In general, RRM is a set of controlling mechanisms supporting intelligent radio access by means of segregation and pooling of radio resources, distribution of traffic, selection of the optimal transmission format using the optimal time, etc. RRM works over multiple dimensions in terms of frequency, space, time, orthogonal codes, and RATs, with various time scales ranging from milliseconds to days, based on the profiles of users, services, networks, and terminals, thereby aiming at optimized usage of radio resource and maximized system capacity.

Advanced RRM can operate in centralized mode, decentralized mode, or hybrid mode. The centralized mode requires intelligence and decisions on the network side; whereas, the decentralized mode primarily depends on the terminal side. The hybrid mode allows the radio network to deploy decentralized functions simultaneously with other centralized mechanisms. More specific examples are given in later sections. When the network infrastructure and signaling permit, centralized approaches show better performances than the others. We focus in this chapter primarily on the centralized mode. The RRM mechanism can decide resource allocation to *mobile stations* (MSs) based on their terminal profile (e.g., multi-antenna in a distributed antenna system). Furthermore, radio resources from different radio networks can be managed jointly to solve the encountered problems more effectively.

10.2.1 Common Functions and System Dimensioning

10.2.1.1 ISO/OSI Dimension

RRM for a radio network is studied by dimensioning its functions. One dimension follows the layers of the radio network. A layer is a form of hierarchical modularity and one of the central principles of network design. Advantages such as simplicity, cooperability, standard, efficiency, and reproducibility motivate and force hierarchical modularization [6].

From an ISO/OSI viewpoint, several layers are of interest to us. The *physical layer* (PHY) functions provide links for transmitting information and control bits between communication peers (e.g., between the MS and the *base station* [BS]).

A layer higher is the *data link control* (DLC) layer, which converts the unreliable bit pipe from PHY to a higher level as blocks of bits. The vulnerability of successful transmissions caused by fading, distortions, and interference will be further reduced by the DLC layer. Inside the DLC layer, normally the *radio link control* (RLC) layer, the *medium access control* (MAC) layer, and other sublayers specific for services such as broadcasting and IP-based data transmissions are defined.

The network layer is the third layer, which is immediately above the DLC layer. At this layer, user packets reassembled from the DLC layer are added with network layer headers and then routed between network entities.

UMTS has termed the *radio resource control* (RRC) layer at the network layer and defined its function inside the control plane, which does not contain information from the user data at the user plane. RRC is actually a specific term for radio systems and terminates in the radio network controller (RNC). This layer controls the usage of resources and as well as all other lower layers and RRC protocols that terminate here [35]. The transport, session, presentation, and application layers are therefore out of the scope of this chapter.

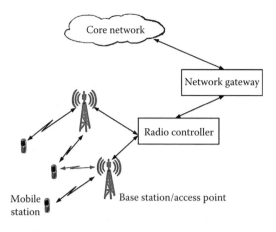

Figure 10.1   Network architecture reference model.

10.2.1.2   Dimension over the System Level and Services

Another dimension is through the network entities viewed at the system level. At this dimension, different hierarchical modules penetrate into the involved network entities. Therefore, RRM function allocation is also intimately related to the network architecture through this dimension. In this chapter we assume a simple model representing classical radio network architecture (Figure 10.1).

Additional to the dimension through the OSI layer and system level, we still can go ahead with the dimension between the switching schemes (e.g., circuit switched and packet switched), along with the multiple access schemes discussed in Section 10.2.2.1.

10.2.1.3   Function Allocation

This section describes the common functions and allocates them according to the aforementioned dimensions. These described functions exist in most wireless systems:

■ *Power control.* Power control is important for systems with radio signal attenuation. There are several kinds of power control mechanisms, namely, open-loop power control and the closed-loop power control. Closed-loop power control includes inner-loop power control and outer-loop power control, differentiated by their update rate and function, respectively. Some faster power control algorithms can even compensate for fast fading. An example is given in Section 10.2.2.2.

■ *Admission control.* The system capacity evaluated by the available radio resources will be reduced after a new call admission. One significant impact is the increase in interference in a highly loaded system. In this case, the system may enter an unstable state and may even lead to call dropping. Hence, admission control is also used to achieve high capacity and maintain system stability. In a CDMA system, the cell coverage reduces due to the increase in interference, which is usually known as the cell breathing phenomenon. Admission control is needed to limit the interference in order to secure the cell coverage. Admission control can vary its controlling parameters according to the radio context (e.g., service level, load, and profile) in order to optimize the system capacity.

■ *Handover.* Because of the default cellular architecture employed to maximize spectrum utilization, handover is extremely important in any ongoing radio links in a mobile cellular system. Both when MS is at the cell border or suffers from a bad connection, such as deep fading or high interference at the current cell, a handover may be triggered. The scenario with more than one radio cell being involved during the handover phase is termed a *soft* or *softer* handover.

   The MS or the network holds information about cells that are served by a connection or might be candidates for a handover. To save the battery life of the MS, only the cells previously indicated by the network (monitor set) are monitored.

■ *Load control.* An overloaded system might still occur even an efficient admission control algorithm and efficient scheduling procedure are implemented. The load problem is particularly severe in the uplink where the high interference is more hazardous to the system. When reaching overload, system performance rapidly degrades. Such a situation will cause a *nervous* system, such as high transmission power of all the MSs. The load control can employ the following measures to reduce system load:
   ■ Downlink fast load control to deny downlink power-up commands received from the MS
   ■ Uplink fast load control to reduce the uplink *signal-to-interference ratio* (SIR) target value
   ■ Slightly degrading the QoS of the users in the overloaded cell during the time it takes to resolve the overload situation, such as by lowering the data rate of one or several services that are insensitive to increased delays
   ■ Performing inter-frequency handovers or intersystem handovers; removing one or several connections

   The load control is activated once the load threshold is exceeded. For parameters such as interference level, SIR distributions can be used to measure the instantaneous load. The same measurements as those in the admission control are used but they must be updated continuously because the considered values change rapidly.

■ *Packet scheduling.* Packet scheduling is a fundamental function of wireless communication systems controlled by the packet scheduler. Its functions are to divide and allocate the available air interface capacity between the packets from different MSs. All systems with packet switched transport must be capable of supporting this function. In the implementation, the network decides the transport channel to be used for each user's packet data transmission. In addition, the network monitors the packet allocations and the system load.

Assume that all users have the same QoS demand, irrespective of the user queue. By assigning the resource (a radio access) to the user with the best channel capacity, multi-user diversity gain can be obtained. The principle is illustrated in Figure 10.2. Theoretically, assuming that all the users to be scheduled have i.i.d. *signal-to-interference plus noise ratio* (SINR) density function as $f_S(s)$, then the complementary *cumulative density function* (CDF) for an SINR threshold S_T is defined as $\bar{F}_{S_T}(s) = \int_{S_T}^{\infty} f_S(s)ds$. With a total of N users controlled by the scheduler, probability of at least the CDF of one user being higher than the threshold value is readily $\sum_{i=1}^{N} \binom{N}{i} \bar{F}_{S_T}^i(s) \left[1 - \bar{F}_{S_T}(s)\right]^{N-i}$. A simple example is given with the assumption of Rayleigh distributed SINR and i.i.d. for all MSs, that is, $f_S(s) = \frac{1}{2\sigma^2} e^{-s/2\sigma^2}$, where σ^2 represents the variance [32]. It is a rather simple assumption without considering co-channel interference and shadowing effects. The probability distribution depicted in Figure 10.3 shows the multi-user diversity gain.

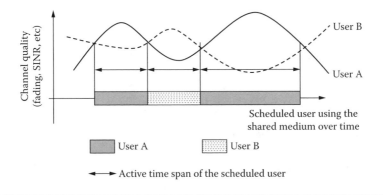

Figure 10.2   Scheduling based on channel quality.

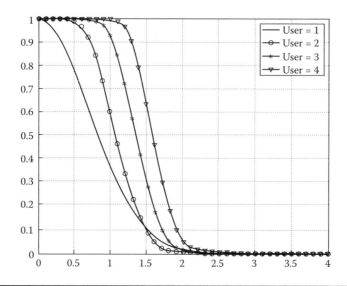

Figure 10.3 **Multi-user diversity gain with Rayleigh fading.**

10.2.2 System-Specific Functions

To deal with wireless communications with multiple involved entities in the radio environment, multiple radio medium access schemes have been developed. Consequently, the RRM mechanisms differentiate from each other according to the basic multiple access schemes that the wireless system selects. In general, radio links with respect to different MSs can be classified into centralized and distributed categories. In the centralized category, multiple access techniques can be differentiated by the basic radio resource units one MS obtains at a time exclusively. Time, frequency, orthogonal code, and even spatial resource are the radio units. Thus, radio systems nowadays have a selection or combination among *Time Division Multiple Access* (TDMA), *Code Division Multiple Access* (CDMA), *Frequency Division Multiple Access* (FDMA), and *Space Division Multiple Access* (SDMA).

In the centralized multiple radio access category, MSs follow the instruction as *Resource Assignment* from the radio network to process an access. In contrast, a number of distributed medium access protocols are necessary for the completion of the centralized radio system or favor one of many new emerging radio access techniques.

Examples of such multiple access protocols included Aloha, slotted Aloha, Carrier Sensing Medium Access (CSMA) with Collision Avoidance (CA) or Collision Detection (CD). A typical example is the ad hoc network using more peer-to-peer communication between MSs and APs. IEEE 802.11 Wireless LAN deploys the CDMA/CA protocol, which is similar to

CD used by the 802.3 Ethernet wired line standard. The CSMA/CA protocol avoids collisions by measuring the detection signal power.

Radio access attempts from MSs that have not obtained the control link in order to get the *resource assignment* normally use the slotted Aloha technology, which has been shown to double the capacity of the pure Aloha system [13,23].

10.2.2.1 Radio Access Scheme Identification

In classical traffic engineering, when radio systems deploy the aforementioned multiple access schemes and allow the admitted call to maintain the assigned resource units until termination, the system capacity can be evaluated using the developed queueing models suitable for circuit switched systems. A typical example is the $M/M/m/m$ model, which is known as the Erlang B formula [22].

More data services penetrate into the radio access network along with the evolution. We are able to assign the resource to MS more temporarily with the help of queueing user plane data without letting the delay be noticed. At the network node (e.g., RNC or gateway), we can identify the correlation in the arrival process with the backlogged processes at the output ports of the switch. The backlog is related to the delay that will be experienced at the output port of the resource controller. To design such a system, the total delay as an important component of the QoS parameters must be specifically considered.

10.2.2.2 Power Control in a WCDMA System

Take the *Wideband CDMA* (WCDMA) as a typical radio system, for example. Fast closed-loop power control is employed both in the uplink and the downlink. This function is the most important aspect in WCDMA for the circuit switched services using dedicated channels, especially for the uplink. Without power control, near-far effects may arise because the power from the MSs near the BS is higher than that of other MSs located further away. The fastest power control rate is 1.5 MHz resulting from power control at each time slot, and the power control step can be varied adaptively according to the MS speed and operating environment. Normally, closed-loop power control is based on SIR values, which requires that the receiver compare the estimated SIR with a target value and commands the transmitter to increase or decrease the power accordingly. The power control command increases or decreases the power of all physical channels on one connection. Fast power control is able to compensate for path loss, shadowing, and also fast fading.

The SIR target value is controlled by outer-loop power control. The outer-loop power control is needed to keep the quality of communication at the required level by setting the target for the fast power control. The outer loop measures the link quality, typically a combination of frame and *bit-error rates* (BERs), depending on the service, and adjusts the SIR target value accordingly. In addition, the outer loop is used to independently control the relative power of different physical channels belonging to the same connection. As an example, the *Dedicated Physical Data Channel* (DPDCH) and *Dedicated Physical Control Channel* (DPCCH) power difference can be controlled by the outer loop while taking into account the variations in DPDCH coding gain for different environments. The frequency of the outer-loop power control is much slower than that of the inner-loop power control and is typically 10 to 100 Hz.

10.2.2.3   Handover in a CDMA System

The *soft handover scheme* is a feature used in the UMTS system. The spectrum efficiency of the system and the quality of the call in progress are significantly affected by handover signaling [25]. In the WCDMA system, the parameters embedded in the radio protocol can be controlled by the network to optimize handover performance resulting from the handover ratio and delay time.

During its ongoing call, an MS can have either one link or multiple radio links established to the network. *Softer handover* means maximum-ratio-combining of two incoming signals and this is always applied in the downlink or at one site in the uplink. In contrast, when two different sites are selected, selection diversity in the uplink (termed soft handover) is adopted. The branch with the best BER is chosen among the active links in the uplink, presuming that the received blocks are error-free.

The candidate set holds all cells that fulfill the requirements for a potential handover. The requirement is that the cell must be good enough to serve as a connection with the MS. The active set contains all cells that are currently serving a connection. If the connection is in soft handover, the active set can contain more than two cells, while in all other cases only one cell. Here we choose pathloss-based handover to simplify our description. This is allowed when we have homogeneously distributed users. In addition to the received signal strength at the MS, the MS decodes the broadcast information and the network can determine the ranking of the received cells. The MS chooses only those cells with lowest attenuation resulting from the distance to the candidate/active BS, operating frequency, antenna gain, and shadowing effects.

10.2.2.4   Packet Scheduling and Function Evolution

WCDMA packet access allows nonreal-time bearers to use common, dedicated, shared channels and high-speed channels dynamically. The network allocates a bit rate for a bearer and possibly changes this bit rate during an active connection.

The packet scheduler is located in the RNC where scheduling can be done efficiently for multiple cells, also taking into account the soft handover connections. The BS provides the measurements of the air interface load for the packet scheduler. If the load exceeds the target, the packet scheduler can decrease the load by decreasing the bit rates of packet bearers. On the other hand, if the load is less than target, the packet scheduler can increase the load by allocating more data.

In Release 99 and Release 4 of 3GPP specifications, all RRM is performed at the *Controlling RNC* (CRNC). And DSCH was the only shared channel in the downlink. However, due to the latencies across the Iub interface between the RNC and BS, and RRC connections from the RNC to the MS, CRNC cannot accurately manage resource allocation based on up-to-date *channel state information* (CSI). In the *high-speed downlink packet access* (HSDPA) system, the CRNC allocates a block of *high-speed DSCH* (HS-DSCH) resources (code tree and transmit power) to the BS, and the BS is allowed to autonomously allocate resources between MSs on a dynamic basis. Because there exists only the Uu interface between the BS and the MS, the BS can collect CSI from the MSs by means of fast signaling from the physical layer and then perform fast HS-DSCH allocation to MSs. HSDPA adopts an advanced adaptive *modulation and coding scheme* (MCS). Upon each radio access instance of an MS, the BS is able to optimize its transmission format according to the CSI. The MS sends to the BS an indication of the radio link quality using the CSI. Based on this, the BS can adapt the MCS in terms of *adaptive modulation and coding* (AMC), where the modulation schemes are changeable between QPSK and 16QAM and coding schemes can be adjusted through an amount of puncturing applied to turbo-coded data by means of varying the amount of data information transmitted in a *transmission time interval* (TTI), that is, 2 milliseconds [5]. During the TTI, the selected coding rate and modulation format remain constant. In addition, the transmission power of the HS-DSCH is kept constant, and the coding and modulation format is chosen to maximize the throughput to the MS. Increasing the coding rate or moving from QPSK to 16QAM increases the size of the transport block that can be sent during the TTI. The link adaptation scheme in HSDPA is extended by the multiple-code transmission option. Under good channel conditions, the BS can exploit multi-codes for a user in order to reach optimum overall throughput.

Following the HSDPA example, we can reconsider or reallocate functions to the new entity as they used to be. Guidelines for us to reconsider the solutions to the function reallocation are principles such as lower latency at the *radio access network* (RAN); high spectral efficiency; high peak rate and average throughput; reduced cost per bit, including transmission, interoperability, high resource utilization, high backward compatibility, reuse of defined control plane; fast market introduction, and protection of investment. With those principles, we have to introduce evolving paths to the current telecommunication network. Such function reallocation from classical commercial central resource control (e.g., RNC) will experience a migration toward the radio front; that is, the BS, in order to further support high demanding data service without losing QoS, additional functions such as flow control and inter BS context management for MSs must be deployed.

10.2.2.5 Trunk Resource Allocation in an OFDMA System

In general, the emerging *Orthogonal Frequency Division Multiplex* (OFDM)-based radio systems attract a large user population, and demand new design methods quite different from those used in conventional systems. OFDM distributes the data over a large number of subcarriers that are spaced apart at precise frequencies. The orthogonality provided by the signal construction prevents the demodulators from receiving interferences caused by carriers other than their own. The IEEE 802.11a standard and HiperLAN 2 use OFDM in the 5-GHz band [15,20]. OFDM systems deploy dynamic resource allocation in the frequency domain, called OFDMA. Currently, WIMAX [7] and 3G evolution [2] are typical examples.

Future systems based on OFDM hold characteristics such as new radio interface and advanced transmission schemes, new network architecture, and new services. One of the main objectives of RRM in such a system is to handle and to avoid interference due to spectrum reuse in order to make efficient use of this scarce natural resource. Of particular interest is the RRM for wireless multimedia systems requiring a multitude of data rates and transmission qualities. Viewing RRM functions at protocol layers in addition to the physical layer, overall system performance highly depends on the mechanisms defined at the MAC, DLC, RLC, and RRC layers.

We encounter the following impacts under the OFDMA system:

- *Scalability of the radio resource.* A basic radio resource unit given by the OFDMA system, called a *chunk*, is determined by not only code and time, but also by frequency and space due to the introduction of *multiple-input-multiple-output* (MIMO) techniques. Therefore, an evaluation of average system load over the entire base band might be biased. This nature of finer grained resource allocation in fact gives addition flexible freedom for the RRM scheme.
- *Sensitivity for system load information.* Optimal indications of the system load (e.g., at the RRC layer) and the system load information selected for the OFDMA-based system would not only be the total transmission power at the downlink or the interference value at the uplink, as for the Universal Mobile Telecommunications System (UMTS) FDD system. The reason behind this is the characteristics of the system blocking nature and the resource allocation flexibility compared to the conventional system.

- *Changes in the network architecture.* Referring to the 3G systems, future system architectures below the RNC might include the absence of an Iub interface. The same tendency can also be applied to the inter-RNC interfaces. Changes in network architecture in the end limit the application of the RRM functions, such as soft handover and slow measurement reports during the handover phase. Combined with the chunk-wise resource allocation scheme, optimal hysteresis values for system-level handover will depend on the chunk size, that is, the scalability of the system (refer to Section 10.2.2.3).

- *Impacts on admission control and load estimation.* Future high bursty traffic with *long-tail distribution* affects admission control and load estimation. For example, the peak number of users at a given time will not be valid for load indication. On the other hand, the admission of new users should depend on the user service profile. For example, rejecting an arrival data user simply by just comparing the peak rate of the service to the remaining capacity degrades the performance of the system.

10.3 Radio Resource Management for Distributed Wireless Systems

A deterministically planned wireless network purely based on the busy-hour traffic cannot match the variant emerging traffic demands. Hence, *extensible* systems, featuring flexible and scalable network elements, are desired. Possible future radio network architectures are equipped with radio controlling mechanisms for multiple APs, which allows for tight interworking among them through *central units* (CUs). A network can contain multiple cells defined by inserted APs that can be immediately controlled by the CU after plug-in. The CU gathers signal processing and RRM tasks from the distributed antenna elements. This philosophy is similar to the *Common Public Radio Interface* (CPRI), which decomposes the BS into two building blocks: (1) the *Radio Equipment Control* (REC) and (2) the *Radio Equipment* (RE) itself. The REC provides access to RE via the interface between REC and RE, whereas RE serves as the APs [12].

The advantage of such an architecture not only eases the integration of newly deployed entities, but also provides deployment flexibility. Following the same principle, the antenna elements can be distributed, so that a *connection-everywhere* and *optimizable* network can be organized. With the continuous increase of media demand and high mobility, future systems will encounter quite temporal-spatial changing traffic, which does not necessarily show a uniform distributed traffic pattern. Following the *Open Wireless Architecture* (OWA) philosophy such as *plug-and-play,* future DWCS will experience differently. The traffic characteristic in each small cell defined by each AP is more sporadic and unpredictable. After deployment of the distributed APs, the spatial-temporal varying traffic must be catered to through advanced RRM mechanisms.

Performance of the DWCS will significantly improve by introducing a novel beamforming technique, termed *organized beam hopping* (OBH). In this case, we do not randomly select a *beamforming vector* (BV) as in the opportunistic or random beamforming techniques; instead, we hop the beamforming vectors in an organized pattern. In addition, the organized pattern is changeable according to the distribution of the MS in the cell. As illustrated [19], the OBH technique has the following advantages. First, using OBH, we could attain good performance for both the small and the large number of MS cases. Second, the OBH technique does not need training sequences for every beam. Moreover, the OBH technique is capable of reducing co-channel interferences by assigning orthogonal beam-hopping patterns to the neighboring cells. Finally, the performance improvement of the OBH technique need not rely on the number of the antennas at the receiver side, so the OBH technique is applicable for both MIMO and *multiple-input-single-output* (MISO) scenarios.

Although OBH is able to hop the BVs in an organized pattern suitable for scenarios with both large and small numbers of MSs, we still encounter challenges such as the in-time reactions of the *beam hopping pattern* (BHP) to the fast changing traffic and the high probability of beam-hitting. When two beams in adjacent cells hit, that is, point to the same place, if we serve different users using a single frequency, the system will experience severe co-channel interference for some of the serving channels. Further, users will experience completely different characteristics compared to today; for example, the user is willing to be always connected to the network. Due to the relative longer OFF period compared to the shorter ON period, many users are in a *dormant* mode, that is, a mode between *idle* and *active*, which is also currently identified by the 3G evolution discussions [4]. In the connected period, some user actions might change the profiles of the MS that are relevant for the RRM functions. One example is that the user might plug a device with an additional power supplier, or embed additional antenna elements to the end device. Therefore, the timing relationship between the natural events will be: *connection time > profile updates > data bursts > channel changes*, which is also different from today's experiences. In addition, unbalanced load between neighboring cells due to relatively smaller cells and high bursty traffic might cause a *catastrophe* problem such as exponentially increased delay time. Therefore, an advanced RRM mechanism is needed, one that is expected to have the profile awareness, efficiency in solving the beam-hitting problem, load-balancing, and providing optimal transmission format and fast reaction to traffic changes.

10.3.1 Multi-loop RRM

A suitable RRM approach involving an efficient multi-loop RRM control protocol specified for a system with fast varying traffic is introduced in this section. The multi-loop RRM approach contains a radio protocol for on-demand uplink reporting and an efficient signaling solution. In the outer loop, a cell-level cooperative power control is applied in order to cater to the spatial changing traffic. In the middle loop, a profile-based scheduling algorithm according to the capabilities of the MS is defined. Cooperative power control is defined in the inner loop. In addition, an enabling mechanism with the capability of beam-hitting awareness to support a *hybrid transmission scheme* is included in the middle-loop control circuit. The outer-loop control function provides the optimal network configuration to the middle-loop and inner-loop control functions. One of the most important configurations is the overall transmission power of an AP, which consequently results in different cell ranges compared to the neighboring cells. Whenever the performance at the outer loop is observed to be not sufficiently good, further network triggered handover, resource request, call dropping, etc., will be required through other RRC functions. At this loop, changes in the BHP take place. Because these changes might cause beam-hitting, profile demands are automatically triggered by the outer-loop function through the interface between the central controller and the AP to AP in order to assist an efficient middle-loop control. At the middle-loop, profiles of the involving MSs are reported to assist proper multi-user scheduling, spatial multiplexing and cooperative power control in the inner-loop. The inner-loop functions adjust the transmission power, physical mode defined by modulation and, coding schemes, and preprocessing of signals for each link.

10.3.1.1 Outer-Loop Power Control

As explained above, the traffic loads are not necessarily equally distributed in neighboring cells. The network therefore must regularly estimate the traffic demands for each deployed AP. The estimated traffic demand includes information such as traffic class, traffic intensity, volume amount, etc. These information items are needed by the CU through the APs to control the cell coverage provided by each AP. By calculating the optimal power setting, cells level power in highly loaded cell is reduced; on the contrary, cell level power in lower loaded cells is increased. Consequently, more users are handovered to lower loaded cells due to their relatively larger coverage, and the traffic load in the neighboring cells is balanced. From a traffic management viewpoint, the possible catastrophic problems caused by high load are significantly reduced.

10.3.1.2 Profile-Based User Scheduling in the Middle Loop

At the middle loop, based on the profiles of MSs reported to the network, the network chooses the proper MSs for scheduling under the situation of beam-hitting. The procedure of potential beam-hitting awareness can be classified by:

- An MS that could detect the APs in the neighboring cell warns its associated APs (in the MS-centric case).
- Otherwise, the BHP between the neighboring cells must be mutually known (through an interface between APs directly or through their associated CU, controlled by the outer loop).

There are two resulting solutions for the beam-hitting case. One is the spatial multiplexing scheme, and the other one is cooperative power control. The spatial multiplexing scheme allows different traffic streams to transmit separately from independent APs to their respectively associated MSs. One example is given in Figure 10.6: MS 1 is associated to Cell 1 and MS 2 is associated to Cell 2, respectively. Beams from Cell 1 and Cell 2 hit each other at the moment. However, based on the profile report from the uplink that both MS 1 and MS 2 have sufficient receiving antenna elements, the network can schedule these two MSs at the same time. The reasoning behind this is the capabilities of both MSs being able to resolve independent radio paths of P 11 (AP in Cell 1 to MS 1), P 12, P 21, and P 22, so that the interference caused by beam-hitting from neighboring APs is avoided. The diversity mode is complementary to the spatial multiplexing scheme, and it has been described. Among selected users, we are able to embed an enabling technique called *hybrid transmission*, as detailed in Section 10.3.2. If the involving MSs do not have multiple receiving antenna elements, or their current services allow lower transmission power by the serving AP, the cooperative power control can be used. Cooperative power control requires additional signaling, which needs to be transmitted much more often than the outer loop and middle loop signaling, we therefore place it at the inner loop in the next section.

10.3.1.3 Cooperative Power Control

Without losing generality, for inner-loop cooperative power control, frequency deployment among APs with reuse 1 is assumed. At the downlink, the AP asks for the QoS target (e.g., SINR target) of the ongoing service from the MS by giving a threshold. Only the MS(s) that is higher than the threshold reports the SINR target, and all the MSs report the measured SINR for scheduling. Based on this information, the AP selects the MS(s) with the lowest power requirement to serve with suitable physical mode (modulation and coding). It may happen that more than one MS can be selected at the beam hopping period whenever the user data of the first MS is completed. In the uplink, the MS measures the received quality from the serving AP (and the ones from neighbor cells if applicable), and reports the AP its measurement and QoS target back to the AP. Whenever it obtains the token indicating the optimal power level (serving rights) from the AP, it adapts its physical mode accordingly. The process is done whenever potential beam-hitting should occur. In case of orthogonal BHP, no cooperative power control is needed because the beam-hitting can be identified by the network by comparing the BHPs for the neighboring cells. When the inter-cell beamformings are not jointly coordinated (e.g., without the presence of signaling between neighbor APs), it is up to the communication peers in a cell to monitor the radio environment and learn about possible upcoming interference situations during their communication process and to predict possible beam-hitting in order to fulfill the proposed procedure. Note that in the inner loop, the RRM mechanism is based on the exploitation of the beam-hitting prediction, that is, on the prediction of the expected interference. Based on this prediction, the suitable MSs with less power requirements are selected, and the corresponding transmission powers are coordinated between the adjacent cells, so that the high interference due to hitting beams is reduced.

10.3.2   Hybrid Transmission

In a DWCS, each MS is no longer limited to connecting to one dedicated AP. Instead, each MS could be flexibly served by one or multiple AP(s). Therefore, we develop a new *hybrid transmission* scheme for the DWCSs, where the MS could follow a separate serving, diversity, or data splitting mode, based on the concurrent SINR from each AP and the QoS requirements of the MSs [18]. In the separate serving mode, the MS transmits the data with only one AP. Separate serving mode typically occurs when the QoS requirement of the MS can be fully satisfied by one AP, or the signal transmitted from other APs cannot help the MS support a higher-order modulation scheme with the required BER. The separate serving mode is simple and easy to implement. Comparably, for the diversity mode, the MS is served by multiple APs at the same time. Diversity mode is applied when the received signal from each AP, respectively, is not powerful enough to support the required QoS of the MS, unless the macroscopic space diversity is exploited. In addition, if the combined signal could support a higher throughput than the summed throughput provided by the separate branches, the diversity mode will be applied. In the case that the data links between the MS and multiple APs are good enough (at least can support BPSK modulated stream with the required BER) and can be separated by, for example, orthogonal codes, it is possible to apply a data splitting mode at the MS. A data splitting mode could also be applied in the case where the MS or AP cannot support a higher modulation scheme, so that the diversity mode could not help to further increase the throughput. In the data splitting mode, the MS communicates with multiple APs simultaneously, by receiving from multiple APs and transmitting to the APs different data streams. Performances of data splitting is detailed in Section 10.4 in the case with multiple radio access techniques.

10.4 Joint Radio Resource Management for Heterogeneous Networks

An extension to RRM as introduced in Section 10.2, *joint radio resource management* (JRRM) controls the resources jointly over multiple radio networks or cell layers with the necessary support of multi-link MSs. In the following, we discuss the JRRM functions based on the network coupling architecture.

10.4.1 Network Coupling Architecture

From the network infrastructure viewpoint, the integration of networks and the inter-communication between them are emerging. As Figure 10.4 shows, the convergence toward an IP (Internet Protocol)-based core network and ubiquitous, seamless access to radio access techniques in different generations, augmented by self-organizing network schemes and short-range connectivity between intelligent communicating apparatuses enforce common MS and network entity platforms. The future evolution of 3G mobile wireless networks is widely accepted and guided by the vision of a concept called *mobile wireless internet* (MWI). In this vision, Internet access is granted by a ubiquitous mobile network anytime and anyplace; access by MSs will be the principle means of Internet access [30]. The MWI aims at an overall network with convergence toward an IP-based network and ubiquitous, seamless access among 2G, 3G, local broadband, short-range and broadcast wireless access schemes [17,30,35]. The interworking between some radio interfaces can be first obtained using the existing interfaces as an intermediate step before *all IP-based 3G* or *beyond 3G* systems are realized. This section provides a high-level summary of the architecture vision in terms of the interworking of radio access networks.

In general, four types of network coupling levels can be classified, as depicted in Figure 10.4. *Very tight coupling* has the coupling point at RNC shown by point ①; it requires high-end RNC or fast interworking between WLAN AP and RNC [14]. *Tight coupling* has the coupling point at the serving

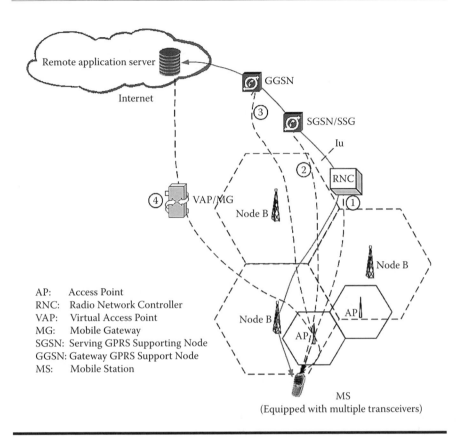

Figure 10.4   Network coupling architecture.

GPRS support node (SGSN) shown by point ②, which generates relatively slower interaction between subnetworks (e.g., handover, load balancing, etc.). There is no common control over radio resources. *Loose coupling* has the coupling point beyond the Gateway GPRS Support Node (GGSN) shown by point ③, *Mobile Gateway* (MG) or *Virtual Access Point* (VAP) shown by point ④. No Iu interface is involved [33] but common *Authentication, Authorization, and Accounting* (AAA) exists. *Open coupling* does not define any interworking between subnetworks, but only maximum off-line billing can be shared between them [14].

Followed by the possible coupling architectures for WLAN and UMTS [16], 3GPP in Reference [1] studies the feasibility and environment of interworking between 3GPP systems and WLANs with different levels. In that document, six different scenarios of 3GPP-WLAN interworking, ranging from common billing to the provisioning of seamless services between the WLAN and the 3GPP system, are given. From the viewpoint of JRRM, broadly, the levels of subnetwork coupling are classified into two categories: *core network coupling* (CNC) and *radio access network coupling* (RANC). The former level assumes the coupling between subnetworks is through the Switching Center, e.g., the *mobile-services switching center* (MSC) or the SGSN; whereas, the latter one is based on an integrated architecture between subnetworks, where the coupling point is a *radio resource controller* (RRC), which controls radio resources from all integrated subnetworks.

Viewing a telecommunication network consisting of a number of coexisting radio networks as a composite network of subnetworks, protocols for handling data, involving entities as well as the segmented from individual subnetworks, are needed. On the one hand, these protocols are able to deal with the heterogeneities inherited by the subnetworks so that higher conductivities are provided. On the other hand, we expect to deploy JRRM running over the protocols to increase the efficiency of the wireless connections. Possible future radio network architectures can be equipped with radio controlling mechanisms over multiple air interfaces, which allows tightly coupled subnetworks [14]. Various wireless standards and air interfaces are processed in a single central office, (RNC or hotel BTS) target at utilizing reconfigurable processing elements for great flexibility with low cost. It is also expected that multiple cells defined by different air interfaces can utilize the same optical link using *Wavelength Division Multiplexing* (WDM) techniques to have the connections directly connected to the CU. Given the broad bandwidth of the optical link, the BTS can be implemented as multi-band/multi-mode, covering many RATs. This step significantly simplifies the implementation of the JRRM mechanisms.

The JRRM algorithms not only span over sub-systems, but also over management layers and service types. The management architecture and strategy are based on the assumption of coexisting different RATs with different profiles, even flexible spectrum allocation [14,28]. The estimated traffic types, mean density, and the estimated volume are used by the CU to optimize the radio resource jointly. To enable a more proper resource allocation, the load information and traffic information between the co-operating subnetworks are also exchanged through the central controller or directly through the cooperating subnetworks. Inside each subnetwork, an efficient interworking between traffic volume, measurement (prediction) function, traffic scheduler, load control unit, and admission control function is also needed. The Traffic Estimation module in each system informs the administrative entity Session/call Admission Control (SAC) on the predicted traffic and planned traffic information to update the priority information for each connection and the admission decision within the subnetwork. The priority information is an input vector for the scheduling algorithm in a lower layer. The subnetworks can also be frequency layers.

The relevant terminal type in this chapter is at least equivalent to the third type of 3GPP MS, that is, one Tx and several Rx chains per MS, one of the Rx chains is used for faster scanning. Along with the common interest of UMTS *Extension Band* (EB) discussions, MSs being able to receive multiple downlink signals are foreseen by 3GPP [3].

10.4.2 Joint Admission Control

The interworking between different subnetworks requires new protocols defined for convergence reasons. An example for subnetwork interworking is through IP, as widely accepted. Functions designed for the common resource controller interfacing the packet-based convergence sublayers are desired to guarantee QoS. Due to the heterogeneity of coexisting different networks, many different policies are conceivable for JRRM, in particular when considering legacy and new network types. Systems in different generations are equipped with different functionalities, protocols, and management requirements. MSs having simultaneous connections to different RATs is one possible operation mode. In general, loose up to very tight coupling schemes between different types of networks must be considered for such multiple connections. For a possible tight coupling between a UMTS subnetwork and a WLAN subnetwork, one must consider the restrictions in each subnetwork; for example, the transport block size and minimum transmission time interval for each are differently defined according to the specifications. Tight coupling allows joint scheduling of traffic streams between involved networks and MSs. The conventional admission control is designed for each access system working independently among coexisting access systems and RATs. In the cooperating environment, a *JOint Session/call Admission Control* (JOSAC), which takes neighbor RAT system load into account, must be defined. The traffic stream can be routed alternatively through the cooperating subnetworks according to the constraints and the capacity of each. The wide coverage can be obtained by the universal cellular system (e.g., GSM, UMTS). In contrast, the high transmission rate can be obtained through a WLAN. With the information of estimated load in all the subnetworks (dynamic network profile), the *JOint Load Control* (JOLDC) entity, located together with the JOSAC, will distribute the traffic based on the characteristic of the coexisting RATs (i.e., the static and dynamic network profile), the QoS requirements for the service, and the number of applicants (e.g., which RAT and channels with committed capacity should be selected). The *JOint Resource Scheduler* (JOSCH) is important for MSs having simultaneous connections to different networks. The JOSCH is responsible for scheduling traffic streams being split over more than one RAT. In case the single traffic is split into two streams carried by two tight cooperating subnetworks, the admissions should be granted to the connections in both systems, and joint scheduling can therefore be applied. In the tight cooperating scenario, queues for each subnetwork will be filled up by the amount of data coming from the corresponding traffic sub-stream.

A typical queueing model adopted by Erlang B is used for two independently operated systems. If we assume the incoming traffic is based on a Poisson distribution with arrival rate λ and one basic channel has the capability to process the incoming call with rate μ, the call blocking probability for subnetwork i is calculated according to the Erlang B formula stated in Section 10.2.2. The blocking probability for the overall system is then calculated as the average of over all available i. If JOSAC is implemented to the cooperating subnetworks with M_i basic channels each (i.e., the incoming calls can be admitted by any subnetwork alternatively and only stay in the SAC level), the call blocking probability is derived as:

$$p_{B,C} = \left[(\lambda/\mu)^{\sum_i M_i} \Big/ \sum_i M_i! \right] \Big/ \left[\sum_{k=0}^{\sum_i M_i} (\lambda/\mu)^k / k! \right] \tag{10.1}$$

with subscript i indexing the subnetworks. It can be easily derived that JOSAC outperforms the noncooperating subnetworks.

10.4.3 Joint Scheduling Mechanism

The Internet Engineering Task Force (IETF) proposed IP *multi-homing* concept [8,29] and this concept is contributed by research projects such as IST MIND [21]. This framework manages IP traffic being routed through different RANs to the same MN. At the IP layer, the multi-homing algorithm allows one to route the traffic for each individual stream through a specific interface according to the type of traffic. An extension to the conventional multi-homing concept is to run simultaneous connections on the radio-frame level, which we call w.r.t. reconfigurable MSs as the *adaptive radio multi-homing* (ARMH) approach. ARMH is an overall management framework extended from the IP MH concept. It provides multiple radio accesses for multi-mod/multi-band MSs in order to allow the MS to maintain simultaneous links with radio networks. It selects the most proper JRRM function based on the identified information from the cooperating subnetworks, terminals, users, and services. To support the selected JRRM functions, proper traffic classification, calibration, interworking between the service application server and RRCR, and the configuration of transmission format as well as MAC protocols are managed by ARMH.

The system capacity gain obtained from the JRRM is, in principle, the enlargement of the number of operational servers from the queueing model viewpoint, which therefore results in a higher trunking gain. On the other hand, by alternatively allocating the resource to call units among the interworking radio networks or frequency layers, the load balancing effect among the radio networks is realized. In a typical soft blocking sensitive radio network, such an effect is very significant. In addition to the capacity gain from a network operation point of view, the advantages of having concurrently parallel streams are manifold. If one bearer service has a high availability in the network (low data rate bearer services result in high coverage, e.g., a 16 Kbps service is available in 99 percent of the cases), this link would be used for transferring important information to the MS. On the other hand, a low data rate service cannot fulfill the requirements for multimedia traffic, resulting in high data rate demands. If traffic is intelligently split into rudimentary and optional information streams, a higher QoS for the user is provided. Whenever possible, the user combines both streams for yielding a higher QoS and, due to the higher availability of a lower data rate service in UMTS, a minimal QoS can be fulfilled to the user.

We study in the following the performance comparison between JOSAC and JOSCH in the packet switched domain. As the M/G/1-PS model [23] shows the relationship between the response time when the processor is being shared and the service time when there is only one customer in the network, the delay factor $f = 1/(1 - \rho)$ can be used to determine the performance degradation due to resource sharing compared to the single user case, where ρ stands for the average system utilization factor. The term "delay factor" was first used by Lindberger [26] to derive the analytical model for multi-server processor sharing system. Suppose now we have r servers; if the number of customers N is less than the number of servers r, and user capacity is constant w.r.t. servers, calls can be processed without delay. In the case $N > r$, i.e., the busy period when all r servers are occupied, the state probabilities can be shown to have geometrical proportions (i.e., $p_k = \rho^{k-r} p_r$), as the procedure to prove the queue length in an $M/M/1$ system, the average queue length is $\rho/(1 - \rho)$. Therefore, the mean delay factor for the customer during the busy period is $1 + \rho/[r(1 - \rho)]$. Because the average system utilization factor is ρ, the proportion of traffic served during the busy period is $p[N \geq r]/\rho$. The average delay factor results from the proportion of traffic being served during the nonbusy period when $N \leq r$ and the proportion of traffic during the busy period:

$$f_R = \left(1 - \frac{p[N \geq r]}{\rho}\right) \cdot 1 + \frac{p[N \geq r]}{\rho}\left(1 + \frac{\rho}{r(1 - \rho)}\right) = 1 + \frac{p[N \geq r]}{r(1 - \rho)}$$

(10.2)

The probability in the busy period is simply the queueing probability resulting from the Erlang C formula [22]. The Erlang C formula is generally valid for an FCFS queueing model where the packets are delayed when all servers are busy. It gives the probability that the call has to wait. In this model, the probability $p[N \geq r]$ is identical to the waiting probability described in Erlang C.

A very important nature holds: for a terminal that receives relatively the same capacity offered by two cooperating radio networks, simply by allowing simultaneous radio access at each scheduled period, the response time for JOSCH will be reduced compared to the one given by the JOSAC algorithm. This nature can be easily proven by comparing Equation 10.3 and the M/G/1-PS delay factor, considering the fact that end user capacity in JOSCH doubles the one in the JOSAC case; in mathematical formulations, the following nature holds:

$$\frac{1}{2(1-\rho)} < 1 + \frac{P[N \geq r]}{r(1-\rho)} \qquad (10.3)$$

10.4.4 Intersystem Handover

Standardization bodies have defined handover requirements and procedures among RATs to allow for fully exploring radio resource. From GSM Release 99 [9,10], it is defined that handover from GSM to UTRA will be supported. It is specified in 3GPP that group vertical handover between UTRA to GSM is intended [9].

According to the introduced JRRM mechanisms in terms of JOSAC and JOSCH (see previous subsections), their proper implementation can be selected based on the knowledge of network coupling architecture. For the JOSAC approach, if the maximum delay constraint by the bearer service is τ_{MC} and the minimum delay caused by the handover between subnetworks is τ_{MH}, JOSAC cannot be applied if the following condition is not fulfilled: $\tau_{MC} \geq \tau_{MH} + T_J$, with the jitter taking place caused by propagation delay and processing delays by network elements during the handover process. As the basic difference between the JOSAC and JOSCH approaches is whether the radio system is able to support simultaneous radio connections in subnetworks, JOSCH by definition requires less signaling delay between subnetworks. The functionality of the coupling point (see Figure 10.4) supporting levels of interworking subnetworks plays an important role. From our intuition, more tightly interworking subnetworks grant less signaling delay and therefore the minimum delay is reduced, which allows more freedom for JRRM implementation. Two basic definitions based on the structure with respect to the signaling flow through the coupling point are therefore given to clarify the constraints of different JRRM schemes.

In this section we go one step further with the proposal, called as *Generalized Processor Sharing* (GPS)-based vertical (intersystem) soft handover [31]. Here, the term "soft" means the link in the old sub network is not released during the handover process; that is, during the *soft handover* process, the terminal has more than two simultaneous connections over different subnetworks. Based on the information of capability in both systems, the current load (grade of service) of both systems and the load brought by the handover process in old system, the resource obtained by the handover call/connections in each subnetwork can be handled by the weighting vector based on the GPS approach.

For voice users, handover results in an audible click interrupting the conversation for each handover. And because of handover, data users may lose packets and unnecessary congestion control measures may come into play. The degradation of the signal level is a random process that makes a very simple handover scheme result in a ping-pong effect. Thus, a completed algorithm with hysteresis, timer, and intelligent decision mechanisms is required [11,34]. Under the soft handover assumption, the simultaneous connection is required. A system deployment is shown in Figure 10.4. As a WLAN subnetwork offers higher system capacity, the probability of mobile handover from the WLAN to UMTS would be rather rare when the MS receives signals from both RATs. Vertical handover often happens upon the coverage border of the WLAN cell or inside a highly loaded WLAN system. Under the assumption from above, a handover process can be triggered in an *event-driven* (ED) manner. It requires the right triggering message resulting from comparing the QoS in each system, with agreeing hysteresis and timer values. The timer value is a system parameter codesigned with the hysteresis value to reduce the ping-pong effect [24,27]. A general handover procedure is shown in Figure 10.5, whereas the JOSCH-based *vertical handover* (VHO) (Flow ①) was discussed in Section 10.4.3. Because complete handover from the WLAN to UMTS for nomadic users under policy-based category is considered to happen rarely (Flow ②), even if it happens, it will be very similar to single system call admission, and this procedure is therefore only shown in Figure 10.5.

10.4.5 Intersystem Load Control

Intersystem load control can be realized by the intelligent admission, resource assignment to the traffic sub-streams, and suitable intersystem handover. Properly fitting the traffic component into the cooperating subnetworks significantly improves system performance.

For example, the dedicated data channels in a CDMA system are power controlled. A power value P_{bi} is assigned from the BS b for an MS i; and the power value that this MS receives is $A_{bi}P_{bi}$, where A_{bi} models the signal attenuation. Similarly, the interference received at the MS i consists of inter-cell interference, intra-cell interference, and noise power. The inter-cell interference is modeled as $\sum_{\bar{b}} A_{\bar{b}i}P_{\bar{b}}$, with $\bar{b}$ the interfering BSs and $P_{\bar{b}}$ the transmission power of the BS $\bar{b}$; the intra-cell interference is modeled as $\beta \sum_{j\neq i} A_{bi}P_{bj}$, with β the system level orthogonality factor; the noise power is denoted by N. The *carrier-to-interference ratio* (CIR) for the MS

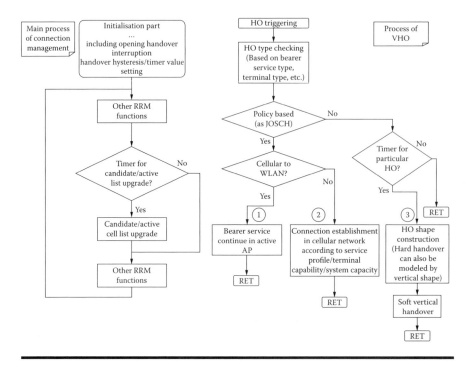

Figure 10.5 Flow diagram for inter-system handover.

i is written as:

$$\xi_i = \frac{A_{bi} P_{bi}}{\sum_b A_{bi} P_b + \beta \sum_{j \neq i} A_{bi} P_{bj} + N} \tag{10.4}$$

The power control for a dedicated channel is executed in each time slot (0.667 msec) by increasing or decreasing the power value in a fixed step in dB (decibel) value periodically in order to satisfy the target *signal-to-interference ratio* (SIR). We will check the effect of adjusting the power value on the CIR in dB value. Equation 10.4 can be derived to

$$\xi_i^{(dB)} = A_{bi}^{(dB)} + P_{bi}^{(dB)} - I_i^{(dB)} \tag{10.5}$$

where $I_i = \sum_b A_{bi} P_b + \beta \sum_{j \neq i} A_{bi} P_{bj} + N$. To estimate the effects of changes in power on the CIR, the derivative of $\xi_i^{(dB)}$ with respect to the power value in dB is taken as:

$$\frac{\partial \xi_i^{(dB)}}{\partial P_{bi}^{(dB)}} = 1 \tag{10.6}$$

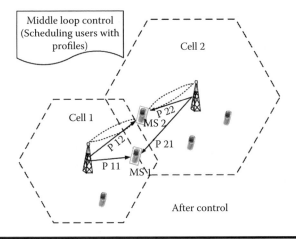

Figure 10.6 Profile-based scheduling.

$$\frac{\partial \xi_j^{(dB)}}{\partial P_{bi}^{(dB)}} = -\frac{\partial I_j^{dB}}{\partial P_{bi}^{dB}} = -\frac{A_{bi} P_{bi}}{I_j} \tag{10.7}$$

$$\frac{\partial \xi_j^{(dB)}}{\partial P_{bi}^{(dB)}} = -\frac{\partial I_j^{dB}}{\partial P_{bi}^{dB}} = -\frac{\beta A_{bi} P_{bi}}{I_j} \tag{10.8}$$

The equations above indicate the effect of power changes on the received signal quality to the existing communication links. If we increase $P_{bi}^{(dB)}$ by 1 dB, we can improve the $\xi_i^{(dB)}$ by 1 dB. However, the expense is the decrease in the CIR to other ongoing radio links, both in intra-cell and the neighbor cells. Taking the nature of mapping from CIR to SIR is roughly a multiplication of the spreading factor; admitting a new call with a higher spreading factor for a cell is preferable compared to a lower one. As JOSCH allows for a traffic split over frequency layers, we allow the user data during connection to shift smoothly between cooperating subnetworks to realize intersystem load balancing. The benefit is proven by the previous analysis.

10.5   Summary

There are many solutions to effectively use the limited radio resources. The RMH concept is able to select and control the proper subnetworks, and the JRRM mechanism to allocate radio resources to the involved mobile terminals is one promising solution. As analyzed in the part of system dimensioning, the system capacity gain can be evaluated through multiple dimensions, that is, switching technique dimension, load elastic (soft blocking) dimension, and the dimension of implementing the advanced RRM functions.

Taking the terminal, network, and user profile into account, the future advanced RRM is expected to have several advantages, such as power saving, interference reduction, and higher system capacity. With the discussed approaches, much finer grained RRM scalability for optimized flow control and system overall spectrum efficiency can be obtained.

To fully exploit the flexibility provided by the multi-mode/multi-band terminals and network capabilities, further research on advanced RRM, joint optimized solution w.r.t. to the spectrum setting and network management is needed. This aims to provide efficient solutions for RRM in a composite radio environment, supporting multiple RATs in different network topologies (hierarchical, decentralized) and, moreover, being potentially managed by the same or different operators. It can also be derived from the traffic splitting approach, that instantaneous big trunk allocation with a number of frequency carriers to a multi-link terminal results in a higher peak rate as well as improved overall system performance. As investigated in this chapter, in case the subnetworks are coupled with each other through different deployment patterns and coupling structures, it will result in significant improved performance. For a given traffic demand, how to jointly reconfigure the network elements and JRRM functions to offer the optimal operational parameters of the network elements is still an open question to be answered.

References

[1] 3GPP TR 22.934, "Feasibility Study on 3GPP Systems to Wireless Local Area Network (WLAN) Interworking," V6.1.0, December 2002.
[2] 3GPP TR 25.892, "Feasibility Study for Orthogonal Frequency Division Multiplexing (OFDM) for UTRAN enhancement," V6.0.0, June 2004.
[3] 3GPP TR 21.910, "Multi-mode UE Issues; Categories, Principles and Procedures," V3.0.0, July 2000.
[4] 3GPP TR 25.913, TS RAN, "Requirements for Evolved UTRA (E-UTRA) and Evolved UTRAN (E-UTRAN)," June 2005.
[5] 3GPP TS 25.308, "High Speed Downlink Packet Access (HSDPA), Overall Description," Rel-5, March 2003.
[6] D. Bertsekas and R. Gallarger, *Data Networks*, Prentice-Hall International Editions, 1992, ISBN 0-13-201674-5.
[7] C.F. Ball, E. Humburg, K. Ivanov, and F. Treml, "Comparison of IEEE 802.16 WiMAX Scenarios with Fixed and Mobile Subscribers in Tight Reuse," *IST Mobile and Wireless Communication Summit*, Dresden, Germany, June 2005.

[8] T. Bates and Y. Rekhter, RFC 2260: "Scalable Support for Multi-homed Multi-provider Connectivity," IETF, January 1998.
[9] BR6.0 FRS0407, "Handover Decision due to BSS Resource Management Criteria."
[10] BR7.0 FRS0490, "Handover (Circuit Switched) from GSM to UMTS."
[11] K.H. Chiang, N. Shenoy, and J. Asenstorfer, "Intelligent Handover and Location Updating Control for a Third Generation Mobile Network," *in Proc. IEEE Globecom'98*, Vol. 4, pp. 1963–1968, November 1998.
[12] CPRI Specification, Developed by Ericsson, Huawei, NEC, Nortel and Siemens, July 2004, (www.cpri.info).
[13] J.N. Daigle, *Queueing Theory with Applications to Packet Telecommunication*, Springer Science, 2005, ISBN 0-387-22857-8.
[14] M. Dillinger et al., *Software Defined Radio Architecture, Systems and Functions*, John Wiley & Sons Inc., 2003, ISBN: 0-470-85164-3.
[15] ETSI, "Broadband Radio Access Network (BRAN); HIPERLAN TYPE 2 Functional Specifications," 1999.
[16] ETSI, "Broadband Radio Access Network (BRAN); Requirements and Architectures for Interworking Between Hiperlan/2 and 3G Generation Cellular Systems," April 2001.
[17] N. Gerlich and H. Becker, "A Function Architecture for 3G IP Based Radio Access Network," in *Proc. IEEE 3G Wireless*, San Francisco, CA, pp. 296–301, 2001.
[18] H. Hu, M. Weckerle, and J. Luo, "Adaptive Transmission Mode Selection Scheme for Distributed Wireless Communication Systems," *IEEE Communication Letters*, July 2006.
[19] H. Hu, M. Weckerle, J. Luo, and E. Schulz, "Organized Beam-Hopping Scheme for Mobile Communication Systems," in *Proc. IEEE/CIC GMC'05*, pp. 156–160, October 2005.
[20] IEEE Standard 802.11a-1999, "Wireless LAN Medium Access Control (MAC) and Physical Layer (PHY) Specifications, Supplement to IEEE Standard for Information Technology," September 1999.
[21] IST-2000-28584 MIND Deliverable D2.2, "MIND Protocols and Mechanisms Specification, Simulation and Validation," November 2002.
[22] L. Kleinrock, *Queueing Systems. Vol I: Theory*, John Wiley & Sons Inc., 1975, ISBN: 0-471-49110-1.
[23] L. Kleinrock, *Queueing Systems. Vol II: Application*, John Wiley & Sons Inc., 1975, ISBN: 0-471-49111-X.
[24] M.D. Kulavaratharasah and A.H. Aghvami, "Teletraffic Performance Evaluation of Microcellular Personal Communication Networks (PCN's) with Prioritised Handoff Procedures," *IEEE Transactions on Vehicular Technology*, Vol. 48, No. 1, pp. 137–152, January 1999.
[25] W. Lee, *Mobile Cellular Telecommunications Systems*, McGraw-Hill, 1989, ISBN 0-070-37030-3.
[26] K. Lindberger, "Balancing Quality of Service, Pricing and Utilization in Multiservice Networks with Stream and Elastic Traffic," in *Proc. ITC 16*, Edinburgh, U.K., pp. 1127–1136, June 1999.

[27] J. Luo, M. Dillinger, E. Schulz, and Z. Dawy, "Optimal Timer Setting for Soft Handover in WCDMA, IEEE 3Gwireless," in *Proc. IEEE 3G Wireless*, San Francisco, CA, June 2000.

[28] J. Luo, R. Mukerjee, M. Dillinger, E. Mohyeldin, and Egon Schulz, "Investigation on Radio Resource Scheduling in WLAN Coupled with 3G Cellular Network," *IEEE Communication Magazine*, June 2003.

[29] A. Mihailovic, Tapio Suihko, and Mark West, "Aspects of Multi-Homing in IP Access Networks," IST Mobile Communications Summit, pp. 115–119, Thessaloniki, Greece, June 2002.

[30] MWIF, Technical Report MTR-007 V1.0.0, "OpenRAN Architecture in 3rd Generation Mobile Systems," September 2001.

[31] A. Parekh and R. Gallager, "A Generalized Processor Sharing Approach to Flow Control in Integrated Services Networks: The Single-Node Case," *IEEE/ACM Transactions on Networking*, Vol. 1, No. 3, pp. 344–357, June 1993.

[32] J.G. Proakis, *Digital Communications*, 3rd edition, McGraw-Hill, 1995, ISBN 0-07-113814-54.

[33] K. Pahlavan et al., "Handoff in Hybrid Mobile Data Networks," *IEEE Personal Communication*, Vol. 7, No. 2, pp. 34–47, April 2000.

[34] E. Del. Re, R. Fantacci, and G. Giambene, "Handover and Dynamic Channel Allocation Techniques in Mobile Cellular Networks," *IEEE Transactions on Vehicular Technology*, Vol. 44, No. 2, pp. 229–237, May 1995.

[35] B. Walke, *Mobile Radio Networks*, second edition, John Wiley & Sons Inc., ISBN: 0471-49902-1, 2001.

[36] J. Zander and S. Kim, *Radio Resource Management for Wireless Networks*, Artech House, 2001, ISBN 1-58053-146-6.

[37] S. Zhou, M. Zhao, X. Xu, J. Wang, and Y. Yao, "Distributed Wireless Communication System: A New Architecture for Future Public Wireless Access," *IEEE Communication Magazine*, Vol. 41, pp. 108–113, March 2003.

MOBILITY
MANAGEMENT

Chapter 11

Fractional Resource Reservation in Mobile Cellular Systems

F.A. Cruz-Pérez and L. Ortigoza-Guerrero

Contents

11.1   Overview

Call admission control (CAC) strategies are required to guarantee that all service types meet their Quality-of-Service (QoS) requirements. The fractional reservation concept allows fine control of the communication service quality by varying the control parameter(s) (i.e., admission thresholds) of the CAC strategies by a fraction of one. This chapter begins with a classification, description, and analysis of fractional resource reservation strategies for handoff prioritization in cellular telecommunication systems with a single service type. The fractional reservation concept is considered in both guard channel (GC)-based and threshold-type priority-based strategies. Then, the generalization of these strategies for cellular systems with multiple service types is presented. A comprehensive comparative study addressing the advantages and disadvantages of each strategy is also included. Additionally, the teletraffic analysis for the performance evaluation of different resource reservation strategies is shown. Algorithms to maximize system capacity subject to QoS constraints are also described for several resource reservation strategies. The chapter also presents simple recursive formulas for the performance evaluation of the GC-based strategies.

11.2   Introduction

This chapter focuses on call-level Quality-of-Service (QoS) measures, and the new call blocking and forced termination probabilities are considered the relevant QoS metrics. The proportion of new call requests that are denied service is known as the new call blocking probability P_b. The forced termination probability, on the other hand, is defined as the probability that a successfully established call is dropped due to the unavailability of radio resources during a handoff attempt. It is important to distinguish between the forced termination P_{ft} and handoff failure P_h probability concepts. The handoff failure probability is the proportion of handoff attempts that fail due to the unavailability of radio resources in the target cell. In general, P_{ft} is a more meaningful system performance metric than the handoff failure rate because it provides better information regarding the QoS experienced by users than P_h. The handoff failure probability only provides information about the QoS experienced by a single handoff attempt, whereas the forced termination probability provides information of the QoS experienced along the duration of calls. As supported by Ghader and Boutaba [1], despite this fact, most research articles focus on the handoff failure probability because its calculation is more convenient and simple. System capacity is defined as the maximum offered traffic for which all the QoS constraints are still satisfied [2] (i.e., P_h and P_{ft} are kept below their respective maximum acceptable values). Thus, the QoS promised to users consists of the new call blocking and forced termination probabilities at the nominal load with the system traffic engineered such that, up to the nominal load, these probabilities are below the guaranteed levels.

11.3   Classification of Resource Reservation Strategies

CAC is a mechanism that manages QoS in cellular networks by restricting access to network resources. Basically, an admission control mechanism accepts new call requests provided their QoS requirements and these for ongoing calls are simultaneously met. In wireless mobile networks, because forced termination of calls in progress is more annoying than blocking of new call requests, handoff call attempts typically require higher priority than new calls while accessing wireless resources. The most popular approach to prioritize handoff call attempts over new call requests is by reserving a portion of available bandwidth in each cell to be used exclusively for handoffs. Depending on which admission control policy is based, the admission control schemes can be classified as guard channel (GC)-based or threshold-type priority (TTP)-based strategies. The admission control policy for new call requests in GC-based strategies depends on the total number of resources being used. On the other hand, in TTP-based strategies, the admission control policy for new call requests is based on the number of ongoing calls at the initial access or, equivalently, on the number of resources used for this type of call. In both GC-based and TTP-based strategies, handoff call attempts are accepted as long as enough resources to serve the call request are free. Fractional resource reservation strategies are more general call admission mechanisms where, provided that sufficient free resources exist, new call requests are accepted

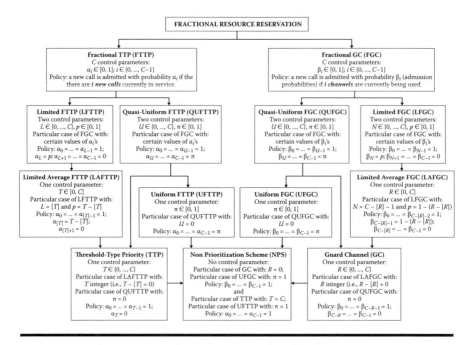

Figure 11.1 Classification of fractional resource reservation strategies.

with a certain probability that, in general, depends on the cell state (i.e., number of resources in use or number of ongoing calls of each type). Figure 11.1 depicts a classification of the fractional resource reservation strategies.

11.3.1 Guard Channel-Based Strategies

The guard channel[1] (GC) priority policy is a classical topic widely studied in the literature [3]. In single-service (i.e., voice) cellular networks, under GC, priority to handoff calls is ensured by reserving a certain number R of channels (known as guard channels). Then, a new call is accepted only if the *total* number of calls in progress (or, equivalently, the total number of channels being used), regardless of their type, is below a cutoff value.

The fractional guard channel (FGC) schemes were first proposed by Ramjee et al. [4] and shown to be more general and more effective than the traditional GC scheme. FGC schemes admit a new call with probability β_i (admission probabilities) that depends on the number i of busy channels.

---

[1] Also referred to in the literature as channel reservation, trunk reservation, or cutoff priority strategy.

However, due to the large number of control parameters, it is difficult to determine the optimum values of the admission probabilities in FGC. With the intention of overcoming the aforementioned problem, the Limited FGC (LFGC) and the Uniform FGC (UFGC) schemes were proposed in References [4] and [5], respectively. GC, LFGC, and UFGC are particular cases of the FGC policy [4]. The LFGC scheme has only two control parameters: (1) an admission threshold (N) and (2) an admission probability (p). Three cases are identified in LFGC. When the number of active users is less than the admission threshold N, LFGC admits new calls with probability one. When the number of active users equals the admission threshold N, LFGC admits new calls with the admission probability p. When the number of active users is greater than the admission threshold N, LFGC does not admit new calls. The LFGC policy effectively reserves a nonintegral number of guard channels for handoff calls by rejecting new calls with some probability. Obviously, it is easier to design, optimize, and operate a CAC strategy with less control parameters. Contrary to the LFGC scheme proposed in Ramjee et al. [4], in Cruz-Pérez [6] a LFGC strategy with a *single* control parameter (the average number R of reserved channels) was proposed, referred to in this chapter as the Limited Average FGC (LAFGC). LAFGC finely controls the communication service quality by effectively varying the average number of reserved channels by a fraction of one. On the other hand, UFGC accepts new calls with an admission probability π independent of channel occupancy. Finally, the Quasi-Uniform FGC (QUFGC) strategy is introduced here as the counterpart for GC-based schemes proposed in Lea and Alyatama [7]. In QUFGC, if the number of channels being used exceeds the admission threshold U, then new call requests are accepted with probability π.

11.3.2 Threshold-Type Priority-Based Strategies

The concept of a single threshold-type priority (TTP) policy was first used in congestion control store-and-forward communication networks [8]. The use of TTP in the context of mobile cellular networks was first investigated in Gavish and Sridhar [9]. Under TTP, a new call is accepted only if a free channel is available and the number of new calls in progress is below a threshold value T. In Fang and Zhang [10], a strategy referred to in this chapter as the Fractional Threshold-Type Priority (FTTP) CAC strategy was proposed. FTTP admits new calls with probability α_i that depends on the number i of new calls currently in service. All calls are blocked if all channels are busy. Additionally, Figure 11.1 shows several particular cases of the FTTP strategy. These strategies (i.e., LFTTP, QUFTTP, LAFTTP, and UFTTP) are introduced here as the counterpart of some GC-based strategies (i.e., LFGC, QUFGC, LAFGC, and UFGC, respectively).

11.4   Modeling of Mobile Cellular Networks

The channel holding time is one of the most important parameters in the performance evaluation of cellular networks and is defined as the time a call, which is accepted in a cell and is assigned resources, will use these resources before completion or handoff to another cell. The call holding time, also known in the literature as the unencumbered service time, is defined as the duration of the requested call connection. On the other hand, the cell dwell time, also known in the literature as the cell residence time, is defined as the time spent by a mobile station in a cell, irrespective of whether it is engaged in a session. Thus, the channel holding time is given by the minimum of the call holding time and the cell residence time.

11.4.1   System and Teletraffic Model

For teletraffic analysis, a homogeneous multi-cellular system with omni-directional antennas located at the center of each cell is assumed. The general guidelines of the model presented in References [11,12] are adopted to cast the system here considered in the framework of birth and death processes. Those guidelines, as well as some additional assumptions include:

1. The number of channels or basic bandwidth units (BBUs)[2] in each cell is represented by C.
2. The new call arrival process is considered a Poisson process with arrival rate λ_n.
3. The handoff call arrival process is considered a Poisson process with arrival rate λ_b. This handoff arrival rate is calculated by the iteration method described in Lin and Noerpel [12].
4. The unencumbered call duration is a random variable with a negative exponential probability density function (pdf) with mean $1/\mu$.
5. The cell dwell time is a random variable with negative exponential pdf with mean $1/\eta$. Then the mean channel holding time is given by $1/(\mu+\eta)$.
6. $\beta_i(i = 0, \cdots, C-1)$ denotes an admission probability (i.e., $0 \leq \beta_i \leq 1$) and $\beta_C = 0$.
7. The number of reserved channels for handoff prioritization in LAFGC and GC is represented by R, and the admission probability in the UFGC scheme is denoted by π.

---

[2] As noted in Santucci et al. [13], because the number of channels in a cell depends on the interference level in a DS/CDMA-based system, it is not fixed. As in Santucci et al. [13], to simplify the problem, we assume that C is the average number of CDMA channels in a cell and, therefore, the performance of a CDMA cellular system should also be treated as an average performance.

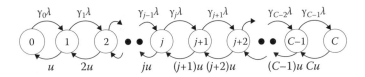

Figure 11.2   State transition diagram for FGC policies, where $u = \mu + \eta$.

8. For the LAFTTP strategy, T represents the average maximum number of simultaneously ongoing new calls in a cell.
9. $\lambda = \lambda_n + \lambda_h$, $A = \lambda/(\mu + \eta)$, and $\alpha = \lambda_h/\lambda$, where λ represents the total call arrival rate, A represents the effective offered traffic, and α represents the proportion of call arrivals that are handoff attempts.
10. As defined in [4], $\gamma_i = \alpha + (1-\alpha)\beta_i$ and $\Gamma = [\gamma_0, \ldots, \gamma_{C-1}]$ (note that, as shown in Figure 11.2, when γ_i is multiplied by λ, the transition rate from state i to state $i + 1$ is obtained).
11. The new call acceptance probability vector is $\mathbf{B} = [\beta_0, \ldots, \beta_{C-1}]$.
12. The offered traffic load per cell[3] is $a = \lambda/\mu$.

11.4.2   Teletraffic Analysis

In this subsection the teletraffic analysis of two representative strategies is developed: (1) FGC and (2) LAFTTP. As noticed, the main difference between the GC-based and TTP-based strategies is the admission policy. In GC-based strategies, the admission policy depends on the total number of resources used; and in the TTP-based strategies, it depends on the number of ongoing new calls. It can be shown that when the admission decision to accept a request of a given type depends only on the number of ongoing calls of that type, a product form solution is produced. On the other hand, when the admission decision depends only on the number of free resource units, no product form solution is produced.

11.4.2.1   GC-Based Strategy

FGC policies use a vector $\mathbf{B} = [\beta_0, \cdots, \beta_{C-1}]$ to determine if new calls can be accepted, and the components of that same vector determine the strategy. For example, when $\beta_0 = \cdots = \beta_{C-R-1} = 1$ and $\beta_{C-R} = \cdots = \beta_{C-1} = 0$, FGC becomes the GC scheme with an integer number R of reserved channels [3]; when $\beta_0 = \cdots = \beta_{C-\lfloor R \rfloor - 2} = 1$, $\beta_{C-\lfloor R \rfloor - 1} = 1 - (R - \lfloor R \rfloor)$, and $\beta_{C-\lfloor R \rfloor} = \cdots = \beta_{C-1} = 0$, FGC becomes the LAFGC scheme with

---

[3] Note that contrary to the effective offered traffic (A), the offered traffic load (a) does not depend on the mobility conditions.

a real average number R of reserved channels [6]. Finally, when $\beta_0 = \cdots = \beta_{C-1} = \pi$, the FGC policy becomes the UFGC scheme with admission probability π [5]. Because a homogeneous system is assumed, the overall system performance can be analyzed by focusing on one given cell. Let us denote the state of the cell as j, where j represents the number of active users in the cell. Describing the birth and death process by "rate up equals rate down" equations [14], the steady-state probabilities are given by:

$$p_j = \frac{A^j}{j!} \prod_{i=0}^{j-1} \gamma_i p_0; \quad 0 \le j \le C \tag{11.1}$$

and

$$p_0 = \frac{1}{\sum_{j=0}^{C} \frac{A^j}{j!} \prod_{i=0}^{j-1} \gamma_i} \tag{11.2}$$

where p_j is the equilibrium probability that the system is in state j. As mentioned before (see Figure 11.2), the transition rate from state j to state $j+1$ is given by the product of γ_j and the total call arrival rate (λ). As explained in Lin and Noerpel [12], the analysis begins with an initial guess of the incoming handoff arrival rate (λ_b). Then, from the equilibrium solution, the outgoing handoff departure rate is calculated as follows [12]:

$$\lambda_b = \frac{(1 - P_b)\lambda_n}{\mu/\eta + P_b} \tag{11.3}$$

The procedure is continued iteratively until the incoming and outgoing handoff arrival rates converge. With the equilibrium solution found, the blocking probabilities can be determined. To facilitate the analysis, the auxiliary variable M_j is defined as:

$$M_j = \prod_{i=0}^{j-1} \gamma_i = \gamma_0 \gamma_1 \cdots \gamma_{j-1} \tag{11.4}$$

The handoff failure probability $P_b(C)$ is given by:

$$P_b(C) = p_C = \frac{\frac{A^C}{N!} M_C}{\sum_{j=0}^{C} \frac{A^j}{j!} M_j} \tag{11.5}$$

The new call blocking probability $P_b(C)^4$ is given by:

$$P_b(C) = \sum_{j=0}^{C}(1 - \beta_j)p_j = \frac{\sum_{j=0}^{C}(1 - \beta_j)\frac{A^j}{j!}M_j}{\sum_{j=0}^{C}\frac{A^j}{j!}M_j} \qquad (11.6)$$

The forced termination probability is given by Lin and Noerpel [12]:

$$p_{ft}(C) = \frac{P_b(C)}{\mu/\eta + P_b(C)} \qquad (11.7)$$

This expression is valid for any of the considered CAC strategies.

11.4.2.2 TTP-Based Strategy

In this subsection, the LAFTTP strategy is analyzed. LAFTTP, on average, limits to T the maximum number of simultaneous ongoing new call in a cell. Let us denote the state of the system as (i, j), where i and j represent, respectively, the number of ongoing new and handoff calls in the cell under study. It is well known that for coordinate convex access policies, the steady-state probabilities can be decomposed into a simple product form [15]. Then, the probability that the analyzed cell is in the state (i, j) when LAFTTP is used is given by:

$$P(i, j) = \begin{cases} \left(\frac{\lambda_n}{\mu+\eta}\right)^i \frac{1}{i!} \left(\frac{\lambda_n}{\mu+\eta}\right)^j \frac{1}{j!} P(0, 0); & 0 \le i \le \lfloor T \rfloor, \quad 0 \le i+j \le C \\ \left(\frac{\lambda_n}{\mu+\eta}\right)^i \frac{(T-\lfloor T \rfloor)}{i!} \left(\frac{\lambda_n}{\mu+\eta}\right)^j \frac{1}{j!} P(0, 0); & i = \lfloor T \rfloor + 1, \quad j = C - (\lfloor T \rfloor + 1) \end{cases}$$
$$(11.8)$$

where P(0,0) is obtained from the normalization condition. The new call blocking probability is given by:

$$P_b = (1 - T + \lfloor T \rfloor)\sum_{j=0}^{C-\lfloor T \rfloor-1} P(\lfloor T \rfloor, j) + \sum_{j=0}^{C-\lfloor T \rfloor-1} P(\lfloor T \rfloor + 1, j)$$

$$+ \sum_{i=0}^{\lfloor T \rfloor-1} P(i, j = C - i) \qquad (11.9)$$

---

[4] Note that both the handoff failure and the new call blocking probabilities are functions of **B**, A, and C. However, for the sake of space and to emphasize the recursive nature of the algorithm, only the dependence on C is explicitly indicated.

and the handoff failure probability is given by:

$$P_b = \sum_{i=0}^{\lfloor T \rfloor + 1} P(i, C - 1) \tag{11.10}$$

11.4.3 Optimization of Control Parameter for Capacity Maximization

It has been shown that there is an optimal value of the (average) number of reserved channels, R, in the GC (LFGC) strategy for which the new call blocking probability is minimized, subject to the hard constraint on the handoff failure probability. The algorithms for finding such optimal values in the GC, LFGC, and LAFGC strategies are given in References [4], [16], and [17], respectively. There is also an optimal value for the admission probability π of the UFGC strategy for which the new call blocking probability is minimized, subject to the hard constraint on the handoff failure probability. The algorithm for finding the optimal value of the control parameter of UFGC is shown in Beigy and Meybodi [5]. To determine the optimum number of reserved channels (admission probability), R (π), in LFGC (UFGC), as well as the maximum system offered traffic, the bisections method can be used. The procedure takes into account that the new call blocking probability is a monotonically increasing function of the number of reserved channels (admission probability) in LAFGC (UFGC) and that both the handoff failure and, therefore, the forced termination probabilities are monotonically decreasing functions of the number of reserved channels (admission probability[5]) [5,16].

11.5 Recursive Formulas for the Performance Evaluation of Guard Channel-Based Strategies

Recursive formulas have been derived in References [16, 18–21] that allow simple and stable computing of (new call or handoff) blocking probabilities, especially when the number of channels is large. These include the Erlang loss system, the Engset model, and the GC scheme for handoff prioritization in cellular networks. More recently, in Vázquez-Ávila et al. [11] recursive formulas for the new call blocking and handoff failure probabilities for FGC policies were derived. To obtain the recursive formulas for the new call blocking and handoff failure probabilities, the following

---

[5] The effective number of reserved channels in the UFGC is increased by decreasing the admission probability π.

definitions apply:

$$\phi(C) = \frac{A^C}{C!} M_C \qquad (11.11)$$

$$G(C) = \sum_{j=0}^{C} \frac{A^j}{j!} M_j \qquad (11.12)$$

$$Y_1(C) = \sum_{j=0}^{C} (1 - \beta_j) \frac{A^j}{j!} M_j \qquad (11.13)$$

Substituting Equations 11.11 and 11.12 into Equation 11.5 and making some simplifications, one obtains:

$$P_h(C) = \frac{\phi(C-1)\frac{A}{C}\gamma_{C-1}}{G(C-1) + \phi(C-1)\frac{A}{C}\gamma_{C-1}} = \frac{P_h(C-1)}{\frac{C}{\gamma_{C-1}A} + P_h(C-1)} \qquad (11.14)$$

Substituting Equations 11.11, 11.12, and 11.13 into Equation 11.6 and making some simplifications, one obtains:

$$P_b(C) = \frac{\frac{Y_1(C-1)}{G(C-1)} + (1 - \beta_C)\frac{\phi(C-1)}{G(C-1)}\frac{A}{C}\gamma_{C-1}}{1 + \frac{\phi(C-1)}{G(C-1)}\frac{A}{C}\gamma_{C-1}}$$

$$= \frac{P_b(C-1) + (1 - \beta_C)\frac{A}{C}\gamma_{C-1}P_b(C-1)}{1 + \frac{A}{C}\gamma_{C-1}P_b(C-1)} \qquad (11.15)$$

Equations 11.14 and 11.15 can be used to calculate, respectively, the handoff failure and new call blocking probability for LFGC, LAFGC, QUFGC, UFGC, and GC by properly setting the corresponding values of admission probabilities in the vector **B**.

11.6 Multi-Services Mobile Cellular Systems

To guarantee acceptable QoS in multi-service mobile environments, network planners must consider certain constraints that provide upper limits for the blocking probability of different service types[6]. An admission control policy, based on certain criteria (i.e., ensuring fair access among services) is also necessary. However, it may also be necessary to protect some delicate

---

[6] We limit the QoS discussion to the issues of call acceptance and dropping to minimize the dimensionality of the problem.

calls, such as handoff calls, to further ensure acceptable QoS. Additionally, calls of different services may need different amounts of system resources, depending on their data rate requirements [22], which cause unfairness because it is more difficult to find a greater amount of available resources for services with high bandwidth[7] requirement. Thus, different service types may have different prioritization level requirements. As noticed in Epstein and Schwartz [23], the level of relative prioritization provided to different service types is specified by relative blocking/dropping probabilities. Hence, in a multi-service mobile environment, it is necessary to provide multiple prioritization levels to efficiently satisfy the QoS of the different traffic classes. This can be achieved by generalizing the GC-based and TTP-based strategies, setting multiple admission thresholds. Several extensions or generalizations for both the GC-based and TTP-based CAC strategies for multi-service systems have been proposed in the literature. Figure 11.3 depicts a classification of the multiple fractional resource reservation strategies.

11.6.1 Description of Multiple Resource Reservation Strategies

Complete sharing (CS) is the simplest admission policy and performs as follows. It blocks an arriving request if and only if the total available bandwidth is less than the requested bandwidth. CS allows the total bandwidth to be shared by all traffic classes or call types (i.e., new and handoff calls) without any restriction. However, when CS is employed, neither call type differentiation nor service differentiation can be provided. Then, among other drawbacks, CS suffers from the fact that it is not fair to users with large bandwidth requirements.

Admission control in the presence of mobility in voice service scenarios or admission control in the presence of multiple services is well studied. However, it has not been until recently that the scenario where multi-service and mobility meet has received attention [24]. In the remainder of this section, the most representative extensions of the GC-based and TTP-based strategies for multi-service mobile cellular networks are described.

11.6.1.1 GC-Based Strategies

Extensions of the GC [3], LAFGC [6], and FGC [4] schemes have been proposed for multi-service mobile cellular networks in [25–27], respectively. In the Multiple GC (MGC) [25] scheme, a request of a given traffic class is admitted if the number of busy channels is less than a certain correspondingly cutoff threshold. That is, different integer numbers of channels or BBUs are reserved for the different call types. On the other hand, in

---

[7] We use a very general term, *bandwidth* (not necessarily meaning a frequency band in Hz), to describe the different resource requirements between two service types.

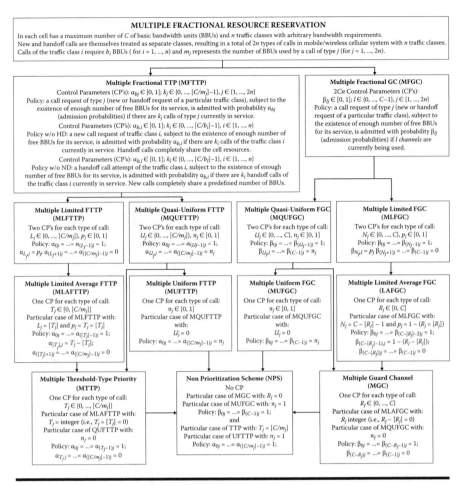

Figure 11.3 Classification of the multiple fractional resource reservation strategies.

the Multiple LAFGC (MLAFGC)[8] [26] strategy, on average, real numbers of channels are effectively reserved for the different call types. Finally, the Multiple FGC (MFGC)[9] [27] scheme admits a service request of type j with a probability $\beta_{l,j}$ (admission probabilities) that depends on both the number l of busy channels and the traffic class of the request.

MLAFGC assumes that the prioritization level assigned to a call type is directly proportional to the amount of resources to which it has access. In multi-service mobile cellular networks with n different services and $2n$ individual Quality-of-Service (QoS) constraints (n for n new call types $+n$ for n handoff call types), MLAFGC reserves $2n-1$ different real numbers of channels, or BBUs, to provide a certain protection level to each of the

---

[8] Called in [26] simply Multiple Fractional Channel Reservation (MFCR).
[9] Called Thinning Scheme I in [27].

diverse call types. This occurs because $2n-1$ independent control variables are needed[10] to differentiate $2n$ call types [26]. A detailed description of MLAFGC is given in Heredia-Ureta et al. [26].

The sorted list of call types based on their relative priorities is known as the "prioritization order" in MLAFGC. In Heredia-Ureta et al. [26], the optimal prioritization order (the one that achieves maximum capacity) was found to be that order for which all $2n$ individual QoS constraints simultaneously equal their respective maximum acceptable values[11]. The reservation of real numbers of channels R_j is achieved because $\lfloor R_j \rfloor + 1$ channels are reserved with probability $R_j - \lfloor R_j \rfloor$ while R_j channels are reserved with probability $1 - (R_j - \lfloor R_j \rfloor)$ [6].

11.6.1.2 TTP-Based Strategies

Coordinate convex policies (to which the optimal policy does not belong to in general) are of particular interest as they lead to product-form solutions of the steady-state probabilities, from which all of the performance measures of interest can be determined[12]. In particular, the TTP-based strategies have proved to be optimal over the class of coordinate convex policies [29]. The Multiple TTP (MTTP) policy[13] concept was first considered in [30], under the context of the stochastic knapsack. A class of state-dependent control policies that produces a product form distribution for steady-state probabilities was shown. The rules described by Equation 3 of Lea and Alyatama [7] are simple but a combination of them can generate complex control policies. Examples of these policies are called the multiple QUFTTP (MQUFTTP) strategy.

The use of MTTP in the context of cellular networks was first investigated in Kwon et al. [31]. A newly arriving call of a given class is blocked if the current number of ongoing calls with the class of the arriving call is equal to or greater than a predefined threshold. An incoming handoff call of a given class is accepted if the available bandwidth is sufficient, and bandwidth adaptation can be performed to accommodate the handoff call if needed. Notice that with this admission criterion, no differentiation of the handoff attempts of the different traffic classes can be performed. For this reason, this strategy is referred here to as MTTP without handoff calls differentiation (MTTP w/o HD). Then, handoff calls with lower bandwidth requirements will have a better chance of occupying the bandwidth than those with higher bandwidth requirements. The performances of two CAC

---

[10] Every service type has two call types (new and handoff calls). Hence, there are n service types and $2n$ call types.

[11] The selection of the optimum prioritization order is addressed in Heredia-Ureta and Cruz-Pérez [26].

[12] Coordinate convex policies were first defined by Aein and Kosovych [28].

[13] Called in Ross and Yao [30] multibandwidth tree network.

strategies — double GC strategy and double TTP strategy — were compared in Deniz and Mohamed [32]. Notice that the double TTP strategy proposed in Deniz and Mohamed [32] is a particular case of the MTTP w/o HD strategy proposed in Kwon et al. [31].

An MTTP-based bandwidth reservation policy for multimedia wireless cellular networks was proposed and mathematically analyzed [33]. The policy gives priority to handoff calls over new calls and prioritizes between different classes of handoff calls according to their QoS constraints by assigning a series of bandwidth thresholds. However, differentiation of the new calls of the different traffic classes is not provided. For this reason, this strategy is referred to here as MTTP without new calls differentiation (MTTP w/o ND).

The so-called thinning schemes, which smoothly reduce the traffic admission rates, were presented in Fang [27]. It was demonstrated that the thinning schemes include the MGC, MFGC, and MTTP schemes as special cases and can be used to obtain new CAC schemes for multimedia networks. In the thinning schemes proposed in Fang [27], to provide the desired QoS for each service in a wireless network supporting multiple types of services, channels are assigned according to priority level. If there are $2n$ traffic types of call arrivals (including new calls and handoff calls), each type of call is assigned one priority level. In the thinning schemes, a call is admitted with certain probability based on the priority and the current traffic situation. The Thinning Scheme I, here referred to as MFGC, uses the information about the total number of busy channels; while the Thinning Scheme II, here referred to as Multiple FTTP (MFTTP), utilizes the numbers of channels occupied by the individual priority traffic streams. In MFTTP, a newly arriving jth priority call, provided there are enough free resources, is admitted with probability $\alpha_{kj,j}$ if there are k_j calls with jth priority currently in service. Notice that for the MFTTP strategy, $2nC$ control parameters (admission probabilities) are required. Because of the large number of control parameters, the tuning of such control parameters becomes an extraordinary complex problem. A multi-cell, multi-class, analytic model to investigate the effects of mobility on bandwidth allocation strategies based on the here-called MQUFTTP policy was presented in Mitchell and Sohraby [34]. For each class of traffic, handoffs have priority over new call arrivals. New and handoff calls are themselves treated as separate classes, resulting in a total of $2n$ types of calls in mobile/wireless cellular systems with n traffic classes. Thresholds are defined for each type of call in each cell of the network. If the number of channels that a type of call occupies in a cell exceeds its corresponding threshold[14], then successive arrivals of

---

[14] Notice that the admission thresholds of the TTP-based strategies can be defined in terms of either the number of ongoing calls of each type or the bandwidth used by each type of call. In this chapter, the first option, was selected.

that type of call are accepted with a predefined probability (here called the admission probability)[15]. Lea and Alyatama [7] generalized on Kelly's analysis [35] and demonstrated that these policies (among these, the here-called MQUFTTP[16] strategy) resulted in product form solutions. Thus, although complex control policies can be generated, they can be easily analyzed.

11.6.2 System and Teletraffic Model

A homogeneous multi-cellular system with fixed cell capacity is assumed, and n different service types with arbitrary bandwidth requirements ($2n$ different call types) are considered as follows: new calls of service 1, $N1$; handoff calls of service 1, $H1$, ...; new calls of service n, Nn; and handoff calls of service n, Hn. For teletraffic analysis, without loss of generality, the prioritization order $N1-H1-N2-H2-\cdots-Nn-Hn$ in terms of increasing importance is assumed. The following mathematical analysis involves the most common assumptions made in the applicable literature [24–26]. Each cell has a maximum number of channels or BBUs (C). Calls of service i require b_i channels (where b_i is considered an integer).[17] New and handoff call arrival processes offered to a given cell are Poisson processes for the n services with arrival rates λ_{Ni} and λ_{hi} (for $i = 1, \ldots, n$), respectively. The values of the handoff arrival rates $\lambda_{h1}, \lambda_{h2}, \ldots$, and λ_{hn} are determined using an iterative procedure, as in Lin and Noerpel[12]. The unencumbered call duration and cell dwell time for service type i (for $i = 1, \ldots, n$) have a negative exponentially pdf with mean $1/\mu_i$ and $1/\eta_i$, respectively.

11.6.3 Teletraffic Analysis

Adding access control often creates state dependent arrival rates that do not always produce a product form distribution. Without a product form, for calculating the blocking probabilities, it may be necessary to numerically solve a set of multidimensional birth and death lineal difference equations. For a system with n different services, the cardinality of the set of allowable states grows roughly as the nth power of the total number of channels or BBUs.

When the admission decision to accept a request of a given type depends only on the number of resource units occupied by the calls or

---

[15] The adjective *quasi-limited* was chosen to differentiate the case when an admission probability is used in only one occupancy state or number of admitted calls (i.e., limited fractional guard channel or limited fractional threshold type policy).

[16] Notice that for the MQUFTTP CAC strategy, two control parameters are required for each type of call.

[17] If there is a contiguousness requirement, packing is assumed. A channel, then, refers to a basic unit of resource.

sessions of the same type in progress, a product form solution is produced. On the other hand, when the admission decision depends only on the number of free resource units (trunk reservation), no product form solution is produced.

To obtain compact expressions, we use the auxiliary variables λ_j and m_j (with $j = 1, 2, \ldots, 2n$). λ_j represents the arrival rate of calls of type j (either new or handoff request) and m_j represents the number of resources used by a call of type j. Assume further that $R_{2n} = 0$ and, because of the prioritization order assumed, $\lambda_{2i-1} = \lambda_{Ni}$, $\lambda_{2i} = \lambda_{hi}$, $m_{2i-1} = m_{2i} = b_i$ (for $i = 1, \ldots, n$). b_i (for $i = 1, \ldots, n$) represents the number of resources used by calls of service i and $\mathbf{b} = (b_1, \ldots, b_n)$. For the sake of space, only the teletraffic analysis for both the MLAFGC and MFTTP strategies are presented.

11.6.3.1 GC-Based Strategy

This subsection shows a teletraffic analysis for MLAFGC developed through the use of multidimensional birth and death process theory. The state of the system can be denoted as $\mathbf{k} = (k_1, \ldots, k_n)$, where $k_1, \ldots, k_n$ represent the number of users of service 1 to n in a given cell, respectively. Describing the process by "rate out equals rate in" equations [14], the equilibrium state lineal difference equations are given by:

$$\left[\sum_{i=1}^{n} (A_{2i-1}(\mathbf{k}) + (A_{2i}(\mathbf{k}) + (B_i(\mathbf{k})))) \right] P(\mathbf{k}) = \sum_{i=1}^{n} \{ [A_{2i-1}(\mathbf{k} - \mathbf{e}_i) + (A_{2i}(\mathbf{k} - \mathbf{e}_i))] P(\mathbf{k} - \mathbf{e}_i) + B_i(\mathbf{k} + \mathbf{e}_i) P(\mathbf{k} + \mathbf{e}_i) \} \quad (11.16)$$

for $k_1 \geq 0, \ldots, k_n \geq 0$, and $\mathbf{k} \cdot \mathbf{b} = \sum_{i=1}^{n} k_i b_i \leq C$, where $\mathbf{e}_i$ is a unity vector of dimension n with a 1 in position i (i.e., $\mathbf{e}_2 = (0, 1, \ldots, 0)$) and $P(\mathbf{k})$ is the probability that the system is in the state $\mathbf{k} = (k_1, \ldots, k_n)$. Now let us define the state transition rates $A_{2i-1}(\mathbf{k})$, $A_{2i}(\mathbf{k})$, and $B_i(\mathbf{k})$. The call birth rate $A_j(\mathbf{k})$ for the call type j (for $j = 1, \ldots, 2n$) is given by:

$$A_j(\mathbf{k}) = \begin{cases} \lambda_j; & \text{if } C - \mathbf{k} \cdot \mathbf{b} - m_j > \lfloor R_j \rfloor \\ \lambda_j [1 - (R_j - \lfloor R_j \rfloor)]; & \text{if } C - \mathbf{k} \cdot \mathbf{b} - m_j = \lfloor R_j \rfloor \quad (11.17) \\ 0; & \text{otherwise} \end{cases}$$

for $0 \leq k_i \leq \lfloor C/b_i \rfloor - 1$ (for $i = 1, \ldots, n$); otherwise, it is zero.

The call death rate $B_i(\mathbf{k})$ for calls with service type i (for $i = 1, \ldots, n$) is given by:

$$B_j(\mathbf{k}) = \begin{cases} \lambda_j(\mu_i + \eta i); & \text{if } \mathbf{k} \cdot \mathbf{b} \le C \\ 0; & \text{otherwise} \end{cases} \quad (11.18)$$

for $0 \le k_i \le \lfloor C/b_i \rfloor$ (for $i = 1, \ldots, n$). Together with the normalization condition:

$$\sum_{k_1=0}^{\lfloor \frac{C}{b_1} \rfloor} \cdots \sum_{k_n=0}^{\lfloor \frac{C}{b_n} \rfloor} P(\mathbf{k}) = 1 \quad (11.19)$$

$$\{ \mathbf{k} | \mathbf{k} \cdot \mathbf{b} \le N_C \}$$

the steady-state probabilities can be calculated by means of the Gauss-Seidel method [14]. The blocking probability for call type j (for $j = 1, \ldots, 2n$), P_{Bj}, is then given by:

$$P_{Bj} = \sum_{k_1=0}^{\lfloor \frac{C}{b_1} \rfloor} \cdots \sum_{k_n=0}^{\lfloor \frac{C}{b_n} \rfloor} P(\mathbf{k}) + (R_j - \lfloor R_j \rfloor) \sum_{k_1=0}^{\lfloor \frac{C}{b_1} \rfloor} \cdots \sum_{k_n=0}^{\lfloor \frac{C}{b_n} \rfloor} P(\mathbf{k}) \quad (11.20)$$

$$\{ \mathbf{k} | \mathbf{k} \cdot \mathbf{b} > C - \lfloor R_j \rfloor - m_j \} \quad \{ \mathbf{k} | \mathbf{k} \cdot \mathbf{b} = C - \lfloor R_j \rfloor - m_j \}$$

For the prioritization order assumed, the new call blocking probability for service type i is $P_{bi} = P_{B(2i-1)}$, and the handoff failure probability service type i is $P_{bi} = P_{B(2i)}$ (for $i = 1, \ldots, n$).

11.6.3.2 TTP-Based Strategy

In this subsection, the MFTTP strategy is analyzed. In MFTTP, a call request of service type j (for $j = 1, \ldots, 2n$), provided there are enough free resources, is admitted with probability $\alpha_{k_j, j}$ if there are k_j calls with jth priority currently in service. For the prioritization order assumed, let the state of the system be denoted as $\mathbf{k} = (k_1, \ldots, k_{2n})$, where k_{2i-1} and k_{2i} represent, respectively, the number of ongoing new and handoff calls of service i (for $i = 1, \ldots, n$) in the analyzed cell. Then, the probability that the analyzed cell is in the state (i, j) when MFTTP is used is given by:

$$P(\mathbf{k}) = \prod_{j=1}^{2n} \prod_{n_j=1}^{k_j} \left(\frac{\alpha_{k_j-1, j}, \lambda_j}{k_j \left(\mu_{\lfloor \frac{j+1}{2} \rfloor} \eta_{\lfloor \frac{j+1}{2} \rfloor} \right)} \right) P(0, \ldots, 0) \quad (11.21)$$

for $\sum_{j=1}^{2n} k_j m_j \le C$. $P(0, \ldots, 0)$ is obtained from the normalization condition. The blocking probability for call type j (for $j = 1, \ldots, 2n$), P_{Bj}, is then given by:

$$P_{Bj} = \sum_{k_1=0}^{\lfloor \frac{C}{m_1} \rfloor} \cdots \sum_{k_{2n}=0}^{\lfloor \frac{C}{m_n} \rfloor} (1 - \alpha_{k_j, j}) P(\mathbf{k})$$

$$\left\{ \mathbf{k} \mid \sum_{t=1}^{2n} k_t m_t \leq C \right\}$$

(11.22)

11.6.3.3 Approximate Analysis of the MLAFGC Strategy through a One-Dimensional Recursive Formula

This subsection presents the one-dimensional recursion formulas derived in Cruz-Pérez et al. [36] to calculate the blocking probabilities of the different call types in MLAFGC. The macro-states technique [14] is used to effectively collapse the n-dimensional system state representation to a one-dimensional "macro-state" sufficient to analyze all quantities of interest. The derivation of Cruz-Pérez et al. [36] is based on the fact that MLAFGC can also be described as follows: when the number of busy channels at a BS is l, an arriving call of type j will be admitted with probability $\beta_{l,j}$ for $l = 0, \ldots, C$ and $j = 1, \ldots, 2n$, with:

$$\beta_{l,j} = \begin{cases} 1; & 0 \leq l \leq C - m_j - \lfloor R_j \rfloor - 1 \\ 1 - (R_j - \lfloor R_j \rfloor); & l = C - m_j - \lfloor R_j \rfloor \\ 0; & \text{otherwise} \end{cases}$$

(11.23)

where, due to the prioritization order assumed,

$$m_{2i-1} = m_{2i} = b_i; \quad \text{for } i = 1, \ldots, n$$

(11.24)

It is important to pay attention to the random variable: $l = \mathbf{k} \cdot \mathbf{b} = $ total number of resource units occupied. Similar to Kaufman [15] and Roberts [37], intuitively it is expected that:

$$(\beta_{l-b_i, 2i-1} \lambda_{Ni} + \beta_{l-b_i, 2i} \lambda_{bi}) q(l - b_i) = E\{n_i | l\}(\mu_i + \eta_i) q(l)$$

(11.25)

for $i = 1, \ldots, n$ and $l = 0, \ldots, C$, where

$$E\{n_i | l\} = \text{mean number of ongoing service type } i \text{ calls,}$$

$$\text{given a total of } l \text{ units of the resource in use.} \quad (11.26)$$

Equation 11.25 arises from interpreting it as a one-dimensional birth-death balance equation with the first factor on the left-hand side of Equation 11.25 as the service type i birth rate and the factors of $q(l)$ on the right-hand side as the service type i death rate.

Then, multiplying Equation 11.25 by b_i and summing over i yields:

$$\sum_{i=1}^{n} \frac{(\beta_{l-b_i, 2i-1}\lambda_{Ni} + \beta_{l-b_i, 2i}\lambda_{bi})}{\mu_i + \eta_i} b_i q(l - b_i) = q(l) \sum_{i=1}^{n} b_i E\{n_i | l\}$$

$$= q(l) E \left\{ \sum_{i=1}^{n} b_i n_i | l \right\} = lq(l)$$

$$(11.27)$$

where $q(x) = 0$ for $x < 0$ and $\sum_{l=0}^{C} q(l) = 1$. Equation 11.27 defines a one-dimensional recursion that trivially generates the distribution of l recursively. Hence, the blocking probability for call type j (for $j = 1, \ldots, 2n$), given by Equation 11.20, can also be obtained using:

$$p_{Bj} = \sum_{r=0}^{m_j + \lfloor R_j \rfloor - 1} q(C - r) + (R_j - \lfloor R_j \rfloor) q(C - m_j - \lfloor R_j \rfloor) \qquad (11.28)$$

Equations 11.27 and 11.28 effectively reduce the calculation of the blocking probabilities of the different call types to one-dimensional recursive expressions instead of the numerical solution of a multidimensional set of lineal difference equations. Note that the MGC (where all the numbers of reserved channels are integer [37]) and CS (when all admission probabilities $\beta_{l,j}$ for $l = 0, \ldots, C$ and $j = 1, \ldots, 2n$ are one [15]) schemes are particular cases of MLAFGC. This type of approximation is accurate if the mean channel holding times do not greatly differ from each other.

11.7 Optimal Configuration of Admission Control Policies in Multi-Service Mobile Wireless Networks

Different CAC policies in various multi-service cellular scenarios were evaluated and an optimization methodology based on a hill climbing algorithm (i.e., gradient method) to find the optimum configuration for most policies was proposed [24]. The configuration of the MLAFGC policy specifies the average amount of resources to which each service has access. The optimal configuration maximizes the offered traffic load that the system can handle while meeting certain QoS requirements (that is, maximizes system capacity). To the best of our knowledge, three algorithms for computing the system capacity of the MLAFGC policy have been proposed in the literature [26,38,39]. A performance comparison of these algorithms is presented in Garcia-Roger, Martinez-Bauset, and Pla [39]. An efficient alternative to overcome the computational effort required is the use of the one-dimensional recursive formula derived in Section 11.6.3.

11.8   Performance Comparison and Numerical Results

11.8.1   Single-Service Scenario

The numerical results obtained with the analysis described in Section 11.5 of this chapter are presented here. The performances of four admission control strategies are evaluated: (1) Non-Priority Scheme (NPS), (2) GC, (3) LAFGC, and (4) UFGC. In the numerical evaluations, it is assumed that each cell has $C = 30$ channels, $1/\mu = 180$ seconds, and the mean cell dwell time varies between 100 and 10,000 seconds. The QoS constraints to be met are assumed to be $P_b \leq P_{b\_max} = 0.02$ and $P_{ft} \leq P_{ft\_max} = 0.002$.

Figure 11.4 plots the system capacity achieved by the different call admission policies versus the mean cell dwell time. The Figure shows that the system capacity decreases as the relative users' mobility increases (i.e., the mean cell dwell time diminishes). This is because the optimum number

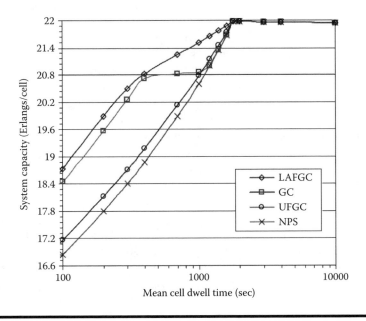

Figure 11.4   System capacity achieved by different call admission policies versus the mean cell dwell time.

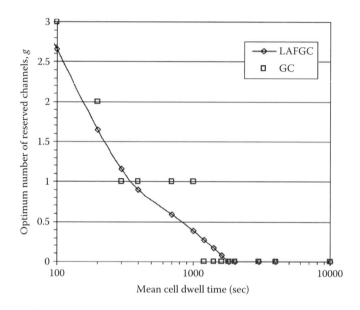

Figure 11.5   Optimum number of reserved channels for the LAFGC and GC schemes versus the mean cell dwell time.

of reserved channels (new call admission probability) in GC and LAFGC (UFGC) increases (decreases) in order to protect handoff calls, as the user mobility increases (see Figure 11.5 and Figure 11.6). Furthermore, it can be seen that system capacity decreases exponentially as the mean cell dwell time[18] decreases. Both figures show that for systems with high mobility users (i.e., small cell dwell time relative to the service time $\eta > \mu$), LAFGC and GC achieve similar performances. This is due to the fact that the fractional part of the optimum number of reserved channels in LAFGC is considerably smaller than its integer part. Results show that LAFGC achieves the highest capacity for the whole range of mean cell dwell times considered.

Figure 11.7 shows the relative capacity gain for the different call admission policies versus the mean cell dwell time. The figure shows the system capacity gain of LAFGC relative to GC (labeled "LAFGC/GC"), to UFGC (labeled "LAFGC/UFGC"), and to NPS (labeled "LAFGC/NPS"). It also shows the system capacity gain of GC relative to UFGC (labeled "GC/UFGC") and NPS (labeled "GC/NPS"). Finally, it shows the system capacity gain of UFGC relative to both GC (labeled "UFGC/GC") and NPS (labeled "UFGC/NPS"). Observe that the relative capacity gain of each of the different schemes

---

[18] Observe that Figures 11.4 through 11.7 are plotted on a logarithmic scale for the x-axis.

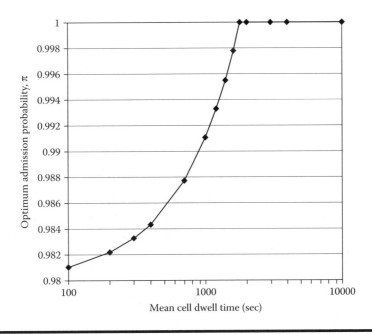

Figure 11.6   Optimum new call admission probability in the UFGC strategy versus the mean cell dwell time.

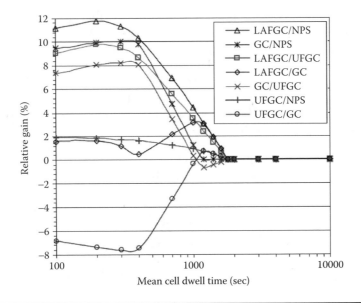

Figure 11.7   Relative capacity gain of the different call admission policies versus the mean cell dwell time.

depends on the value of the mean cell dwell time. LAFGC can achieve as much as 3 percent capacity gain over GC, 10 percent over UFGC, and 12 percent over NPS. There is a brief interval in which UFGC achieves a higher capacity than GC (at most 1 percent relative capacity gain) when user mobility is moderated[19]. Finally, it is shown that for low mobility conditions (i.e., $\eta \ll \mu$), handoff prioritization is unnecessary due to the fact that there are practically no handoff attempts.

11.8.2 Multiple Services Scenario

As noticed in Beigy and Meybodi [41], there have been relatively few studies that investigate the performance of different bandwidth allocation schemes for multi-service wireless cellular networks. A comprehensive comparative evaluation of CAC policies in cellular multi-service networks is presented [24]. For each of the studied CAC policies, the maximum system capacity is obtained. In Garcia et al. [24], it was shown that MLAFGC achieves practically the same maximum system capacity that the optimum CAC policy achieves (relative difference smaller than 1 percent and it decreases when the number of total channels increases). In Garcia et al. [24], it was shown that policies of the trunk reservation class outperform policies that produce a product form solution, and the improvement ranges approximately between 5 and 15 percent.

In summary, even when it has been shown that, in general, (1) the optimal CAC policy does not belong to the coordinate convex policies [42], (2) policies of the trunk reservation class outperform policies that produce a product form solution [24,32] (the improvement ranges approximately between 5 and 15 percent [24]), and (3) trunk reservation policies achieve practically the same performance as the optimum CAC policy in multiservice mobile cellular networks [24,43], it is important to consider the TTP-based strategies for the following reasons:

1. They lead to product form solutions of the steady-state probabilities (which are well known, stable, efficient, and give accurate results) [34].
2. The TTP-based strategies have proven to be optimal over the class of coordinate convex policies [29].
3. The handoff dropping probability in the TTP-based strategies may be lower than that of the trunk reservation CAC when the arrival rate of wideband users is higher than the arrival rate of narrowband users [32].

---

[19] This is similar to the observation in Li et al. [40] that UFGC performs better than GC under the low handoff/new traffic ratio.

4. As noticed in Fang and Zhang [10], the TTP-based strategies may work best when the call arrivals are bursty (for example, before or after a football game) by spreading the potential bursty calls (users will try again when the first few tries fail)[20]. As explained in Fang and Zhang [10], wireless network traffic will tend to be self-similar considering more data services will be supported and the TTP-based strategies could be useful in future wireless multimedia networks.

References

[1] M. Ghader and R. Boutaba, "Call admission control in mobile cellular networks: a comprehensive survey," *Wireless Communications and Mobile Computing*, Vol. 6, No. 1, pp. 69–93, 2006.

[2] J.G. Markoulidakis, J.E. Dermitzakis, G.L. Lyberopoulos, and M.E. Theologou, "Optimal system capacity in handover prioritised schemes in cellular mobile telecommunication systems," *Computer Communications*, Vol. 23, pp. 462–475, March 2000.

[3] D. Hong and S.S. Rappaport, "Traffic model and performance analysis for cellular mobile radio telephone systems with prioritized and nonprioritized handoff procedures," *IEEE Trans. Veh. Technol.*, Vol. 35, pp. 77–92, August 1986.

[4] R. Ramjee, R. Nagarajan, and D. Towsley, "On optimal call admission control in cellular networks," *Wireless Networks*, Vol. 3, No. 1, pp. 29–41, 1997.

[5] H. Beigy and M.R. Meybodi, "Uniform fractional guard channel policy," in *Proc. 6th SCI'2002*, Vol. 15, Orlando, FL, July 2002.

[6] F.A. Cruz-Pérez, D. Lara-Rodríguez, and M. Lara, "Fractional channel reservation in mobile communication systems," *IEE Elect. Letters*, pp. 2000–2002, November 1999.

[7] C.-T. Lea and A. Alyatama, "Bandwidth quantization and states reduction in the broadband ISDN," *IEEE/ACM Trans. on Networking*, Vol. 3, pp. 352–360, June 1995.

[8] S.S. Lam and M. Reiser, "Congestion control of store-and-forward networks by input buffer limits an analysis," *IEEE Trans. Commun.*, Vol. 27, No. 1, pp. 127–133, January 1979.

[9] B. Gavish and S. Sridhar, "Threshold priority policy for channel assignment in cellular networks," *IEEE Trans. on Computers*, Vol. 46, No. 3, pp. 367–370, March 1997.

---

[20] In the trunk reservation schemes, the number of total resource units occupied is used as a decision variable for CAC. However, it may well happen that too many new calls are accepted into the system, which can result in congestion in neighboring cells due to the handoffs of these new calls in the future. Then, the use of the number of channels that are currently occupied by new calls as a decision variable for CAC seems reasonable [44].

[10] Y. Fang and Y. Zhang, "Call admission control schemes and performance analysis in wireless mobile networks," *IEEE Trans. Veh. Technol.*, Vol. 51, pp. 371–382, March 2002.

[11] J.L. Vázquez-Ávila, F.A. Cruz-Pérez, and L. Ortigoza-Guerrero, "Performance analysis of fractional guard channel policies in mobile cellular networks," *IEEE Trans. Wireless Commun.*, Vol. 5, No. 2, pp. 301–305, February 2006.

[12] Y.-B. Lin and A. Noerpel, "The sub-rating channel assignment strategy for pcs hand-offs," *IEEE Trans. Veh. Technol.*, Vol. 45, pp. 122–130, February 1996.

[13] F. Santucci, W. Huang, P. Tranquilli, and V.K. Bhargava, "Admission control in wireless systems with heterogeneous traffic and overlaid cell structure," in *Proc. IEEE VTC-Fall*, Boston, MA, September 2000, pp. 1106–1113.

[14] R.B. Cooper, *Introduction to Queueing Theory*. Washington: CEEPress Books, 1990.

[15] J.S. Kaufman, "Blocking in a shared resource environment," *IEEE Trans. Commun.*, Vol. 29, pp. 1474–1481, October 1981.

[16] G. Haring, R. Marie, R. Puigjaner, and K. Trivedi, "Loss formulas and their application to optimization for cellular networks," *IEEE Trans. Veh. Technol.*, Vol. 50, pp. 664–673, May 2001.

[17] J.L. Vázquez-Ávila, F.A. Cruz-Pérez, and L. Ortigoza-Guerrero, "Performance comparison of fractional guard channel policies in mobile cellular networks," in *Proc. IEEE PIMRC'04*, Barcelona, Spain, pp. 1476–1480, September 2004.

[18] D.L. Jagerman, "Some properties of the Erlang loss function," *Bell Syst. Tech. J.*, Vol. 53, pp. 525–551, 1974.

[19] A.A. Jafari, T.N. Saadawi, and M. Schwartz, "Blocking probability in two-way distributed circuit-switched CATV," *IEEE Trans. Commun.*, Vol. 34, pp. 977–984, 1986.

[20] E. Pinsky, A. Conway, and W. Liu, "Blocking formulae for the Engset model," *IEEE Trans. Commun.*, Vol. 42, pp. 2213–2214, June 1994.

[21] F. Santucci, "Recursive algorithm for calculating performance of cellular networks with cutoff priority," *IEE Elect. Lett.*, Vol. 33, pp. 662–664, April 1997.

[22] W. Huang and V.K. Bhargava, "Performance evaluation of a DS/CDMA cellular system with voice and data traffic," in *Proc. IEEE PIMRC*, Taiwan, ROC, pp. 588–592, 1996.

[23] B. Epstein and M. Schwartz, "Reservation strategies for multi-media traffic in a wireless environment," in *Proc. IEEE VTC*, Chicago, IL, July 1995, pp. 165–169.

[24] D. Garcia, J. Martínez, and V. Pla, "Comparative evaluation of admission control policies in cellular multi-service networks," in *Proc. Wireless'04*, Calgary, AB, July 2004.

[25] B. Li, C. Lin, and S. Chanson, "Analysis of a hybrid cutoff priority scheme for multiple classes of traffic in multimedia wireless networks," *Wireless Networks*, pp. 279–290, 1998.

[26] H. Heredia-Ureta, F.A. Cruz-Pérez, and L. Ortigoza-Guerrero, "Capacity optimization in multi-service mobile wireless networks with multiple

fractional channel reservation," *IEEE Trans. on Veh. Technol.*, Vol. 52, pp. 1519–1539, November 2003.

[27] Y. Fang, "Thinning schemes for call admission control in wireless networks," *IEEE Trans. Comput.*, Vol. 52, No. 5, pp. 685–687, May 2003.

[28] J. Aein and O.S. Kosovych, "Satellite capacity allocation," *Proc. IEEE*, Vol. 65, pp. 332–342, March 1977.

[29] K.W. Ross and D.H.K. Tsang, "The stochastic knapsack problem," *IEEE Trans. Commun.*, Vol. 37, No. 7, pp. 740–747, July 1989.

[30] K.W. Ross and D.D. Yao, "Monotonicity properties for the stochastic knapsack," *IEEE Trans. Theory*, Vol. 36, pp. 1173–1179, September 1990.

[31] T. Kwon, S. Kim, Y. Choi, and M. Naghshineh, "Threshold-type call admission control in wireless/mobile multimedia networks using prioritised adaptive framework," *IEE Elect. Lett*, Vol. 36, pp. 852–854, April 2000.

[32] D.Z. Deniz and N.O. Mohamed, "Performance of CAC strategies for multimedia traffic in wireless networks," *IEEE J. Select. Areas Commun.*, Vol. 21, pp. 1557–1565, 2003.

[33] N. Nasser and H. Hassanein, "Bandwidth reservation policy for multimedia wireless cellular networks and its analysis," in *Proc. IEEE ICC'04*, France, pp. 3030–3034, 2004.

[34] K. Mitchell and K. Sohraby, "An analysis of the effects of mobility on bandwidth allocation strategies in multi-class cellular wireless networks," in *Proc. IEEE Infocom'01*, Anchorage, Alaska, April 2001, pp. 1005–1011.

[35] F.P. Kelly, *Reversibility and Stochastic Networks*. New York: Wiley, 1979.

[36] F.A. Cruz-Pérez, J.L. Vázquez-Ávila, and L. Ortigoza-Guerrero, "Recurrent formulas for the multiple fractional channel reservation strategy in multiservice mobile cellular networks," *IEEE Commun. Lett.*, Vol. 8, No. 10, pp. 629–631, October 2004.

[37] J.W. Roberts, "A service system with heterogeneous user requirements — application to multi-services telecommunications systems," *Performance of Data Communications Systems and Their Applications*, G. Pujolle, Ed. The Netherlands: North-Holland, 1981, pp. 423–431.

[38] V. Pla, J. Martínez, and V. Casares-Giner, "Algorithmic computation of optimal capacity in multi-service mobile wireless networks," *IEICE Trans. Commun.*, Vol. 88, pp. 797–799, February 2005.

[39] D. Garcia-Roger, J. Martinez-Bauset, and V. Pla, "Efficient algorithm to determine the optimal configuration of admission control policies in multiservice mobile wireless networks," in *Proc. HET-NETs'05*, Ilkley, West Yorkshire, U.K., July 2005.

[40] L. Li, Bin Li, Bo Li, and X.R. Cao, "Performance analysis of bandwidth allocations for multi-services mobile wireless cellular networks," in *Proc. IEEE WCNC'03*, New Orleans, LA, March 2003, pp. 1072–1077.

[41] H. Beigy and M.R. Meybodi, "A new fractional channel policy," *Journal of High Speed Networks*, Vol. 13, No. 1, pp. 25–36, Spring 2004.

[42] K.W. Ross and D.H.K. Tsang, "Optimal circuit access policies in an ISDN environment: a Markov decision approach," *IEEE Trans. Commun.*, Vol. 37, pp. 934–939, September 1989.

[43] D. Garcia, J. Martínez, and V. Pla, "Admission control policies in multi-service cellular networks: optimum configuration and sensitivity," *Wireless Systems and Mobility in Next Generation Internet, LNCS*, Vol. 3427, pp. 121–135, Springer-Verlag, 2005.

[44] J. Hou and Y. Fang, "Mobility-based call admission control schemes for wirelesss mobile networks," *Wireless Commun. Mob. Comput.*, Vol. 1, pp. 269–282, July 2001.

Chapter 12

Mobility Management
for Mobile IP Networks

Chang Woo Pyo and Jie Li

Contents

12.1 Introduction

During the past decade, the technologies of computing and communication networking have evolved significantly, and Internet and wireless mobile communications have radically revolutionized our communication environment. In the near future, more and more Internet services of both conventional and novel types will be smoothly accessed with various mobile devices through widely deployed wireless communication networks. The boundary between the Internet and wireless mobile communications is disappearing, through the convergence of the Internet and wireless mobile communications. With the rapid improvement in both networks and mobile terminals, significant advances have emerged in all fields of wireless mobile communications, including the number of mobile subscribers, the deployment of mobile communication systems, new advances in mobile techniques, along with the miscellaneous mobile services.

With the convergence of the Internet and wireless mobile communications and with the rapid growth in the number of mobile subscribers, *mobility management* emerges as one of the most important and challenging problems for wireless mobile communication over the Internet. Typically, mobility management is one of the major functions of cellular networks such as Global System for Mobile Communications (GSM) [1] and Code Division Multiple Access (CDMA) [2] that enable the cellular networks to locate roaming terminals for call delivery and to continue seamless connection as the terminal is moving into a new service area [3]. Thus, mobility management supports mobile terminals, allowing users to roam while simultaneously offering them incoming calls and supporting calls-in-progress.

The original Internet Protocols (IPs) do not support node mobility. At the time they were designed, it was assumed that nodes did not travel around the networks at all. That is, a node's point of attachment to the network remains unchanged at all times, and an IP address identifies a particular network. With IP nodes roaming between different portions of an IP network, such as wireless-enabled laptop computers and personal digital assistants (PDAs), being greatly popularized, the traditional viewpoint on the Internet began to change with the new scenarios of the mobile applications. There need to be mechanisms to ensure that mobile nodes have a relevant address for the network subnet. To support mobility for these nodes with the current IPs, reconfiguration is necessary any time they move.

IP extensions for mobility solution [5–11] — Mobile IPs (Mobile IPv4 and Mobile IPv6) — mainly carried out by the Working Groups in the Internet Engineering Task Force (IETF) [4] were developed to allow IP nodes using either IPv4 or IPv6 to seamlessly "roam" among IP subnets. Mobile IPs provide ways to support global node mobility that allow a node to continue to use its permanent IP address without changing its IP address as it moves around the Internet. When a mobile user leaves the network with which his device is associated (his home network) and enters the domain of a foreign network, the foreign network uses the Mobile IP protocols to inform the home network of a care — of an address to which all packets for the user's device should be sent.

The concerns of high mobility management overhead and poor efficiency in standardized global mobility management protocols of Mobile IPs led to the introduction of micro-mobility management methods, such as regional micro-mobility management [24], hierarchical mobile IPv6 [25], Cellular IP [26,27], HAWAII [28], IDMP [29,30], and user-based micro-mobility management [32]. These micro-mobility management methods are intra-domain mobility solutions, focusing mainly on fast and efficient mobility support within a restricted network domain.

This chapter focuses on mobility management for Mobile IPs. We first provide an overview of Mobile IP (also called Mobile IPv4 in this chapter) and Mobile IPv6. Then, features of typical micro-mobility management methods are described. Furthermore, we introduce two agent-based micro-mobility management methods for Mobile IP. For the purpose of distributing the overall network load, the proposed agent-based micro-mobility management methods distribute the functions of the central gateway into all routers. The routers configure their own networks (domains) consisting of several routers, and each of them acts as a gateway on its domain. Performance evaluations and comparisons conducted via computer simulation are also presented.

12.2   Mobile IP

With IP nodes moving around different IP networks rapidly becoming popularized, mobility support of IP mobile nodes is an essential technology and there need to be mechanisms to ensure that the mobile nodes have a relevant address for the network subnet. Mobile IP (i.e., Mobile IPv4) proposed by the IETF [4] is a mechanism for maintaining transparent network connectivity to mobile nodes [5–11]. Mobile IP enables a mobile node to be identified by the IP address it uses in its home network, regardless of the network to which it is currently attached physically. Therefore, ongoing network connections to a mobile node can be maintained even as the mobile node is moving from one subnet to another.

12.2.1 Mobile IP Components

Mobile IP introduces the following entities:

- **Mobile Node (MN):** a host that may change its point of attachment from one network or subnetwork to another over the Internet. An MN continues to communicate with other nodes at any location using its preassigned home address.
- **Home Agent (HA):** a router on an MN's home network that maintains current location information for the MN. An HA tunnels the addressed datagrams to the MN located away from home.
- **Foreign Agent (FA):** a router on an MN's visited network that provides routing services to the MN while registered. The FA detunnels and delivers datagrams to MNs. For datagrams sent by an MN, the FA may serve as a default router for registered MNs.
- **Correspondent Node (CN):** a node that sends the packets addressed to the MN.
- **Home Network:** a network having a network prefix matching that of an MN's home address.
- **Foreign Network:** any network other than the MN's home network.
- **Home Address:** a permanent IP address assigned to an MN. It remains unchanged regardless of where the MN is attached to the Internet.
- **Care-of Address (CoA):** an address that identifies the MN's current location on the visited network. It can be viewed as the end of a tunnel directed toward an MN. It can be either assigned dynamically or associated with its FA.
- **Tunnel:** the path taken by encapsulated packets. That is, it is the path that leads packets for an MN from the HA to the FA.
- **Encapsulation/Decapsulation:** *Encapsulation* is the process of enclosing an IP datagram within another IP header that contains the CoA of the MN. The IP datagram itself remains intact and untouched throughout the enclosing process [9]. *Decapsulation* is the process of stripping the outermost IP header of the incoming packets so that the enclosed datagram can be accessed and delivered to the proper destination. Decapsulation is the reverse process of encapsulation.

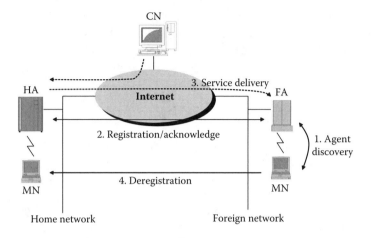

Figure 12.1   Protocol operations of Mobile IP.

12.2.2   Mobile IP Protocol Operations

To support mobility of roaming MNs as shown in Figure 12.1, an IP network will perform the following operations:

1. **Agent discovery:** At first, an MN should determine whether its current attachment point is in the home network or any foreign network. When an MN is away from home, it will find an FA for attachment to the Internet. There are two ways of finding agents, which use ICMP router discovery extension messages [12]: an *agent advertisement message* and an *agent solicitation message.* One way is by receiving agent advertisement messages periodically advertised from an FA, and another way is by sending out from an MN a periodic agent solicitation message until it receives a response from an FA. The MN thus gets the CoA, being dynamically assigned or associated with its FA.

2. **Registration:** The MN will register the obtained CoA with its HA to continue service. The registration process can be performed directly from the MN or relayed by the FA to the HA, depending on whether the CoA is dynamically assigned or associated with its FA.

3. **Service delivery:** Service continues while the MN stays in the service area, that is, during the period after the registration process and before the service time expiration. During the service period, the MN gets forwarded packets from its FA that were originally sent from the MN's HA. Tunneling is the method used to forward packets from HA to FA and finally to MN [9].

4. **Deregistration:** After the MN returns home, it deregisters its registered CoA in its HA. That is, the MN sets its CoA back to its

home address. The MN achieves this by sending a registration request directly to its HA with the lifetime being set to zero. There is no need to deregister with the FA because the service expires automatically when the service time expires.

12.2.3   Problems of Mobile IP

There are some well-known concerns in standardized Mobile IPv4. If an MN changes the point of attachment rapidly, as would occur when it is in a fast-moving vehicle, the MN will report its new attachment to its HA frequently. This may cause high signaling traffic.

Due to the long registration process during handoff, the packet loss rate will be quite high.

The way to encapsulate the datagram is to put the original datagram inside another IP envelope. The fields in the outer IP header add too much overhead to the final datagram, such that several fields are duplicated from the inner IP header. This waste of unnecessary space is not economical.

Even if an MN moves to the same subnet, the CN will send the datagram all the way to the MN's HA, and its HA will then forward the datagram to its CoA, which might take a half-second to reach if the datagram is sent directly from the CN. This kind of *triangle routing* is inefficient and undesirable.

12.3   Mobile IPv6

Mobile IP version 6 (Mobile IPv6) [8] is based on the same principles as Mobile IPv4, enabling IP nodes to move from one IP subnet to another, but has been designed to allow transparent routing of IPv6 packets to IPv6 MNs and appears to be much more powerful. This is well integrated into the new IPv6 protocol and exploits its facilities for overcoming several deficiencies encountered in Mobile IPv4.

12.3.1   IP version 6

A brief overview of the basic characteristics of IP version 6 (IPv6) is given here. IPv6 is a new version of the Internet Protocol, designed by the IETF [4] as the successor to IP version 4 (IPv4). Currently, the numbers and kinds of electronic devices and machines are explosively increasing. As the devices and machines are trying to add to the Internet, it brings about a shortage of IPv4 addresses. IPv6 solves the problems with IPv4, such as the limited number of available IPv4 addresses. With 128-bit-long address, IPv6 will be able to cope with the eventuality of providing Internet access for all kinds of information devices.

IPv6 also adds many improvements to IPv4, such as optimal header format [15], reasonable addressing architecture [19], neighbor discovery mechanism [17], stateless auto-configuration [16], and security and Quality of Service (QoS) support. New extensions for supporting authentication and data integrity are optionally provided. There is a new capability to enable the labeling of packets belonging to particular traffic flows for which the sender requests special handling, such as QoS or real-time service. IPv6 has provision for automatic configuration, which promises reduced complexity of network deployment and administration.

12.3.2 Mobile IPv6 Protocol Operations

The basic idea for both Mobile IPv4 and Mobile IPv6 is always similar in the way that an MN needs to notify its HA of its current location in order to route packets to the node's current point of attachment. An MN maintains a home address, which serves as its identification to higher layers. It is reserved for a link-local address, which is not routable but which is guaranteed to be unique on a link (i.e., on a local network). Also, it is configured with a CoA on each visiting link, which serves as a routable address for being able to receive packets.

In Mobile IPv6, an MN is always addressable by its home address, and packets can be routed to it using this address, regardless of the MN's current point of attachment to the Internet. Thus, the MN communicates with other nodes after changing its link-layer point of attachment from one IP subnet to another, yet without changing the MN's IPv6 address.

More detailed protocol operations for Mobile IPv6 are given as follows. An MN, while attached to its home network, is able to receive packets addressed to its home address and that are forwarded by means of conventional IP routing mechanisms.

When an MN moves from one IP subnet to another, the MN will configure its CoA by stateless address auto-configuration, or alternatively by some means of stateful address auto-configuration. In the former case, the address can be generated easily by combining the network prefix of the visited network and an interface identifier of the MN. In the latter case, a DHCPv6 server [23] is required for making such address assignment to the MN. The decision about which manner of automatic address configuration to use is made according to IPv6 neighbor discovery mechanisms [17]. In addition, an MN can have more than one CoA at a time, for example, if it is link-level attached to more than one wireless network at a time or if more than one IP network prefix is present on a network to which it is attached but only one of them is registered to the MN's HA. The association between the MN's home address and its CoAs, known as *binding*, is maintained by the node.

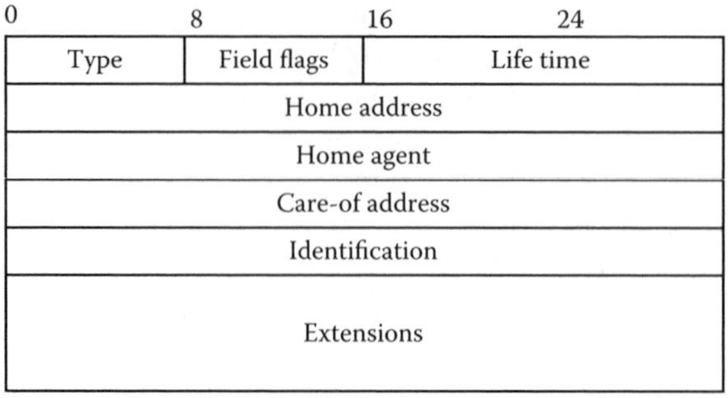

0	8	16	24
Type	Field flags	Life time	
Home address			
Home agent			
Care-of address			
Identification			
Extensions			

Figure 12.2   Registration request format for Mobile IPv4.

While away from the home network, an MN registers one of its bindings with a router in its home network, requesting this router to function as the HA for the MN. The MN's HA retains this entry in a cache, known as a binding cache, until its lifetime expires. Figure 12.2 and Figure 12.3 show the binding update datagram formats of Mobile IPv4 and Mobile IPv6, respectively [7,8]. There are several differences between Mobile IPv4 and Mobile IPv6. Contrary to the fixed set of address fields in Mobile IPv4, the field of mobility option in Mobile IPv6 is a variable-length field and could include various information to support the mobility of MNs. The field of mobility options contains zero or more mobility options, which are the binding authorization data option and the alternate CoA option. If no options are present in this message, the CoA is specified either by the source address field in the IPv6 header or by the alternate CoA option.

While the HA has a home registration entry in its binding cache, it uses IPv6 neighbor discovery to intercept any IPv6 packets addressed to the MN's home address on the home subnet [17]. The HA will send a

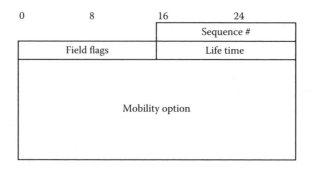

Figure 12.3   Binding update format for Mobile IPv6.

neighbor advertisement message to the all-nodes multicast address on the home link so it can associate its own link-layer address with the node's home IP address. Certainly, the HA needs to reply to any *neighbor solicitation messages* addressed to the MN after checking that it maintains a binding cache for it. In this way, all packets destined to the node's home address will be delivered to the HA, which will tunnel each intercepted packet to the MN's CoA indicated in the binding cache entry. To tunnel the packet, the HA encapsulates it using IPv6 encapsulation [18].

Moreover, Mobile IPv6 provides a mechanism for IPv6 CNs communicating with an MN to dynamically learn the MN's binding. The CN adds this binding to its binding cache. When sending a packet to any IPv6 destination, a CN checks its binding cache for an entry for the packet's destination address; and if a cached binding for this address is found, the CN routes the packet directly to the MN at the CoA indicated in this binding. In this routing optimization mechanism, it uses an IPv6 routing header instead of IPv6 encapsulation, which is able to add less overhead to the size of the packet. If it does not find the binding cache entry, the CN sends the packet to the destination (the MN's HA), and the packet is then intercepted and tunneled by the MN's HA.

12.3.3 Mobile IPv4 versus Mobile IPv6

The design of Mobile IP support in IPv6 (Mobile IPv6) benefits both from the experiences gained from the development of Mobile IP support in IPv4 (Mobile IPv4) and from the opportunities provided by IPv6. Mobile IPv6 thus shares many common features with Mobile IPv4, and offers many other improvements. This section summarizes the major differences between Mobile IPv4 and Mobile IPv6.

One major difference with Mobile IPv6 is that no FAs are needed to support the MNs. To save the limited IPv4 address resource, an IPv4 MN at the new attachment point is asked to use the FA's CoA as its temporary address. However, the IPv6 functionality of stateless or stateful auto-configuration is sufficient for letting MNs acquire their CoA without the FA's assistance. In this case, the IPv6 MN will process most of the FA's function instead. Thus, Mobile IPv6 operates in any location without any special support required from the local router.

Each IPv4 MN has just one CoA at a time. In contrast, an IPv6 MN is allowed to acquire multiple CoAs, for example, if the link-level is attached to more than one wireless network at a time or if more than one IP network prefix is present on a network to which it is attached. This multiple-CoA functionality is particularly important for supporting smooth handovers. In Mobile IPv4, when an MN moves to another foreign network, the packets addressed to the MN's previous CoA are silently discarded. To support smooth handoff for Mobile IPv4, an additional design could be required such that the previous FA acts as a forward pointer and the previous FA is responsible to retunnel the packets to the MN's current CoA. Otherwise, the support of smooth handoffs is originally provided in Mobile IPv6. The previous default router is requested to temporarily play the role of an HA in the previous network when the MN just moved to another network. When receiving a binding update of the MN, the previous default router acts as an HA, and starts to multicast proxy neighbor advertisement among the links to inform other routers of receiving packets on behalf of the MN. Thereafter, the previous default router encapsulates the intercepted packets and tunnels them to the MN's current CoA.

In Mobile IPv4, the MN's HA has the potential to perform service delivery. Thus, the well-known triangle routing problem happens in Mobile IPv4. Therefore, routing optimization is an additional design to help Mobile IPv4 to efficiently deliver packets [22]. In Mobile IPv4 routing optimization, any CN maintains a binding cache. A binding update message will be sent to the source correspondent to inform the MN's new CoA by the FA or the MN itself. The correspondent then updates the binding cache and tunnels any later packets directly to the MN's current CoA without bypassing the HA anymore. In Mobile IPv6, however, routing optimization is one of the fundamental functions. The IPv6 routing header, instead of encapsulation, uses routing optimization for direct delivery of packets to the MN without the intervention of the HA.

IPv6 neighbor discovery, instead of an address resolution protocol (ARP), is used by the HA to intercept packets addressed to the MN's home address. This mechanism is used for determining the link-layer address of neighbors known to reside on the attached links. This decouples the Mobile IPv6 functionality from a particular link layer and therefore enhances the protocol's robustness. IPv6 neighbor discovery is also used in Mobile IPv6, with some modifications when required, to allow move detection by MNs and to enable home address discovery that an MN might not discover the IP address of its own HA in such a case where the home subnet prefixes can change over time.

In Mobile IPv4, it is not always possible for an MN to send packets directly to the correspondent without being routed by the HA due the router's effect of ingress filtering [13]. In solving this, reverse tunneling must be used in Mobile IPv4 [14], with the sacrifice of routing efficiency optimization. In Mobile IPv6, because the correspondent can cache MN's binding, the problem of reverse tunneling is solved without affecting the operation of ingress filtering.

Because it is assumed that all IPv6 nodes are to implement strong authentication and encryption features to improve Internet security [20,21], the Mobile IPv6 protocol is then simplified so that it need not specify the security procedures by itself.

12.4   Micro-Mobility Management for Mobile IPs

Mobile IPv4 and Mobile IPv6 standards offer global mobility management mechanisms, allowing users to move between different wide wireless access networks connected to the Internet. However, these global mobility management protocols lack many essential features to satisfy the requirements of IP mobile networks, for example, low signaling delay and low network overhead. Many solutions have been proposed to overcome the problems of global mobility management within IP networks; they are often called *micro-mobility management protocols.* Micro-mobility covers the management user movement at a local area level, inside a particular wireless network. A quick review of several well-known IP micro-mobility management protocols, including regional micro-mobility management, hierarchical Mobile IPv6, Cellular IP, HAWAII, IDMP, and user-based micro-mobility management, is given.

12.4.1   Regional Micro-Mobility Management

Regional micro-mobility management [24] intends to minimize control traffic overhead, by which movements within a foreign network do not inform the MN's HA of the new attachment. That is, regional micro-mobility management is a solution for performing registrations locally in a regional network consisting of one or more FAs, as shown in Figure 12.4. Regional

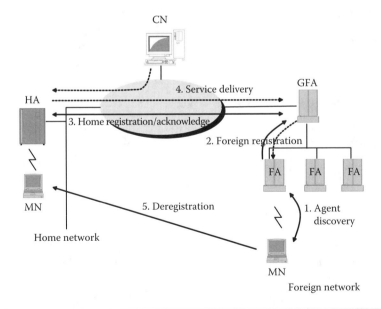

Figure 12.4   Protocol operations of regional micro-mobility management.

micro-mobility management uses a gateway foreign agent (GFA) acting as a local HA while roaming within a region. The GFA maintains the local point of attachment of MNs, receives all packets on behalf of the MNs, and encapsulates and forwards them to their current point of attachment.

When an MN first arrives at a visited domain, the MN finds an FA by getting the CoA that may be associated with its FA (agent discovery).

The MN then performs a foreign registration with the GFA and it registers the obtained CoA with the GFA (foreign registration). The GFA keeps a visitor list of all the MNs currently registered with it. Thereafter, it performs a *home registration* with its HA. At this home registration, the HA registers the CoA of the MN. The CoA that is registered at the HA is the address of a GFA. Because the CoA registered at the HA is the GFA address, it will not change when the MN changes FA under the same GFA. Thus, the HA does not need to be informed of any further MN movements within the visited domain.

When the MN moves from one FA to another FA within the same visited domain, that is, the advertised GFA address that the MN receives from the new FA is the same as the one the MN has registered as its CoA during its last home registration, the MN registers the new FA's CoA with the GFA instead of HA (foreign registration).

When CNs send data packets to the MN's home address, these packets are intercepted and then encapsulated by the MN's HA. The encapsulated packets are forwarded to the registered GFA. After the GFA receives the packets from the mobile's HA, it will tunnel them to the FA of the target mobile. After the FA receives the packets from the GFA, it will send them to the mobile, and the mobile will process them in the usual manner (service delivery).

12.4.2   Hierarchical Mobile IPv6

The basic idea of Hierarchical Mobile IPv6 [25] is the same as for regional micro-mobility management in the sense that the MN's HA need not be informed of every movement that the MN performs inside the foreign network. Hierarchical Mobile IPv6 introduces a new Mobile IPv6 node called the Mobility Anchor Point (MAP). The MAP can be located at any level in a hierarchical network of routers, including the access router (AR). The AR is the MN's default router and it aggregates the outbound traffic of the MN.

In Hierarchical Mobile IPv6, an MN is assigned two CoAs, called Regional CoA (RCoA) and On-Link Care-of Address (LCoA). An MN obtains the RCoA from the visited network and it is an address on the MAP's subnet. The LCoA is the CoA that is based on the prefix advertised by AR.

When an MN enters a new site, it gets a router advertisement containing information on one or more local MAPs. The router advertisement will inform the MN of the available MAPs and their distances from the MN. After selecting a MAP, the MN gets the RCoA on the MAP's domain and the LCoA from the AR. The MN sends an update of binding between the RCoA and LCoA to the MAP. The MAP records the binding and inserts it in its binding cache (foreign registration). The MAP sends binding update messages also to the MN's HA and to the CNs with which the MN is interacting (home registration).

For the MN outside its home network, the global incoming data goes through MAP hierarchy. When CNs or HA send messages to the MN's RCoA, they are received by MAP, which in turn tunnels the messages to the MN's local address LCoA using IPv6 encapsulation. The function of the MAP is basically the same as that of the HA. MAP receives all data packets coming from external networks and forwards them to the MN (service delivery).

As the MN roams locally, it gets a new LCoA from its new AR. The RCoA remains the same as long as the MN is within the site. When an MN sends a binding update to the CNs that are situated within the site, these CNs are able to send data packets directly to the MN without MAP intervention (routing optimization).

12.4.3 Cellular IP

Cellular IP, which basically supports the mobility of nodes in cellular networks, is based on hierarchical micro-mobility management [26, 27]. In contrast to Hierarchical Mobile IPv6, Cellular IP uses mobile-originated data packets to maintain reverse path routes, as shown in Figure 12.5. Access nodes or base stations (BSs) in Cellular IP monitor mobile originated packets and maintain a distributed hop-by-hop location database that is used to route packets to MNs. BSs on the path of routing forward incoming packets to the MN by referring their routing information.

A gateway periodically broadcasts a beacon packet, depending on which BS can form the uplink to the gateway and then route the uplink packets from the MN to the gateway hop-by-hop. Each BS is also responsible for maintaining a routing cache. An entry in the routing cache binds the MN's HA with the interface through which the MN can arrive. The routing information of the entry is generated and refreshed by monitoring regular packets sent by an MN in order to minimize control messaging. The downlink path can then be formed by reversing the path. Because the bindings in the routing caches have timeout values, an MN can keep the entry valid by sending empty packets (route-update packets) to the gateway at regular intervals.

In particular, Cellular IP supports two types of handoff schemes: (1) hard handoff and (2) semi-soft handoff. Cellular IP hard handoff is based on a simple approach that trades off some packet loss in exchange for minimizing handoff signaling rather than trying to guarantee zero packet

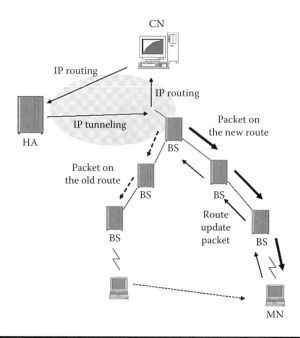

Figure 12.5   Protocol operations of Cellular IP.

loss. Cellular IP semi-soft handoff prepares for handoff by proactively notifying the new access point before actual handoff. Semi-soft handoff minimizes packet loss, providing improved Transmission Control Protocol (TCP) and User Datagram Protocol (UDP) performance over hard handoff.

Cellular IP also supports IP paging and is capable of distinguishing between active and idle MNs. Paging systems help minimize signaling in support of better scalability and reducing the power consumption of MNs. Cellular IP tracks the location of idle nodes in an approximate and efficient manner. Therefore, MNs do not have to update their location after each movement. This extends battery life and reduces air interface traffic. When packets need to be sent to an idle MN, the node is paged using a limited scope broadcast and in-band signaling. An MN becomes active upon receiving a paging packet and starts updating its location until it moves to an idle state again.

12.4.4   HAWAII

The Handoff Aware Wireless Access Internet Infrastructure (HAWAII) protocol [28] is proposed by Lucent Technologies Bell Labs as a separate routing protocol to take care of the micro-mobility inside the visited domain. A gateway in each domain is called the Domain Root Router (DRR). No HA is involved when an MN's movement is within the home domain, where the MN is identified by its IP address. When an MN is moving within a foreign domain, the MN is assigned a CoA. Packets can then be tunneled to the MN by the HA in its home domain. This CoA remains unchanged as long as the MN is moving within the foreign domain; thus, the HA does not need to be notified of these movements unless the mobile host moves to a new domain. This is achieved by enabling any mobile host to register with a BS while using a CoA and then locally handling the registration by the corresponding BS.

The processing and generation of the registration messages are split into two parts: (1) between the mobile host and the base station, and (2) between the base station and the home agent. Nodes in a HAWAII network execute a generic IP routing protocol and maintain mobility-specific routing information. Location information (i.e., mobile-specific routing entries) is created, updated, and modified by explicit signaling messages sent by MNs.

HAWAII defines different alternative path setup methods to control handoff between the points of attachment. An appropriate method is selected, depending on the service level agreement or operator's priorities. Two scenarios are considered by HAWAII: (1) power up and (2) following handoff. When the MN first powers up and tries to attach to the domain root router, it sends a path power-up message to the router. All intermediate routers between the MN and the domain root router are informed and set up their routing tables without knowing the MN's IP address. When the MN moves from one place to another inside a domain, it sends path setup update messages to establish a new route. The MN also informs the routers of the old route.

Soft handoff states are maintained in intermediate routers for MNs. The MN sends periodic path refresh messages to the base station to which it is attached. The base station and intermediate routers also send periodic aggregate hop-by-hop refresh messages to the domain root router.

Similar to Cellular IP, HAWAII also uses paging to search MNs when incoming data packets arrive at an access network and no recent routing information is available.

12.4.5 IDMP

The Intra-Domain Mobility Management Protocol (IDMP) uses a hierarchical structure with a mobility agent (MA) at the top of the hierarchy and several child sub-network FAs interconnected to it [29,30]. The top-level MA functions as a gateway to the Internet.

In IDMP, there are global and local addresses for handling mobility. The global address points toward the current administrative top-level MA. This address remains unchanged as long as the MN remains in the MA's domain. In contrast, the local address is a pointer toward the visiting FA and changes once an MN moves to a different FA. No global registration is necessary as long as nodes move within the MA's administrative domain.

Similar to Cellular IP and HAWAII, IDMP also uses IP multicasting to page idle MNs when incoming data packets arrive at an access network. In contrast of Cellular IP and HAWAII, IDMP supports dynamic paging area configuration, which changes the number of FAs on a subnetwork. Thus, it could reduce the overhead of multicasting to page MNs.

12.4.6 User-Based Micro-Mobility Management

The method of user-based micro-mobility management by Xie and Akyildiz [32] is an extension of regional micro-mobility management, in which each FA can function either as an FA or a GFA. Whether an agent should act as an FA or a GFA depends on node mobility. When an MN enters a regional network, the first FA of the subnet the MN visits will function as a GFA of this network. If an agent acts as a GFA, it needs to maintain a visitor list and keep entries in the list updated according to the registration requests sent from other FAs within the regional network. The GFA also relays all the home registration requests to the HA. Other agents in the regional network act as the general FAs.

In contrast to regional micro-mobility management, the number of FAs under a GFA for user-based micro-mobility management is not fixed but, rather, optimized for each MN to minimize the total signaling traffic, as shown in Figure 12.6. Each MN has its own optimized system configuration from time to time. To determine the regional network for MNs, each MN

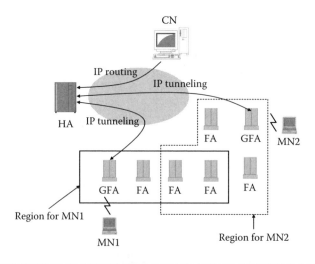

Figure 12.6   Protocol operations of user-based micro-mobility management.

keeps a buffer for storing the IP addresses of FAs. An MN records the address of the GFA into its buffer when it enters a new regional network and then performs a home registration through the new GFA. After the home registration, the optimal number of FAs for a regional network is computed based on the up-to-date parameters (e.g., packet arrival rate and the number of registrations) of the MN. This optimal value is dynamically set for the buffer length threshold of the MN.

When the MN detects that it enters a new subnet, it performs a regional registration by sending a regional registration request to the recorded IP address of the GFA. The mobile then compares the IP address of the FA in the new subnet with the addresses recorded in its buffer. If the address of the current FA has not been recorded in the buffer, then the MN records it — otherwise, ignores it. If the total number of addresses in the buffer as well as the address of the current FA exceeds the threshold, it means the MN is in a new regional network. The MN deletes all the addresses in its buffer, saves the new one, and requests a home registration. Thus, there is no strict regional network boundary for each MN.

Ma and Fang [33] show that registration signaling can be reduced by registering the new CoA to the previous FA when an MN moves from one subnet to another. However, forwarding through multiple FAs will cause some service delivery delay, which may not be appropriate when there is delay restraint for some Internet applications such as video or voice services. To avoid excessive packet transmission delay, they set a threshold on the level number of the subnet hierarchy. When the threshold is reached, the MN will register to its HA. Otherwise, it will register to the previous FA. The threshold is dynamically adjusted based on every MN's traffic load and mobility.

12.5 Proposed Agent-Based Micro-Mobility Management

To improve system performance, we previously proposed agent-based micro-mobility management methods for Mobile IPv4 [34,35], but they can be extended to Mobile IPv6. Similar to user-based micro-mobility management, agent-based micro-mobility management also allows all FAs in a domain to act as a GFA. Unlike user-based micro-mobility management, agent-based micro-mobility management allows each agent to configure its own domain and to act as a GFA when mobiles reside in the domain.

There are two types of GFA: (1) the permanent GFA (P-GFA) and (2) the temporary GFA (T-GFA). The P-GFA performs as a local HA for mobiles residing in the P-GFA's domain, referred to as a permanent domain (P-domain). In addition, every FA acts as a T-GFA on its sub-domain, referred to as a temporary domain (T-domain).

12.5.1 Domain Configuration

Each FA produces the list of FAs, namely a domain list, to indicate the range of the T-domain where the FA functions as a T-GFA. The domain list of an FA is the set of neighboring FAs, including itself, within a distance threshold D from itself, which is given as:

$$D = \begin{cases} \frac{H_{FA\leftrightarrow P-GFA}-1}{2}, & \text{if } H_{FA\leftrightarrow P-GFA} \text{ is odd,} \\ \frac{H_{FA\leftrightarrow P-GFA}}{2}, & \text{if } H_{FA\leftrightarrow P-GFA} \text{ is even,} \end{cases} \qquad (12.1)$$

where $H_{FA\leftrightarrow P-GFA}$ is the number of hops from an FA to the P-GFA.

For the sample network shown in Figure 12.7, Table 12.1 shows the domain list of each FA with which distance threshold D is obtained by (Equation 12.1). The P-GFA has both P-domain and T-domain. The P-domain for

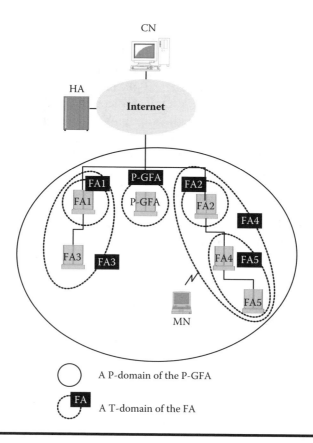

Figure 12.7 A sample network for agent-based micro-mobility management.

Table 12.1   Domain Lists for the Network Shown in Figure 12.7

	P-GFA	T-GFAs				
		FA_1	FA_2	FA_3	FA_4	FA_5
Distance threshold (D)	0	0	0	1	1	1
Domain List	P-GFA	FA_1	FA_2	FA_1	FA_2	FA_4
(T-domain)				FA_3	FA_4	FA_5
					FA_5	

the P-GFA consists of FA_1, FA_2, ..., FA_5 and P-GFA. The T-domain for the P-GFA consists of P-GFA only, as shown in Table 12.1.

In Figure 12.7, when the mobile moves from FA_4 to FA_5, registering its location with FA_4 (i.e., T-GFA) instead of the P-GFA is more beneficial for reducing the registration delay. This is because the distance between FA_5 and FA_4 (i.e., T-GFA) is shorter than that between FA_5 and P-GFA. All other cases where a mobile changes a point of attachment within a T-domain satisfy that the proposed domain configuration provides a distance for performing registrations that is shorter than or equal to that of conventional micro-mobility management methods. Thus, the agent-based micro-mobility management method is useful for reducing packet loss during registration. Although the proposed method imposes an extra overhead to configure a T-domain (i.e., to compute the distance from an FA to the P-GFA and its neighboring FAs), it is not significant because it is enough to establish the T-domain once for the first time.

12.5.2   Domain Detection

Whenever a mobile moves from an FA's service area to another, it must always verify whether it enters a new domain. For domain detection, it requires mobiles as well as FAs to maintain the domain list. For this purpose, every T-GFA sends router advertisement messages, which are specified in Mobile IP message format, containing its domain list to mobiles residing in its T-domain. Then the mobiles obtain the domain lists of their T-GFAs.

See Figure 12.7 and Table 12.1. Let FA_3 be the current point of a mobile, and it also acts as a T-GFA for the mobile. Initially, after receiving a router advertisement message from FA_3 containing FA_3's domain list, <FA_1, FA_3>, the mobile stores the domain list of FA_3 in its buffer. When the mobile moves into FA_1 from FA_3, it must verify whether FA_1 belongs to the domain list. After the mobile has received a router advertisement message from FA_1, it knows that FA_1 is one of the routers in the buffered list, thereby confirming that it still resides in the same domain. When the mobile moves to FA_2, it does not have any location information of FA_2 in its buffered list.

In this case, the mobile confirms that it moves into a new domain, and it then refreshes the old list and stores the new domain list of FA$_2$, <FA$_2$>, which will be the new T-GFA for the mobile.

12.5.3 Registration

In the agent-based micro-mobility management method, registrations occur in the following three cases. The first is the case where a mobile moves out of its home network or moves from one P-domain to another. The second is the case where a mobile moves from a T-domain to another within the same P-domain. The third is the case where a mobile moves from an FA's service area to another within the same T-domain.

For location registration in the first case, when a mobile moves out of its home network or moves from one P-domain to another, it gets an IP address of the first FA on the P-domain. The mobile obtains the FA's domain list through domain detection. The mobile registers the obtained FA's address with its P-GFA. The mobile then registers the P-GFA's address with its HA. The HA sends an acknowledgment message to the mobile via the P-GFA and the FA. After updating the mobile's location in both the P-GFA and HA, the FA will become the new T-GFA for the mobile.

For location registration in the second case, when a mobile moves from one T-domain to another within the same P-domain, it gets an IP address of the first FA on the new T-domain. The mobile obtains the FA's domain list through domain detection. The mobile registers the obtained FA's address with its P-GFA. The P-GFA sends an acknowledgment message to the mobile via the FA. After updating the address in the P-GFA, the FA will become the new T-GFA for the mobile.

For location registration in the third case, when a mobile moves from one FA's service area to another within the same T-domain, it does not need to update the domain list in the buffer, and also does not need to take a new T-GFA. It gets an IP address only from the visiting FA. The mobile registers the obtained address with its T-GFA. The T-GFA sends an acknowledgment message to the mobile.

12.5.4 Packet Delivery

When a CN (or CNs) or a home network sends data packets to the MN's home address, these packets are intercepted and then encapsulated by the MN's HA. The encapsulated packets are forwarded to the registered P-GFA. After the P-GFA receives the packets from the mobile node's HA, it will tunnel them to the T-GFA of the target mobile, after the T-GFA receives the packets from the P-GFA, it will send them to the mobile, and the mobile will process them in the usual manner.

12.5.5 An Extension of Agent-Based Micro-Mobility Management

In the agent-based micro-mobility management method, the single P-GFA remains essential to service registration and packet delivery. High registration messages and packet arrivals may incur high signaling overhead at the P-GFA. Also, the failure of the P-GFA will lead packets to be unreachable to the mobiles managed by it. For system robustness and performance improvement purposes, an extension of agent-based micro-mobility management without implementing the P-GFA is proposed [34,35].

Both agent-based micro-mobility management methods are the same as the following. Every FA produces its domain list by its distance threshold. Each FA functions as a GFA in its domain determined by the domain list. Unlike the agent-based micro-mobility management method, in the extended method, a mobile will register its GFA's address with the HA whenever it changes its associated domain. This is because it does not allow for locating the P-GFA.

The range of the domain for each FA is determined by the distance threshold of the FA. Figure 12.8 shows a sample network for describing the extended method. Table 12.2 shows the domain list determined by the sample distance threshold of each FA. The extension supports that the distance threshold of each FA can be changed dynamically, and thus the FA must be decided to minimize the signaling load and packet loss.

For registration, when a mobile moves out of its home network or moves from one domain to another, the mobile gets an IP address of the first FA in the domain. The mobile obtains the FA's domain list through domain detection. After the mobile registers the obtained FA's address with its HA, the FA will be the new GFA for the mobile.

On the other hand, when a mobile moves from one FA's service area to another within the same domain, it does not need to update the domain list in the buffer, and also does not need to take a new GFA. It gets an IP address only from the visiting FA. In turn, the mobile registers the obtained address with its GFA. The GFA sends an acknowledgment message to the mobile.

For service delivery, if CNs or a home network send data packets to the MN's home address, these packets are intercepted and then encapsulated by the MN's HA. The encapsulated packets are forwarded to the registered GFA. After the GFA receives the packets coming from the HA, it will send them to the mobile. The mobile will process them in the usual manner.

12.5.6 Performance Studies

In simulation studies, each mobile stays in an FA's service area for a random period of time and then moves into another. The average time during

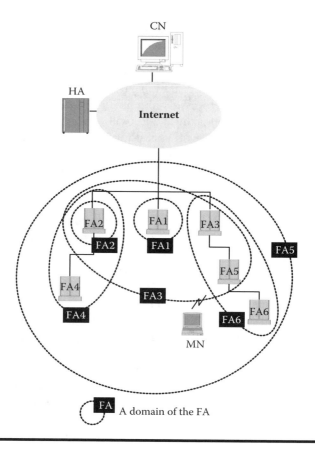

Figure 12.8 A sample network for the extension of agent-based micro-mobility management.

Table 12.2 Domain Lists for the Network Shown in Figure 12.8

	Distributed GFAs					
	FA_1	FA_2	FA_3	FA_4	FA_5	FA_6
Distance threshold (D)	0	0	1	1	3	2
Domain List	FA_1	FA_2	FA_1	FA_2	FA_1	FA_3
(Domain)			FA_2	FA_4	FA_2	FA_5
			FA_3		FA_3	FA_6
			FA_5		FA_4	
					FA_5	
					FA_6	

which a mobile resides in an FA's service area follows an exponential distribution. To generate packet arrivals at mobiles, ten CNs send data packets to mobiles, and each CN sends packets with a Poisson process to a randomly selected mobile.

Most packet loss usually occurs during registration in Mobile IP networks. High packet loss imposes high retransmission overhead on the system or degrades the QoS. Thus, for a high QoS and high packet transmission rate, it is necessary to reduce the delay of registration. In micro-mobility management, registration delay is greatly affected by network traffic. To study the effect of network traffic on registration delay, the number of mobiles roaming in a local network is changed from 10 to 50, where each mobile has an average residence time in an FA's service area of 60 seconds, and it receives packets with an average arrival rate of 1/60. A large number of mobiles will cause high network traffic.

Figure 12.9 shows the registration delay of the various micro-mobility management methods — regional micro-mobility management method (RMMM), a hierarchical micro-mobility management method (HMMM), and an enhanced agent-based micro-mobility management method (EAMMM). A large number of mobiles in a network result in an increase in the registration delay of each method. This result can be easily understood, in that high network traffic will cause high registration delay. EAMMM has a lower registration delay than the others. In particular, EAMMM is more beneficial in reducing the registration delay as the network traffic increases, whereas HMMM has a high delay of registration, in particular for high network traffic. This is because registration processes should occur at all hierarchy GFAs located on the path to registration. This performance degradation problem will also occur in Cellular IP and HAWAII, which are based on the hierarchy network configuration.

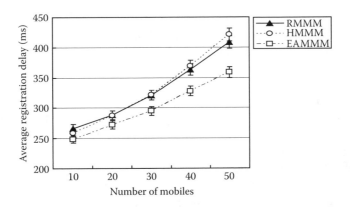

Figure 12.9 Registration delay of micro-mobility management methods.

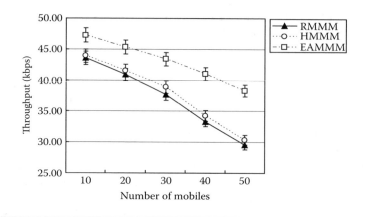

Figure 12.10 Throughput of micro-mobility management methods.

Figure 12.10 shows the throughput as a function of changing the number of mobiles on the network. Each mobile has an average residence time in an FA's service area of 60 seconds, and it receives packets with an average arrival rate of 1/60. As the number of mobiles increases, (i.e., as network traffic increases), the throughput of each method decreases. A large number of mobiles will cause the processing traffic of delivering packets at the GFA to increase. EAMMM has a higher throughput than others. Compared to RMMM and HMMM, in particular, the throughput decreases more slowly as the network traffic increases.

12.6 Summary

This chapter discussed global mobility management and micro-mobility management for both Mobile IPv4 and Mobile IPv6. For the global mobility solution for Mobile IP, we discussed the standard Mobile IPv4 and Mobile IPv6, including the basic principles, enhancement, differences, and problems associated with Mobile IPs. In terms of solutions for overcoming the problems with global Mobile IP mobility management, we discussed micro-mobility management methods, including Regional Micro-Mobility Management, Hierarchical Mobile IPv6, Cellular IP, HAWAII, IDMP, and user-based micro-mobility management.

Furthermore, we introduced two agent-based micro-mobility management methods that provide distributed registration and dynamic network configuration. The performance studies revealed that the agent-based micro-mobility management methods have better performance in terms of signaling delay of registration and throughput.

References

[1]   M. Mouly and M.B. Pautet, "The GSM System for Mobile Communications," M. Mouly, 49 rue Louise Bruneau, Palaiseau, France, 1992.

[2]   EIA/TIA, "Cellular Radio-Telecommunications Intersystem Operations," Tech. Rep. IS-41 Revision B, EIA/TIA, 1991.

[3]   A. Bar-Noy, I. Kessler, and M. Sidi, "Mobile Users: To Update or Not to Update?" *ACM-Baltzer J. Wireless Networks*, Vol. 1, No. 2, pp. 175–186, July 1995.

[4]   IETF Mobile IP Working Group, http://www.ietf.org/.

[5]   C. Perkins, "Mobile IP," *IEEE Commun. Mag.*, Vol. 35, No. 5, pp. 84–99, May 1997.

[6]   C. Perkins, "IP Mobility Support", RFC 2002, October 1996.

[7]   C. Perkins, "IP Mobility Support for IPv4," RFC 3344, August 2002.

[8]   D. Johnson, C. Perkins, and J. Arkko, "Mobility Support in IPv6," RFC 3775, June 2004.

[9]   C. Perkins, "IP Encapsulation within IP," RFC 2003, October 1996.

[10] G. Montenegro, "Reverse Tunneling for Mobile IP," RFC 2344, May 1998.

[11] C. Perkins, "Minimal Encapsulation within IP," RFC 2004, October 1996.

[12] S. Deering, "ICMP Router Discovery Messages," RFC 1256, September 1991.

[13] P. Ferguson and D. Senie, "Network Ingress Filtering: Defeating Denial of Service Attacks which Employ IP Source Address Spoofing," RFC 2267, January 1998.

[14] G. Montenegro, Ed., "Reverse Tunneling for Mobile IP, Revised," RFC 3024, January 2001.

[15] S. Deering and R. Hinden, "Internet Protocol, Version 6 (IPv6) Specification," RFC 2460, December 1998.

[16] S. Thomson and T. Narten, "IPv6 Stateless Address Autoconfiguration," RFC 2462, December 1998.

[17] T. Narten, E. Nordmark, and W. Simpson, "Neighbor Discovery for IP Version 6 (IPv6)," RFC 2461, December 1998.

[18] A. Conta and S. Deering, "Generic Packet Tunneling in IPv6 Specification," RFC 2473, December 1998.

[19] R. Hinden and S. Deering, "Internet Protocol Version 6 (IPv6) Addressing Architecture," RFC 3513, April 2003.

[20] S. Kent and R. Atkinson, "IP Authentication Header," RFC 2402, November 1998.

[21] S. Kent and R. Atkinson, "IP Encapsulating Security Payload (ESP)," RFC 2406, Work in Progress, November 1998.

[22] C. Perkins and D. Johnson, "Route Optimization in Mobile IP," Internet Draft, draft-ietf-mobileip-optim-10, Work in Progress, November 2000.

[23] R. Droms, J. Bound, B. Volz, T. Lemon, C. Perkins, and M. Carney, "Dynamic Host Configuration Protocol for IPv6 (DHCPv6)," RFC 3315, July 2003.

[24] E. Gustafsson, A. Jonsson, and C. Perkins, "Mobile IPv4 Regional Registration," draft-ietf-mip4-reg-tunnel-00, November 2004.

[25] H. Soliman, C. Castelluccia, Karim El-Malki, and L. Ludovic Bellier, "Hierarchical Mobile IPv6 Mobility Management," draft-ietf-mobileip-hmipv6-08.txt, June 2003.

[26] A. Valkó, "Cellular IP: A New Approach to Internet Host Mobility," *ACM SIGCOMM Comp. Commun. Rev.*, Vol. 29, No. 1, pp. 50–65, January 1999.

[27] A.T. Campbell, J. Gomez, S.K. Kim, A. Valkó, and C.Y., Wan, "Design, Implementation, and Evaluation of Cellular IP," *IEEE Personal Communications*, Vol. 7, pp. 42–49, August 2000.

[28] R. Ramjee, "HAWAII: A Domain-Based Approach for Supporting Mobility in Wide-Area Wireless Networks," *Proc. IEEE Int. Conf. Network Protocols*, 1999.

[29] S. Das, A. McAuley, A. Dutta, A. Misra, K. Chakraborty, and S.K. Das, "IDMP: An Intra-Domain Mobility Management Protocol for Next Generation Wireless Networks," *IEEE Wireless Communications*, Special Issue on Mobile and Wireless Internet: Architectures and Protocols, Vol. 9, No. 3, pp. 38–45, June 2002.

[30] A. Misra, S. Das, A. Dutta, A. McAuley, and S.K. Das, "IDMP-Based Fast Hand-offs and Paging in IP-Based 4G Mobile Networks," *IEEE Communications*, Special Issue on 4G Mobile Technologies, Vol. 40, No. 3, pp. 138–145, March 2002.

[31] A.T. Campbell, J. Gomez, S. Kim, C.-Y. Wan, Z.R. Turanyi, and A.G. Valko, "Comparison of IP Micromobility Protocols," *IEEE Wireless Communications*, pp. 72–82, February 2002,

[32] J. Xie and I.F. Akyildiz, "A Distributed Dynamic Regional Location Management Scheme for Mobile IP," *Proc. IEEE INFOCOM 2002*, pp. 1069–1078, 2002.

[33] W. Ma and Y. Fang, "Dynamic Hierarchical Mobility Management Strategy for Mobile IP Networks," *IEEE Journal on Selected Areas in Communications*, Vol. 22, No. 4, pp. 664–676, May 2004.

[34] C. Pyo, J. Li, and H. Kameda, "Dynamic and Distributed Domain-based Mobility Management for Mobile IPv6," *Proceedings of the IEEE Vehicular Technology Conference*, October 6–9, 2003, Orlando, FL.

[35] C. Pyo, J. Li, and H. Kameda, "Performance Study for Domain-Based Mobility Management with a Dynamic and Distributed Manner for Mobile IPv6," *Proceedings of the 11th IEEE International Conference on Networks* (IEEE ICON 2003), pp. 239–244, September 28–October 1, 2003.

Chapter 13

Location Management in Wireless Networks: Issues and Technologies

Christos Douligeris, Dimitrios D. Vergados, and Zuji Mao

Contents

13.1   Introduction

With the increasing demands for new data services, wireless networks should support calls with different traffic characteristics and different Quality-of-Service (QoS) guarantees. For cellular networks to be able to efficiently cope with mobile user requirements, new mechanisms are needed, mechanisms that will help the network know the location of every mobile user it serves, and thus, properly track the user. Knowing where the user resides, the cellular network will be capable of creating suitable connections (creation of a suitable path) between two users and deliver the call efficiently. Moreover, the cellular network should guarantee low call delivery latency and all time connectivity, especially when a user moves from one cell to another, providing at the same time the appropriate amount of bandwidth and avoiding call termination. The mechanism responsible for carrying out the above procedures is generically called *mobility management*. This mechanism is divided into the following two procedures: (1) the *location management procedure* and (2) the *handover procedure*. The scope of the handover operation is to preserve active users' connections while they transact from one registration area to another. For that reason, several schemes have been implemented. On the other hand, location management is an operation that frequently informs the network where the users reside. Several strategies have been proposed and implemented for the location management procedure. The main ideas behind these strategies and a detailed analysis of some of them are presented in this chapter. The location management procedure can be further subdivided into the following two categories:

1. *Location registration procedure:* When a mobile terminal (MT) happens to move, either from one cell to another or from a certain region (RA, registration area) (that contains a certain number of cells)

to another, it sends specific signaling messages to the network to inform it of its current location.

2. *Call delivery procedure (paging)*: The call delivery procedure takes place after the completion of the location registration procedure. Based on the information that has been registered in the network during the location registration procedure, when an incoming call is initiated by an MT, the call delivery procedure interrogates the network as to the exact location of the called MT so that the call can be delivered.

An important problem in designing a cellular network lies in how capable and how flexible is the network in correctly managing the information that a mobile user creates. For this purpose, several schemes for efficiently and effectively managing this information have been proposed. In the following sections we analyze several such schemes and discuss their advantages and disadvantages. We also present how these procedures are performed in real networks, as well as how one can obtain a valid cost model to analyze performance.

13.2   General Background Information

Through location management procedures, the PLMNs (public land mobile networks) keep track of where the MTs are located in the system area. The location information for each MT is stored in functional units called *location registers*. Functionally, there are two types of location registers:

1. *Home location register (HLR):* where all subscriber parameters of an MT are permanently stored and where the current location may be stored.
2. *Visitor location register (VLR):* where all the relevant data concerning an MT is stored as long as the station is within the area controlled by that visitor location register.

A mobile service center (MSC) area is composed of the areas covered by all the base stations controlled by an MTC. An MTC area can consist of several location areas (LAs). An LA is an area in which, after having performed a location update once, MTs can roam without being required to perform subsequent location updates for reasons of location change. An LA consists of one or more cells. The LA identification (LAI) plan is part of the base station identification plan to uniquely identify the base stations. To minimize the costs involved in the location management procedure, the LAs must be designed very carefully. A large number of analytical studies consider that all the cells have the same size and shape [1], all LAs have the same structure, and the movement of the users is homogeneous. Obviously, this situation does not hold in real-world environments where the geography of the area, the type of antennas used, and the distribution of buildings and other obstacles create irregularly shaped cells and result in irregular LAs. To model the traffic movements of the MTs, several traffic models have been proposed. Two traffic models that are widely used are the Markov and the Fluid Flow traffic models. Using a simplified Fluid Flow traffic model [2] derives that the appropriate number of cells that an LA must have is given by:

$$N_{opt} = \sqrt{\frac{vC_C}{\pi RC_R}} \qquad (13.1)$$

where R is the cell radius, v the velocity of the mobile user/terminal, C_C is the paging cost for every incoming call, and C_R the location registration cost. An algorithm was proposed that considers the type of cell and the associated clusters for calculating the perimeter of an LA [2]. Once the LAs have been defined, one needs to identify algorithms that can be used to move the control of the call from one user to another. These algorithms can be divided into static and dynamic. The GSM (Global System for Mobile Communications) standard proposes a static method that uses a timer and the location of an MT to trigger and update. As an MT moves to a new LA, it creates a location update message. Moreover, there is also a timer attached to the MT, which, whenever a user accepts an incoming call, resets and starts counting from the beginning. When this time expires, the MT transmits a location update message. The location update messages are sent according to the time-based scenario and their purpose is to make the databases of the current network refresh. In this way, the transmission of paging messages (when the MT is out of range, the base location update messages of the Time-Based scenario are not received from the corresponding database) is avoided. Dynamic strategies have been proposed to improve over static ones in terms of the required signaling, the delay incurred, and the size of the required databases. In the *dynamic management of the location update area* scenario, an irregular cell architecture is considered, allowing the calculation of the appropriate size of the location update area so as to reduce the location management cost. The MT is located in an area that has a number of $k \times k$ cells. The parameter k is calculated separately for every mobile user, according to the way the user moves and the rate with which the user receives calls [1]. This method performs better than the static time-based registration scenario used in GSM. However, it is not easy to implement. For example, during the paging of the mobile user, the databases of the system must store the location of the users that are in the RA (it consists of a certain number of cells), which results in consuming considerable storage space and requires sizeable computation power for scanning the area. Moreover, for every location update message, the corresponding database that serves the user must send to the MT the RA in which it is located. The MT stores this information in its database. We observe that the databases of the system store not only the location of the MT (in which cell it is located), but also in which RA it is located. Combinations of the movement-based and the distance-based scenarios (defined in following section) have also been proposed [1] to search for the MT, when it is located in a number of cells, which are at distance d from the central cell. The user stores in its memory the number of cells that are at distance b from the central cell. If the user stores in its memory all the cells that are at distance d from the central cell, then we have the movement-based scenario; whereas if it stores those that are at distance b, then the distance-dased scenario is used. An intermediate scenario can be implemented, where $b \leq d$ and the movement threshold can be $m = d - b + 1$.

13.3 Location Update Procedures

In this section we present some location update procedures that have been proposed in the scientific literature in order to understand the basic trade-offs and issues involved.

13.3.1 *Location Update According to Distance*

In this scheme, the location update takes place whenever the distance (counted in number of cells) between the current cell where the MT resides and the cell where the last location update occurred is greater than a certain value d [4, 5]. Therefore, we can outline an LA as the area defined by a central cell, which is the last cell where the location update occurred the last time, surrounded by cells that are at distance (d) from it. Figure 13.1 shows schematically an LA, that is, the central cell and the cells at a distance d from the control cell. Note that the area between the cells is not part of the LA in this strategy.

13.3.2 *Location Update According to Movement*

This location update procedure takes place whenever a mobile user completes d transitions between cells. The value d is called the *threshold value* of movement. When a call arrives, the network detects the LA that encloses all the cells that are at distance d from the cell where the last location update took place [3, 4, 6–9]. The LA consists of all the cells that are at

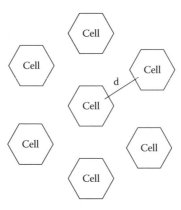

Figure 13.1   Location area in the distance-based strategy.

distance *d* from the central cell, including all the cells enclosed by them, as shown in Figure 13.2.

13.3.3   Location Update According to Time

In this strategy, the location update happens every *t* units of time. The size of the location area is defined according to the mobility of the user in the scenario implemented [4, 10].

The three above strategies present simple and efficient techniques for location management. Many other techniques have been proposed that add more complexity, but achieve better results under a wide array of traffic parameters and movement scenarios.

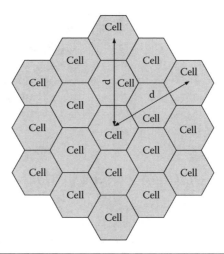

Figure 13.2   Location area in the movement-based strategy.

13.3.4 Local Anchoring Strategy

In this scenario, the signaling cost created by the registration procedure is minimized by reducing the need to send location update messages to the HLR often. The VLR, which serves the mobile user most of the time, is considered the local anchor of this user. The HLR stores the pointer to that VLR [1, 11, 12]. The location of the mobile user is stored in the local anchor instead of the HLR. The registration cost can be reduced because the location of the user is no longer stored in the HLR. When an incoming call arrives, the HLR is queried. The result of this is to create a pointer to the local anchor. If the location of the mobile user is in the local anchor, the paging procedure of the user begins. If the local anchor has a pointer to the VLR already stored, the procedure for locating the user is performed at the VLR. There are two more ways with which this strategy can be used. In the static scenario, the VLR to which an incoming call arrived for the last time becomes the local anchor for the called MT. In the dynamic scenario, the local anchor changes when an incoming call arrives. The system itself has the capability to decide whether it will change the location of the local anchor. This decision depends on the user's mobility. Depending on the mobility of the user and the rate at which incoming calls arrive, this strategy might offer better performance than other strategies.

13.3.5 Reporting Cells Tracking Strategy

The basic idea behind this strategy is to allow the location update only in certain cells (the number of which is small compared to the total number of cells in the network). In this way, the detection of a mobile terminal is limited to a small subset of cells. It has been shown that this scenario restricts the total networkwide cost, while other techniques pay more attention to every user separately [13]. In a simple cellular network, there is one base station in every cell. In this scenario, a chosen number of cells are selected and identified as *reporting cells*. The remaining cells in the network are called *nonreporting cells*. The reporting cells periodically transmit messages to define their role. Therefore, an MT will know whether it is served by a reporting cell. The neighbors of a reporting cell will be the total number of nonreporting cells that are accessible from this reporting cell without the need to pass through another reporting cell. For this reason, two functions are defined:

1. *Renewal:* A location update is performed only when a mobile user arrives at a new reporting cell.
2. *Detection:* When a call arrives for an MT, the system searches for it in the neighboring cells (nonreporting cells) of the reporting cell in which the location update has been performed for the last time. This procedure is shown in Figure 13.3.

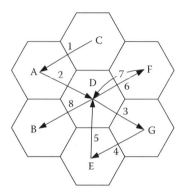

Figure 13.3   Example of the reporting cell tracking strategy.

In Figure 13.3, cells B, C, D, and G are reporting cells, whereas the remainder are nonreporting cells. The neighboring cells of C are A and F; of D are A, F, and E; of G are F and E; and of B are A and E. Consider now that the user follows the route: C, A, D, G, E, D, F, D, B. When the mobile user moves from cell C to cell A, a location update procedure will take place. The same procedure is followed from cell A to cell D because the mobile user comes from a nonreporting cell. On the other hand, when the user moves from cell D to cell G, this procedure will not be followed because the mobile user comes from a reporting cell. The entire procedure continues until the mobile user reaches its final destination, following the same algorithm. During the tracking procedure, when the user is allocated, for example, in cell A, the system will search in cells A, F, and C because cells A and F are the neighboring cells of reporting cell C. When the number of reporting cells increases, the cost of location updates increases as well, whereas the location tracking cost of the MT decreases. On the other hand, when the number of reporting cells is small, the location update cost is higher in this strategy, whereas the tracking cost of the MT will continue to decrease. Therefore, the way in which the cells are chosen as reporting cells and their exact number are very important issues to be addressed in the design of this strategy. To decide in this tracking strategy which cells will be the reporting cells, an algorithm based on the reselection scenario is used. The rate at which MTs enter a cell relates to the location update cost value because the larger this rate is, the higher the location update cost becomes. If a cell with a high arrival rate of mobile terminals is selected as a nonreporting cell, it will not be required to update the mobile terminals' location in the network as they enter the network. In this way, the total location update cost for the entire network is reduced. Thus, that particular cell among the remaining cells presenting the greater arrival rate of mobile terminals is selected. If the location update cost created in this cell is equal to a value Z, the selection of this cell as a nonreporting cell is successful and the algorithm continues to the next step. If this is not the case, the selection is unsuccessful and this specific cell is defined as a reporting cell. The algorithm of the cell reporting tracking strategy continues until all the cells in the network are examined.

13.3.6   Alternative Strategy (AS)

This strategy is based on the fact that the traffic behavior of most mobile users can be predicted. In this way, the signaling messages involved in mobility management procedures can possibly be reduced, achieving better management of network resources. It has been observed that an MT resides in a certain geographical region, for example, in the house or at the workplace for considerable amounts of time. The vehicle location registers or mobile switching centers (VLRs/MSCs) that serve this area have been chosen in such a way as to be the local anchors of these users. In this strategy, it is assumed that there are two local anchors: one in the workplace area of the user and one in the region of the user's residence. Between these local anchors, there are forwarding pointers. The local anchor, where the user is located at a certain moment, is called the *current local anchor*. Depending on the traffic category of the MT, a location update cost and a call delivery cost are attributed to this MT. The traffic categories can be classified as [11, 14, 16, 17]:

■ Daily movement with public means of transportation
■ Workplace
■ Region of residence
■ Personal transportation
■ Social transportation
■ Transportation for pleasure trips

The main advantage of this scenario is the minimization of location update messages because of the storage of the mobility profiles of every user in the local anchors. However, this strategy has some drawbacks, such as the increased delay that the paging messages experience, the high requirements in memory, and the frequent updates of the user profiles, as many of them frequently change their traffic behaviors.

13.3.7   Forwarding Pointer Strategy

In this strategy, a chain of pointers is created (Figure 13.4) from the old MSC/VLR to the new MSC/VLR whenever a subscriber leaves the old MSC/VLR. For that reason, the forwarding pointer strategy decreases to a great

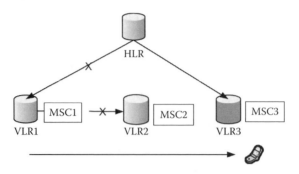

Figure 13.4   Forwarding pointer strategy architecture.

extent the cost created because of the movement of the user (location update cost) [15, 16]. However, there is an additional cost because of the access to many VLR databases for the purpose of locating mobile users. Therefore, this strategy is suitable when the call-to-mobility ratio (CMR) is small, that is, below a certain value. The forwarding pointer strategy is *not suitable* for networks where the users present intense mobility.

13.3.8   Eager Caching Strategy

On the opposite side of the forwarding pointer strategy is the eager caching strategy, which is more suitable for networks where the subscribers present intense mobility [19, 20]. In the forwarding pointer strategy, because many VLRs are queried during user movement, a high location update cost is incurred. When the user presents high mobility, these queries to the VLRs occur more often, so the location update costs become larger. In the case of the eager caching strategy, due to the use of special bypass pointers the queries to the databases that are in a higher level than the level of the bypass pointers are avoided. The basic idea of this strategy is the storage of the location of the call recipient in the database of the caller. Therefore, each time a call is created from the same caller to the same call recipient, there will be a direct delivery of the call with the help of the previous call. A pair of bypass pointers uses the forward and reverse pointers created at the databases s and t, respectively. In this way, the number of databases queried in order to locate the mobile user during the next call (after the next call) is reduced. When a mobile user located in the sub-tree that has at its top the database s makes a call to an MT located in the sub-tree of the database t, then the pair of the bypass pointers can be used to find the call recipient and complete the call connection. It is obvious that the call delivery time is reduced significantly because no queries will be made to the databases located at a higher level than that of s and t. When the CMR becomes larger, this strategy does not present

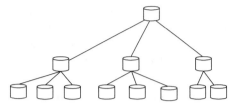

Figure 13.5 Hierarchical strategy architecture.

any problems to the network (e.g., high delay, excessive consumption of network resources, etc.).

13.3.9 Hierarchical Architecture

This architecture is based on the architecture of metropolitan area networks (MANs) and usually has three levels of databases: (1) the MAN access network, (2) the MAN backbone network, and (3) the MAN itself (Figure 13.5). The top of the tree is the database of the metropolitan area that manages the subscribers located in that area. Multiple metropolitan areas can be interconnected through the top root. For every user, there is a home access MAN and a home MAN address [19]. For locating a call recipient, the ripple search scenario has been proposed. This scenario consists of a sequence of detections to the distributed databases. The tracking procedure is first certified in the local address MT of the caller and afterward in the home MT address of the caller. If this tracking procedure fails to locate the call recipient, then the directory database is searched. This database contains all the home access MAN addresses of every user. At the moment the home access MAN is found, the user is also located.

13.4 Location Management in Third-Generation Wireless Networks (3G UMTS)

The way in which the location management procedure is implemented has to do with the current network topology as well as with the components comprising the cellular network. Because the topology and the characteristics of the network are different between the 2.5G and 3G networks, the implemented location management procedures are different.

13.4.1 Architecture of Intercooperating 2.5G–3G Networks

A UMTS (Universal Mobile Telecommunications System) network comprises three intercooperating locations: (1) the core network (CN),

(2) the terrestrial radio access network (UTRAN), and (3) the user equipment (UE). The main operation of the CN location is to provide switching, routing, and transit functionalities for the data exchange of the user. Moreover, the CN includes all the functionalities necessary for managing the network. The main architecture of the location CN in UMTS networks is based on that of the GSM and GPRS (General Packet Radio Services) networks. Therefore, all the equipment of these networks must be altered to provide UMTS services and functionalities. The UTRAN location provides the access mechanism for the mobile user through the air interface. The base station is called Node-B and the control equipment for Node-B is called the radio network controller (RNC).

13.4.1.1   Core Network

The core network (CN) is divided into circuit switching and packet switching areas. Some elements of the circuit switching area are the MSCs, the VLR, and the gateway MSC. The elements in the packet switching area consist of the serving GPRS support node (SGSN) and the gateway GPRS support node (GGSN). Other elements, such as the EIR (equipment identity register), HLR, VLR, and AuC (authentication center), are shared in both locations. The data transmission method for UMTS networks is Asynchronous Transfer Mode (ATM). The CN location architecture can change when new services and characteristics are introduced. The Number Portability Data Base (NPDB) mechanism is used to allow the user to change networks and at the same time preserve the same telephone number. The gateway location register (GLR) can be used to optimize the management of the subscriber at the boundaries of the network.

13.4.2   3G/UMTS Architecture

The UMTS architecture is presented in Figure 13.6. It is a pure UMTS architecture, without intercooperating with GPRS, GSM networks. The dotted lines carry signaling messages, while the continuous lines carry signaling messages and data. The creation of the UMTS aims at serving more services in comparison to GSM and GPRS. Real-time services are envisioned, as well as the ability to make wireless cellular networks cooperate with IP networks. The main differences between UMTS and the preexisting systems (GSM, GPRS) are presented in the following paragraphs. *RNC:* It is the same as the base station controller (BSC) for the GSM network. It provides the interface between the mobile users and the base stations; this interface is called Iub. The interface between the RNCs is called Iur. The special characteristic of the RNC is that it runs the WCDMA (Wideband Code-Division Multiple-Access) protocol, which gives the ability to the

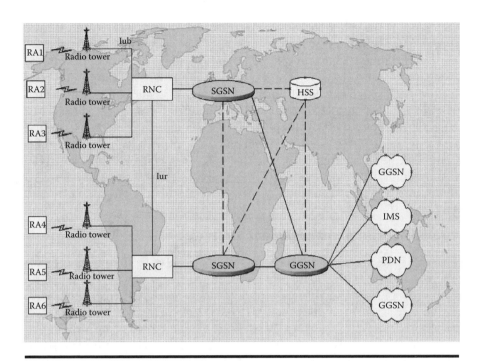

Figure 13.6　UMTS network architecture.

mobile user to accept and send data communicating with two base stations, which are located under different RNCs concurrently. In this case,
one RNC is called the *source RNC* and the other the *target RNC*. Each RNC
is connected to the SGSN. One SGSN can be connected to several RNCs.
The SGSN helps mainly in the intercooperation between the UMTS network and the GPRS network, as well as in providing the necessary services
to the mobile user. All SGSNs are connected to the GGSN, which provides
the necessary connections of the UMTS network to other networks such
as PDN (packet data network) and IMS (IP multimedia system) networks.
Both the SGSN and the GGSN are connected and communicate with the
Home Subscriber Server (HSS), so as to maintain the information regarding the mobility of the user and the management of the connection of the
user. Therefore, all information regarding the location of the user in the
entire UMTS network is recorded in the HSS. As a result, when a user is
located in a network with a large number of users and of high mobility,
too many registrations and too many queries take place. Therefore, there
is a large overhead created in the HSS and a vast consumption of network
resources. Moreover, when a user moves from a UMTS network to another
UMTS network, the information-routing (packet data protocol, PDP) will
pass through the GGSN of the first UMTS network and afterward through

the GGSN of the other UMTS network. Thus, an update will be made to the first HSS and another one to the other HSS. Therefore, we have additional overhead on the HSS. Consequently, there is a need to create a distributed database architecture to relieve the load of the HSS. This proposed architecture consists of two levels, at the top level of which there will be the HSS and at the next level the databases attached to the RNCs. We must take advantage of the fact that there is a direct communication among the RNCs, something that does not happen in GSM or in GPRS, where the BSCs communicate through the MSC and the SGSN, respectively. Thus, we avoid having multiple queries between the databases (for instance, if we had a three-level tree architecture).

13.4.3 Location Management Strategies

The anticipated increase in signaling traffic on "short-haul" and "long-haul" international links has led the standardization organization "3rd Generation Partnership Project" (3GPP) to introduce a new level in the hierarchy of location management entities for UMTS/IMT-2000 networks. This new level consists of the GLR (Figure 13.7), which is a node between the VLR and/or SGSN and the HLR. The GLR can be used to optimize location updating and the handling of subscriber profile data across network boundaries. When a subscriber is roaming, the GLR plays the role of the HLR toward the VLR and SGSN in a visited public land mobile network (VPLMN) and the role of the VLR and SGSN toward the HLR in a home public land mobile network (HPLMN). The following figures that are originally presented in 3GPP TS 23.119 (Figures 13.8 through 13.11) illustrate the location management procedures used in 3G networks.

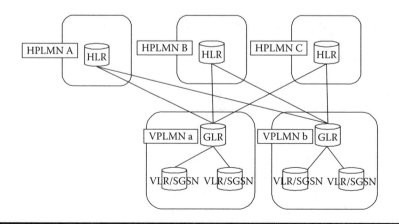

Figure 13.7   Possible location of GLR.

As illustrated in Figure 13.7, the GLR must be deployed at the edge of the visited network, contains the roamer's subscriber profile and location information, and handles mobility management within the visited network. When a node enters the visited network, the first location update procedure under the GLR is initiated, causing the subscriber information to download from the HLR to the GLR. This information permits the GLR to handle the Update Location message from VLR, as if it is the HLR of the subscriber at the second and further location updating procedures. The GLR enables the procedure to be invisible from the home network so that this hierarchical location management can reduce the internetwork signaling for location management. The information is kept at the GLR until the reception of a Cancel Location message from HLR. A GLR interacts with multiple HLRs, which may be located in different PLMNs. The relationship between the GLR and the HLR is the same as that between the VLR and the HLR. The implication of supporting multiple HPLMNs is that the GLR will need to store a large amount of profile data. A GLR interacts with multiple VLRs. Figures 13.8 through 13.11 show the flow of information in different roaming cases.

Figure 13.8 illustrates the sequence of operations when an MT, which was located in a VPLMN that did not have a GLR, roams into a new VPLMN with a GLR. The MT sends a location update to the VLR, which is forwarded to the GLR, and then forwarded to the HLR. Then, the subscriber data is downloaded to the GLR and forwarded to the VLR. Figure 13.9 illustrates

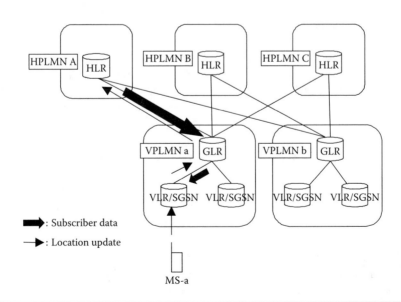

Figure 13.8   Roaming to VPLMN with the GLR from HPLMN without the GLR.

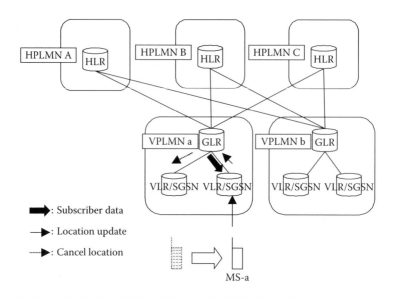

Figure 13.9   Intra-VPLMN with GLR roaming.

the sequence of operations when the MT roams from an old VLR to a new VLR, both of which belong to the same VPLMN and, consequently, are associated to the same GLR. The MT sends a location update message to the new VLR, which is forwarded to the GLR. Because the GLR already has the appropriate subscriber information for the MT, the HLR is notified. The GLR transmits this information to the new VLR and a Cancel Location message to the old VLR.

Figure 13.10 illustrates the sequence of operations when the mobile terminal roams from a VPLMN that has a GLR to a new VPLMN that also has a GLR. The MT sends a location update to the new VLR that is forwarded to the new GLR. The GLR does not have the subscriber's information, so it forwards the message to the HLR. Then, the Information is downloaded to the new GLR, and then forwarded to the new VLR. The HLR sends a Cancel Location message to the old GLR, which is forwarded to the old VLR. Finally, when the MT roams out of a VPLMN with a GLR to enter a VPLMN that does not have a GLR, the location update is transmitted directly from the new VLR to the HLR, and the subscriber information is downloaded to the new HLR (Figure 13.11). The Cancel Location message, on the other hand, is forwarded from the HLR to the old GLR, and then forwarded to the old VLR. Although the problem of location management is not different in context for the 3G/UMTS network compared to GSM, the addition of the GLR in the location management hierarchy causes some modifications to the LA algorithms that were initially designed for second-generation (2G) networks [21]. More specifically, the location

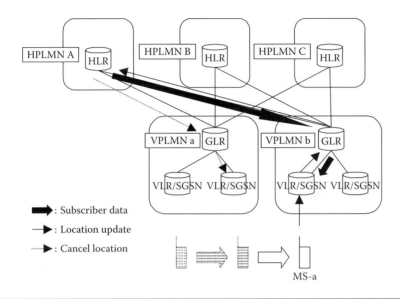

Figure 13.10   Inter-VPLMN with GLR roaming.

management procedures that affect the VLR/HLR interaction in 2G networks can be applied to the VLR/GLR interaction. The new level of GLR/HLR interactions is similar to the functionality as in the above case, although they are different in scale. Therefore, the advantages and disadvantages of the presented static, dynamic, anchor, hierarchical, etc. strategies algorithms

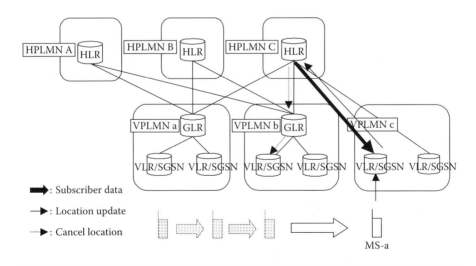

Figure 13.11   Roaming to VPLMN without GLR from VPLMN with GLR.

proposed for 2G networks have similar advantages and disadvantages in the 3G case.

13.5 Location Management Specifications for GSM/UMTS

In this section, we discuss specific problems in location management procedures as they present in the relevant standards.

13.5.1 International Mobile Subscriber Identity (IMSI) Detach/Attach Operation

The *attach/detach operation* is used for monitoring the presence of MTs in the network. The attach operation is initiated by an MT when it joins the network, to let the network be aware of its presence. The attach information is registered to the VRL. On the contrary, when an MT is not present in the network, the detach operation is used for determining which MTs have left the network. The attach/detach operation is crucial for the effectiveness of location management strategies because it can limit the number of unnecessary paging messages in the network. Support of the IMSI detach/attach operation is mandatory in MTs. The facility is optional in the fixed infrastructure of the PLMN.

The *explicit IMSI detach operation* is the action taken by an MT to indicate to the PLMN that the station has reentered an inactive state (e.g., the station is powered down). The *explicit IMSI attach operation* is the action taken by an MT to indicate that the station has reentered an active state (e.g., the station is powered up).

The *implicit IMSI detach operation* is the action taken by the VLR to mark an MT as detached when there has been no successful contact between the MT and the network for a time determined by the implicit detach timer. The value of the implicit detach timer is derived from the periodic location updating timer. During an established radio contact, the implicit detach timer will be prevented from triggering implicit detach. At the release of the radio connection, the implicit detach timer will be reset and restarted. Implicit IMSI detach will also be performed in the case of a negative response to an IMEI (International Mobile Equipment Identity) check.

13.5.2 General Procedures in the Network Related to Location Management

The MSC will pass messages related to location updating between the MT and the VLR. When receiving a Location Updating Request or an IMSI detach/attach message from an MT, the MSC will convey the message to its associated VLR. Any response from the location register will similarly be conveyed to the MT. The MT will identify itself by either the IMSI or the TMSI (Temporary Mobile Subscriber Identity) plus the Location Area Identification of the previous VLR. In the latter case, the new VLR will attempt to request the IMSI and authentication parameters from the previous VLR by the methods defined in Escale and Giner [20]. If this procedure fails, or if the TMSI is not allocated, the VLR will request that the MT identifies itself by use of the IMSI.

13.5.3 Information Transfer between Visitor and Home Location Registers

- *Procedures for location management.* Detailed procedures for the exchange of location updating information between VLRs and HLRs are given in Reference [22]. An overview of these procedures follows below.
- *Location updating procedure.* This procedure is used when an MT registers with a VLR. The VLR provides its address to the HLR. The VLR can also allocate an optional identity for the MT at location updating: the Local Mobile Station Identity [23].
- *Downloading of subscriber parameters to the VLR.* As part of the location updating procedure, the HLR will convey the subscriber parameters of the MT that need to be known by the VLR for proper call handling. This procedure is also used whenever there is a change in subscriber parameters that need to be conveyed to the VLR (e.g., change in subscription, a change in supplementary services activation status). If the HPLMN applies the multi-numbering option, different MSISDNs (Mobile Station International ISDN Numbers) are allocated for different Basic Services [24] and stored in the HLR. Among these MSISDNs, the Basic MSISDN Indicator as part of the HLR subscriber data marks the "Basic MSISDN" to be sent to the VLR at location update. It is used in the VLR for call handling as calling party and as line identity. If the HPLMN applies the Administrative Restriction of Subscribers' Access feature, the HLR will convey the subscriber access restriction parameter (Access Restriction Data) to the VLR. The VLR will check this subscription parameter against the radio access technology that supports the LA/RA in which the MT is roaming to decide whether the location update should be allowed or rejected.
- *Location cancellation procedure.* This procedure is used by the HLR to remove an MT from a VLR. The procedure will normally be used when the MT has moved to an area controlled by a different location register. The procedure can also be used in other cases (e.g., when an MT ceases to be a subscriber of the Home PLMN).

13.5.4 Mobile Subscriber Purging Procedure

A VLR can purge the subscriber data for an MT that has not established radio contact for a period determined by the network operator. Purging means to delete the subscriber data and to "freeze" the TMSI that has been allocated to the purged MT, so as to avoid double TMSI allocation. The VLR will inform the HLR of the purging. When the HLR is informed of the purging, it will set the flag "MT purged" in the IMSI record of the MT concerned. The presence of the "MT purged" flag will cause any request for routing information for a call or short message to the MT to be treated as if the MT were not reachable. In the VLR, the "frozen" TMSI is freed for usage in the TMSI allocation procedure by location updating for the purged MT in the same VLR, location cancellation for the purged MT, or, in exceptional cases, by O&M (Operation and Maintenance). In the HLR, the "MT purged" flag is reset by the location updating procedure and after reload of data from the nonvolatile backup that is performed when the HLR restarts after a failure.

13.6   Location Management Cost Modeling

Estimating the cost of location management procedures is an important issue because an efficient, according to some performance objective, strategy can create heavy signaling and increase the load in several parts of the network. Moreover, a careful cost estimation often reveals overloaded parts of the network or mobility patterns, and the call arrival rates that make some strategies efficient and effective while other strategies are shown to have a very limited implementation horizon. The network topology, the mobility patterns, and the call arrival patterns are key issues in defining a valid cost model and performing an analysis. In the literature one finds several definitions of location management costs. In this section, we present the main components that are common to most of the published works. According to Ho and Akyildiz [1], the location management cost can be divided into two components:

1. *Registration cost (movement):* defined as the cost associated with the fact that the mobile terminal moves around in the network.
2. *Call delivery cost (paging):* defined as the cost incurred in completing the call delivery procedure.

A network model is necessary not only to understand the function of the location management operation, but also to provide the basis for the specification of the individual cost components. A model that has been used extensively in the literature and that describes to a large extent what happens in the GSM network is shown in Figure 13.12. Users roaming in registration

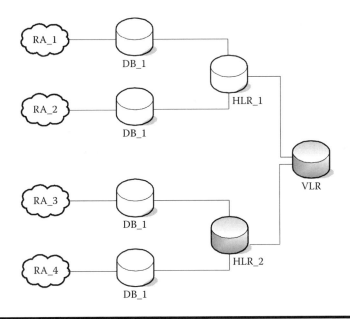

Figure 13.12   GSM network topology.

area i (RA_i) refer for information purposes to Database i (DB_i). Several DB_i's are associated with a certain HLR_j and several HLR_j's report to a VLR. A detailed analysis of the registration and call delivery procedures is necessary to identify the individual components that contribute to the total cost.

13.6.1   Registration Procedure

- *Under the same HLR.* Suppose that an MT moves from RA_1 to RA_2. A registration request message is sent to DB_1 of RA_2. An entry of the MT is recorded in DB_1 of RA_2. Then a message is sent to HLR_1 to inform it that the MT has left RA_1 and that it now resides in RA_2. Moreover, HLR_1 sends a message to VLR to inform it that the MT resides in RA_2. Finally, DB_1 of RA_1 sends an acknowledgment message to DB_1 of RA_2 through HLR_1, including the service profile of the MT. Then it deletes the service profile of the MT from DB_1 of RA_1.
- *Under different HLRs.* When an MT moves from RA_2 to RA_3, the following registration messages are exchanged between the corresponding databases. A registration message is sent to DB_1 of RA_3. An entry is recorded in DB_1 of RA_3 and then it sends a message to HLR_1 through HLR_2 and VLR to inform it that the MT has left RA_2 and that it now resides in RA_3. Then, DB_1 of RA_2 sends an

acknowledgment message to DB_1 of RA_3 through HLR_1, HLR_2, and VLR, including the service profile of the MT. Finally, it deletes the service profile of the MT from DB_1 of RA_1.

13.6.2 Call Delivery Procedure

1. When a call for an MT is originated, a location request message is sent to the VLR that serves the corresponding MT. The VLR is queried and the HLR (HLR_1 or HLR_2) that serves the calling MT is found.
2. The HLR that serves the called MT is queried to find out under which DB_1 the MT resides (DB_1 of RA_1, RA_2, RA_3, and RA_4).
3. After finding the RA in which the called MT resides, a call connection is set up from the calling to the called MT, and the call delivery procedure ends.

To evaluate the total cost of the above presented strategy, the following cost notation must be introduced:

- C_U: cost for a query or an update to DB_1.
- C_H: cost for a query or an update to HLR.
- C_V: cost for a query or an update to VLR.
- C_{D-H}: cost for transmitting a signaling message between DB_1 and HLR.
- C_{H-H}: cost for transmitting a signaling message between two HLRs.
- C_{V-H}: cost for transmitting a signaling message between VLR and HLR.
- p_1: the probability that an MT moves from one RA to another, under the same HLR.

Using the above defined costs and the presented registration procedure, the Registration Cost can be calculated for two distinct cases:

1. Movement under the same HLR:

$$C_{R1} = p_1(C_U + 5C_{D-H} + C_H + C_{V-H} + C_V) \tag{13.2}$$

2. Movement under different HLRs:

$$C_{R2} = (1 - p_1)(5C_{D-H} + 6C_{V-H} + 3C_U 2C_H) \tag{13.3}$$

The total location registration cost is given by:

$$C_R = C_{R1} + C_{R2} \tag{13.4}$$

The Call Delivery Cost (CDC) can be also calculated as:

$$C_C = C_V + 1/2C_H + 1/4C_U \qquad (13.5)$$

The total location management cost is the sum of the Registration Cost and the CDC; that is,:

$$LMC_{TOTAL} = C_R + C_C \qquad (13.6)$$

The above equations present a macroscopic view of the involved costs. To find the individual costs in detail, one must consider the CMR. The CMR shows whether we are in an environment with heavy call arrivals and low movements, or in an environment where the MT moves very fast compared to the arrival rate of the calls and thus many changes occur in the responsible databases and VLRs. In the example mentioned above, we did not apply the CMR parameter because the derived equations would be very complex. Suppose that the velocity of an MT increases while the calls originating to the MT are constant, then the CMR value decreases. On the other hand, if the calls to an MT are increasing while the MT travels with a specific velocity, the CMR value increases. In real life, neither the velocity nor the number of calls to a user is constant. For example, a taxi driver can travel for a long period of time without receiving any calls, or receiving only a small number of calls. In that case, the CMR value is low. On the other hand, a public servant who stays in a specific place for a long period of time, while receiving a large number of calls, experiences a large CMR value. Even for a specific user, the number of received calls depends on the time of day [15, 25, 26]. As one can see, it is not easy to select a particular CMR value that will represent the characteristics of all subscribers of the network. Authors have looked at a wide range of CMR values. For example in Mao and Douligeris [12], the CMR parameter was allowed to take values over the interval [0.1, 100]. Several mobility models/patterns and call arrival processes have been used in the modeling and cost evaluation of location management schemes. The mobility models are needed to calculate the cell residence time as well as the intra-cell movement probability distributions. The most commonly used mobility models included:

- A two-dimensional random walk model where the directions of the mobile units are assumed to be independent and identically distributed
- A two-dimensional Markov walk model with six states (hexagonal pattern) that is able to describe the dynamic behavior of the intercell
- A trip model in which the mobile user can follow a particular path to a destination

- A proximity-based model where the position of the MT's mobility is extrapolated by its proximity to neighboring cells; this model requires knowledge of the exact trajectory of the mobile user and allows for exact estimations of the residence time estimation [27]
- A uniform movement model where the MTs are moving at a fixed or uniformly distributed velocity and the direction of movement is uniformly distributed over $[0, 2\pi]$

There are also mobility models that distinguish between users in buildings, pedestrians, users in cars moving with low speed (e.g., in a town), and users in cars moving with high speed (e.g., highways). The velocity of the vehicles or the residence time of the users in a building can follow several distributions, i.e., Gaussian with different mean values and variances, i.e., 110 km/h and 25 percent, respectively, for high-speed moving vehicles. The most commonly used call arrival patterns are Poisson with time- or location-based dependent arrival rates. Uniform or self-similar arrival patterns have also been examined in simulation studies but have proven difficult to evaluate analytically. Once the mobility models have been defined, the estimation of the total location management cost is evaluated by varying the CMR. Beyond the total cost versus CMR analysis, one finds in the literature several other performance measures, including: the call delivery latency versus CMR (the expected call delivery latency of an incoming call of the proposed strategy, where the cost of a network element represents the delay incurred by its operation); the call arrival time versus relative cost; the call residence time versus relative cost; call update cost versus relative cost; and the CMR versus relative cost.

13.7 Conclusions

The rapid growth of users in wireless networks has automatically resulted in the need for capable and flexible location detection strategies. The management of mobility techniques is useful for maintaining the location of the MT, namely in making the network able to know every time in which location in the wireless network our user is. When a call arrives at an MT, the network automatically, by maintaining the Location Area in which the mobile user has made for the last time a Location Update and with the help of a strategic paging, is capable of finding out the exact location of the mobile terminal and delivering the call. In GSM networks, the overall costs of the location management defined as the sum of registration cost and the cost of call delivery are significantly reduced in comparison to the rate of incoming calls and user mobility. The increasing numbers of subscribers in wireless networks has led to the need for efficient location tracking strategies. Thus, efficient location management schemes should be developed and applied in 3G wireless networks to minimize location management costs.

Another emerging area in wireless networks where the above ideas might be applicable is that of *ad hoc* and *sensor* networks. The lack of a wired infrastructure, the scarce energy of mobile terminals, the low memory and processing capabilities of nodes, and the fact that nodes of the network are not steady and the topology of ad hoc networks is dynamic make the location management schemes designed for mobile cellular networks inefficient for ad hoc and sensor networks. Therefore, new location management strategies should be introduced that take into account the limitations arising from such networks. The application of a group registration strategy with distributed databases in ad hoc and sensor networks remains a challenging research issue [28].

References

[1] J.S.M. Ho and I.F. Akyildiz, "Local anchor scheme for reducing signaling costs in personal communication networks," *IEEE/ACM Trans. Networking*, Vol. 4, No. 5, October 1996, pp. 709–725.

[2] G.M. Andres and M.V. Altamirano, "An Approach to Modeling Subscriber Mobility in Cellular Radio Networks," in *Proceedings of the Telecom Forum '87*, Geneve, Switzerland, 1987, pp. 185–189.

[3] E. Alonso, K.S. Meier-Helsten, and G.P. Pollini, "Influence of Cell Geometry on Handover and Registration Rates in Cellular and Universal Personal Telecommunications Networks," in *Proc. of the 8th International Telecommunications Congress, Special Seminar on Universal Personal Telecommunications*, Santa Margherita Ligure, Genova (Italy), October 1992, pp. 261–270.

[4] A. Bar-Noy, I. Kessler, and M. Sidi, "Mobile users: to update or not to update?," *ACM-Baltzer, J. Wireless Networks*, Vol. 1, No. 2, July 1995, pp. 175–186.

[5] V. Wong and V. Leung, "An adaptive distance based location update algorithm for next generation PCS networks," *IEEE J. Select. Areas Commun.*, Vol. 19, No. 10, October 2001, pp. 1942–1952.

[6] I.F. Akyildiz, J. Mcnair, J.S.M. Ho, H. Uzunalioglou, and W. Wang, "Mobility management in next generation wireless systems," *Proceedings of the IEEE*, Vol. 87, No. 8, August 1999, pp. 1347–1348.

[7] J. Li, Y. Pan, and X. Jia, "Analysis of dynamic location management for PCS networks," *IEEE Trans. Networking*, Vol. 5, No. 1, February 1997, pp. 25–33.

[8] Z. Mao and C. Douligeris, "A location based mobility tracking scheme for PCS networks," *Computer Commun.*, Vol. 23, No. 18, December 2000, pp. 1729–1739.

[9] Y. Xiao, "Optional fractional movement based scheme for PCS location management," *IEEE Commun. Lett.*, Vol. 7, No. 2, February 2003, pp. 67–69.

[10] C. Rose, "Minimizing the average cost of paging and registration: a timer based method," *ACM-Baltzer, J. Wireless Networks*, Vol. 2, No. 2, 1996, pp. 109–116.

[11] Z. Mao and C. Douligeris, "Group registration with local anchor for location tracking in mobile networks," to appear in *IEEE Transactions on Mobile Computing*, Vol. 5, No. 5, May 2006, pp. 583–595.

[12] Z. Mao and C. Douligeris, "Location tracking in mobile networks: a scheme and an analysis framework," *Electronics Lett.*, Vol. 39, No. 19, 2003, pp. 33–35.

[13] A. Hac and X. Zhou, "Location strategies for personal communication networks: a novel tracking strategy," *IEEE J. Select. Areas Commun.*, Vol. 15, No. 8, October 1997, pp. 1415–1424.

[14] S. Tabbane, "An alternative strategy for location tracking," *IEEE J. Select. Areas Commun.*, Vol. 13, No. 5, June 1995, pp. 880–892.

[15] Q.O. Garcia and S. Pierre, "An alternative strategy for location update and paging in mobile networks," *Computer Communications*, Vol. 27, No. 15, 2004, pp. 1509–1523.

[16] Gateway Location Register (GLR) Stage2, 3GPP TS 23.119 V6.0.0 (December 2004).

[17] R. Jain and Y.B. Lin, "An auxiliary user location strategy employing forwarding pointers to reduce network impacts of PCS," *ACM-Baltzer, J. Wireless Networks*, Vol. 1, No. 2, July 1995, pp. 197–210.

[18] E. Pitoura and I. Fudos, "An effective hierarchical scheme for location highly mobile users," *The Computer Journal*, Vol. 44, No. 2, pp. 75–91, 2001.

[19] J.K. Wey, L.-F. Sun, and W.-P. Yang, "Using multilevel hierarchical registration strategy for mobility management," *Information Sciences*, Vol. 89, No. 1–2, February 1996.

[20] P.G. Escale and V.C. Giner, "A Hybrid Movement-Distance-Based Location Update Strategy for Mobility Tracking," in *Proc. of the 5th European Wireless Conference Mobile and Wireless Systems beyond 3G*, Barcelona, Spain, February 2004.

[21] Y. Xiao, Y. Pan, and J. Li, "Design and analysis of location management for 3G cellular networks," *IEEE Trans. on Parallel and Distributed Systems*, Vol. 15, No. 4, April 2004, pp. 339–349.

[22] ETS 300 599 (GSM 09.02): "European Digital Cellular Telecommunications System (Phase 1); Mobile Application Part (MAP) specification."

[23] ETS 300 523 (GSM 03.03): "European Digital Cellular Telecommunications System (Phase 2); Numbering, addressing and identification."

[24] ETS 300 604 (GSM 09.07): "European Digital Cellular Telecommunications System (Phase 2); General Requirements on Interworking between the Public Land Mobile Network (PLMN) and the Integrated Services Digital Network (ISDN) or Public Switched Telephone Network (PSTN)."

[25] A. Quintero, O. Garcia, and Samuel Pierre, "An alternative strategy for location update and paging in mobile networks," *Computer Communications*, Vol. 27, 2004, pp. 1509–1523.

[26] Y.-B. Lin, "Reducing location update cost in a PCS network," *IEEE/ACM Trans. Networking*, Vol. 5, No. 1, February 1997, pp. 25–33.

[27] C. Wu, H. Lin, and L. Lan, "A new analytic framework for dynamic mobility management of PCS networks," *IEEE Transactions on Mobile Computing*, Vol. 1, No. 3, 2002, pp. 208–220.

[28] L. Galluccio and S. Palazzo, "A taxonomy of location management in mobile ad-hoc networks," *Journal of Communications and Networks*, Vol. 6, No. 4, 2003, pp. 397–402.

Chapter 14

Network Mobility

Eranga Perera, Aruna Seneviratne, and Vijay Sivaraman

Contents

14.1   Overview

Providing seamless Internet connectivity to mobile hosts has been studied in the Internet Engineering Task Force (IETF) for some years now, and protocols such as Mobile IP and Mobile IPv6 have been developed. We are now witnessing the emergence of mobile networks, namely a set of hosts that move collectively as a unit, such as on ships, aircrafts, and trains. The protocols for mobility support therefore need to be extended from supporting an individual mobile device to supporting an entire mobile network. In this chapter the state-of-the-art in supporting the mobility of entire networks is examined. The problem is first motivated by considering typical network mobility scenarios and by identifying characteristics that require new solutions. The design requirements of the protocols that support network mobility are explored. Furthermore, current approaches for network mobility support and their strengths and weaknesses in addressing the design requirements are presented. The chapter concludes by identifying some open research issues in the realization of mobile networks[1].

14.2   Introduction

The prediction that most devices will be connected to a network is fast becoming reality with the almost ubiquitous availability of computing and wireless communication capabilities in most electrical and electronic devices. Two emerging forms of this ubiquitous connectedness are (1) personal area networks (PANs) that interconnect a user's devices together, and (2) vehicle networks, especially in public transport systems, that will enable groups of people to access network services while on the move. In such situations, employing one device, namely a mobile router (MR) [1] for mobility management of the entire network, would be a lucrative solution in terms of performance and costs. This architecture is advocated, in particular, by the IETF's NEtwork MObility (NEMO) Working Group [2], and its popularity is evidenced by an increasing number of commercial and research projects. The NEMO Working Group has standardized a protocol, namely NEMO Basic Support Protocol [3], that ensures uninterrupted connectivity to nodes within a mobile network via a mobile router. This protocol extends the mechanisms utilized in the host mobility management protocol Mobile IPv6 [4]. In this chapter, the state-of-the-art for supporting network mobility is presented by extending the survey work carried out in Perera et al. [5]. First, the need for a new network mobility architecture is motivated, and then the characteristics and design goals of such an architecture are presented. This is followed by presenting network mobility schemes in an IPv4 and IPv6 setting. Open research issues pertaining to network mobility and some of the solutions proposed to address these challenges are then discussed.

14.3   Network Mobility: Why a New Architecture?

Many researchers have been working toward developing mechanisms that provide permanent Internet connectivity to all mobile network nodes (MNNs) via their permanent IP addresses as well as maintaining ongoing sessions as the mobile network changes its point of attachment to the Internet. Host mobility protocols, such as MIP and MIPv6, are not sufficient to handle network mobility due to two reasons. First, not all devices in a

---

[1] ©ACM, 2004. This is a minor revision of the work published in *ACM SIGMOBILE Mobile Computing and Communications Review*, Volume 8, Issue 2 (April 2004) http://doi.acm.org/10.1145/997122.997127.

mobile network, such as the sensors on an aircraft, may be sophisticated enough to run these complex protocols. Second, once a device has attached to the MR on a mobile network, it may not see any link-level handoffs even as the network moves. Thus, host mobility protocols, such as MIP and MIPv6, do not get triggers indicating link-level handoffs and, as a result, will not initiate handover and would necessitate support from an MR. The NEMO Working Group identifies three types of nodes that would be supported by an MR [6]. *Local fixed nodes* (i.e., nodes that belong to the mobile network and cannot move with respect to the MR) would typically not be able to achieve global connectivity without the support of the MR. There could also be nodes that can move with respect to the MR, namely local mobile nodes (home link belongs to the mobile network) and visiting mobile nodes (home link does not belong to the mobile network). By employing an MR to act as a gateway, all of these node types within the network should be able to achieve global connectivity, irrespective of their capabilities. In the following subsections, we elaborate on some of the benefits of having a network mobility architecture that relies on an MR.

14.3.1 Reduced Transmission Power

The radio transmission distance from an MNN (such as on ships and aircraft) to an on-board MR is potentially much shorter than to another access router on the Internet. Thus, by employing the MR as an access router, the nodes within the mobile network need only communicate with the MR using minimal power, and these nodes need not be equipped with specialized high-power communication capabilities. For small devices running on battery power, this reduction in power consumption could potentially be quite significant.

14.3.2 Reduced Handoff Signaling

Once the MNNs have established a link with the MR, the link does not need to be torn down, even as the mobile network moves. Because all communications beyond the scope of the mobile network are via the MR, only the MR needs to handle link layer handoffs. This allows unsophisticated (i.e., handoff-unaware) devices to be deployed in the mobile network, potentially yielding low-cost mobile networks.

14.3.3 Reduced Complexity

Once nodes join a mobile network, these nodes would not have to keep changing their address because this functionality would be performed by the MR. When the mobile network changes its point of attachment to the Internet, only the MR needs to auto-configure a location-specific address. This reduces the need for the MNNs having to perform link layer handoffs as well as the need for auto-configuring a new address. By having the MR perform these actions on behalf of the network nodes, the software and hardware complexity on the MNNs can be greatly reduced.

14.3.4   Increased Manageability and Scalability

If protocol updates or additional features were to become necessary in the future, it is much easier to update software or policies on the MR than on each of the network nodes. Thus, an MR network mobility architecture offers an easy central point in managing the mobility features of the entire network and also increases the scalability of mobile networks.

14.3.5   Economic Incentive

From the point of view of transportation systems, it is often commercially lucrative to provide and charge for global connectivity to passengers' mobile devices through an MR installed in the vehicle, as is being currently done by the airlines.

14.4   Network Mobility Characteristics and Design Requirements

Figure 14.1 depicts a typical mobile network operational scenario of a moving vehicle — for example, an aircraft carrying passengers. The aircraft may be equipped with various devices, such as information panels in each cabin that provides information to the passengers, fuel sensors in the engine, and embedded sensors that gather information such as temperature, pressure, and wind velocity. These devices together constitute the vehicular area

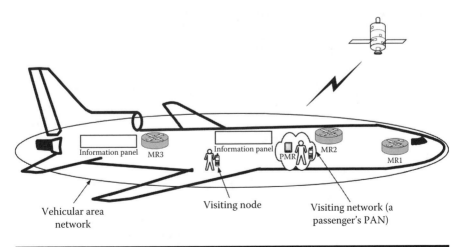

Figure 14.1   Mobile network scenario.

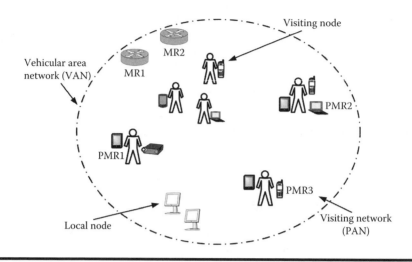

Figure 14.2   Abstract view of a vehicular area network.

network (VAN). Furthermore, passengers might carry personal wireless devices or even an entire PAN of devices, which enter and exit the VAN as and when passengers embark and disembark the aircraft. A designated node in the PAN (such as the PDA denoted in Figure 14.1 as a PAN Mobile Router [PMR]) may act as the MR that helps connect the PAN to the VAN. The VAN is equipped with one or more MRs designated as MR1, MR2, and MR3 in Figure 14.1, (one MR in each cabin), that provide Internet connectivity to the nodes within the VAN. Using the above environment as an example, in the following sections we describe some of the characteristics of mobile networks and discuss how they influence the protocol design. Figure 14.2 illustrates an abstract view of a VAN.

14.4.1   Set of Nodes Moving as a Unit

The defining characteristic of network mobility — the notion of a set of nodes moving as a unit — is evident in the above scenario. The aircraft can be viewed as a single node changing its point of attachment to the Internet. A network mobility protocol should be able to provide global access to all the nodes within the mobile network.

14.4.2   Local Versus Visiting Nodes

The mobile network in the above example includes visiting nodes (such as a passenger's wireless device, which is in a network different from its home network), as well as local nodes, such as the information panel, which is within its home network. The network mobility protocol should cater to both types of nodes.

14.4.3 Mobility Aware Versus Unaware Nodes

The mobile network may contain sophisticated nodes that are "mobility aware," namely, run protocols such as MIP or MIPv6 and are able to perform link layer handoffs. However, it is quite conceivable that the mobile network may also contain nodes that are "mobility unaware," that is, nodes that are not sophisticated enough to handle handoffs. In the case of the aircraft considered above, the temperature, pressure, and wind velocity information may need to be conveyed to a central database located outside the aircraft's network. Running sophisticated protocols on the low-cost sensors may be infeasible, and the network mobility support protocol should therefore be able to handle mobility on behalf of such nodes without requiring any special support from them. This goal, often termed "Network Mobility Support Transparency," is being strongly advocated by the IETF NEMO Working Group as very desirable when providing global connectivity to nodes within a mobile network [7].

14.4.4 Nested Mobility

In the scenario considered above, the VAN of the aircraft carries within it a PAN belonging to a passenger. Where a smaller mobile network could be contained in a larger one is known as *nested mobility*. The network mobility protocol needs to allow for at least two levels of recursive nesting. As we will see in a subsequent section, nested networks have implications for routing and route optimization.

14.4.5 Multi-Homed Mobile Networks

The VAN in the aircraft described above can be considered to be multi-homed if it has more than one active interface connected to the global Internet; these interfaces could be through one MR or via the different MRs in the different cabins. It is desirable that the MNNs are reachable even if one of the active interfaces fails. It would also be desirable to allow the MNNs to attach to the active interface that suits the application it is running. For example, a passenger downloading MP3 files and listening to music may best be supported over the least congested, high-bandwidth link.

It is likely that VANs such as the aircraft VAN would use different access technologies: satellites while flying and WLAN/UMTS while on the ground. An MR offering connectivity to the nodes sitting behind it should be able to ensure reachability to those nodes, irrespective of the access technology used.

Ernst et al. [8] highlight the importance of multi-homing by illustrating the benefits and goals of multi-homing considering real-life scenarios.

14.4.6 Different Sizes of Mobile Networks

A passenger's PAN on its own with a single MR and few devices such as a PDA and a mobile phone is a very small-scale network. The entire aircraft network, on the other hand, with a collection of subnets and a few hundred IP devices can be categorized as a large-scale network. The network mobility solution should scale from small PANs to large VANs.

14.4.7 Disparate Handoff Rates

Wireless cells have a limited coverage area due to the limited transmission range of base stations. Mobile networks can vary significantly in the speed at which they move. For example, the aircraft would be stationary or taxiing at very low speeds while on ground and when airborne would be moving at high speed. In addition, the passengers themselves may be moving within the VAN. This would result in distinct handoff frequencies and network mobility solutions having to handle this wide range of handoffs.

14.4.8 Mobile Devices from Different Administrative Domains

A passenger's PAN might come from an entirely different domain than that of the mobile network. The passenger's mobile devices joining the aircraft network need to *trust* the MR to obtain Internet connectivity. This characteristic of mobile networks, which requires interaction and trust among nodes from different domains, has given rise to security issues unique to network mobility as opposed to host mobility in an MIPv6 network. For example, Johnson et al. [4] eliminated the need for a local routing proxy (foreign agent), which was a feature of the MIP in designing the MIPv6 protocol. This enabled the end node to handle its own routing identifier, the *care-of-address.* In the network mobility architecture, this desirable feature of MIPv6 is again compromised by having the MR as an intermediary node in the end-to-end communication. This requires network mobility support solutions to address specific security issues as well as comply with the standard IETF securities policies and recommendations.

14.5 Network Mobility in IPv4 and IPv6

Advantages of the MR architecture were recognized as early as the 1990s. Hager et al. [10], in their article "MINT — A Mobile Internet Router," describe a router with sufficient computational power to perform all necessary communication protocol operations and enable connectivity for nodes. The MINT router provides communication software transparency, and the nodes connecting to the Internet via such a router need not be modified with basic mobility support software. The use of a mobile router for network mobility has been specified even in the very early Request For Comments [11] on IP mobility support. In the next subsections we present some of the research that has taken place on network mobility in an IPv4 and IPv6 setting.

14.5.1   IPv4 and Network Mobility

Here we present a summary of how mobile networks are handled with the use of the IP Mobility Support Protocol (MIP). This summary follows from the IP Mobility Support RFC 3220 [9].

The MR would act as the foreign agent and provide a foreign agent care-of-address to the mobile nodes. Packets addressed to the mobile nodes within the mobile network go through the MR's home agent (HA) as well as the mobile node's HA. If the nodes are fixed with respect to the mobile network, then MIP specifies two mechanisms that enable global connectivity for these nodes. One method is to have a permanent registration with an HA to reflect the MR's home address as the fixed host's care-of-address. Usually, the MR's HA would be used for this purpose. The second method requires the MR to advertise connectivity for the entire mobile network using normal IP routing protocols through a bi-directional tunnel to its own HA.

14.5.2   IPv6 and Network Mobility

Although it has been claimed that MIP can support mobile networks as single mobile nodes, experiments conducted in Motorola Planete Inria labs [12] have shown that this is not the case with the MIPv6 protocol. These experiments have shown that the HA fails to redirect packets destined for the local fixed nodes sitting behind an MR. If a packet is addressed to a local fixed node, the border router in the home network would attempt to forward it the MR because the MR is the next hop toward the local fixed node. The HA acting on behalf of the MR (assuming the MR has registered its new care-of-address with the HA by means of binding updates) would intercept the packet. Although the HA has a binding update for the MR, it has no binding update for the destination address on the packet (local fixed node's home address). The HA being unable to handle this packet would reroute it to the border router. The border router would once again attempt to route this packet, causing a repetition of the above process and paving the way to a routing loop. The MIPv6 protocol's inability to handle mobile networks and the network mobility characteristics and design goals have paved the way for the NEMO Basic Support Protocol.

14.5.2.1 NEMO Basic Support Protocol

The NEMO Basic Support Protocol is a natural extension of the host mobility protocol, MIPv6. It specifies a mechanism that enables all nodes within a mobile network to be reachable via permanent IP addresses, as well as maintain ongoing sessions as the MR changes its point of attachment within the topology. This protocol, which runs on the MR and its HA, ensures uninterrupted connectivity to the mobile network nodes, without considering issues such as route optimization.

When an MR attaches to an access router in a foreign network, it acquires a care-of-address from the visited link. Upon obtaining the new address, it informs its HA of its current location by way of a registration message, namely a binding update. If the MR is acting as a router as opposed to a mobile host, then this is indicated to the MR's HA by way of enabling a mobile router flag (R) in the Binding Update (BU) message. Further, the MR informs the HA of the mobile network prefixes. (Note that in some scenarios, the HA would already know which prefixes belong to an MR by an alternate mechanism such as static configuration. In these cases, the MR does not include any prefix information in the binding update.) A positive Binding Acknowledgment from the HA with the MR flag set received by the MR indicates the completion of the binding process and the establishment of a bi-directional tunnel between the MR's care-of-address and the HA's address. Any packet addressed for an MNN would then be intercepted by the HA and would be forwarded to the MR at its care-of-address. The MR then decapsulates the packets and forwards them to the mobile nodes. Packets in the reverse direction are also tunneled via the HA to overcome ingress filtering restrictions [13]. In this case, the HA decapsulates the packets and forwards them to the correspondent nodes (CNs).

14.5.2.2 Nested Mobility Management with the NEMO Protocol

So far we have considered only local fixed nodes, which are not capable of managing their own mobility. However, a mobile node managing its own mobility, such as a passenger with an MIPv6-capable mobile device, can enter the mobile network, treating it as a foreign network. Such a *visiting mobile node* will send a BU to its own HA, informing it to deliver all traffic to its new care-of-address using IPv6 tunneling. This results in two, nested levels of mobility management, because the MR manages the mobility of the mobile network. The mobility of the MR is hidden from MNNs, even if they are capable of handling their own mobility. Thus, all packets would traverse via the MR's HA as well as the visiting mobile node's HA with the NEMO protocol. This indirect routing path significantly hinders the performance of the mobile network and this issue is discussed in a subsequent section.

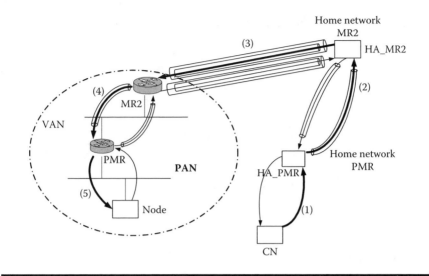

Figure 14.3   Nested bi-directional tunneling.

In the case of nested mobile networks, there would be an overhead of nested bi-directional tunnels. To illustrate the communication path and the nested bi-directional tunnel overhead in the case of nested mobile networks, consider a CN that sends a packet to a mobile device in the passenger's PAN traveling within a train's VAN. This operation of the nested mobility scenario is depicted in Figure 14.3. Again, it is assumed that the PAN's MR (PMR) has registered the care-of-address it obtained from MR2, with its HA (HA-PMR). Furthermore, as before, MR2 has registered the VAN prefixes and the care-of-address with its HA (HA-MR2). A packet from a CN is initially addressed to the home address of the mobile device and gets routed to its home network [(1) in Figure 14.3]. In the home network, it is intercepted by the HA-PMR. A route lookup there indicates that the mobile node's prefix is at the VAN's care-of-address. The HA-PMR then encapsulates and tunnels the packet to the VAN's care-of-address (2). At the VAN's home network, this gets intercepted by VAN MR's HA (HA-MR2). Its route lookup of the VAN's care-of-address determines that the packet should be encapsulated and tunneled yet again to the care-of-address in the network to which the VAN is currently attached (3). When this doubly encapsulated packet reaches MR2, which recognizes the outer care-of-address as its own, it strips it off to reveal the care-of-address of the PMR. It sends the packet to the PMR (4), which again recognizes the (only remaining) care-of-address as its own, strips it off, and sends the original packet to the mobile device in the PAN (5). Packets from the mobile device back to the CN traverse the same path in the reverse direction.

14.6   Network Mobility Open Research Challenges

As described in the previous section, the NEMO Basic Support Protocol defines a mechanism that enables session continuity to all MNNs by means of bi-directional tunneling between the MR and its HA. In defining this solution, to cater to mobility-unaware nodes within the mobile network, the Working Group has assumed that all nodes within the network are mobility unaware. Although this assumption guarantees complete mobility transparency to the mobile network nodes, it significantly hinders the overall performance of the system. This indirect routing issue, and some other challenges that should be addressed for network mobility to be a truly ubiquitous experience, are examined below.

14.6.1   Indirect Routing

According to the NEMO Basic Support Protocol, all packets to and from MNNs traverse through the bi-directional tunnel between the MR and its HA. This indirect routing mechanism leads to issues that degrade the overall performance of the network. There are many undesirable side effects of indirect routing pertaining to the NEMO protocol that have been identified by Ng et al. [14], including:

- *Increased packet overhead.* Tunneling requires that packets be encapsulated, thus increasing the per-packet overhead. This overhead reduces the portion of bandwidth available for application data. For example, real-time traffic with a packet size of 100 bytes would experience a 40 percent overhead. On a high-cost satellite link, such a magnitude of overhead would not be economically viable.
- *Increased processing delay.* The encapsulation and decapsulation processes would potentially involve tasks such as encryption/decryption, which would incur an increased processing delay.
- *Increased chances of packet fragmentation.* The size of a packet increases due to encapsulation, and this could lead to potential packet fragmentation. Packet fragmentation would result in further delays and inefficiencies in bandwidth usage.
- *Increased susceptibility to link failure.* Indirect routing creates a longer path for each packet traversing to and from an MNN to CNs. The chances of link failure are thus higher due to the longer path.
- *Potential bottleneck in the home network.* Because all the packets to and from each and every MNN are via the mobile router's HA, this increased traffic load would lead to congestion at the home network.

In Figure 14.3, tracing the path from a CN to a device in the PAN nested within the VAN illustrates how the route departs more and more from the optimal as the number of nested levels increases. This multi-angular routing issue, which is referred to as *pinball routing*, is highly undesirable because it amplifies the adverse effects of sub-optimal routing with each level of nested mobility. Each added level of nested mobility requires an additional tunnel encapsulation, and these extra IPv6 headers increase the packet size and the associated overheads. Solutions that optimize the routes also reduce the levels of indirection, thereby overcoming the side effects of non-optimal routing such as an increase in packet size. This multi-angular routing issue, referred to as pinball routing, could hinder the deployment of mobile networks because it gives rise to a negative impact not only on the mobile network, but also on the Internet as a whole.

Providing optimal routing in a network mobility setting is not an easy task. This is due to the introduction of an intermediary node in the communication between a node inside a mobile network and a correspondent node. This has raised an issue in using MIPv6 route optimization mechanisms for network mobility, namely that the nodes within the mobile network are unable to perform the MIPv6 Return Routability (RR) test, which is needed to verify to the correspondent nodes that the home address and the care-of-address are collocated. Even if the MNNs are MIPv6 capable, these nodes do not have their own care-of-addresses to perform route optimization. If the MR performs the MIPv6 route optimization procedure on behalf of the nodes sitting behind it, then this would require extensions to the MIPv6 operation of correspondent nodes and would also severely affect the scalability of MRs. Scalability is affected because the MR would be required to keep account of correspondent nodes and would need to send binding updates to them on behalf of the nodes within the network. In a typical mobility scenario such as on public transportation systems, the number of correspondent nodes communicating with mobile network nodes could reach up to several hundreds. Recognizing the impracticability of requiring the MR to send binding updates to CNs, several route optimization solutions have been proposed and these are presented in a subsequent section.

14.6.2 Mobile Router Handoff Performance

A NEMO handoff, which is similar to a MIPv6 handoff, is preceded by a link layer handoff and an IP layer network attachment procedure. In the case of MR handoffs, significant breakages in connectivity would potentially have an impact on a large number of nodes. Perera et al. [15] have implemented and analyzed the handoff performance of MRs with the NEMO protocol and have shown that the NEMO approach results in significant breakages in connectivity due to handoffs. Their results of handoff latency [16] show that, due to the NEMO handoff process, all nodes within the mobile network depending on the MR for their mobility management incur severe end-to-end TCP performance degradations and packet losses.

14.6.3 Mobile Network Prefix Delegation

One or more mobile network prefixes need to be assigned to a mobile network, either dynamically or statically, for the MR to use on the links within the mobile network. The NEMO Basic Support Protocol does not provide a means for provisioning the MRs with essential parameters such as home network prefixes. With the current NEMO protocol, there is no mechanism that enables the MR to obtain a mobile network prefix dynamically; only static configuration at the initial setup allows the MR to obtain mobile network prefixes. Static configuration could lead to binding inconsistencies because after the initial setup of associations between the MRs and their mobile network prefixes, there is no mechanism to maintain the matching states. This could lead to routing loops and unreachable prefixes. Further, if the mobile network prefixes need to be modified, there is no mechanism to do so.

14.6.4 Multi-Homing Issues

A multi-homed mobile network would have more than one point of attachment between the Internet and the MR. With the NEMO protocol, multi-homing translates to simultaneous multiple bi-directional tunnels. Although the NEMO Basic Support Protocol neither prevents nor explicitly specifies mechanisms to handle multi-homed mobile networks, there could be issues pertaining to deploying such networks. Some of the issues pertaining to deployment of multi-homed mobile networks, as analyzed by Ng et al. [17] are listed below:

- *Unbroken connectivity.* If one or more of the active connections fail, then there should be a mechanism by which the MNNs that were served by the failed connection can be served by an alternate one.
- *Path selection.* In a multi-homed mobile network because there would be more than one bi-directional tunneling path available, there should be a dynamic mechanism to select the most suitable path. The HAs, MRs, MNNs, and applications or a combination of these entities should be able to select the path. In the case of a hybrid mechanism, these entities need to cooperate in making the decision.

- *Failure detection.* Failures could happen at the MRs, HAs, ARs, etc., and mechanisms to detect such failures in time to make use of alternate options available in a multi-homed network would allow a more robust system.
- *Multi-homing in nested mobility.* There could be many real-life scenarios where a multi-homed mobile network would be nested within another mobile network. For example, in Figure 14.1 in the aircraft scenario, the passenger's PAN could potentially be a multi-homed network attached to the vehicular area network (VAN). Such cases of network mobility could lead to complex configurations and could possibly create infinite routing loops.

14.6.5   Security and Reliability Issues

In NEMO, the MR should allow subscribers from different domains to get Internet connectivity through it. In such settings where static trust relationships are lacking, a variety of security threats arise. In the NEMO Basic Support Protocol, the use of IPSec to protect signaling messages is advocated. The Protocol itself does not specify any mechanisms to handle security- or reliability-related issues. Petrescu et al. [18] have described security threats related to the NEMO protocol. They have identified signaling between the MR and the HA and nested mobility configurations as two main sensitive points of the protocol. Jung et al. [19], in their threat analysis draft on NEMO, have highlighted issues related to IPSec and other tunneling mechanisms between the MR and the HA. Perera et al. [20] have identified failures of mobile routers in different mobility scenarios and have shown how a new mobility architecture, namely the Ambient Networking architecture [21] with its in-built enhanced failover management functionality, has the potential for creating resilient networks. Until a new mobility architecture such as the Ambient Networking architecture becomes a reality, it is evident that security mechanisms for network mobility will remain an extremely difficult challenge.

14.6.6   AAA (Authentication, Authorization, and Accounting) Issues

Providing Internet access to passengers of public transportation systems would necessitate serious consideration of Authentication, Authorization, and Accounting (AAA) issues. Access control is vital in such wireless public access networks in order for any NEMO solution to be viable.

Currently within the IETF MIPv6 Working Group there is much interest in adapting IPv6 AAA mechanisms for host mobility. Mechanisms to adapt the designated AAA protocol Diameter for an MIPv6 network is currently being studied [22]. But network mobility introduces new AAA issues due to mobile network nodes relying on a previously unknown entity to take care of their mobility management tasks. Recognizing such issues, Barz et al. [23] have outlined a network access control model for vehicular mobile networks. Because the introduction of network access control via AAA entities causes handover delays, Barz et al. [23] advocate distributing these entities to minimize such delays.

Billing mechanisms for passengers of public transportation systems for Internet services is a very important issue that should be tackled from the business point of view.

14.7   Route Optimization

In large mobile networks, requiring the MR to send each CN an individual binding update causes a binding update implosion. To overcome this scalability issue, Ernst et al. [24] have proposed a method in which the MR sends a *prefix scope binding update* to a multicast address to which CNs would have subscribed. A mobile network needs to have a multicast address, which it registers with the Domain Name Server (DNS). The MR sends a periodic binding update containing the mobile network prefix and the MR's care-of-address to this multicast address. CNs can join the multicast group using IPv6 multicast mechanisms. Although this solution is beneficial for large mobile networks with many local fixed nodes, it requires major changes to MIPv6 and also to the already widely deployed DNS system.

The Optimized Route Cache Management Protocol (ORC) [25] relies on scattering a route of a mobile network to portions of the Internet by means of *binding routes* (BRs) (an association between the mobile network prefix and the care-of-address) and ORC routers. Some Interior Gateway Protocol (IGP) routers called ORC routers are used to maintain a BR to the mobile network persistently. Whenever the MR moves, ORC routers receive a BR notification that will be cached in their routing tables. The ORC routers will advertise a proxy route to the mobile network using IGP and will capture packets destined to the mobile network. The packets will be forwarded to the care-of-address on the BR, thus avoiding routing via the MR's home network. Because it is not possible to make every router on the Internet an ORC router, it has been suggested that these routers be deployed in networks where there are CNs for the mobile network. This scheme would only provide optimal routing if ORC routers are available on the CN's networks.

Jeong et al. [26] proposed an optimization mechanism for MIPv6-enabled nodes by requiring that the MR be a *neighbor discovery proxy*. This technique requires the MR to act as a bridge to the visiting mobile nodes in order for these nodes to be virtually connected to the foreign link, while requiring the MR to act as a router for the local fixed nodes. Perera et al. [27] have proposed and implemented [16] a route optimization technique, OptiNets RO, which requires the MR to act as an access router to the visiting mobile nodes and deliver the foreign network prefix on its Ingress interface. This enables the MIPv6-capable nodes within the mobile network to auto-configure a location-specific care-of-address. This OptiNets RO technique exploits the desirable characteristics of both NEMO and MIPv6 protocols. By employing this route optimization technique, it is evident that the mobile nodes need not perform a link-level handoff as in the NEMO Basic Support Protocol, and these nodes are able to achieve optimal routing as in MIPv6.

Several route optimization techniques were also proposed in the context of nested mobile networks. Kang et al. [28] suggested the use of a bi-directional tunnel between the HA and the top-level MR (TLMR) to improve the routing latency. In their scheme, a tunnel is set up between the TLMR of the nested mobile network and the HA of each MR. The TLMR advertises its existence using an extended Router Advertisement message. Each MR relays the extended Router Advertisement message of the TLMR transmitted from its parent MR to its child MR. After receiving the extended Router Advertisement message, the MR registers with the TLMR and notifies the address of the TLMR to its own HA. When the TLMR receives a registration message from the MR, it detects a relationship between the MR and its access router and uses this information to create a tunnel to each MR. Ohnishi et al. [29] proposed an HMIP-based approach for route optimization. They suggested employing the TLMR as the MAP (as in the context of HMIPv6 [30]). In their approach, an MNN registers its location within the nested mobile network once to the TLMR, and also registers the location of the nested mobile network (the TLMR's subnet address) to its HA. This hierarchical series of registrations allows the HA of MRs to directly send packets to the TLMR. Thubert and Molteni [31] have proposed the use of a new routing header called the Reverse Routing Header (RRH) to build a nested mobile network that avoids the nested tunnels overhead. The RRH, which is similar to MIP Loose Source Routing, records the route out of the nested mobile network. This can be converted to a routing header for packets destined to the mobile network. To further illustrate the RRH solution, consider the nested mobile network bi-directional tunneling depicted in Figure 14.3. The packets originating from a node in the PAN must go through two HAs before reaching the CN, leading to very inefficient pinball routing. If an RRH is used, the PMR would, in addition to tunneling the packet to its home agent, add a routing header with a predetermined number of slots. In the first slot, the PMR puts its home address. This packet then has the PMR's care-of-address as source address and PMR's home agent (HA_PMR) as the destination address.

The next router (in this case, last router) on the path, which is MR2, notices that the packet already contains an RRH, so it overwrites the source address of the packet with its own address (MR2 care-of-address) and puts the PMR's care-of-address in the second slot. The outer packet now has MR2_COA as source address and HA_PMR as destination address. When the packet reaches HA_PMR, this home agent uses the information on the RRH in entering a binding update for the PMR's home address. Now when a packet arrives for PMR to its home network, HA_PMR can trivially construct a routing header with MR2_COA and PMR_COA. This allows bypassing the home agent of MR2. It is evident even if there are n levels of nested mobility, the packet would go through only a single home agent, bypassing $(n-1)$ home agents.

14.8   Seamless Handoffs

A NEMO handoff can occur in two ways: (1) MR first loses connectivity with its current access router and becomes unreachable at its current care-of-address, and then establishes connectivity with a new access router and acquires a new care-of-address. This is known as a *break-before-make handoff.* (2) The MR establishes connectivity with a new AR while still being connected to the old AR and performs a so-called make-before-break handoff after connectivity is established. In the first case, the connectivity between MNNs and CNs is broken for the duration of the handoff; therefore, the handoff latency translates directly to packet loss. In the second case, the communications are not affected by the handoff latency.

Localized mobility management schemes, such as Hierarchical Mobile IPv6 [30] and Cellular IP [32], aim to reduce the effects of network latency between the mobile node and its home agent. Another large factor in Mobile IPv6 handoff latency is the network attachment latency. The use of optimized procedures such as Optimistic DAD [33] and Fast Router Advertisements [34] for network attachment can reduce the latency to half to two round-trip times between the mobile node and its access router. Link layer handoffs from an old base station or access point to a new one and authentication procedures are other factors affecting the handoff latency that can be improved using access-technology-specific optimizations [35].

As long as a mobile router connects to only one network at a time and is limited to performing break-before-make handoffs, handoffs are likely to cause some degree of disruption to traffic. Fast Mobile IPv6 [36] provides an access-technology-independent way of emulating make-before-break handoffs by buffering and using a localized forwarding scheme that could be used in a network mobility setting. However, the potential performance benefits from the emulated make-before-break handoffs are offset by the increased complexity required in the access network. Petander et al. [37] have addressed this issue, and proposed and implemented an access-technology-independent make-before-break handoff scheme for mobile routers. They have shown that by using a dual interface MR, where one interface is utilized for data communications and the other interface for scanning, it is possible to achieve virtually loss-free handoffs. Because the scanning and the handoff are performed on a different interface than that of the active interface, this approach enables make-before-break handoffs and minimizes the packet losses incurred due to handoff latencies without any network infrastructure support. A further benefit of this approach is that it alleviates the periodic interruption to the active data communication that would occur if scanning was performed on the same interface.

14.9   Prefix Delegation Mechanisms for NEMO

Droms and Thubert [38] have extended the DHCPv6 protocol to handle the assignment of dynamic prefix delegation for NEMO. In doing so, they extend the home agent capabilities to perform the duties of a DHCPv6 delegating router and the mobile router to perform the capabilities of a requesting router. The DHCPv6 protocol messages would flow through the NEMO bi-directional tunnel between the home agent and the mobile router. Kniveton and Thubert [39] have argued that this DHCPv6-based solution is not desirable due to reasons such as the need for the home agent to have a collocated delegating router functionality, operational overhead, etc. Thus, they advocate extending the NEMO Basic Support Protocol in order for the MR to synchronize its mobile network prefixes and obtain new ones dynamically instead of using another protocol such as the DHCPv6 Prefix Delegation protocol. They propose extensions to the current NEMO protocol to handle dynamic prefix delegation, resynchronization at binding creation after loss of states, mobile network prefix renumbering, and configuration checking for loop avoidance.

14.10   Multi-Homing Solutions

A mobile network, being multi-homed due to the MR being multi-homed, is one of the least complicated configurations. This case can be handled by the solutions proposed for host mobility because the MR is seen as a single node in the global topology. On the other hand, much research work needs to be done to handle multi-homed mobile networks due to multiple MRs.

The Inter Home Agents Protocol (HAHA) proposed by Wakikawa et al. [40] allows a mobile router/mobile node to utilize multiple home agents simultaneously. The adopted mechanism is to place multiple home agents serving the same home prefix on different links that coordinate with each other to provide home agent redundancy and load balancing. Chung and Mahbub [41] addressed the issue of, how to "best" distribute user traffic among the set of available access networks in a multi-homed on-board mobile network. Two traffic distribution schemes were proposed that optimize the profit and performance of the mobile network operator. For policy-driven mechanisms supported by multi-homed mobile networks to be realized, it is necessary to consider accounting issues that arise inadvertently. Providing QoS (quality-of-service) for mobile network nodes by exploiting the redundancy provided in multi-homed mobile networks is a research area that needs further study.

14.11   Network Mobility Projects

Along with the IETF NEMO Working Group, network mobility research projects influence the evolvement of network mobility protocols. We consider some of these projects in the following subsections.

14.11.1   eMotion (Network in Motion) Child Project of OCEAN (On-board Communication, Entertainment And iNformation)

OCEAN [42], which is a University of New South Wales and Australian Research Council sponsored project with collaborators such as the National ICT Australia (NICTA) [43] and Boeing Airline Company, was founded in 2003. This project recognizes the need for the extension of Internet services for public transportation systems and encompasses two research areas, namely networking and data management. The child project eMotion [44] handles networking issues pertaining to providing global Internet access to passengers via mobile routers and wide area wireless access systems. This project has implemented a network mobility test bed [45], has studied the NEMO protocol extensively, and has proposed many optimizations for network mobility [16].

14.11.2   Ambient Networks Project

The Ambient networks (AN) project [21] is geared toward increasing competition and cooperation in an environment populated by a multitude of user devices, wireless technologies, network operator, and business entities. This architecture aims to extend all IP networks with three fundamental requirements of today's networking world. These requirements include dynamic network composition, mobility, and heterogeneity. By encompassing these notions the AN project strives to achieve horizontally structured mobile systems that offer common control functions to a wide range of different applications and air interface technologies. The Ambient networking approach includes the flexibility of every end system to be not just a node, but also an entire network. Therefore, network mobility is inadvertently a central focus of this architecture.

14.11.3 OverDRiVE (Over Dynamic Multi-Radio Networks in Vehicular Environments)

OverDRiVE [46] is an Information Society Technologies (ISTs) project and is ongoing work of the DRiVE [47] project. The main objective of this project is to enable high-quality wireless communication to vehicular networks in multi-radio access environments. This project also uses the mobile router architecture for network mobility in an IPv6 setting.

14.11.4 Nautilus6 (WIDE Project)

The WIDE (Widely Integrated Distributed Environment) project, which was launched in 1988, established the Nautilus6 Working Group [49] to deploy mobile Internet. To do so, the Nautilus6 Working Group uses the IETF standards whenever appropriate. The Nautilus6 NEMO Working Group was established in November 2003 and specifically looks at issues concerned with mobile networks. This group has a demonstration test bed as well as an indoor NEMO test bed.

14.11.5 InternetCAR (Internet Connected Automobile Researches)

The demand for intelligent transportation systems (ITSs) has influenced the need for projects such as the InternetCAR [50]. This project, which is also a WIDE project, was launched in 1996. Its main aim is to view a car as a node on the Internet and to provide Internet connectivity permanently and in a transparent manner, regardless of the underlying access medium.

The InternetCAR project implemented the Prefix Scoped Binding Update approach for network mobility proposed by Ernst et al. [51]. The details of this implementation and the requirements for connecting vehicles to the Internet are described in Ernst et al. [52].

14.11.6 FleetNet: Internet on the Road

The FleetNet [53] project also aims to connect vehicles to the Internet. However, the MR-HA bi-directional tunneling architecture for network mobility is not the approach adopted when providing Internet connectivity to the devices in a vehicle in FleetNet. This project has introduced a novel network mobility architecture called MOCCA (MObile CommuniCation Architecture).

The MOCCA architecture relies on a new entity — namely, an Internet gateway installed on roadsides that plays the role of a gateway router for devices requiring Internet connectivity. These gateways can be regarded as FleetNet radio nodes because they run the same communication system employed on the vehicles. The Internet gateways have another interface that connects them to the Internet. The devices in the vehicles are able to obtain connectivity through the Internet gateways transparently. This MOCCA architecture was developed by Bechler et al. [54].

14.12 Conclusion

This chapter presented the basic network mobility architecture using mobility scenarios to highlight the benefits of such an architecture. As shown, providing basic network mobility support is relatively simple and requires only minimal extensions to the MIPv6 operations of the MR and its HA. On the other hand, providing route optimization for the nodes within the mobile network is quite challenging, given the MR-HA bi-directional tunneling architecture. The IETF NEMO Working Group is chartered only to standardize solutions for basic network mobility; nevertheless, numerous members within the community are currently working on optimal routing solutions for a network mobility setting. The future of network mobility at the IP level at a large scale relies on the ability to provide advanced mobility support that can coexist with the currently deployed protocols.

References

[1] Manner, J. and Kojo, M., "Mobility Related Terminology," RFC 3753, IETF, June 2004.

[2] http://www.ietf.org/html.charters/nemo-charter.html

[3] Devarapalli, V., Wakikawa, R., Petrescu, A., and Thubert, P., "Network Mobility (NEMO) Basic Support Protocol," RFC 3963, IETF, January 2005.

[4] Johnson, D., Perkins, C., and Arkko, J., "Mobility Support in IPv6," RFC 3775, IETF, June 2004.

[5] Perera, E., Sivaraman, V., and Seneviratne, A., "Survey on Network Mobility Support," *ACM SIGMOBILE Mobile Computing and Communications Review*, Volume 8, Issue 2, April 2004.

[6] Ernst, T. and Lach, H.-Y., "Network Mobility Support Terminology," (draft-ietf-nemo-terminology-01), Internet Draft, IETF, February 2004, work in progress.

[7] Ernst, T., "Network Mobility Support Goals and Requirements," (draft-ietf-nemo-requirements-04.txt), IETF, Internet Draft, February 2005, work in progress.

[8] Ernst, T., Montavont, N., Wakikawa, R., Paik, E., Ng, C., Kuladinithi, K., and Noel, T., "Goals and Benefits of Multihoming," (draft-multihoming-generic-goals-and-benefits-01), IETF, Internet Draft, February 2005.

[9] Perkins, C., "IP Mobility Support for IPv4," RFC 3220, IETF, January 2002.

[10] Hager, R., Klemets, A., Maguire, G.Q., Reichert, F., and Smith, M.T., "MINT — A Mobile Internet Router," *1st International Symposium on Global Data Networking*, Cairo, Egypt, December 13–15, 1993.

[11] Perkins, C., "IP Mobility Support for IPv4," RFC 2002, IETF, October 1996.

[12] Ernst, T., "Network Mobility Support in IPv6," Ph.D. thesis, University Joseph Fourier Grenoble, France, October 2001.

[13] Ferguson, P. and Senie, D., "Network Ingress Filtering: Defeating Denial of Service Attacks which Employ IP Source Address Spoofing," RFC 2267, IETF, January 1998.

[14] Ng, C., Thubert, P., Watari, M., and Zhao, F., "Route Optimization Problem Statement," (draft-ietf-nemo-ro-problem-statement-02), Internet Draft, IETF, December 2005, work in progress.

[15] Perera, E., Petander, H., Lan, K., and Seneviratne, A., "Implementation and Evaluation of a Mobile Hotspot," *The Third ACM International Workshop on Wireless Mobile Applications and Services on WLAN Hotspots 2005*, Cologne, Germany, September 2005.

[16] Petander, H., Perera, E., Lan, K., and Seneviratne, A., "Measuring and Improving Performance of Network Mobility Management in IPv6 Networks," NICTA Technical Report, http://nicta.com.au/director/research/publications/technical_reports/2005.cfm

[17] Ng, C., Paik, E., and Ernst, T., "Analysis of Multihoming in Network Mobility Support," (draft-ietf-nemo-multihoming-issues-02), Internet Draft, IETF, February, 2005, work in progress.

[18] Petrescu, A., Olivereau, A., Janneteau, C., and Lach, H.-Y., "Threats for Basic Network Mobility Support (NEMO Threats)," (draft-petrescu-nemo-threats-01.txt), IETF, Internet Draft, January 2004, work in progress.

[19] Jung, S., Zhao, F., Wu, F., Kim, H., and Sohn S., "Threat Analysis for NEMO Basic Operations," (draft-jung-nemo-threat-analysis-02.txt), Internet Draft, IETF, February 2004, work in progress.

[20] Perera, E., Seneviratne, A., Boreli, R., Eyrich, M., Wolf, M., and Leinmuller, T., "Failover for Mobile Routers: A Vision of Resilient Ambience," IEEE International Conference on Networking (ICN'05), Reunion Island, April 2005.

[21] Ambient Networks, http://www.ambient-networks.org/

[22] Faccin, S., Le, F., Patil, B., and Perkins, C., "Mobile IPv6 Authentication, Authorization and Accounting Requirements," (draft-le-aaa-mipv6-requirements-02.txt), Internet Draft, IETF, April 2003, work in progress.

[23] Barz, C., Frank, M., Lach, H.-Y., Maihoefer, C., Petrescu, A., Pilz, M., and Zombik, L., "Network Access Control in OverDRIVE Mobile Networks," IST Mobile Summit 2003, Aveiro, Portugal, 16–18 June 2003.

[24] Ernst, T., Castelluccia, C., and Lach, H., "Extending Mobile-IPv6 with Multicast to Support Mobile Networks in IPv6," ECUMN'00, Colmar, France, October 2–4, 2000.

[25] Wakikawa, R., Koshiba, S., and Uehara, K., "ORC: Optimized Route Cache Management Protocol for Network Mobility," *Proceedings of the 10th Interantional Conference on Telecommunication (ICT)*, Tahiti Papeete, French Polynesia, February 2003.

[26] Jeong, J., Lee, K., Park, J., and Kim, H., "Route Optimization Based on ND-Proxy for Mobile Nodes in IPv6 Mobile Networks," IEEE VTC 2004-Spring, Milan, Italy, May 17–19, 2004.

[27] Perera, E., Seneviratne, A., and Sivaraman, V., "OptiNets: An Architecture to Enable Optimal Routing for Network Mobility," *International Workshop on Wireless Ad-Hoc Networks 2004 (IWWAN 2004)*, Oulu, Finland, 31 May–3 June, 2004.

[28] Kang, H., Kim, K., Han, S., Lee, K., and Paik, J., "Route Optimization for Mobile Network by Using Bi-directional between Home Agent and Top Level Mobile Router," (draft-hkang-nemo-ro-tlmr-00.txt), Internet Draft, IETF, June 2003.

[29] Ohnishi, H., Sakitani, K., and Takagi, Y., "Route HMIPv6 Based Route Optimization Method in a Mobile Network," (draft-ohnishinemo-ro-hmip-00.txt), October 2003.

[30] Soliman, H., Castelluccia, C., Malki, K., and Bellier, L., "Hierarchical Mobile IPv6 Mobility Management (HMIPv6)," RFC 4140, IETF, August 2005.

[31] Thubert, P. and Molteni, M., "IPv6 Reverse Routing Header and Its Application to Mobile Networks," (draft-thubert-nemo-reverse-routing-header-05.txt), Internet Draft, IETF, June 2004, work in progress.

[32] Valko, A., "Cellular IP: A New Approach to Internet Host Mobility," *ACM SIGCOMM Computer Communication Review*, Volume 29, Issue 1, January 1999.

[33] Moore, N., "Optimistic Duplicate Address Detection for IPv6," (draft-ietf-ipv6-optimistic-dad-05.txt), February 2005.

[34] Kempf, J., Khalil, M., and Pentlan, B., "IPv6 Fast Router Advertisement," (draft-mkhalil-ipv6-fastra-05.txt), January 2005.

[35] Ramani, I., and Savage, S., "Syncscan: Practical Fast Handoff for 802.11 Infrastructure Networks," in *Proceedings of IEEE INFOCOM*, Miami, FL, March 2005.

[36] Koodli, R., "Fast Handovers for Mobile IPv6," (draft-ietf-mobileip-fast-mipv6-07.txt), Internet Draft, IETF, September 19, 2003, work in progress.

[37] Petander, H., Perera, E., and Seneviratne, A., "Multiple Interface Handoffs: A Practical Method for Access Technology Independent Make-Before-Break Handoffs," NICTA Technical Report, http://nicta.com.au/director/research/publications/technical_reports/2005.cfm.

[38] Droms, R. and Thubert, P., "DHCPv6 Prefix Delegation for NEMO," (draft-ietf-nemo-dhcpv6-pd-00.txt), Internet Draft, IETF, August 2004, work in progress.

[39] Kniveton, T. and Thubert, P., "Mobile Network Prefix Delegation," (draft-ietf-nemo-prefix-delegation-00.txt), Internet Draft, IETF, August 2005, work in progress.

[40] Wakikawa, R., Devarapalli, V., and Thubert, P., "Inter Home Agents Protocol (HAHA)," (draft-wakikawa-mip6-nemo-haha-01.txt), Internet Draft, IETF, February 2004, work in progress.

[41] Chung, A. and Mahbub, H., "Traffic distribution algorithm for multi-homed mobile hotspots," in *Proceedings of IEEE Vehicular Technology Conference (VTC)*, Stockholm, Sweden, Spring 2005.

[42] http://www.ocean.cse.unsw.edu.au/

[43] http://www.nicta.com.au/

[44] http://www.ocean.cse.unsw.edu.au/emotion/index.html

[45] Lan, K., Perera, E., Petander, H., Libman, L., Dwertman, C., and Hassan, M., "MOBNET: The Design and Implementation of a Network Mobility Testbed for NEMO Protocol," *14th IEEE Workshop on Local and Metropolitan Area Networks*, September 18–21, 2005, Crete, Greece.

[46] OverDRiVE: http://www.ist-overdrive.org

[47] DRiVE, http://www.ist-drive.org/index2.html

[48] Ronai, M., Petrescu, A., Tönjes, R., and Wolf, M., "Mobility Issues inOver-DRIVE Mobile Networks," IST Mobile Summit 2003, Aveiro, Portugal, 16–18 June 2003.

[49] http://www.nautilus6.org/

[50] InternetCAR, http://www.sfc.wide.ad.jp/InternetCAR/

[51] Ernst, T., Olivereau, A., Bellier, L., Castelluccia, C., and Lach, H. "Mobile Networks Support in Mobile IPv6," (draft-ernst-mobileip-v6-network-03) Internet Draft, IETF, March 2002, work in progress.

[52] Ernst, T., Mitsuya, K., and Uehara, K., "Network Mobility from the Internet-CAR Perspective," *Journal Of Interconnection Networks (JOIN)*, June 2003.

[53] FleetNet, http://www.fleetnet.de/

[54] Bechler, M., Franz, J., and Wolf, L., "Mobile Internet Access in FleetNet," *13. Fachtagung Kommunikation in Verteilten Systemen (KiVS 2003)*, Leipzig, February 2003.

SECURITY
MANAGEMENT

Chapter 15

Key Management in Wireless Sensor Networks: Challenges and Solutions

Yun Zhou, Yanchao Zhang, and Yuguang Fang[1]

Contents

[1] This work was supported in part by the U.S. Office of Naval Research under Young Investigator Award N000140210464 and by U.S. NSF Career Award ANI-0093241.

15.1   Introduction

Wireless sensor networks (WSNs) [1] have been seen as a promising network infrastructure for many military applications, such as battlefield surveillance and homeland security monitoring. In those hostile tactical scenarios and important commercial applications, security mechanisms are necessary to protect WSNs from malicious attacks.

Key management is very critical to security protocols because most of the cryptographical primitives, such as encryption and authentication, in those protocols are based on the operations involving keys. Encryption requires that a key be fed into an algorithm so that plaintexts can be transformed into ciphertexts. To authenticate a packet, a *Message Authentication Code* (MAC) can be attached to the packet. However, MACs are usually computed by hashing the concatenations of packets and keys. Hence, key management is of paramount importance for establishing security infrastructures for WSNs.

However, to establish keys is a very challenging task in WSNs due to their unique characteristics, which are different from conventional wired networks such as the Internet and other wireless networks such as *mobile ad hoc networks* (MANETs). The openness of wireless channels renders adversaries' capability to analyze the eavesdropped packets transmitted between sensor nodes such that some key information can be exposed, from which adversaries can derive keys between sensor nodes. A sensor node is usually built with constrained resources in terms of memory, radio bandwidth, processing capability, and battery power. Strong security algorithms may not be supported by sensor platforms due to their complexity. In a hostile environment, it is infeasible to provide constant surveillance on a WSN after deployment, and sensor nodes can be captured so that all their keying materials are compromised. A WSN may have to scale up to thousands of sensor nodes; the WSN simple, flexible, and scalable security protocols. However, to design such security protocols is not an easy task. Higher-level security and less computation and communication overhead are contradictory requirements in the design of security protocols for WSNs. In most cases, a trade-off must be made between security and performance.

A general approach to establishing keys in WSNs includes two steps. Before sensor nodes are deployed, each node is configured with some key materials. After those nodes are deployed into a designated terrain, they perform several rounds of communications to agree on pairwise keys associated with their key materials. Based on the algorithms used to establish pairwise keys, current solutions can be classified into symmetric key schemes and asymmetric key (or public key) schemes. The symmetric key schemes of the early stage are probabilistic, in that two nodes can use their key materials to establish a pairwise key with a certain probability. Therefore, some pairs of nodes can establish keys directly and some pairs of nodes have to establish pairwise keys through multi-hop paths. However, their probabilistic nature makes some nodes isolated, in that they are not able to establish keys with their neighboring nodes. Then, a deterministic approach is proposed so that every pair of nodes can establish a pairwise key directly or through a multihop path. The probabilistic approach and the deterministic approach uniformly distribute key materials, and thus each node can establish direct keys with a small portion of its neighbors and must establish indirect keys with other neighbors through a multi-hop path, which may cost large power consumption. Then, location infomation is used so that nodes close to each other are configured with correlated key materials. This location-based approach can increase the probability that each pair of neighboring nodes establishes a pairwise key, thus saving energy consumption on multi-hop routing. Public key schemes are mainly based on Diffie-Hellman and RSA. However, algorithms in the field of elliptic curves have drawn much attention recently because of their efficiency.

In this chapter, some design issues and challenges are first introduced in Section 15.2. Then, symmetric key schemes are described in Section 15.3, including probabilistic, deterministic, and location-based solutions. Public key schemes are discussed in Section 15.4. Section 15.5 sheds light on some open issues. Finally, the chapter offers a conclusion in Section 15.6.

15.2   Design Issues and Challenges

15.2.1   Cryptographical Issues

In his classic paper "Communication Theory of Secrecy Systems" [2], Shannon, who had established information theory, developed the theoretical framework for *symmetric key*-based cryptography. In his cryptographical system model, there are two information sources (i.e., a message source and a key source) at the transmission end. The key source produces a particular key K from among those that are usable in the system. This key K is transmitted by some means, supposedly *not interceptible*, for example by a messenger, to the receiving end. The message source produces a message M (in the "clear"), which is enciphered by the encipherer T_K. The resulting ciphertext E is sent to the receiving end by possibly *interceptible* means, for example radio. At the receiving end, the ciphertext E and the key K are combined in the decipherer T_K^{-1} to recover the message M. The transformation T_K and its inverse T_K^{-1} are possibly known to the public.

The Diffie-Hellman [3] and Rivest, Shamir, and Adleman (RSA) [4] algorithms mark the establishment of *asymmetric key*-based cryptography. Unlike a single key used in symmetric key systems, there are two keys in asymmetric key systems. The transmission end encrypts a message M into a ciphertext E by an encryption key K. The receiving end decrypts the ciphertext E to get the message M by a decryption key K^{-1}. Here, the encryption key K and the decryption key K^{-1} are different. Although the decryption key is kept secret, the encryption key is usually known to the public. Asymmetric key systems, therefore, are also called *public key* systems.

In a cryptographical system, the message source and the ciphertext space are usually accessible to an attacker. The encryption and the decryption transforms are also seen as accessible to the attacker. Although in some specific systems the cryptographical algorithms can be kept secret, this approach may increase system vulnerability because an algorithm that is not inspected carefully by critical experts may have some potential defects that can be utilized by hackers. Therefore, most "secure" algorithms are public so that they can be carefully inspected. In this case, the security of the entire system primarily relies on the secrecy of the keys it uses.

If an attacker can find the key, the entire system is broken. The attacker can achieve this goal by cryptanalysis. Most cryptographical systems are vulnerable to cryptanalysis due to the existence of the redundancy of message source in the real world. The redundancy can always provide the attacker with a possible tool to do cryptanalysis over intercepted ciphertexts during their transmission. Moreover, the attacker knows the system being used, that is, the message space, the transformation T_i, and the probabilities of choosing various keys, and has unlimited time and manpower available for the analysis of ciphertexts. The attacker thus can use all these resources to find the key if time is not important to him. Another way is to directly intercept the key during its transmission between the message source and receiving end. Therefore, how to securely achieve key agreement between the source and sink is a very important issue.

Generally, establishing keys involves two steps. First, the source and sink should be configured with some key materials. Second, those materials are used to establish a shared symmetric key between the source and sink. In symmetric key systems, those key materials can be the shared symmetric key or parameters used to calculate a symmetric key. In asymmetric key systems, they are parameters associated with the chosen asymmetric key algorithm (e.g., Diffie-Hellman or RSA), and the source and sink can negotiate a shared symmetric key using the asymmetric key algorithm. After those nodes are deployed into a designated terrain, they perform several rounds of communications to agree on pairwise keys associated with their key materials.

15.2.2 Challenges

Although the key management problem has been investigated thoroughly in conventional wired networks such as the Internet and wireless networks such as cellular networks, WLANs, or ad hoc networks, the existing solutions can hardly be transplanted into WSNs due to their unique characteristics.

In wired networks, the key materials transmitted over shielded wired lines during the negotiation phase between the source and sink are more difficult to intercept. But a wireless channel is open to eavesdroppers. In addition to eavesdropping key materials to expose corresponding keys, adversaries can also intercept the encrypted ciphertexts so that adversaries can analyze the eavesdropped packets to get some key information, from which adversaries can derive keys between sensor nodes.

In cellular networks and WLANs, the communication pattern is one-hop, that is, between the base station or the access point and the mobile node; but in WSNs, all the nodes are involved into multi-hop communications. Most centralized secure protocols cannot be directly applied in distributed

WSNs. Although ad hoc networks bear more similarities to WSNs, the nodes in ad hoc networks are more powerful than those in WSNs, and are thus able to support more secure, more complex protocols.

Moreover, a wireless channel is very dynamic. Key establishment protocols can endure frequent interruptions when channel conditions vary. Although link layer protocols may have some error control mechanisms, the cost of establishing keys is inevitably increased.

The openness of a wireless channel also renders adversaries the ability to launch the *denial-of-service* (DoS) attack [5]. Constant or random jamming interferences can be introduced by adversaries to corrupt normal communications, thus leading to the failure of key establishment protocols.

To save cost, sensor nodes are usually built with constrained resources. For example, the latest MICA motes [6] only use 8-bit processors. Their memory size is measured in units of kB (kilobytes). The radio interface can support approximately 40 kbits/sec, The entire mote platform is only powered by 2 AA batteries. The constrained processing capability makes the implementation of strong security algorithms a very challenging task. Considering the power limit, it is impratical to support complex protocols if a long lifetime of the WSN is desirable.

WSNs are usually deployed in hostile environments, such as battlefields or disaster locations, where fixed infrastructures are not available. After deployment, it is infeasible to provide constant surveillance of and protection for a WSN. In many security-critical scenarios, adversaries may have ability to access sensor nodes without being detected. Adversaries can use proper devices to dig into sensor hardware and find key materials. Due to cost constraints, it is also unrealistic and uneconomical to employ tamper-resistant hardware to secure the cryptographic materials in each individual node. Even if tamper-resistant devices are available, they are still not able to guarantee perfect security of secret materials [7]. Hence, adversaries can capture any node and compromise the secrets stored in that node. Furthermore, adversaries can use the compromised secrets to derive more secrets shared between other noncompromised nodes. This means that the *node compromise* attack is unavoidable in WSNs. What we can do is to reduce the impact on other normal nodes as much as possible.

A compromised node can be used as a platform to launch other tricky attacks. The adversary can let the compromised node impersonate another normal node to establish secure communications with other normal nodes. Therefore, node authentication should be considered during the key establishment procedure. If the compromised node is involved as a router between a pair of source and sink nodes, the key negotiation procedure may fail just because the compromised node intentionally drops some packets for the negotiation between the source and sink.

Scalability is another important issue. According to different application scenarios, a WSN can have from tens to thousands of sensor nodes.

Moreover, during the lifetime of the WSN, some nodes can run out of power, and some new nodes can be inserted to increase the network processing capability. Therefore, the number of nodes can vary from time to time. These node dynamics demand simple, flexible, and scalable security protocols. However, to design such security protocols is not an easy task. The reason mainly lies in that the memory cost per node increases quickly when the network size enlarges. For example, conventional key establishment schemes [8,9] cannot support large networks due to their memory cost of $N - 1$ units in a network of N nodes. To increase the scalability, many works partition the entire network into several portions and apply conventional schemes in each portion, but the security performance is reduced.

Generally, higher-level security and less computation and communication overhead are contradictory requirements in the design of security protocols for WSNs. In most cases, a trade-off must be made between security and performance.

15.3   Symmetric Key Management

Due to its computational efficiency, it is commonly believed that symmetric key technology is more viable for resource-constrained, low-end devices than public key technology. Therefore, most of the security protocols developed thus far for WSNs are based on symmetric key technology. We discuss this approach first.

15.3.1   Global Key

The simplest symmetric key scheme is to use a *global key* [10], which is shared by all the sensor nodes. Data traffic is protected by the global key. The global key is updated periodically. In each period, a sensor node is elected as a key manager that generates and distributes a new global key to all the other sensor nodes. If all the sensor nodes are trustful, this scheme can effectively prevent external adversaries from accessing critical information that is secured by the global key. However, the scheme is very vulnerable to *node compromise* in that adversaries can get the global key by compromising only one node and thereby break into the entire sensor network.

15.3.2   Key Server

A WSN is usually connected to external wired networks through a *base station* (BS). The BS can act as a key server to distribute keys for any pair of sensor nodes [11]. In particular, each sensor node shares a unique key with the BS. When two nodes need to establish a shared key, one can generate the key and let the BS forward the key to the other node. This procedure can be secured by the keys shared between the sensor nodes and BS.

This centralized approach can reduce the impact of node compromise because compromising one node does not result in the exposure of keys between noncompromised nodes. However, the communication overhead is high because two close nodes may have to carry out handshakes through the central key server at a distant place. There is still a concern for security, in that the key server may become a potential point of failure even if the server is under careful protection. The entire network can break down if adversaries successfully corrupt the key server.

15.3.3 Full Predistribution

Usually, a WSN belongs to an authority. The authority can do configurations before deploying the network such that any pair of nodes is preloaded with a unique shared key. This scheme is the most resilient to node compromise because adversaries will not know each key unless they compromise one of the two nodes sharing the key. Moreover, each key is preloaded, so no negotiation is required between nodes. However, the memory cost is very high. In a WSN of N nodes, each node needs to store $N - 1$ keys, and the overall number of keys in the WSN is $\frac{N(N-1)}{2}$. It is unaffordable on memory-constrained sensor platforms when N is very large. Therefore, this approach works only in small sensor networks.

15.3.4 Blom Scheme

A similar approach was proposed by Blom [8]. His method is based on $(N, t + 1)$ *maximum distance separable* (MDS) linear codes [12]. Before a WSN is deployed, a central authority first constructs a $(t + 1) \times N$ public matrix P over a finite field $\mathbb{F}_q$. Then the central authority selects a random $(t + 1) \times (t + 1)$ symmetric matrix S over $\mathbb{F}_q$, where S is secret and only known to the central authority. An $N \times (t + 1)$ matrix $A = (S \cdot P)^T$ is computed, where $(\cdot)^T$ denotes the transpose operator. The central authority preloads the i-th row of A and the i-th column of P to node i, for $i = 1, 2, \ldots, N$. When nodes i and j need to establish a shared key, they first exchange their columns of P, and then node i computes a key K_{ij} as the product of its own row of A and the j-th column of P and node j computes K_{ji} as the product of its own row of A and the i-th column of P. Because S is symmetric, it is easy to see that:

$$K = A \cdot P = (S \cdot P)^T \cdot P = P^T \cdot S^T \cdot P$$

$$= P^T \cdot S \cdot P = (A \cdot P)^T = K^T \tag{15.1}$$

Therefore, nodes pair (i, j) will use $K_{ij} = K_{ji}$, as a shared key.

The Blom scheme has a *t-secure* property, in that in a network of N nodes, the collusion of less than $t + 1$ nodes cannot reveal any key shared by other pairs of nodes. This is because as least t rows of A and t columns of P are required to solve the secret symmetric matrix S. The memory cost per node in the Blom scheme is $t + 1$. To guarantee perfect security in a WSN with N nodes, the $(N - 2)$-secure Blom scheme should be used, which means the memory cost per node is $N - 1$. Hence, the Blom scheme can provide strong security in small networks.

15.3.5 Polynomial Model

A polynomial-based key establishment scheme was described by Blundo et al. [9]. It is a special case of Blom's scheme, in that a Vandermonde matrix is used as the generator matrix of MDS code. They used a *t*-degree bivariate polynomial, which is defined as:

$$f(x, y) = \sum_{i=0}^{t} \sum_{j=0}^{t} a_{ij} x^i y^j \qquad (15.2)$$

over a finite field $\mathbb{F}_q$, where q is a prime that is large enough to accommodate a cryptographic key. By choosing $a_{ij} = a_{ji}$, we can get a symmetric polynomial in that $f(x, y) = f(y, x)$. Each sensor node is assumed to have a unique, integer-valued, non-zero identity. For each sensor node u, a *polynomial share* $f(u, y)$ is assigned, which means the coefficients of univariate polynomials $f(u, y)$ are loaded into node u's memory. When nodes u and v need to establish a shared key, they broadcast their IDs. Subsequently, node u can compute $f(u, v)$ by evaluating $f(u, y)$ at $y = v$, and node v can also compute $f(v, u)$ by evaluating $f(v, y)$ at $y = u$. Due to the polynomial symmetry, the shared key between nodes u and v has been established as $K_{uv} = f(u, v) = f(v, u)$. Similar the Blom scheme, a *t*-degree bivariate polynomial is also $(t + 1)$-secure, meaning that adversaries must compromise no less than $(t + 1)$ nodes holding shares of the same polynomial to reconstruct it. But the memory cost is also the same as that of the Blom scheme. Hence, the polynomial model is also suitable in small networks.

15.3.6 Random Key Predistribution

In ideal cases, every pair of nodes in a network should have a unique shared key. Although full predistribution, the Blom scheme, and the polynomial model can achieve this goal, the cost is that each node needs to store $N - 1$ keys in a network of N nodes. This is impractical for WSNs due to the memory constraints of sensor nodes when the network scale is very large. Instead, most recent research articles in this field lose the security requirement and follow a *partial predistribution* approach, wherein key materials are predistributed such that some sensor nodes can establish shared keys directly and they can help to establish indirect shared keys between other sensor nodes.

A pioneer work following this approach is *random key predistribution* (called RKP hereafter) [13]. In RKP, each node is preloaded with a subset of *m* keys randomly selected from a global pool of *M* keys such that any pair of neighboring nodes can share at least one key with a certain probability, that is:

$$p = 1 - \frac{\binom{M-m}{m}}{\binom{M}{m}} \tag{15.3}$$

RKP is developed based on an observation that it is unnecessary to guarantee full connectivity for a sensor node with all its neighbors, as long as multi-link paths of shared keys exist among neighbors that can be used to set up a path key as needed. The rationale behind this observation is the random graph theory. When the probability p that a link exists between two nodes increases, the probability P_c that the entire graph is connected also increases. There is a required p such that P_c is almost 1. Hence, we can choose m and M such that the entire network is almost connected. In RKP, two neighboring nodes can have a shared key *directly* if their key subsets have an intersection or negotiate an *indirect* key through a *secure path*, along which every pair of neighboring nodes has a direct shared key.

A major concern of RKP is node compromise. By tampering or cryptanalysis, adversaries can compromise a node and expose its key subset. Because each key is reused by many sensor nodes, those exposed keys can be used to corrupt links between other noncompromised nodes if those links happen to be secured by the exposed keys.

Another concern is the communication overhead. Due to the memory constraint, each sensor node cannot keep too many keys. Hence, the value of p is rather small, which means that each sensor node needs to negotiate keys with a large portion of neighboring nodes through multi-link paths.

15.3.7 Q-Composite RKP

To mitigate the impact of node compromise, Chan et al. [14] suggested to improve RKP such that any pairs of neighboring nodes are required to share at least q keys with a certain probability. Let $p(i)$ be the probability that two nodes share i keys; then:

$$p(i) = \frac{\binom{M}{i}\binom{M-i}{2(m-i)}\binom{2(m-i)}{m-i}}{\binom{M}{m}^2} \tag{15.4}$$

and the probability that two nodes share at least q keys is:

$$p = 1 - (p(0) + p(1) + \cdots + p(q-1)) \tag{15.5}$$

This scheme achieves greatly strengthened security under small-scale attacks while trading off increased vulnerability in the face of large-scale attacks on network nodes. As the amount of required key overlap increases, it becomes exponentially more difficult for an attacker with a given key set to break a link. However, to preserve the given probability p of two nodes sharing sufficient keys to establish a secure link, it is necessary to reduce the size of the global key pool M. This allows the attacker to gain a larger sample of the global key pool by breaking fewer nodes.

15.3.8 Random-Pairwise Key

Authentication is necessary to provide assurance for the identities of communicating parties. This can be achieved through the normal *challenge-response* approach based on the unique key shared by the communicating parties. In particularly, one verifier node can send an encrypted random number, called a challenge, to the other node, and that node can prove its identity by returning the decrypted result to the verifier node. The identity of the verifier node can be authenticated in the same way.

RKP and q-composite RKP can hardly provide authentication because of the reuse of the same key in many sensor nodes. To solve the problem, Chan et al. [14] also proposed the *random-pairwise key* (RPK) scheme. For each node, a set of M nodes is randomly selected from the entire network, and a unique pairwise key is assigned for the pair of the node and each of the nodes in the set. When the network consists of N nodes, any pair of nodes can share a pairwise key with a probability

$$p = \frac{M}{N} \tag{15.6}$$

Based on the random graph theory, the entire network can be connected as long as the probability is larger than a threshold. Hence, any pair of nodes can either share a direct key or negotiate an indirect pairwise key through a multi-link secure path. The uniqueness of direct keys can be used to authenticate node identities.

In RPK, each direct key is uniquely generated and shared by a unique pair of nodes, so RPK is resilient to node compromise in terms of the secrecy of direct pairwise keys. However, the negotiation of indirect pairwise keys can introduce a lot of communication overhead due to the discovery of secure paths and handshakes between two end nodes.

15.3.9 Random Key Assignment

To discover whether the key sets of two nodes have an intersection, usually both nodes need to broadcast their key indices or find common keys through the challenge-response procedure. Such methods are not communication efficient. Pietro et al. [15] improved the RKP scheme by associating the key indices of a node with the node identity. For example, each node is assigned a pseudo-random number generator $g(x, y)$, and the key indices for the node are calculated as $g(ID, i)$ for $i = 1, \ldots, M$, where ID is the node identity. In this way, other nodes can quickly find out which key is in its key set by checking its node identity.

15.3.10 Multiple-Space Key Predistribution

In the face of node compromise, the security level of RKP and its derivatives described above will deteriorate quickly, in that each time adversaries compromise one more node, more secrets such as keys are exposed.

To improve the resilience to node compromise, Du et al. [16] developed the *multiple-space key predistribution* (MSKP) scheme based on the Blom scheme. Specifically, a public matrix P and a global pool of symmetric matrices S_i for $i = 1, \ldots, w$ are constructed. Each tuple (S_i, P) is called a *key space*. For each sensor node, v spaces are randomly selected, and the node is configured with parameters derived according to the Blom scheme. Obviously, if two neighboring sensor nodes have a space in common, they can calculate a direct pairwise key according to the Blom scheme. The merit of MSKP is that it can tolerate up to a certain number of compromised nodes before the security level of the network begins to deteriorate. This is due to the threshold-based security of the Blom scheme.

15.3.11 Polynomial Pool–Based Key Predistribution

Another scheme, called *polynomial pool–based key predistribution* (PPKP) [17], is basically the same as the MSKP scheme, but each Blom matrix is replaced by a t-degree bivariate polynomial. Each sensor node randomly selects v polynomials from a global pool of w polynomials. Any pair of neighboring nodes can calculate a direct pairwise key if they share the same polynomial. Like MSKP, PPKP can also provide threshold-based resilience to node compromise.

Another merit of MSKP and PPKP is that each direct key is tied to the identities of the nodes sharing it. Hence, they can provide authentication like RPK.

15.3.12 Hwang-Kim Scheme

The schemes discussed above require each sensor node keep many key materials such that two nodes share a key with a probability that can guarantee that the entire network is almost connected. The requirement may be too harsh in memory-constrained sensor networks. Hwang and Kim [18] revisited the RKP scheme and its derivatives, and proposed to reduce the amount of key materials that each node keeps while still maintaining a certain probability of sharing a key between two nodes. Their idea is to guarantee that the largest giant component of the network, instead of the entire network, is almost connected. Hence, each sensor node can keep less key materials. The probability that two nodes have a key in common is reduced, but it is still large enough for the largest network component to be connected. The trade-off is that more nodes can be isolated because they do not share keys with their neighbors.

15.3.13 PIKE Scheme

The probabilistic nature of the random distribution of key materials cannot guarantee that two neighboring nodes establish a shared key. To facilitate key establishment between every pair of neighboring nodes, a deterministic approach can be taken. In the *peer intermediaries for key establishment* (PIKE) scheme [19], all N sensor nodes are organized into a two-dimensional space (Figure 15.1), where the coordinate of each node is (x, y) for $x, y \in \{0, 1, 2, \ldots, \sqrt{N} - 1\}$. Each node shares unique pairwise keys with $2(\sqrt{N} - 1)$ nodes that have the same x or y coordinates in the two-dimensional space. For two nodes with no common coordinate, an intermediate node, which has common x or y coordinates with both nodes, is used as a router to forward a key for them. However, the communication overhead is rather high because the secure connectivity is only $\frac{2}{\sqrt{N}}$, which

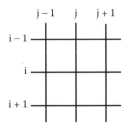

Figure 15.1   PIKE scheme. Sensor nodes are organized into a two-dimensional space.

means that each node must establish a key for almost each of its neighbors through multi-link paths.

15.3.14 Grid-Based Key Predistribution

The *grid-based key predistribution* (GBKP) scheme [17] uses the same two-dimensional space as PIKE. Instead of pairwise keys in PIKE, GBKP assigns a bivariate symmetric polynomial for each set of nodes with the same x or y coordinate. Hence, direct keys can be established between nodes with the same x or y coordinate according to the polynomial model. Indirect keys can be negotiated in the same way as that in PIKE. PIKE and GBKP can guarantee that any pair of nodes shares a direct key or negotiates an indirect key through a third node. Moreover, a node can find whether it has a direct shared key with another node based on the coordinate of that node. This can provide an authentication service, in that the identity associated with the coordinate of a node can be challenged based on its keys that are related to its identity.

15.3.15 Scalable Key Agreement

Zhou and Fang [20] developed a scalable key agreement. They use a t-degree $(k+1)$-variate symmetric polynomial to establish keys in a deterministic way.

A t-degree $(k+1)$-variate polynomial is defined as:

$$f(x_1, x_2, \ldots, x_k, x_{k+1}) = \sum_{i_1=0}^{t} \sum_{i_2=0}^{t} \cdots$$

$$\sum_{i_k=0}^{t} \sum_{i_{k+1}=0}^{t} a_{i_1, i_2, \ldots, i_k, i_{k+1}} \, x_1^{i_1} x_2^{i_2} \cdots x_k^{i_k} x_{k+1}^{i_{k+1}} \tag{15.7}$$

All coefficients are chosen from a finite field $\mathbb{F}_q$, where q is a prime that is large enough to accommodate a cryptographic key.

A $(k+1)$-tuple permutation is defined as a bijective mapping:

$$\sigma : [1, k+1] \longrightarrow [1, k+1] \tag{15.8}$$

By choosing all the coefficients according to

$$a_{i_1, i_2, \ldots, i_k, i_{k+1}} = a_{i_{\sigma(1)}, i_{\sigma(2)}, \ldots, i_{\sigma(k)}, i_{\sigma(k+1)}} \tag{15.9}$$

for any permutation σ, a symmetric polynomial can be obtained in that

$$f(x_1, x_2, \ldots, x_k, x_{k+1}) = f(x_{\sigma(1)}, x_{\sigma(2)}, \ldots, x_{\sigma(k)}, x_{\sigma(k+1)}) \tag{15.10}$$

Each node is identified by an ID $(n_1, n_2, \ldots, n_k)$, which is the coordinate of a point in a k-dimension space $\mathcal{S}_1 \times \mathcal{S}_2 \times \cdots \times \mathcal{S}_k$, where $n_i \in \mathcal{S}_i \subset \mathbb{Z}$ for $i = 1, \ldots, k$ and $\mathcal{S}_i \bigcap \mathcal{S}_j = \phi$, for $i \neq j$.

For a node $(n_1, n_2, \ldots, n_k)$ in the network, a *polynomial share*

$$f_1(x_{k+1}) = f(n_1, n_2, \ldots, n_k, x_{k+1})$$

$$= \sum_{i_{k+1}=0}^{t} b_{i_{k+1}} \, x_{k+1}^{i_{k+1}} \qquad (15.11)$$

is calculated using the node ID as inputs to the t-degree $(k+1)$-variate symmetric polynomial.

If two nodes u with ID $(u_1, u_2, \ldots, u_k)$ and v with ID $(v_1, v_2, \ldots, v_k)$ have only one mismatch in their IDs, say $u_i \neq v_i$ for some i but $u_j = v_j = c_j$ for other $j \neq i$, then nodes u and v can calculate a shared key as:

$$K_{uv} = f(c_1, \ldots, u_i, \ldots, c_k, v_i) = f(c_1, \ldots, v_i, \ldots, c_k, u_i) \qquad (15.12)$$

If two nodes have more than one mismatch in their IDs, they cannot calculate a direct key. In this case, they can negotiate an indirect key over a multi-hop path, along which every pair of neighboring nodes has already calculated a direct key.

An example of a three-dimensional ID space is given in Figure 15.2. For any edge, the pair of nodes at its two ends can calculate a direct key because the two nodes have only one mismatch in their IDs. Suppose node

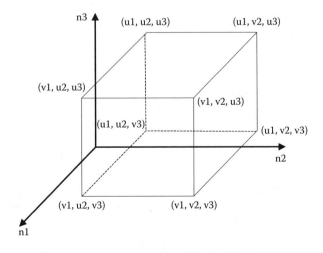

Figure 15.2 Multidimension key graph. Nodes (u_1, u_2, u_3) and (v_1, v_2, v_3) can negotiate a key through any path consisting of connected edges of the cube.

(u_1, u_2, u_3) needs to establish a shared key with node (v_1, v_2, v_3), where all three indices in their IDs are mismatching. There are three disjoint paths from node u to node v. For example, three disjoint paths are:

$$(u_1, u_2, u_3) \rightarrow (v_1, u_2, u_3) \rightarrow (v_1, v_2, u_3) \rightarrow (v_1, v_2, v_3)$$

$$(u_1, u_2, u_3) \rightarrow (u_1, u_2, v_3) \rightarrow (v_1, u_2, v_3) \rightarrow (v_1, v_2, v_3)$$

and

$$(u_1, u_2, u_3) \rightarrow (u_1, v_2, u_3) \rightarrow (u_1, v_2, v_3) \rightarrow (v_1, v_2, v_3)$$

Obviously, the above set of disjoint paths is not unique.

The dimension of the ID space k is a parameter to be controlled to achieve the trade-off between memory cost per node and scalability. To guarantee each direct key unsolvable by adversaries, no matter how many nodes are compromised, the memory cost per node is less than

$$M \leq k \log k + \log N + \left(\sqrt[k]{N} \sqrt[k+1]{\frac{k(k+1)!}{2}} + 1 \right) \log q \qquad (15.13)$$

where N is the total number of nodes, q is the field size, and all the subspace S_i have the same cardinality. Obviously, the scheme has good scalability, in that the memory cost is only on the order $\mathcal{O}(\sqrt[k]{N})$ when k is fixed.

15.3.16 Location-Based Key Predistribution

In the aforementioned schemes, key materials are uniformly distributed in the entire terrain of a network. The uniform distribution makes the probability that two neighboring nodes share a direct key, called *secure connectivity*, rather small. Therefore, a lot of communication overhead is inevitable for the establishment of indirect keys. If some location information is known, two nearby sensor nodes can be preloaded intentionally with the same set of key materials. In this way, we can expect improvement in secure connectivity.

In the *location-based key predistribution* (LBKP) scheme [21], the entire sensor network is divided into many square cells. Each cell is associated with a unique t-degree bivariate polynomial. Each sensor node is preloaded with shares of the polynomials of its home cell and four other cells horizontally and vertically adjoining its home cell. After deployment, any two neighboring nodes can establish a pairwise key if they have shares of the same polynomial.

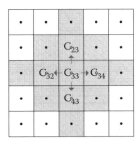

Figure 15.3   LBKP scheme. The polynomial of cell C_{33} is also assigned to cells C_{32}, C_{34}, C_{23}, and C_{43}. A node in C_{33} has some common polynomial information with other nodes in the shadow areas.

For example, in Figure 15.3, the polynomial of cell C_{33} is also assigned to cells C_{32}, C_{34}, C_{23}, and C_{43}. The polynomials of other cells are assigned in the same way. As a result, a node in C_{33} has some polynomial information in common with other nodes in the shadow areas.

15.3.17   Key Establishment Using Deployment Information

Du et al. [22] divide the entire network into many square cells. Each cell is assigned a subset key pool S_{ij}, $i = 1, \ldots, u$ and $j = 1, \ldots, v$ out of a global key pool S. Those subset key pools are set up such that the key pools of two neighboring cells will share a portion of keys. In each cell, the RKP [13] scheme is applied.

Using deployment knowledge to achieve the same connectivity, the size of the key ring that one node holds in this scheme is much less than that in the RKP scheme. This can significantly save the memory of sensors. However, it still inherits the same weakness as RKP, in that it does not provide authentication.

15.3.18   Location-Aware Key Management

Huang et al. [23] also used square cells. To provide intra-cell connectivity, the MSKP scheme [16] is applied in each cell such that any pair of nodes having a common space in one cell can establish a shared key. The RPK scheme [14] is applied between neighboring cells in that for each sensor node, a node from each of its neighboring cells is selected and a unique key is assigned to the pair of nodes. This provides inter-cell connectivity.

The MSKP scheme provides strong resilience to node compromise. However, network connectivity is low because the randomly distributed key spaces in each cell and the randomly assigned pairwise keys between cells can only guarantee that each node has shared keys with a portion of its neighbors.

15.3.19 Neighboring Cell–Based Predistribution Model

Threshold-based schemes such as the Blom [8] and polynomial schemes [9] can tolerate only a certain number of compromised nodes. The more nodes sharing a Blom matrix or a polynomial, the more likely they are exposed due to node compromise. Therefore, if we can reduce the number of nodes sharing a Blom matrix or a polynomial, the security then increases. Following this idea, Zhou et al. investigated a neighboring-cell-based predistribution model based on hexagon [24] and triangle [25] grid models.

Zhou et al. [24] divided the entire network terrain into non-overlapping hexagon cells. The polynomial model [9] is used here. Unlike LBKP [21], which assigns a polynomial to each cell and its four adjacent cells, [24] assigns a polynomial to each pair of neighboring cells. For example, in Figure 15.4, cell $c0$ is assigned six polynomials, each of which is uniquely shared with one of its neighboring cells; therefore, the nodes in cell $c0$ can establish direct pairwise keys with the nodes in the shadow area. The reduced number of nodes sharing one polynomial means less chance the polynomial can be exposed by collusion of compromised nodes. Hence, this new predistribution method can improve the resilience to node compromise. At the same security level, [24] requires less memory cost for each node compared with LBKP [21]. Due to the use of deployment knowledge, the security connectivity is very high.

Later, Zhou et al. [25] improved the hexagon grid model to the triangle grid model, in which the entire network is divided into non-overlapping triangle cells. The same key materials predistribution method is used as in

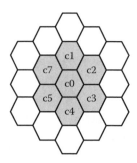

Figure 15.4 Hexagon grid-based key predistribution. Cell *c* 0 is assigned six Blom matrices or polynomials, each of which is uniquely shared with one of its neighboring cells.

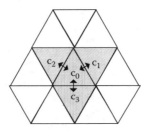

Figure 15.5 **Triangle grid-based key predistribution. Cell** *c* **0 is assigned three Blom matrices or polynomials, each of which is uniquely shared with one of its neighboring cells.**

[24], in which each pair of neighboring cells is associated with a unique *t*-degree bivariate polynomial [9] or a Blom matrix [8]. For example, in Figure 15.5, cell *c0* is assigned three Blom matrices or polynomials, each of which is uniquely shared with one of its neighboring cells, and the nodes in cell *c0* can establish direct pairwise keys with the nodes in the shadow area. This scheme further reduces the memory cost of each node at the same security level compared to the hexagon or the square grid model because each node needs to keep only three polynomials or matrices. Although the security connectivity is smaller than that of hexagon grid model, it is still much larger than conventional schemes that do not use location information.

15.3.20 *Group-Based Key Predistribution Framework*

Liu et al. [26] established a group-based key predistribution framework that may incorporate previous schemes. They divide all the sensor nodes into many *deployment groups*. In each group, a specific keying material distribution scheme, which could be one of the schemes discussed above, is applied to provide in-group connectivity. Also, each group picks one node and all those picked nodes form a *cross-group*. There is also a specific keying material distribution scheme for each cross-group. Therefore, two nodes from different deployment groups can establish a shared key through a path in a cross-group.

15.3.21 *LEAP*

Zhu et al. [27] proposed a *Localized Encryption and Authentication Protocol* (LEAP) for sensor networks. LEAP supports the establishment of four types of keys for each sensor node: (1) an *individual key* shared with the base station, (2) a *pairwise key* shared with another sensor node, (3) a *cluster key* shared with multiple neighboring nodes, and (4) a *group key* shared by all the nodes in the sensor network.

In this protocol, each node has an individual key that is only shared with the base station. This key is generated and preloaded into each node prior to its deployment. The individual key K_u^m for a node u is generated as $K_u^m = f_{K_s^m}(u)$ (assuming that each node has a unique ID). Here, f is a pseudo-random function and K_s^m is a master key known only to the base station.

A pairwise key refers to a key shared only between the node and one of its direct neighbors (i.e., one-hop neighbors). At first, each node u is preloaded with an initial key K_I. Then sensor node u derives a master key $K_u = f_{K_I}(u)$. When it is deployed, node u initiates a timer T_{min} and then broadcasts a *HELLO* message that contains a nonce N to each neighbor v. The reply from each neighbor v is authenticated using its master key K_v. Because node u can compute K_v using K_I, it is able to verify node v's identity independently. The handshake is as follows:

$$u \rightarrow * : \quad u, N_u$$

$$v \rightarrow u : \quad v, MAC(K_v, N_u|v)$$

Then node u computes its pairwise key with v as $K_{uv} = f_{K_v}(u)$. Node v can also compute K_{uv} independently. When a preset timer T_{min} expires, node u erases K_I and all the keys K_v it computed in the previous phase.

A cluster key is a key shared by a node and all its neighbors, and is primarily used for securing locally broadcast messages, for example, routing control information, or securing sensor messages that can benefit from passive participation. Consider the case that node u wants to establish a cluster key with all its immediate neighbors $v_1, v_2, \ldots, v_m$. Node u first generates a random key K_u^c, then encrypts this key with the pairwise key of each neighbor, and then transmits the encrypted key to each neighbor v_i:

$$u \rightarrow v_i : \quad \left(K_u^c\right)_{K_{uv_i}}$$

Node v_i decrypts the key K_u^c and stores it in a table. When one of the neighbors is revoked, node u generates a new cluster key and transmits to all the remaining neighbors in the same way.

A group key is a key shared by all the nodes in the network, and is necessary when the base station is distributing a secure message (for example, a query on some event of interest or a confidential instruction) to all the nodes in the network. To tolerate node failures, the group key should be updated occasionally. They employ the μTESLA protocol [11] to distribute group keys.

Unlike the schemes discussed above, LEAP assumes the network is safe during its initialization phase. Otherwise, the pairwise key establishment procedure can be attacked by adversaries.

15.3.22 Key Infection

The schemes described above assume a strong adversary model in which adversaries are able to compromise any node in a network. In such a case, security protocols are usually heavyweighted such that they can deal with worst-case attacks. However, in many scenarios, adversaries may not have enough resources or the ability to launch such worst-case attacks. Therefore, Anderson et al. [28] assumed a weak adversary model where adversaries can only eavesdrop a small portion of communications between sensor nodes during node deployment phase. Hence, a node wishing to communicate securely with other nodes simply generates a symmetric key and sends it in the clear to its neighbors.

15.4 Public Key Management

A symmetric key algorithm is efficient on low-end devices, but the schemes described in the previous section are rather complex for sensor networks in terms of communication overhead and memory cost. As a contrast, public key technology is easier to manage and more resilient to node compromise, although it is much more computationally expensive. Each node can keep its private key secret and only publish its public key; therefore, compromised nodes cannot provide any clue to the private keys of non-compromised nodes.

15.4.1 RSA Algorithm

RSA [4] is a very popular public key algorithm that can provide authentication and encryption services. At first, a node generates two large random and distinct primes p and q, each roughly the same size. Then it computes $n = pq$ and $\phi = (p-1)(q-1)$. A random integer e is selected, such that $1 < e < \phi$ and $\gcd(e, \phi) = 1$. After that, a unique integer d is calculated such that $1 < d < \phi$ and $ed = 1 \pmod{\phi}$. Hence, the node's public key is (n, e) and its private key is d.

When a node B needs to secretly send a message to node A, B can represent the message as an integer m in the interval $[0, n-1]$, compute $c = m^{e_A} \bmod n_A$ using A's public key, and send c to A. Node A can use its private key d_A to decrypt c to get $m = c^{d_A} \bmod n_A$.

To authenticate itself to node A, node B can generate a public message m and calculate a signature $s = m^{d_B} \bmod n_B$, then send a certificate $\langle m, s \rangle$

to A. Node A can verify node B's identity by checking whether m is equal to $s^{e_B} \bmod n_B$.

However, the exponential operation in RSA is very expensive, especially for large exponents. Hence, the focus of applying RSA in sensor networks is to develop efficient implementations on resource-constrained sensor platforms. RSA is rarely used to provide encryption in sensor networks. The private key operation of RSA is more expensive than public key operation because d is usually rather larger than e. *TinyPK* [29] uses RSA to provide the authentication service for external parties when they access sensor networks. In particular, each external party carries an RSA-based certificate and shows it to the sensor network. Sensors can verify the certificate and authenticate the external party. To simplify the signature verification on the sensor side, the RSA public key was chosen as $e = 3$. To avoid the secret key operation on the external party side, the certificate is calculated by a central network manager and preloaded to the external party.

15.4.2 Diffie-Hellman Algorithm

The Diffie-Hellman algorithm [3] is usually used to achieve key agreement between communication parties. At first, two nodes A and B agree on two parameters g and p, where g is a generator of $\mathbb{Z}_p$ and p is a large prime. Then node A chooses a secret integer x_A and sends $g^{x_A} \bmod p$ to B, and B also chooses a secret integer x_B and sends $g^{x_B} \bmod p$ to A. The shared key between A and B can be calculated as $K_{AB} = (g^{x_A})^{x_B} = (g^{x_B})^{x_A} = g^{x_A x_B} \bmod p$.

To make the Diffie-Hellman algorithm viable on sensor platforms, *TinyPK* chooses the base of exponentiation operation as $g = 2$ [29].

15.4.3 Elliptic Curve Cryptography

Recently, *elliptic curve cryptography* (ECC) [30,31] has become a very hot topic in academia and industry, and is seen as the basis for the next-generation security infrastructure. The reason is that algorithms based on ECC are more efficient than RSA and Diffie-Helman at the same security level. The fundamental operation underlying RSA is the modular exponentiation in integer rings. Its security stems from the difficulty in factorizing large integers. Currently, there only exist sub-exponential algorithms to solve the integer factorization problem[2]. ECC operates on groups of points

---

[2] Given a positive integer $n = pq$, where p and q are large pairwise distinct primes, find p and q.

over elliptic curves and derives its security from the hardness of the *Elliptic Curve Discrete Logarithm Problem* (ECDLP)[3]. The best algorithms known for solving ECDLP are exponential. Hence, ECC can achieve the same level of security as RSA with smaller key sizes. For example, 163-bit ECC can provide comparable security to conventional 1024-bit RSA [32]. Under the same security level, the smaller key sizes of ECC offer merits of faster computational efficiency, as well as memory, energy, and bandwidth savings; thus, ECC is better suited for resource-constrained devices.

In Malan et al. [32], Diffie-Hellman over ECC is suggested to achieve key agreement between sensor nodes. An elliptic curve E with a generator G is chosen as a global public parameter. Node A chooses a secret integer x_A and sends $x_A G$ to node B, and B as well chooses a secret integer x_B and sends $x_B G$ to A. The shared key between A and B can be calculated as $K_{AB} = x_B(x_A G) = x_A(x_B G) = x_A x_B G$.

Huang et al. [33] consider a sensor network consisting of some secure managers and many sensor nodes. An ECC-based authenticated key establishment protocol is proposed for the key establishment between secure managers and sensor nodes. To reduce the computational overhead of sensor nodes, most computationally expensive public key operations are put on the secure manager side.

15.4.4 Efficient Implementations

Computational efficiency is a critical issue in applying public key technology to sensor platforms. Gaubatz et al. [34] showed the feasibility of implementing public key technology with the right selection of algorithms and associated parameters, optimization, and low-power techniques. The conceptual implementations of the Rabin scheme [35] and NtruEncrypt scheme [36] were described as examples. In many cases, high-level programming languages cannot be optimized for specific hardware platforms, and therefore assembly languages are required to further reduce computing time. Gura et al. [37] evaluated the assembly language implementations of ECC and RSA on the Atmel ATmega128 processor [38], which is popular for sensor platforms such as Crossbow MICA Motes [6]. In their implementation, a 160-bit point multiplication of ECC requires only 0.81 seconds, while 1024-bit RSA public key operation and private key operation require 0.43 and 10.99 seconds, respectively.

---

[3] Given a generator G of a finite cyclic point group $\mathbb{G}$ over an elliptic curve $E(\mathbb{F}_q)$ and a point Q in the group, find an element $x \in \mathbb{F}_q$ such that $xG = Q$.

15.4.5 Authentication of Public Keys

Another critical issue in applying public key technology is the authenticity of public keys. A public key should be really owned by the node that claims to have the public key. Otherwise, adversaries can easily impersonate any node by claiming its public key and launch a *man-in-the-middle* attack. For example, a malicious node C can impersonate node B to node A and also impersonate A to B if A and B cannot verify the public key of each other. In this way, node C can act as an invisible router and learn all the messages between A and B. The conventional solution to public key authentication is to use a certificate signed by a trustful *certificate authority* (CA). Therefore, node B can send its public key with corresponding certificate to node A such that A can verify the correctness of the certificate with the well-known public key of CA. Node B can verify the authenticity of A's public key by following the same procedure.

Although technical advances have made the usage of public key technology viable in WSNs, public key algorithms are still more expensive than symmetric key algorithms. The authentication of public keys can incur high energy consumption because it is likely to be performed many times. Du et al. [39] developed public key authentication scheme based on a symmetric key technique, the *Merkle tree* [40]. In the Merkle tree, each parent is a hash of the concatenation of its children, and each leaf is corresponding to a node and is calculated as a hash of the node ID and its public key. When a sensor wants to authenticate its public key, it attaches its public key with the siblings of the tree nodes along the path from the leaf of the sensor to the root. Other sensors verify whether they can recover the root and decide the authenticity of the public key.

15.4.6 Location-Based Keys

Based on identity-based cryptography [41], where the publicly known identity information of a node is used as its public key, Zhang et al. [42,43] proposed the notion of *location-based keys* by binding the private keys of individual nodes to both their IDs and locations.

The following parameters are chosen and preloaded to each node: a q-order cyclic group G_1 of points on an elliptic curve; a pairing over the elliptic curve $e : G_1 \times G_1 \to G_2$ that satisfies the bilinear property, that is,

$$e(P + Q, R + S) = e(P, R)e(P, S)e(Q, R)e(Q, S) \qquad (15.14)$$

a hash function maps a bit string to a point in G_1 (i.e., $H_1 : \{0, 1\}^* \to G_1$); and another hash function maps a bit string to an integer, $H_2 : \{0, 1\}^* \to \mathbb{Z}_q^*$. For a node A, an identity-based key (IBK), $IK_A = kH_1(A)$, is preloaded to A, where k is a networkwide secret parameter. After deployment, the position of A is used to derive a location-based key (LBK), $LK_A = kH_1(pos_A)$. The LBK is encrypted by the IBK and securely transmitted to A.

When node A needs to communicate with node B, A first sends an authenticate request, including its position pos_A and a nonce n_A to B. If node A is in the transmission range of node B, B returns a reply including its own location pos_B, a random nonce n_B, and an authenticator V_B calculated as:

$$V_B = H_2(e(LK_B, H1(pos_A)) \parallel n_A \parallel n_B \parallel 0) \qquad (15.15)$$

If node A finds that node B is in its transmission range, A proceeds to compute a verifier V'_B as:

$$V'_B = H_2(e(H_1(pos_B), LK_A) \parallel n_A \parallel n_B \parallel 0) \qquad (15.16)$$

If and only if both A and B have the authentic LBKs corresponding to their claimed locations, they can have:

$$e(LK_B, H_1(pos_A)) = e(H_1(pos_B), LK_A)$$
$$= (e(H_1(pos_B), H_1(pos_A)))^k \qquad (15.17)$$

After verifying the equality of V'_B and V_B, A can ascertain that B is an authentic neighbor with the claimed location pos_B. Node A, in return, should send to B its own authenticator V_A computed as:

$$V_A = H_2(e(H_1(pos_B), LK_A) \parallel n_A \parallel n_B \parallel 1) \qquad (15.18)$$

Then, node B can determine whether A is an authentic neighbor with the claimed location pos_A. Based on this three-way handshaking, nodes A and B can achieve mutual authentication and establish an authentic link between them. Meanwhile, a pairwise key between A and B is established as $(e(H_1(pos_B), H_1(pos_A)))^k$.

The location-based keys have perfect resilience to node compromise in that no matter how many nodes are compromised, the location-based keys of noncompromised nodes as well as their pairwise keys always remain secure. It has been shown in Zhang et al. [42,43] that the solution can defend against a wide range of attacks, such as the Wormhole attack, the Sybil attack, and the node replication attack, in sensor networks.

15.5 Open Issues

In this section we discuss some problems that need to be fully studied.

15.5.1 Memory Cost

High security and lower overhead are two objectives that a key management protocol needs to achieve. Although there have been several proposals for key establishment in sensor networks, they can hardly address these two requirements, as discussed in this chapter. Strong security protocols usually require large amounts of memory cost, as well as high-speed processors and large power consumption. However, they cannot be easily supported due to the constraints on hardware resources of the sensor platform.

It is well known that in wireless environments, transmission of one bit can consume more energy than computing one bit. Therefore, communication overhead can dominate the entire power consumption. In key management protocols, direct key establishment does not require communication or only a few rounds of one-hop communications, but indirect key establishment is performed over multi-hop communications. To reduce the multi-hop communication overhead, high secure connectivity, which is the probability of direct key establishment between a pair of nodes, is desirable. However, highly secure connectivity requires more key materials in each node, which is usually impratical, especially when the network size is large.

Considering the above two issues, memory cost can be the major bottleneck in designing key management protocols. How to reduce memory cost while still maintaining a certain level of security and overhead is a very important issue.

15.5.2 End-to-End Security

The major merit of symmetric key technology is its computational efficiency. However, most current symmetric key schemes for WSNs aim at the link layer security — not the transport layer security — because it is impractical for each node to store a transport layer key for each of the other nodes in a network due to the huge number of nodes.

However, end-to-end communication at the transport layer is very common in many WSN applications. For example, to reduce unnecessary traffic, a fusion node can aggregate reports from many source nodes and forward a final report to the sink node. During this procedure, the reports between source nodes and the fusion node and the one between the fusion node and the sink node should be secured. In hostile environments, however, any node can be compromised and become malicious. If one of the intermediate nodes along a route is compromised, the message delivered along the route can be exposed or modified by the compromised node. Employing end-to-end security can effectively prevent message tampering by any malicious intermediate node.

Compared with symmetric key technology, public key technology is expensive but has flexible manageability and supports end-to-end security. A more promising approach to key establishment in WSNs is to combine the merits of both symmetric key and public key techniques, in that each node is equipped with a public key system and relies on it to establish end-to-end symmetric keys with other nodes. To achieve this goal, a critical issue is to develop more efficient public key algorithms and their implementations so that they can be widely used on sensor platforms.

How to prove the authenticity of public keys is another important problem. Otherwise, a malicious node can impersonate any other normal node by claiming its public key. Identity-based cryptography is a shortcut to avoid the problem. Currently, most identity-based cryptographical algorithms operate on elliptic curve fields, and pairing over elliptic curves is widely used in the establishment of identity-based symmetric keys. However, the pairing operation is very costly, comparable to or even more expensive than RSA. Therefore, fast algorithms and implementations are the major tasks of researchers.

15.5.3   Efficient Symmetric Key Algorithms

There is still a demand for the development of more efficient symmetric key algorithms because encryption and authentication based on symmetric keys are very frequent in the security operations of sensor nodes. For example, in the link layer security protocol TinySec [44], each packet must be authenticated, and encryption can also be triggered if critical packets are transmitted. Therefore, fast and cost-efficient symmetric key algorithms should be developed.

15.5.4   Key Update and Revocation

Once a key has been established between two nodes, the key can act as a master key and be used to derive different sub-keys for many purposes (e.g., encryption and authentication). If each key is used for a long time, it may be exposed due to cryptanalysis over the ciphertexts intercepted by adversaries. To protect the master key and those sub-keys from cryptanalysis, it is wise to update keys periodically. The period of update, however, is difficult to choose. Because the cryptanalysis capability of adversaries is unknown, it is very difficult to estimate how long it takes for adversaries to expose a key by cryptanalysis. If the key update period is too long, the corresponding key may also be exposed. If it is too short, frequent updates can incur large overhead.

A related problem is key revocation. If one node is detected to be malicious, its key must be revoked. However, key revocation has not been thoroughly investigated. Although Chan et al. [45] proposed a distributed revocation protocol, it is only based on the random-pairwise key scheme [14], and cannot easily be generalized into other key establishment protocols.

15.5.5 Node Compromise

Node compromise is the most detrimental attack on sensor networks. Because compromised nodes have all the authentic key materials, they can result in very severe damage to WSN applications and cannot be detected easily. How to counteract node compromise remains under investigation.

Most current security protocols try to defend against node compromise through careful protocol design such that the impact of node compromise can be restricted to a small area. However, a hardware approach is more promising. With advances in hardware design and manufacturing techniques, much stronger, tamper-resistant, and cheaper devices can be installed on the sensor platform to counteract node compromise.

15.6   Conclusion

Key management is the most critical component in the design of security protocols for WSNs, and has been drawing intensive interest from both academia and industry. In this chapter we surveyed current solutions to the key management issue in WSNs and shed light on future directions of the issue in WSNs. There are many challenges in the design of key management schemes due to various resource limitations and salient features of WSNs. Secure and efficient key management schemes are still under exploration.

References

[1]   I.F. Akyildiz, W. Su, Y. Sankarasubramaniam, and E. Cayirci, "A survey on sensor networks," *IEEE Communication Magazine*, Vol. 40, No. 8, pp. 102–114, August 2002.

[2]   C.E. Shannon, "Communication theory of secrecy systems," *Bell Sys. Tech. J.*, Vol. 28, pp. 656–715, October 1949.

[3]   W. Diffie and M.E. Hellman, "New directions in cryptography," *IEEE Transactions on Information Theory*, Vol. IT-22, No. 6, pp. 644–654, 1976.

[4]   R.L. Rivest, A. Shamir, and L. Adleman, "A method for obtaining digital signatures and public-key cryptosystems," *Communications of the ACM*, Vol. 21, No. 2, pp. 120–126, February 1978.

[5]   A. Wood and J. Stankovic, "Denial of service in sensor networks," *IEEE Computer*, pp. 54–62, October 2002.

[6]   Crossbow Technology, http://www.xbow.com/

[7]   R. Anderson and M. Kuhn, "Tamper resistance — a cautionary note," *Proc. 2nd USENIX Workshop on Electronic Commerce*, Oakland, CA, November 18–21, 1996, pp. 1–11.

[8]   R. Blom, "An optimal class of symmetric key generation systems," in *Proc. of EUROCRYPT '84*, 1985, pp. 335–338.

[9]   C. Blundo, A. De Santis, A. Herzberg, S. Kutten, U. Vaccaro, and M. Yung, "Perfectly-secure key distribution for dynamic conferences," in *Advances in Cryptology C CRYPTO 92, LNCS 740*, 1992, pp. 471–486.

[10] S. Basagni, K. Herrin, D. Bruschi, and E. Rosti, "Secure pebblenets," *ACM Mobihoc'01*, Long Beach, CA, 2001.

[11] A. Perrig, R. Szewczyk, J.D. Tygar, V. Wen, and D.E. Culler, "SPINS: security protocols for sensor networks," *Wireless Networks*, Vol. 8, pp. 521–534, 2002.

[12] F.J. MacWilliams and N.J.A. Sloane, *The Theory of Error Correction Codes*, North-Holland, New York, 1977.

[13] L. Eschenauer and V. Gligor, "A key management scheme for distributed sensor networks," in *ACM CCS2002*, Washington, D.C., 2002.

[14] H. Chan, A. Perrig, and D. Song, "Random key predistribution schemes for sensor networks," in *Proceedings of the 2003 IEEE Symposium on Security and Privacy*, May 11–14, 2003, p. 197.

[15] R.D. Pietro, L.V. Mancini, and A. Mei, "Random key-assignment for secure wireless sensor networks," in *Conference on Computer and Communications Security (CCS'03)*, 2003.

[16] W. Du, J. Deng, Y.S. Han, and P.K. Varshney, "A pairwise key pre-distribution scheme for wireless sensor networks," in *CCS'03*, Washington, D.C., October 27–30, 2003.

[17] D. Liu and P. Ning, "Establishing pairwise keys in distributed sensor networks," *CCS'03*, Washington, D.C., 2003.

[18] J. Hwang and Y. Kim, "Revisiting random key pre-distribution schemes for wireless sensor networks," in *Proceedings of the 2nd ACM Workshop on Security of Ad Hoc and Sensor Networks (SASN'04)*, October 25, 2004, Washington, D.C.

[19] H. Chan and A. Perrig, "Pike: peer intermediaries for key establishment in sensor networks," in *IEEE INFOCOM'05*, March 2005.

[20] Y. Zhou and Y. Fang, "A scalable key agreement scheme for large scale networks," *2006 IEEE International Conference on Networking, Sensing and Control (ICNSC'06)*, Fort Lauderdale, FL, April 23–25, 2006.

[21] D. Liu, and P. Ning, "Location-based pairwise key establishments for relatively static sensor networks," in *ACM Workshop on Security of Ad Hoc and Sensor Networks (SASN'03)*, October 2003.

[22] W. Du, J. Deng, Y.S. Han, S. Chen, and P.K. Varshney, "A key management scheme for wireless sensor networks using deployment knowledge," in *IEEE INFOCOM 2004*, Hong Kong, March 2004.

[23] D. Huang, M. Mehta, D. Medhi, and L. Harn, "Location-aware key management scheme for wireless sensor networks," in *Proceedings of the 2nd ACM*

Workshop on Security of Ad Hoc and Sensor Networks (SASN'04), October 25, 2004, Washington, D.C.

[24] Y. Zhou, Y. Zhang, and Y. Fang, "LLK: a link-layer key establishment scheme in wireless sensor networks," *IEEE WCNC'05*, New Orleans, LA, March 2005.

[25] Y. Zhou, Y. Zhang, and Y. Fang, "Key establishment in sensor networks based on triangle grid deployment model," in *Proc. IEEE MILCOM'05*, Atlantic City, NJ, October 17–20, 2005.

[26] D. Liu, P. Ning, and W. Du, "Group-based key pre-distribution in wireless sensor networks," *ACM WiSe'05*, September 2005.

[27] S. Zhu, S. Setia, and S. Jajodia, "LEAP: efficient security mechanism for large-scale distributed sensor networks," in *ACM CCS'03*, Washington, D.C., October 27–31, 2003.

[28] R. Anderson, H. Chan, and A. Perrig, "Key infection: smart trust for smart dust," in *Proceedings of the 12th IEEE International Conference on Network Protocols (ICNP'04)*, 2004.

[29] R. Watro, D. Kong, S. Cuti, C. Gardiner, C. Lynn, and P. Kruus, "TinyPK: securing sensor networks with public key technology," *SASN'04*, Washington, D.C., October 25, 2004.

[30] N. Koblitz, "Elliptic curve cryptosystems," *Mathematics of Computation*, Vol. 48, pp. 203–209, 1987.

[31] V. Miller, "Uses of elliptic curves in cryptography," *Lecture Notes in Computer Science 218: Advances in Cryptology — CRYPTO'85*. Springer-Verlag, Berlin, 1986, pp. 417–426.

[32] D.J. Malan, M. Welsh, and M.D. Smith, "A public-key infrastructure for key distribution in TinyOS based on elliptic curve cryptography," *First IEEE International Conference on Sensor and Ad Hoc Communications and Networks* (SECON'04), Santa Clara, CA, October 2004.

[33] Q. Huang, J. Cukier, H. Kobayashi, B. Liu, and J. Zhang, "Fast authenticated key establishment protocols for self-organizing sensor networks," *ACM WSNA'03*, San Diego, CA, 2003.

[34] G. Gaubatz, J. Kaps, and B. Sunar, "Public key cryptography in sensor networks — revisited," *ESAS'04*, 2004.

[35] A. Menezes, P. van Oorschot, and S. Vanstone, *Handbook of Applied Cryptography*, CRC Press, Boca Raton, FL, 1996.

[36] J. Hoffstein, J. Pipher, and J.H. Silverman, "NTRU: a ring based public key cryptosystem," *Lect. Notes in Comp. Sci.*, Springer-Verlag, Berlin, Vol. 1433, pp. 267–288, 1998.

[37] N. Gura, A. Patel, A. Wander, H. Eberle, and S.C. Shantz, "Comparing elliptic curve cryptography and RSA on 8-bit CPUs," *CHES'04*, 2004.

[38] Atmel Corporation, *http://www.atmel.com/*

[39] W. Du, R. Wang, and P. Ning, "An efficient scheme for authenticating public keys in sensor networks," *ACM MobiHoc'05*, May 2005.

[40] R. Merkle, "Protocols for public key cryptosystem," *Proceedings of the IEEE Symposium on Research in Security and Privacy*, April 1980.

[41] D. Boneh and M. Franklin, "Identity-based encryption from the weil pairing," in *Proc. CRYPTO'01*, Ser. LNCS, Vol. 2139, Springer-Verlag, 2001, pp. 213–229.

[42] Y. Zhang, W. Liu, W. Lou, and Y. Fang, "Securing sensor networks with location-based keys," *IEEE WCNC'05*, March 2005.

[43] Y. Zhang, W. Liu, W. Lou, and Y. Fang, "Location-based compromise-tolerant security mechanisms for wireless sensor networks," *IEEE Journal on Selected Areas in Communications, Special Issue on Security in Wireless Ad-Hoc Networks*, Vol. 24, No. 2, pp. 247–260, 2006.

[44] C. Karlof, N. Sastry, and D. Wagner, "TinySec: a link layer security architecture for wireless sensor networks," in the *Second ACM Conference on Embedded Networked Sensor Systems (SensSys'04)*, Baltimore, MD, November 2004.

[45] H. Chan, V. Gligor, A. Perrig, and G. Muralidharan, "On the distribution and revocation of cryptographic keys in sensor networks," *IEEE Transactions on Dependable and Secure Computing*, Vol. 2, No. 3, July–Sept. 2005.

Chapter 16

Secure Routing for Mobile Ad Hoc Networks

Joo-Han Song, Vincent W.S. Wong,
Victor C.M. Leung, and Yoji Kawamoto

Contents

16.1   Introduction

Most of the current ad hoc routing protocols [1,2] proposed for mobile ad hoc networks (MANETs) assume that there is an implicit trust-your-neighbor relationship in which all the neighboring nodes behave properly. However, in practice, many MANETs are subject to attacks by rogue users who try to paralyze the network by manipulating the messages (e.g., dropping all data or control packets, sending incorrect route advertisement messages). This problem is further complicated by a lack of centralized management control, error-prone multi-hop wireless channels, and the dynamic changes in network topology due to node mobility.

A number of secure ad hoc routing protocols have been proposed [3–8] with the aim of preventing various attacks. We now summarize several of these protocols.

The SAODV [3] is a secure extension to the original Ad hoc On-demand Distance Vector (AODV) [1] routing protocol. SAODV uses the digital signature to authenticate nonmutable fields and uses the hash chains to authenticate the hop-count field in both RREQ (Route REQuest) and RREP (Route REPly) messages. During the route discovery process, the source node sets a maximum hop-count (MHC) to the time-to-live (TTL) value in the Internet Protocol (IP) header, and generates a one-way hash chain. During the route maintenance process, SAODV uses the digital signature to protect the Route ERRor (RERR) message. Both originating and forwarding nodes of the RERR sign the whole message, and thus its neighboring nodes can verify the signature of its previous forwarding node. However, because SAODV does not have a mechanism for authenticating intermediate nodes, malicious attackers can easily join a path and launch various malicious attacks.

The Ariadne [4] is a secure routing extension for an on-demand source routing protocol. It authenticates routing messages using a shared secret key and a broadcast authentication protocol. The source node includes an HMAC (Hash-based Message Authentication Code) computed over non-mutable fields in an RREQ. To ensure that each intermediate node cannot remove the existing nodes from or add extra nodes to the node list in the RREQ, each intermediate node authenticates new information in the RREQ by appending an HMAC of the entire RREQ. When the destination determines that the RREQ is valid, it returns an RREP by appending an

HMAC computed over nonmutable fields in the RREP. Each intermediate node along the source route appends its TESLA key in the RREP. When the source receives the RREP, it verifies that the end-to-end HMAC and hop-by-hop HMACs are valid. If all these tests give positive results, the source will accept the RREP.

The Authenticated Routing for Ad hoc Networks (ARAN) proposed in Sanzgiri et al. [5] is a secure routing protocol for AODV. It prevents message tampering attacks by appending the time stamp, certificate, and digital signature in the routing messages. A preliminary certification process is assumed. Route discovery in ARAN is accomplished by a broadcast route discovery message from the source node. The destination node sends a reply to the source node by unicast. The routing message is authenticated at each hop by the previous node's certificate and signature from source to destination. Using an unalterable physical metric such as time delay, ARAN avoids attacks against the hop-count field in routing messages. The main limitation of ARAN is that each node must verify multiple signatures for both RREQ and RREP messages. The use of multiple digital signatures on networkwide broadcast messages can be expensive.

SRP (Secure Routing Protocol) [6] is a secure source routing protocol that provides correct routing and authentic connectivity information between end-to-end nodes. SRP requires a symmetric security association between the communicating nodes. It attempts to guarantee that the node initiating the route discovery can detect the replies that provide false topological information and can discard these malevolent replies. HMAC is calculated using the shared key between the source and destination and two identifiers. Not only can it validate the integrity of RREQ messages, but it can also authenticate the origin of the packet to the destination. However, this is realized through the existence of a security association between source and destination without the intermediate nodes having to cryptographically validate the control traffic. To limit flooding, each node records the rate at which each neighbor forwards an RREQ and gives a high priority to an RREQ sent through neighbors that less frequently forward RREQs. Because SRP does not provide hop-by-hop authentication, a malicious user can join a path and modify the contents of routing messages.

In the Secure Efficient Ad hoc Distance vector routing protocol (SEAD) proposed in Hu et al. [7], the receiver of the routing update authenticates the sender. Computationally efficient one-way hash functions are used to secure the routing update messages. As it is impossible to invert a one-way hash function, intermediate nodes can only increase the metric in the routing update but cannot decrease it. Therefore, SEAD can authenticate the lower bound on the hop-count metric in each update message. However, SEAD needs either a shared secret key among each pair of nodes or a broadcast authentication mechanism with synchronized clock to authenticate the source of each routing update message.

SLSP (Secure Link-State Protocol) [8] is a secure proactive link-state routing protocol. Link-state information is managed using both the Neighbor Lookup Protocol (NLP) and Link-State Update (LSU) messages. NLP's HELLO and LSU messages are signed by the sender's private key. Thus, all receivers can verify those messages using the sender's public key. A hash chain is used to authenticate the hop-count field of LSU messages, similar to SAODV.

Although various secure routing protocols have been proposed to prevent different attacks (e.g., message tampering, message dropping, message replay), the possible threats of routing table tampering attacks [9, 10] have not been resolved in MANETs. Routing table tampering attacks can be launched by a compromised user who is authenticated by the network and trusted by other users, but is behaving maliciously. These attacks include physical deletion, alteration, or falsification of information stored in the routing tables in a node.

In AODV, the fields in each routing table entry include destination IP address, destination sequence number, hop-count (number of hops needed to reach destination), next hop address, and lifetime (expiration or deletion time of the route) [1]. A table tampering attacker can manipulate the route update mechanism by modifying the fields in the corresponding route entry. For example (see Figure 16.1), when the table tampering attacker in node *C* receives an RREQ from its neighbor *B*, node *C* can redirect traffic to itself by unicasting to node *B* an RREP that contains a higher destination sequence number. Node *C* will then become one of the intermediate nodes of the path. Moreover, in node *C*'s routing table module, the attacker can alter the next hop address *D* of the corresponding destination entry *F* to an unreachable or nonexisting address *U*. As a result, data packets passing through node *C* will never reach their intended destination *F*. This attack can also cause an increase in control traffic via the transmission of RERRs. In node *C*, routing loops can be created by replacing the

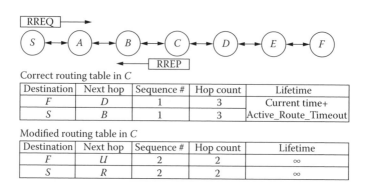

Correct routing table in *C*

Destination	Next hop	Sequence #	Hop count	Lifetime
F	*D*	1	3	Current time+
S	*B*	1	3	Active_Route_Timeout

Modified routing table in *C*

Destination	Next hop	Sequence #	Hop count	Lifetime
F	*U*	2	2	∞
S	*R*	2	2	∞

Figure 16.1   Correct and modified routing tables in AODV.

next hop address *B* to destination *S* either by node *D* or by another rogue node *R*.

Because an attacker can allow its mobile node to be selected as the intermediate node of a chosen path, other possible ways for the attacker to exploit the table tampering attack include discarding rather than forwarding all data packets, thereby creating a permanent denial-of-service (DoS) attack, or selectively discarding data packets. The devastating impacts of table tampering attacks on the performance of AODV have been investigated in [9,10] based on simulations. For example, a single attacker can drop up to 75 percent of the packets by manipulating destination sequence numbers in some scenarios [10].

The table tampering attack cannot be avoided by simply checking the authentication and integrity of the control packets because the modification is performed in the routing table's entry. Although the table tampering attack results in message tampering, we distinguish this attack from other message tampering attacks by malicious users, such as impersonating other nodes or modifying one or several fields in control packets.

The objectives of this chapter are to provide a security mechanism to prevent routing table tampering attacks, and to secure both control and data messages in MANETs. We focus on AODV as the basis of the design of our security mechanisms. The main contributions of this chapter are as follows:

1. We propose the Secure Table Entry Protection (STEP) scheme to detect routing table tampering attacks from compromised users. STEP provides authentication for both the destination sequence number and hop-count fields in the routing table entry. The receiving node can confirm the correctness of messages by verifying the signatures of two consecutive upstream nodes and the source node that creates the packet.

2. We propose SeAODV, a secure routing extension to the original AODV protocol. SeAODV aims to prevent attacks by malicious users. It uses the double signature for both RREQ and RREP packets to prevent routing message tampering attacks. We also provide the modifications of RREQ and RERR packets to prevent data dropping attacks.

3. We propose a Secure Data Forwarding (SDF) scheme based on SeAODV for secure transmissions of data packets over wireless links. SDF uses the keyed Hash-based Message Authentication Code (HMAC) [11] to maintain the integrity of data messages.

4. We conduct simulation experiments to determine the performance between STEP, SeAODV with SDF, and the original AODV.

The remainder of this chapter is organized as follows. The rationale for the assumptions made in our framework are explained in Section 16.2. Our proposed STEP mechanism is described in Section 16.3. Our proposed

SeAODV protocol with SDF mechanism is described in Section 16.4. Performance comparisons between STEP, SeAODV with SDF, and AODV through simulations are presented in Section 16.5. Conclusions are given in Section 16.6.

16.2   Network Environments

The secure protocols proposed in this chapter aim to prevent and detect routing table tampering attacks, message tampering attacks, and message dropping attacks in the network layer. Attacks in other layers (e.g., physical, transport, application) are beyond the scope of this chapter. Our proposed schemes work under several assumptions:

1. The network links are bi-directional. That is, if node A is able to transmit to node B, then node B is also able to transmit to node A.
2. There exists a public key infrastructure in the MANET. Each mobile node stores the trusted certification authority's (CA) public key.
3. Elliptic curve cryptography (ECC) [17] is used to generate the digital signature.

We now explain the rationale behind these assumptions. The first assumption is common in practice. Many wireless medium access control protocols require bi-directional links to exchange several link-layer frames between a source and destination to avoid collisions.

In the second assumption, the CA's public key is used to authenticate the public keys of other nodes. Certificate-based public key distribution is feasible in MANETs. The key distribution can be performed either through a noncryptographic approach [14] or some other distributed key management schemes [6,15]. There are also several offline methods to exchange the secret information [16]. These mechanisms allow each node to obtain the public keys from other nodes in the network.

The third assumption aims to reduce the overhead in terms of delay and bandwidth. The use of ECC reduces the time to generate and verify the signature when compared with the RSA (Rivest, Shamir, Adleman) algorithm. For example, for a 1024-bit RSA key, the delay for signature generation and verification is about 60 milliseconds. On the other hand, for an ECC 160-bit key, the delay is only 6 milliseconds [18].

16.3   Secure Routing Table in AODV

In this section, we describe a security mechanism that protects the routing table in AODV against routing table tampering attacks. Our proposed scheme can also be used to prevent message tampering attacks. We use the following notations in this chapter:

1. $Sign_A(M)$ denotes that a message M is digitally signed by the private key of node A.
2. $<M_1||M_2>$ denotes that message M_1 is concatenated with message M_2.

16.3.1 Secure Table Entry Protection (STEP)

Recall that AODV is vulnerable to routing table tampering attacks because node B cannot verify the correctness of routing control message received from its neighboring node A. That is, node B cannot detect whether node A has maliciously changed either the destination sequence number or the hop-count field in a control message. To address this problem, node A must prove the correctness of message to its downstream neighbor B.

Our proposed STEP mechanism provides authentication for both the *destination sequence number* and *hop-count* fields in the routing control messages. The receiving node can confirm the correctness of control messages by verifying the digital signatures of the two consecutive upstream nodes and the source node that creates the packet.

As shown in Figure 16.2, whenever node S sends a message to its neighbor, it needs to compute the digital signature for the *sequence number* (sn_S), *hop-count* (hc_{SS}), and *timestamp* (ts_S) fields. The digital signature is then appended at the end of the message. Therefore, node S sends a message $M_S = <MESSAGE, ts_S, Sign_S(sn_S||hc_{SS}||ts_S)>$, where MESSAGE denotes either RREQ or RREP, and $hc_{SS} = 0$. Note that although a control message has several different fields (e.g., source and destination addresses, sequence number, hop-count), for simplicity only the fields that are relevant to STEP are shown in Figure 16.2.

When one of its neighbors, node A, receives this message, it verifies the digital signature and forwards the modified message M_A, which is signed by node A, to its neighbor B. The value of hc_{AS} (or hc_{BS}) denotes the hop distance from node A (or B) to S [i.e., 1 (or 2)]. In this way, every node except the destination node attaches two additional fields [i.e., *digital signature* (40 bytes) [17] and *timestamp* (4 bytes)] in each packet, and forwards it to its neighbor.

Each intermediate node stores the digital signature [i.e., $Sign_S(sn_S||hc_{SS}||ts_S)$] of *destination sequence number*, *hop-count*, and *timestamp* of

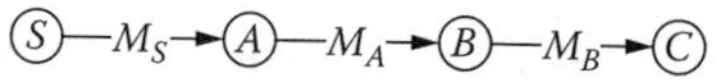

$$\text{Ⓢ}—M_S→\text{Ⓐ}—M_A→\text{Ⓑ}—M_B→\text{Ⓒ}$$

$M_S = <\text{Message}, ts_S, Sign_S(sn_S|hc_{SS}|ts_S)>$
$M_A = <\text{Message}, ts_S, Sign_S(sn_S|hc_{SS}|ts_S), ts_A, Sign_A(hc_{AS}|ts_A)>$
$M_B = <\text{Message}, ts_S, Sign_S(sn_S|hc_{SS}|ts_S), ts_A, Sign_A(hc_{AS}|ts_A), ts_B, Sign_B(hc_{BS}|ts_B)>$

Figure 16.2 Example for STEP with digital signature.

source node S in its routing table. Thus, it can prove to its neighbors that it has the valid destination sequence number upon S by attaching the digital signature generated by S. Any malicious change in the destination sequence number on S will be detected by the attacker's one-hop downstream node by verifying this signature.

Moreover, because each node's signature about hop-count field is forwarded up to two-hop downstream nodes, the receiving node can verify the correctness of the hop-count field. For example, if the previous node has not increased the value in the hop-count field by one, the signature of the two-hop upstream neighbor cannot be verified correctly by substituting (*hop-count*−1) for the value of hop-count field. If there are no colluding nodes in the network, then the two digital signatures from the two-hop upstream nodes are sufficient to verify the correctness of the hop-count field in the received message. The scenario with colluding attackers is discussed in Section 16.3.4.

16.3.2   *Route Discovery with STEP*

We now present the secure AODV route discovery scheme based on STEP. The intuition behind secure route discovery is to make both destination sequence number and hop-count fields verifiable. Thus, routing table tampering attacks can be detected in the network.

Before node S sends an RREQ to node D, it needs to compute the digital signature for the sequence number, hop-count, and timestamp fields, and then append those fields to the message (see Figure 16.3). When an intermediate node A receives this message, it needs to verify the digital signature. If the digital signature is valid, it will update the reverse path to

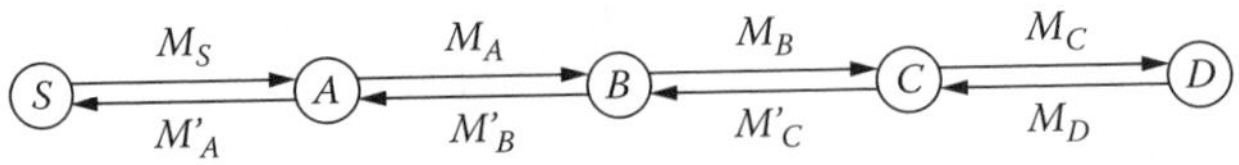

$M_S = \langle \mathrm{RREQ\_}, ts_S, Sign_S(sn_S\|hc_{SS}\|ts_S)\rangle$

$M_A = \langle \mathrm{RREQ\_}, ts_S, Sign_S(sn_S\|hc_{SS}\|ts_S), ts_A, Sign_A(hc_{AS}\|ts_A)\rangle$

$M_B = \langle \mathrm{RREQ\_}, ts_S, Sign_S(sn_S\|hc_{SS}\|ts_S), ts_A, Sign_A(hc_{AS}\|ts_A), ts_B, Sign_B(hc_{BS}\|ts_B)\rangle$

$M_C = \langle \mathrm{RREQ\_}, ts_S, Sign_S(sn_S\|hc_{SS}\|ts_S), ts_B, Sign_B(hc_{BS}\|ts_B), ts_C, Sign_C(hc_{CS}\|ts_C)\rangle$

$M_D = \langle \mathrm{RREP\_}, ts_D, Sign_D(sn_D\|hc_{DD}\|ts_D)\rangle$

$M'_C = \langle \mathrm{RREP\_}, ts_D, Sign_D(sn_D\|hc_{DD}\|ts_D), ts'_C, Sign_C(hc_{CD}\|ts'_C)\rangle$

$M'_B = \langle \mathrm{RREP\_}, ts_D, Sign_D(sn_D\|hc_{DD}\|ts_D), ts'_C, Sign_C(hc_{CD}\|ts'_C), ts'_B, Sign_B(hc_{BD}\|ts'_B)\rangle$

$M'_A = \langle \mathrm{RREP\_}, ts_D, Sign_D(sn_D\|hc_{DD}\|ts_D), ts'_B, Sign_B(hc_{BD}\|ts'_B), ts'_A, Sign_A(hc_{AD}\|ts'_A)\rangle$

Figure 16.3   Example for route discovery (RREP from node D) where RREQ and RREP denote the original AODV's RREQ and RREP, respectively. Note that original AODV messages have the fields of the destination sequence number and hop-count.

node *S* and broadcast this packet with its signature and timestamp to its neighbors again. Eventually, an RREQ will reach destination *D*. Node *D* can then verify the digital signatures of *S*, *B*, and *C* (see Figure 16.3). If this message is valid, node *D* will store both the routing information with ID and its digital signature in its routing table.

When the RREQ arrives at the destination or to an intermediate node (which has an entry in its cache), an RREP is being sent. Depending on whether the node is the destination *D* or an intermediate node (which has a fresh route to node *D*), the processing is slightly different. These two scenarios are described in the following subsections.

16.3.2.1 RREP from the Destination

Figure 16.3 shows an example of secure route discovery. In this example, only destination *D* sends an RREP to the corresponding RREQ. Once created, the RREP is sent to the next hop toward the originator of the RREQ according to its local routing table entry for that originator. When the source *S* or intermediate nodes *A*, *B*, and *C* receive an RREP, they will verify the digital signatures as described in Section 16.3.1. If the message is valid, they will store those signatures and update their forward route to the destination *D* using the neighbor from which they receive the RREP.

16.3.2.2 RREP from the Intermediate Node

Consider the example in Figure 16.4 where the RREP packet is generated by an intermediate node *B*. In the RREP packet, node *B* includes the destination sequence number and the distance from the destination *D*. In addition, node *B* attaches messages (which are signed by both node *D* and its upstream node *C*) in the RREP packet. Node *E* can confirm that the received RREP from *B* is correct, up-to-date, and has not been modified by table tampering attackers.

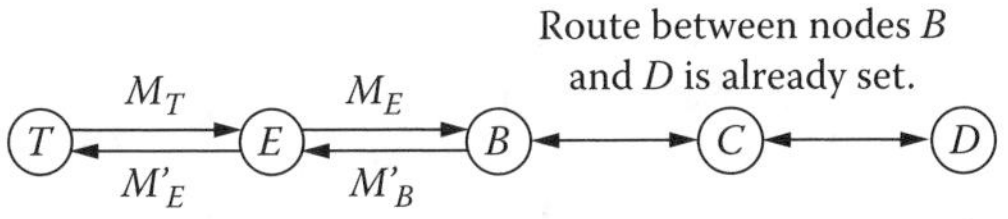

$M_T = \langle \text{RREQ}\_, ts_T, Sign_T(sn_T | hc_{TT} | ts_T) \rangle$

$M_E = \langle \text{RREQ}\_, ts_T, Sign_T(sn_T | hc_{TT} | ts_T), ts_E, Sign_E(hc_{ET} | ts_E) \rangle$

$M_B = \langle \text{RREP}\_, ts_D, Sign_D(sn_D | hc_{DD} | ts_D), ts'_C, Sign_C(hc_{CD} | ts'_C), ts'_B, Sign_B(hc_{BD} | ts'_B) \rangle$

$M'_E = \langle \text{RREP}\_, ts_D, Sign_D(sn_D | hc_{DD} | ts_D), ts'_B, Sign_B(hc_{BD} | ts'_B), ts'_E, Sign_E(hc_{ED} | ts'_E) \rangle$

Figure 16.4   Example for route discovery (RREP message from node *B*).

16.3.3 Route Discovery with Efficient STEP (ESTEP)

The main limitation of STEP is that each node must verify multiple signatures for both RREQ and RREP packets. Moreover, the use of multiple digital signatures on networkwide broadcast RREQ packets can be expensive. To this end, we propose an *Efficient STEP (ESTEP)* that can avoid the use of expensive multiple digital signatures on an RREQ using a single signature.

When node *S* sends an RREQ to node *D*, it needs to compute the digital signature for *sequence number* and *timestamp*, and then appends this message at the end: $MS = < \text{RREQ}_-, ts_S, Sign_S(sn_S||ts_S) >$. When one of its neighbors, node *A*, receives this message, it can verify the digital signature to check the integrity of the message. If the digital signature is valid, it will update the reverse path to node *S* and broadcast this packet M_A to its neighbor *B* again and so on, where the message $M_A = M_S$. Note that the other fields are updated according to the AODV protocol. Eventually, the RREQ will reach the destination *D*. If this message is valid, node *D* will create an RREP packet. Because the intermediate nodes (i.e., *A*, *B*, *C* in Figure 16.3) cannot verify either the *hop-count* or *next hop address* fields in its routing entry corresponding to node *S*, they are not allowed to generate an RREP.

After receiving a valid RREQ, the destination node *D* creates an RREP. The RREP is sent to the next hop toward the originator of the RREQ. Unlike an RREQ, the RREP must be relayed according to the STEP mechanism described in Section 16.3.1 with multiple signatures. There is one additional field in an RREP: the hop-count field from node *S* to *D*. This field is signed by destination node *D* and verified by source node *S*. If this value matches that of the hop-count field in the RREP, this unidirectional path from source *S* to destination *D* can be trusted as the path without routing table tampering attackers. If it is necessary to have a secure bi-directional path between two end-to-end nodes, the source node may have to send its first data packet according to the STEP mechanism.

16.3.4 Extension against Colluding Attackers

STEP is effective against routing table tampering attacks from noncolluding users (i.e., individual compromised users). However, it cannot prevent colluding attackers (i.e., a set of compromised attackers) from cheating the hop-count field. To prevent *n* serial colluding nodes from decreasing or not increasing a hop-count to node *S* in a message, each node must have $(n+1)$ signatures in the hop-count field, starting from its one-hop neighbor to $(n+1)$ hops upstream nodes toward node *S*. For example, suppose nodes *A* and *B* are colluding routing table tampering attackers (i.e., $n = 2$). When node *A* receives a message from node *S*, it will not increase the hop-count field and forward to its neighbor *B*. Node *B* can increase the hop-count field and relay again to its neighbor node *C*. Because node *C* verifies three (i.e., $n + 1 = 3$) signatures from its three upstream nodes S, A, and B, it can prove that both nodes A and B are tampering the message from S. This signature can also be used for non-repudiation purposes.

16.3.5 Integrating STEP with Secure Routing Protocol

Due to the use of multiple signatures, STEP can introduce significant routing overhead relative to the AODV protocol. It can create a scalability problem and degrade network performance. However, STEP can be invoked only when necessary. For example, when STEP is used together with a secure on-demand routing protocol such as SAODV [3] or our proposed SeAODV, a node can find a route using those routing protocols. If the source node cannot find a valid route due to malicious changes in either destination sequence number or hop-count, it will set a flag indicating that it activates the use of STEP (or ESTEP). In this way, STEP can still be considered efficient in large and dense networks.

16.3.6 Discussion

In this chapter we do not consider the protection of the source and destination addresses explicitly. However, when the source and destination pair can set up the shared secret key using any key distribution schemes [6,14–16], both source and destination addresses can be protected with the efficient symmetric operations such as HMAC [11]. Although both sequence number and timestamp are nonmutable fields, the end-to-end shared symmetric key cannot be used to prevent routing table tampering attacks.

STEP can use the efficient HMAC instead of computationally expensive digital signatures. However, using HMAC may have two limitations. First, the setup of $n(n-1)/2$ secret keys among n nodes can be an expensive operation in a MANET. Second, the HMAC value for both destination sequence number and hop-count fields can only be verified with the same secret key used in calculating this HMAC. Therefore, each sender must include in the packet a separate HMAC for all other nodes, thus making the packet size large. If all nodes can share the same key, the need to include separate HMACs in the packet can be avoided. However, this allows any node to impersonate others with the same key.

16.4 Secure AODV (SeAODV)

In this section we first propose SeAODV, a secure routing extension to the original AODV protocol. SeAODV aims to prevent routing message tampering attacks by malicious users. We then propose a Secure Data Forwarding (SDF) scheme based on SeAODV for secure transmission of data packets

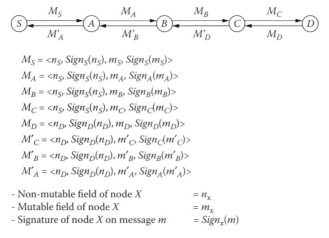

$M_S = <n_S, Sign_S(n_S), m_S, Sign_S(m_S)>$
$M_A = <n_S, Sign_S(n_S), m_A, Sign_A(m_A)>$
$M_B = <n_S, Sign_S(n_S), m_B, Sign_B(m_B)>$
$M_C = <n_S, Sign_S(n_S), m_C, Sign_C(m_C)>$
$M_D = <n_D, Sign_D(n_D), m_D, Sign_D(m_D)>$
$M'_C = <n_D, Sign_D(n_D), m'_C, Sign_C(m'_C)>$
$M'_B = <n_D, Sign_D(n_D), m'_B, Sign_B(m'_B)>$
$M'_A = <n_D, Sign_D(n_D), m'_A, Sign_A(m'_A)>$

- Non-mutable field of node X $= n_X$
- Mutable field of node X $= m_X$
- Signature of node X on message m $= Sign_X(m)$

Figure 16.5 **Transmissions with double signatures.**

over wireless links. SDF aims to maintain the integrity of data messages and prevent data dropping attacks.

16.4.1 Secure Route Discovery against Message Tampering

For either RREQ or RREP packets, we propose the use of *double signatures*; that is, two digital signatures are applied. The first one is for the non-mutable part (e.g., destination address field) and the second one is for the mutable part (e.g., hop-count field). An example is shown in Figure 16.5. Suppose source node *A* begins route discovery to destination node *D* by broadcasting an RREQ. When node *B* receives the RREQ, it increases the hop-count field in the packet's header by one before rebroadcasting it to neighboring node *C*, and so on. To protect the integrity and to provide end-to-end authentication of the nonmutable part of RREQ, the digital signature of the nonmutable part is signed by source node *A*. On the other hand, to protect the integrity and to guarantee hop-by-hop authentication of the mutable part of RREQ, the digital signature of the mutable part is updated and signed by intermediate node *B* with its own private key. Because malicious users cannot sign the RREQ with a valid private key, the altered content will result in digital signature mismatch. Thus, it will be detected by the receiving node. In this way, the integrity and authentication of RREQ can be maintained using double signatures.

When destination node *D* receives an RREQ, it generates an RREP. Once generated, it is unicast to the next-hop toward node *A*. As the RREP is sent toward *A*, the hop-count field is incremented by one at each hop. The double signature is used again in the same way as shown in Figure 16.5.

Note that when an intermediate node receives an RREQ and has a routing cache for that destination, the AODV standard allows an intermediate node to send an RREP to the source node directly. To this end, we propose to store the previously received valid digital signature and the corresponding nonmutable part of routing message in its routing cache. This feature is inspired by SAODV [3]. In SAODV, whenever a node generates an RREQ, it includes several additional fields with the digital signature that can be used to send an RREP by any intermediate node that creates a reverse route to the originator of the RREQ.

16.4.2 Secure Route Maintenance

In AODV, a node can initiate the generation of an RERR packet in several situations [1]. Consider the case when a node detects a link breakage for the next-hop of an active route in its local routing table while transmitting data. That is, when an upstream node does not receive the link layer ACK after several retransmissions, it will declare this link as broken and send an RERR to the source node by signing the whole message.

However, a problem arises when packet dropping attack occurs by the downstream node. The upstream node cannot distinguish whether the link breakage is due to mobility of the downlink node or intentional packet dropping. We resolve this issue by proposing an extension of the RERR to include the addresses of the two suspicious end nodes of the broken link.

When the source node receives the RERR, it will first reinitiate the route discovery process. We propose the use of a cache to store the addresses of these two end nodes of the broken link (Figure 16.6). We also propose that the source node maintains a parameter called the *suspicious value* for each node. The suspicious values of the two end nodes of the broken link are increased by one. This suspicious value will not be informed to any other nodes. This avoids the blackmail attack, which refers to the false report of a good node as a bad one. The suspicious values of these two nodes are decreased by one when the source node has not received an

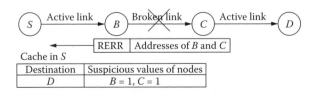

Figure 16.6   Example for suspicious node detection.

RERR with the addresses of these two suspicious nodes for a certain period of time.

The suspicious value is initially set to zero. The source node will append the addresses of nodes that have reached the maximum suspicious value θ in the RREQ for route discovery later. Those suspicious nodes will not be chosen as part of a new route from this source node to the destination. Note that this avoidance mechanism is managed by each node independently. Only the node that creates the RERR packet is required to sign this message for authentication and integrity. Note that SeAODV does not allow changing the destination sequence number when receiving an RERR. This is because the destination does not sign the destination sequence number in RERR.

16.4.3   Consideration of Control Message Dropping Attacks

There are three different control messages in AODV: (1) RREQ, (2) RERR, and (3) RREP. Because an RREQ dropping attack has no benefit for the attackers as they will not be able to join (or attack) the communication session as an intermediate node. Thus, RREQ dropping attacks will not be considered. Because each routing entry, which is in a soft state, will be removed after a certain time period, an RERR dropping attack has no benefits to the attackers either. An RREP dropping attack may have limited impact. Because the intermediate nodes can generate an RREP when they have a route to the destination, a source node can receive multiple RREPs. The source will then select one of the available routes. Thus, an RREP dropping attack may not always prevent the source from discovering a route to the destination.

16.4.4   Consideration of Replay Attacks

Attackers can replay an RREQ that is eavesdropped. To prevent this attack, we can use the idea of temporal leashes proposed in Hu et al. [19] by assuming that all nodes have loosely synchronized clocks.

16.4.5   Consideration of DoS Attacks

In general, DoS attacks cannot be avoided completely from MANET. By injecting a large amount of RREQs into the network, attackers can reduce the network's throughput. To reduce the power of a DoS attack, each node can measure the rates at which RREQs are received from each of its neighbor. If the measured rate of incoming RREQs from node A is greater than a certain threshold value, all incoming messages from node A will be discarded without being verified.

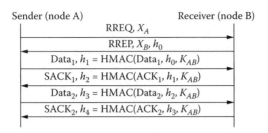

Figure 16.7 showing:

Sender (node A) — Receiver (node B)

RREQ, X_A

RREP, X_B, h_0

$Data_1, h_1 = HMAC(Data_1, h_0, K_{AB})$

$SACK_1, h_2 = HMAC(ACK_1, h_1, K_{AB})$

$Data_2, h_3 = HMAC(Data_2, h_2, K_{AB})$

$SACK_2, h_4 = HMAC(ACK_2, h_3, K_{AB})$

Figure 16.7   Data integrity check.

16.4.6   Secure Data Forwarding (SDF) Based on SeAODV

Applying an asymmetric authentication method for data packets such as the digital signature may not be suitable due to the high computational power required to generate and verify the digital signature for each packet. Therefore, we propose the use of a shared symmetric key between neighbors on the path.

Because each participating node on the route from source to destination must exchange RREQ and RREP during the route discovery phase, we take advantage of this interaction for key exchange, using the Elliptic Curve Diffie-Hellman (ECDH) [20] key exchange method to generate the symmetric keys (e.g., K_{AB} in Figure 16.7). The detailed operations for ECDH key exchange method can be found in Reference [20].

Considering the example in Figure 16.7, the upstream node A is the sender while the downstream node B is the receiver. Note that a generated key is different on each link along the route. The data packet integrity is maintained using HMAC [11]. Each data packet includes an HMAC value h_i (20 bytes in SHA-1 [21]) that is a function of the data, the symmetric key K_{AB}, and the previous HMAC value h_{i-1}. Each $SACK_i$ (Secure Acknowledgment) packet also includes an HMAC value h_{j+1} that is a function of the ACK_i, the symmetric key K_{AB}, and the previous HMAC value h_j. The detailed operations for HMAC can also be found in Krawczyk et al. [11]. With the use of HMAC, each node can verify the integrity of data messages in the path.

16.4.7   Extension of SeAODV

SeAODV can be extended, depending on its working environment. When power consumption is at a premium, the use of double signatures for RREQ may be too expensive. We can optimize the operation of SeAODV using a single signature (or HMAC, depending on the key management scheme). This scheme is almost the same as ESTEP described in Section 16.3.3.

Another extension is related to multipath routing. As mentioned in Section 16.4.3, multiple RREPs can be generated from several nodes corresponding to the same RREQ. However, RREP dropping can possibly result in multiple route discoveries when none of intermediate nodes have a routing cache to the destination. To reduce the power of an RREP dropping attack, we can also extend SeAODV with a path disjoint multipath routing protocol, such as AOMDV [23]. There are two different path disjoint routes: (1) link-disjoint and (2) node-disjoint routes. Path disjointness has the property that paths fail independently. In general, node-disjointness is a more strict condition than link-disjointness, and thus presents a lower number of disjoint routes. However, node-disjoint multipath is advantageous when there are attackers in the networks. This is because the link-disjoint path cannot avoid rogue users having several independent links with multiple neighbors.

16.4.8   Comparison with Other Schemes

Comparing our proposed SeAODV with the SAODV [3] and ARAN [5], we note that all three protocols work in the managed-open scenario [5] where the compromised users are excluded from route discovery and maintenance, and there is a key management subsystem as described in Section 16.2. To protect the integrity and to provide end-to-end authentication, all three protocols use the digital signature for the nonmutable field of messages. On the other hand, they use different schemes to protect the mutable part. SAODV [3] uses hash chains to authenticate the hop-count field of RREQ and RREP to verify that an attacker has not decreased the hop-count. ARAN [5] removes the field of hop-count in the routing entry, and does not allow RREP generation from an intermediate node to the source node. It avoids routing message tampering attacks in mutable fields at the cost of higher latency and higher overall routing load in route discovery. In SeAODV, the digital signature of the mutable part is updated and signed by intermediate nodes with their own private keys. Because malicious users cannot sign the mutable part with a valid private key, the altered content will result in a digital signature mismatch at their neighbors. Although SeAODV can protect the hop-count field in control messages, it is still vulnerable to the manipulation of the hop-count field in its routing table. As mentioned in Section 16.3.5, STEP (or ESTEP) can be integrated with SeAODV to protect routing table entries.

16.5   Performance Comparisons

We use the network simulator (ns-2) to compare the performance among STEP, ESTEP, and the original AODV routing protocol. For simplicity, the extensions described in Sections 16.3.4 through 16.3.6 are not included in this section. We compare the performance between our proposed SeAODV with SDF and the original AODV routing protocol. In this section, we refer to SeAODV with SDF simply as SeAODV without the extension described in Section 16.4.7.

The network topology consists of 50 nodes randomly placed over a 1000×1000 square-meter flat-grid, and each simulation run takes 900 simulated seconds (Figures 16.9 and 16.10). We assume that 15 of these nodes are constant bit-rate data sources, each sending fixed-size 512-byte packets at a rate of 4 packets/second. A random waypoint model is used for the mobility model at a speed uniformly distributed from 0 to 20 meters/second. The physical characteristics of each mobile node's radio interface approximate the Lucent WaveLAN, operating as a shared-medium radio with a nominal bit rate of 2 Megabits/second and a nominal radio range of 250 meters. The propagation model combines a free space propagation model and a two-ray ground reflection model. Other parameters are the same as in ns-2 b8a [24].

For SeAODV, we use ECC with a 20-byte key, 40-byte signature, and 60-byte certificate [17]. The digital signature generation delay and verification delay are assumed to be 2.0 and 4.0 milliseconds, respectively. These delay values are based on the measurements in Brown et al. [18]. Our results are based on simulation over 16 runs, and the error bars represent the 95 percent confidence interval of the mean in Figures 16.8 through 16.10. The packet sizes of RREQ, RREP, and RERR in the original AODV [1] are 24, 20, and 20 bytes, respectively. The packet sizes of RREQ, RREP, and RERR in SeAODV are 244 bytes (i.e., two signatures [80 bytes], two certificates [120 bytes], and domain parameter X_A [20 bytes] in Figure 16.7); 260 bytes (with an additional HMAC [20 bytes]); and 132 bytes (i.e., signature [40 bytes], certificate [60 bytes], and two physical addresses [12 bytes]), respectively. In this simulation, the maximum suspicious value θ is set to 2.

We evaluate six different metrics for performance evaluations:

1. Delivery ratio of data packets
2. Average end-to-end delay of transferred data packets
3. Normalized routing overhead (packets)
4. Normalized routing overhead (bytes)
5. Average route acquisition latency
6. Average path length

Transmission at each hop along the path counted as one transmission in the calculation.

16.5.1 STEP versus AODV without Attackers

In this subsection, the network topology consists of 100 randomly placed over a 1500×1500 square-meter flat-grid, and each simulation run takes 300 simulated seconds. Figure 16.8(a) shows that the packet delivery ratios

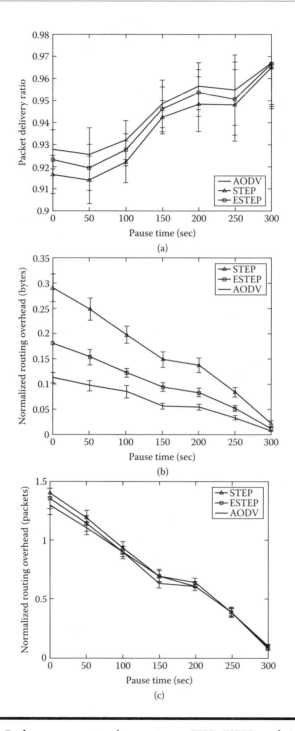

Figure 16.8 Performance comparisons among STEP, ESTEP, and AODV without attackers.

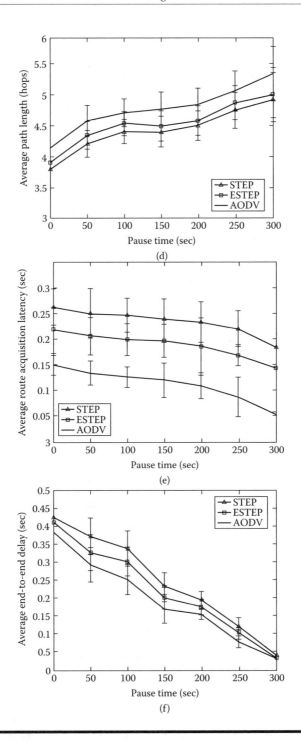

Figure 16.8   (Continued).

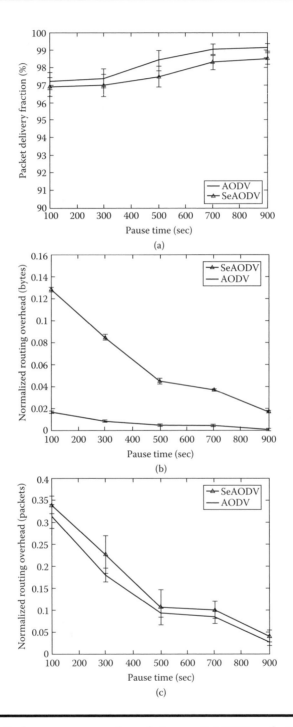

Figure 16.9 Performance comparison between SeAODV and AODV without attackers.

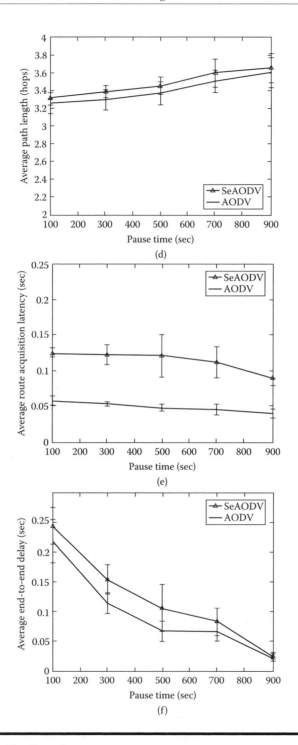

Figure 16.9   (Continued).

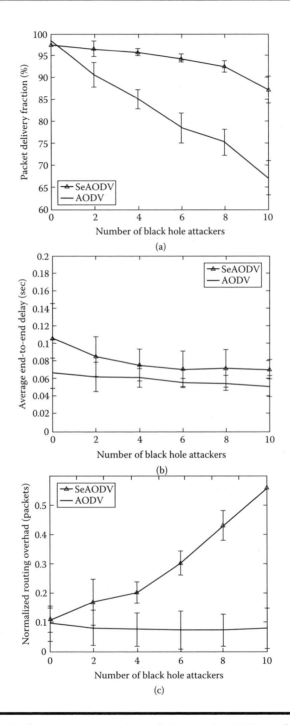

Figure 16.10 **Performance comparison between SeAODV and AODV with blackhole attackers in the network (500-second pause time).**

among STEP, ESTEP, and AODV are very close and above 90 percent in all scenarios. This suggests that both STEP and ESTEP are effective in discovering and maintaining routes for delivery of data packets even in relatively high mobility scenarios. Figure 16.8(b) shows that STEP's routing overhead is higher than that of AODV in terms of bytes. That is due to the increase in size of routing control packets with digital signatures in STEP. As mobility increases, the amount of control overhead of STEP increases linearly. On the other hand, ESTEP remains efficient, as compared to STEP, by avoiding multiple signatures in RREQ broadcast messages. Figure 16.8(c) shows the average number of control packets transmitted per data packet delivered. The three routing protocols demonstrate nearly the same amount of routing overhead. AODV has the advantage of smaller control packets; smaller packets have a higher probability of successful reception at the destination.

Figure 16.8(d) shows that AODV selects a slightly longer path when compared to STEP and ESTEP. Because all three protocols find the shortest path explicitly based on the same route discovery mechanism of AODV, the length of paths should not differ significantly. However, it is possible that due to a higher contention or queueing delay along the shortest path, a sub-optimal path is being used instead. In STEP, due to the additional delay of signature verification and generation in each hop, the possibility of finding a longer path instead of the shortest one decreases as compared to AODV. Figure 16.8(e) shows that the average route acquisition latency for STEP is approximately double that for AODV. This is due to the additional processing delay for multiple digital signature generations and verifications at each node for control packets. Because multiple digital signatures are not used for RREQ messages in ESTEP, ESTEP can find a route with a smaller additional delay when compared to STEP. Figure 16.8(f) shows that the average end-to-end delays for both STEP and ESTEP are slightly higher than that of AODV. Although STEP and ESTEP have higher average route acquisition delays than AODV, the number of route discoveries performed is a small fraction compared with the number of data packets delivered. Therefore, the effect of route acquisition latency on the average end-to-end delay of data packets is not significant. Note that the processing delay of data packets is identical in all three protocols.

16.5.2 SeAODV versus AODV without Attackers

We now compare AODV and SeAODV without attackers. Figure 16.9(a) shows that the packet delivery fractions between AODV and SeAODV are very close. This suggests that SeAODV is effective in discovering and maintaining routes for delivery of data packets, even with relatively high node mobility. Figure 16.9(b) shows that SeAODV's routing overhead is higher than that of AODV. This is due to the increase in the size of routing control packets in SeAODV. Figure 16.9(c) shows the average number of control

packets transmitted per data packet delivered. The two protocols demonstrate nearly the same amount of routing packet overhead. AODV has the
advantage of small control packets; smaller packets have a higher probability of successful reception at the destination. However, due to the IEEE
802.11 medium access control for unicast transmissions (i.e., Request-to-
Send, Clear-to-Send messages), a portion of time is spent in acquiring the
channel not in transmission.

Figure 16.9(d) shows that SeAODV selects a slightly longer path when
compared to AODV. Because both protocols find the shortest path explicitly based on the same route discovery mechanism of AODV, the length
of paths should not differ significantly. However, due to the secure route
maintenance mechanism with the suspicious value, any link breakage in
the path will be regarded as a dropping attack. Therefore, sometimes, the
sub-optimal path is being used instead of the shortest path in SeAODV. Figure 16.9(e) shows that the average route acquisition latency for SeAODV is
approximately double that for AODV. This is due to the additional processing delay for digital signature generation and verification for control packets. Figure 16.9(f) shows that the average end-to-end delay for SeAODV
is slightly higher than AODV. Note that in SeAODV, the number of route
discoveries performed is a small fraction of the number of data packets
delivered. Hence, the effect of route acquisition latency on the average
end-to-end delay of data packets is not remarkable.

16.5.3 SeAODV versus AODV with Blackhole Attackers

Figure 16.10 shows the performance metrics as a function of the number
of blackhole attackers in the network. We assume that blackhole attackers
only drop data packets, not routing packets. For AODV, the network will
not be able to detect the presence of attackers. In Figure 16.10(a), SeAODV
maintains a higher packet delivery fraction than AODV without security
mechanisms. It shows that our proposed mechanism can effectively isolate
blackhole attackers. Results in Figure 16.10(b) indicate that AODV has lower
average end-to-end delay when compared with SeAODV. Because SeAODV
incurs more routing control packets to the network, the average end-to-
end delay for data packets increases. Note that the average end-to-end
delay of AODV decreases slightly as the number of blackhole attackers
increases. The rationale is that blackhole attackers drop data packets at the
intermediate nodes. Because the dropped packets are not counted in the
end-to-end delay calculation, the average end-to-end delay will decrease.
Results in Figure 16.10(c) show that AODV incurs a lower routing control
overhead than SeAODV. This is due to the fact that SeAODV can detect
blackhole attackers and initiate a new route discovery to avoid those nodes.
The increase in routing control overhead is due to the increase in the
number of RRER, RREQ, and RREP control packets in the network.

16.6   Conclusions

In this chapter we proposed the Secure Table Entry Protection (STEP) mechanism to prevent routing table tampering attacks. STEP provides authentication for both the destination sequence number and hop-count fields in the routing table entry. The receiving node can confirm the correctness of messages by verifying the signatures of two consecutive upstream nodes and the source node that creates the packet. We described how STEP can be incorporated in the AODV routing protocol. We proposed the Efficient STEP (ESTEP), which can avoid the use of expensive multiple digital signatures on RREQ broadcast messages. We also proposed several techniques to improve the security of the AODV routing protocol against malicious users. We proposed a secure routing extension (SeAODV) and secure data forwarding (SDF) mechanism for the AODV routing protocol. Double signature is used for both RREQ and RREP to prevent routing message tampering attacks. Both RREQ and RERR are modified to prevent data dropping attacks. For secure data transmission, we introduced the use of HMAC to maintain the message integrity. Simulation results showed that both STEP and ESTEP continue to maintain a high packet delivery fraction and a small end-to-end delay at the expense of slightly higher route acquisition latency and control overhead in route discovery. In the presence of data packet dropping attackers, SeAODV continues to maintain a high packet delivery fraction and a small end-to-end delay. As future work, we plan to implement STEP and SeAODV on mobile devices and study them in real-world scenarios by taking into account the energy issues.

Acknowledgments

This work was supported by the University of British Columbia Graduate Fellowship, and by the Canadian Natural Sciences and Engineering Research Council under grants RGPIN 261604-03 and 44286-00.

References

[1] C. Perkins, E.M. Belding-Royer, and S. Das, "Ad hoc on-demand distance vector (AODV) routing," *IETF RFC 3561*, July 2003.
[2] X. Hong, K. Xu, and M. Gerla, "Scalable routing protocols for mobile ad hoc networks," *IEEE Network*, Vol. 16, No. 4, pp. 28–39, July/August 2002.
[3] M. Zapata and N. Asokan, "Securing ad hoc routing protocols," *Proc. of ACM Workshop on Wireless Security*, Atlanta, GA, September 2002.
[4] Y.-C. Hu, A. Perrig, and D.B. Johnson, "Ariadne: a secure on-demand routing protocol for ad hoc networks," *Proc. of ACM Mobicom*, Atlanta, GA, September 2002.

[5] K. Sanzgiri, B. Dahill, B.N. Levine, C. Shields, and E.M. Belding-Royer, "A secure routing protocol for ad-hoc networks," *Proc. of International Conference on Network Protocols*, Paris, France, November 2002.

[6] P. Papadimitratos and Z.J. Haas, "Secure routing for mobile ad hoc networks," *Proc. of SCS Communication Networks and Distributed Systems Modeling and Simulation Conference*, San Antonio, TX, January 2002.

[7] Y.-C. Hu, A. Perrig, and D.B. Johnson, "SEAD: secure efficient distance vector routing for mobile wireless ad-hoc networks," *Proc. of IEEE Workshop on Mobile Computing Systems and Applications,* June 2002.

[8] P. Papadimitratos and Z.J. Haas, "Secure link state routing for mobile ad hoc networks," *Proc. of IEEE Workshop on Security and Assurance in Ad hoc Networks,* Orlando, FL, January 2003.

[9] P. Ning and K. Sun, "How to misuse AODV: a case study of insider attacks against mobile ad hoc routing protocols," *Proc. of IEEE Information Assurance Workshop,* West Point, NY, June 2003.

[10] W. Wang, Y. Lu, and B. Bhargava, "On vulnerability and protection of ad hoc on-demand distance vector protocol," *Proc. of the International Conference on Telecommunication,* Papeete, France, February/March 2003.

[11] H. Krawczyk, M. Bellare, and R. Canetti, "HMAC: keyed-hashing for message authentication," IETF RFC 2104, February 1997.

[12] J. Kohl and B.C. Neuman, "The Kerberos network authentication service (V5)," IETF RFC 1510, September 1993.

[13] R. Rivest, A. Shamir, and L. Adleman, "A method for obtaining digital signatures and public key cryptosystems," *Communications of the ACM*, Vol. 21, No. 2, pp. 120–126, February 1978.

[14] F. Stajano and R. Anderson, "The resurrecting ducking: security issues for ad hoc wireless networks," *Security Protocols International Workshop*, Springer-Verlag, Berlin, Germany, 1999.

[15] H. Yang, X. Meng, and S. Lu, "Self-organized network-layer security in mobile ad hoc networks," *Proc. of ACM Workshop on Wireless Security*, Atlanta, GA, September 2002.

[16] B. Schneier, *Applied Cryptography: Protocols, Algorithms, and Source Code in C*, John Wiley & Sons, New York, 1996.

[17] D.B. Johnson, "ECC, future resiliency and high security systems," Certicom White Paper, March 1999.

[18] M. Brown, D. Cheung, D. Hankerson, J.L. Hernandez, M. Kirkup, and A. Menezes, "PGP in constrained wireless devices," *Proc. of the USENIX Security Symposium*, Denver, CO, August 2000.

[19] Y.-C. Hu, A. Perrig, and D.B. Johnson, "Packet leashes: a defense against wormhole attacks in wireless networks," *Proc. of IEEE Infocom*, San Francisco, CA, March/April 2003.

[20] Certicom Research, "SEC1: elliptic curve cryptography," *Standards for Efficient Cryptography Group*, September 2000.

[21] C. Madson and R. Glenn, "The use of HMAC-SHA-1-96 within ESP and AH," IETF RFC 2404, November 1998.

[22] S. Buchegger and J.-Y. Le Boudec, "A robust reputation system for mobile ad hoc networks," EPFL Technical Report, No. IC/2003/50, July 2003.

[23] M.K. Marina and S.R. Das, "On-demand multipath distance vector routing in ad hoc network," *Proc. of the International Conference for Network Procotols*, Riverside, CA, November 2001.

[24] The Network Simulator — NS-2 notes and documentation.

Chapter 17

Security and Privacy in Future Mobile Networks

Geir M. Køien

Contents

17.1 Overview

The topic of cellular access security is limited to security schemes and protection of the link layer from the mobile station to a terminating point in the access network. Access security can be split into two main areas: (1) the Authentication and Key Agreement (AKA) protocol and (2) the link layer protection mechanisms. In this chapter we investigate the requirements for access security in future mobile networks and look specifically into new schemes for how to provide effective and efficient authentication and key agreement for cellular access security.

17.2   Introduction

The security services normally covered by access security are authentication and authorization for access on the link layer, including cryptographic protection of the over-the-air access link. Access link protection will normally consist of cryptographic confidentiality protection and cryptographic integrity protection. The protection will be subscriber specific and it will normally cover both subscriber data and subscriber control signaling data. A prerequisite for subscriber authentication is to have a subscriber identity. For access security, the identity (or identities) will be link layer identities, and these identities are logically distinct from higher layer identities in the same manner as a link layer address (message authentication code or MAC address) is distinct from the network address (Internet protocol or IP address). Subscriber privacy is a growing concern. Subscriber data and subscriber-related control data must be protected against eavesdropping. For cellular access security, we also have the privacy issues of identity confidentiality and location confidentiality for the public air interface. These two issues are closely connected with subscriber identity handling. In the following, we examine the requirements for access security in future mobile networks. Performance is always an issue with real systems. We therefore look at how to integrate selected mobility management and radio resource allocation procedures with the access security procedures to get improved setup performance and still achieve improved security and user privacy. We also briefly discuss how best to protect the link-layer connection.

17.3   Access Security

This section introduces some common concepts and defines some security services used for access security purposes. The section is not intended to be a tutorial. For an introduction to mobile access security, the reader may want to look at the 3G/WLAN-related discussions in references [1–4]. Access security consists of two main parts: (1) security context establishment, also known as the authentication and key agreement part; and (2) protection of the access link.

17.3.1 Authentication and Key Agreement (AKA)

AKA is the set of procedures and protocols necessary for authenticating the principal entities and for establishing a *security context*. The security context consists of the established key material, together with other relevant security parameters. The security context is bounded to a specific run of the AKA protocol. The AKA protocol will be executed in a potentially hostile environment. The protocol must therefore be designed to be robust and able to withstand active attacks. We do not require the AKA protocol to always succeed, but we do require that a failure should never compromise legitimate users or the system in any way. An adversary may thus be able to disrupt the AKA protocol, but it should not be able to achieve more than that (e.g., impersonate as a legitimate user or get access to secret key material, etc). The AKA execution is in some ways analogous to a remote "login" on a computer system, and it may be useful to think of it in this way.

17.3.2 Access Link Protection

Access link protection is the set of procedures and protocols that ensures a sufficient level of (cryptographically based) protection of the access link. We require the access link to be protected against eavesdropping, willful manipulations of messages, etc. In practice, this requires the link to be encrypted (confidentiality protection) and to be cryptographically integrity protected. Another issue to deal with is to decide exactly how deep into the network the access link protection should reach. The "link" part suggests that the protection should extend for exactly one link segment, but the issue is more complex than that. We return to this subject later.

17.3.3 Security Services

We now define some access-related security services. A security service by itself is an abstract entity and it must be realized by a concrete security mechanism. The security mechanisms are, in turn, devised from profiled use of cryptographic primitives or functions. Note that one security mechanism can be configured to provide multiple security services, and that one may need several mechanisms to provide one service.

17.3.3.1 Entity Authentication

When communicating over a network or a wireless link, the communicating parties cannot physically assert each other's respective identities. We must therefore devise some means of asserting that the communicating parties are who they claim to be. We note that entity authentication presumes that the principal entities have recognized identities. In many wireless or cellular systems, the subscriber "identity" is taken to be the link address. However, an address is not strictly the same as an identity; and for security and privacy purposes, it will generally be better if the two are cleanly separated. Entity authentication is primarily used during security context establishment. The security context will subsequently be the basis for protected communication.

17.3.3.2   Data Authentication and Data Integrity

By *data authentication* we tend to mean data origin authentication. To protect against forgery, one often needs assurance that a message is indeed from the claimed sender. To this end, we must guarantee that the address or identity information has not been tampered with. The normal way of realizing this is by a Message Authentication Code (MAC)[1] function.

Data integrity is a service complementary to data authentication, and it is almost always provided by the same mechanism. The data integrity service provides us with assurance that the message content has not been manipulated in any way. Data integrity, as a security service, is different from the normal transmission-related notion of integrity. The security service must be able to withstand willful manipulations by an adversary, and then the standard checksum-based mechanisms will not suffice. As for data authentication, the normal realization of the integrity service is by a MAC function. Data integrity (Section 5.2.4 in [5]) can be subdivided into more specific services. For link-layer access security purposes, we normally only require per-message (connectionless) integrity; hence, there is no integrity protection over the message sequence. This has the implication that one cannot detect missing messages or duplicated messages. Such capability may be added at the higher layers if considered necessary. For control signaling, we only require connectionless integrity protection. At the link layer, the frames will contain channel state information, including channel number, frame number, etc. Provided that the cryptographic checksum is computed over the frame/packet content *and* the channel state information, one does achieve a higher level of data integrity protection. Thus, one will effectively have replay protection. The replay protection is indirect because it is provided by the link-layer frame/packet scheduler, but this should be considered a feature and not a shortcoming.

17.3.3.3   Confidentiality

Confidentiality comes in several flavors (Section 5.2.3 in [5]). In addition to message content protection against eavesdropping (data confidentiality),

---

[1] In this chapter we reserve the MAC abbreviation for this purpose.

we also have address/identity confidentiality and traffic flow confidentiality. To provide identity confidentiality, one must protect the user identity and other elements of the message that might allow an adversary to monitor and track users. In a mobile context, location confidentiality is also an issue. The terms "identity confidentiality" and "location confidentiality" are also known as identity privacy and location privacy, respectively.

17.3.3.4 Availability and Non-repudiation

To guarantee availability in a wireless environment is difficult. Not only is the wireless medium a shared resource, but it also represents an adverse environment with complex radio transmission problems and, adding to this, the potential presence of malicious and capable adversaries. The only feasible line of defence is to ensure that the security services are as robust as possible. This means that it should be infeasible for an adversary to turn the security procedures against the user or network.

One cannot, in general, prevent an adversary from disrupting wireless services; denial-of-service (DoS) attacks are relatively easy to mount and we can at best hope to make them too inconvenient or too costly to be considered practical. The Transmission Control Protocol (TCP)-syn flooding attack [6] is a classical example of what happens when an attack is cheap to execute. In the TCP-syn case, the target servers had to maintain state for the semi-alive sessions, and this either exhausted the available resources (memory, etc.) or one reached a limit on the maximum number of open connections. In either case, legitimate requests were rejected due to the inability of the server to respond. We must therefore seek to prevent attacks from scaling in our design or to otherwise limit the effects of the attack.

Non-repudiation is a service normally associated with higher-layer services and applications. We do not discuss non-repudiation here, but we observe that charging and billing for the access service depends on the credibility of the entity authentication (AKA) protocol. It is therefore prudent that the authorization procedure is trustworthy and that none of the parties can easily deny actual (chargeable) events. For high-value transactions (money transfer between banks, etc.), one needs high-level assurance; but for ordinary mobile services, this is not necessary.

17.3.4 Principals, Interfaces, Trust, Threats, and Adversaries

To discuss and analyze security requirements and solutions for future mobile systems, we need to introduce a few abstractions and definitions.

17.3.4.1 Principal Entities

To simplify our analysis, we only define three principal entities and one adversary/intruder.

1. *User entity (UE)*. The UE is a subscriber entity, and the subscription is registered with the HE (home entity). The HE has security jurisdiction over the UE. The HE assigns the permanent HE–UE security credentials and unilaterally decides on the permanent UE identity (*UEID*).
2. *Home entity (HE)*. The HE is the *home operator*. It is responsible for service provision to the UE. It manages macro-mobility and handles charging and billing on behalf of the UE when it consumes access services.
3. *Serving network (SN)*. The SN provides a core network (CN) with a nonempty set of access networks (ANs) attached to it. Through HE-SN *roaming agreements*, the SN provides access services to UEs. The roaming agreements are mutual in nature.
4. *Dolev-Yao intruder (DY)*. The DY intruder is our proverbial adversary. The DY intruder [7] can selectively read, store, delete, inject, and otherwise manipulate all transmitted messages (in real-time). The DY intruder may be a legitime principal, and it may try to impersonate any of the other principals.

Most cellular operators today own and operate both HE and SN services. Virtual operators will only have HE functionality. In our model, the HE has security jurisdiction over the UE, and it has full authority over UE identity assignment. It will therefore not make much sense to assume that the HE will willfully try to deceive the UE with respect to impersonation. We therefore rule out the possibility that the HE will impersonate its own subscribers.

17.3.4.2 The User Device

The *user device* (UD) can include a powerful general-purpose computing platform, or it may be a dedicated device with scarce resource. A UD consists of one or more *mobile termination* (MT) units and one or more *security module* (SM) units. An SM is owned or controlled by exactly one administrative entity, the HE operator. Each SM contains one or more subscriptions, and we assume that all subscriptions on a single SM are issued by the controlling HE. The SM can be implemented as a smartcard, as is common in 2G and 3G systems, but the actual type of implementation is not important. However, we do require the SM to provide secure storage and a computation engine for processing cryptographic primitives and protocols. Each user subscription is represented by exactly one UE. The UE, which physically is an application running on the SM, is uniquely referenced by the UE identity (*UEID*). Figure 17.1 illustrates the possibilities. We normally refer only to the UE, but observe that sometimes one implicitly includes SM or MT capabilities in the discussion.

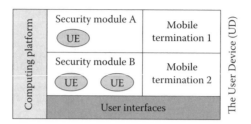

Figure 17.1   The user device (UD).

17.3.4.3   Interfaces

Modern mobile systems are complex and contain a number of physical nodes, interfaces, and protocols. However, for our purposes we can simplify this picture. As shown in Figure 17.2, we need only consider the interfaces between the principals when considering the AKA protocol. Figure 17.3 shows the Universal Mobile Telecommunications System (UMTS) AKA and some of the nodes involved.

- *A-interface.* This is the over-the-air interface. It has comparatively low bandwidth and relatively long and variable latency.
- *B-interface.* This is the fixed interface between the SN and the HE. We assume no bandwidth restrictions here. The signal propagation delay is constant.

17.3.4.4   Trust Relationships

Trust can be defined as the intersection between beliefs and dependence. The UE clearly depends on the HE to get service. The UE also has reason to believe that the HE is benevolent. Still, the UE should only trust the HE to the extent needed to obtain the required services. The relationship between the HE and the SN is normally well balanced; and while neither has reason to distrust the other, they too should only trust their partner to the extent necessary. The relationship between the UE and the SN also requires some trust between the parties, and here one will have the HE as

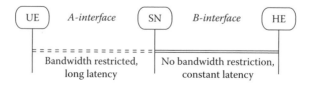

Figure 17.2   Interfaces.

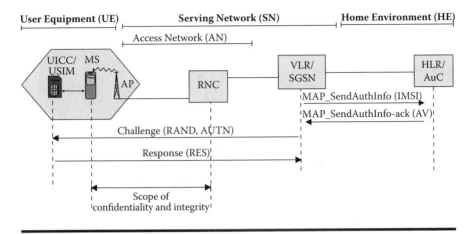

Figure 17.3   UMTS authentication and key agreement (AKA).

a mediator (proxy). Figure 17.4 depicts the trust relationships between HE, SN, and UE. We have the following:

- *Long-term HE–UE relationship:*
  - *Security.* The relationship is based on the subscription contract, which gives the HE security jurisdiction[2] over the UE. The UE must therefore necessarily trust the HE.
  - *Privacy.* The HE assigns the *UEID* and hence the UE cannot have *identity privacy* from the HE. However, the UE can insist on *location privacy* from the HE.
- *Long-term HE–SN relationship:* This is a mutual trust relationship. The roaming partners are likely to be competitors in other markets, and they will not want to divulge more information than strictly necessary to the opposite party. The HE operators will want to retain some control (home control) over the UE while it is roaming. Charging and billing is, of course, the central issue, and one will want to limit the trust. For future systems, one should expect requirements for (near) real-time billing control for all services.
- *Temporary SN–UE relationship:* The SN–UE trust relationship is derived from the HE–UE and HE–SN relationships. While one may assume trust to be transitive, the derived relationship obviously has a weaker basis. Given the weaker basis, it is advisable that the scope of the derived SN-UE relationship be limited. It is also reasonable to require online confirmation from the proxy (the HE) to assert the validity of the derived relationship.

---

[2] *Control* and *jurisdiction* can be used interchangeably in this context.

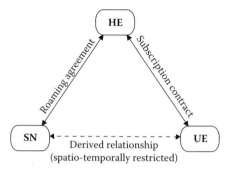

Figure 17.4 Security trust relationships.

- *Privacy.* Whenever the SN is in radio contact with the UE, the SN will know the approximate UE position. During idle periods, the SN will only know the location area, but the subscriber cannot rely on this because the subscriber cannot know or control when the SN and the MT may be in contact. We must therefore conclude that the UE cannot attain *location privacy* from the SN. However, as we will see later, it is possible to provide UE *identity privacy* from the SN.

17.3.4.5 Security Architecture and Adversaries

It is very difficult to assess all present and future threats to a system, but even an imperfect risk analysis will have value in clarifying the needs of the security architecture. Based on the analysis, one can decide on security service requirements. The security architecture will likely be in place for the lifetime of the system, and it must therefore be flexible enough to allow for modifications and upgrades. Requirements for upgrading means that one must devise secure negotiation schemes. The negotiation capabilities must be present in the original design. We note that backward compatibility and fallback issues are notoriously difficult to tackle in a secure way. Over time, these aspects have a tendency to compromise the consistency of the security architecture.

We already mentioned the DY intruder and its formidable capabilities. While powerful, the DY intruder also has some limitations. The DY intruder is not capable of breaking the cryptographic primitives and it will not be able to physically compromise the principals. A DY intruder in a mobile environment must be physically distributed and, as such, the DY intruder is an unrealistic assumption today. On the other hand, we note that sensor technology is getting cheaper and that a system that withstands the DY intruder will withstand less-capable intruders. The DY intruder is also commonly assumed when devising security proof for AKA protocols, and much of the methodology is designed with the DY intruder in mind. So we advise to play it safe and assume that the full DY intruder is present. While we do not assume that the DY intruder is able to break the cryptography, many cryptosystems have been broken. The Global System for Mobile Communications (GSM) system is a point in case. Barkan, Biham, and Keller (BBK) [8] have demonstrated how various design flaws in the GSM system and one weak cipher algorithm (the A5/2 cipher) seriously compromised GSM security. The BBK article is quite readable and is recommended reading for anyone wishing to design access security schemes. Another example of broken cryptography is the WEP scheme for IEEE 802.11 WLAN systems [9].

In real-life, we will also have intruders that are more capable than the DY intruder. For example, a UD may get compromised by malware at the higher layers. To maintain UD integrity will be difficult, but future HE operators should prepare to take responsibility in this area. The issue is outside the scope of access security and we do not discuss it further here. Another aspect is physical security in the access network. Access points (APs) will be geographically distributed and difficult to protect. Because APs should be inexpensive devices, there will be a limit as to how much effort one puts into protecting them. The APs will therefore have comparably little physical protection and they will inevitably be exposed to intrusion. One should therefore make sure that the access security architecture design is not too vulnerable to the effects of compromised APs. One way of doing this is to avoid terminating security at the APs; that is, one extends the link layer security protection to a controller or server further into the access network. For large systems, we cannot readily assume that we will be able to protect all entities and nodes at all times. Even with a "perfect" security architecture, there will be security failures. Security on the large-scale system level should be viewed as a statistical property and not in absolute terms. There will be security breaches, and we should recognize this and prepare for how to handle incidents. The security architecture must therefore aim at minimizing the effects of security breaches.

17.4 Principles for Security and Privacy in Mobile Systems

How should one design a security architecture for a mobile system? There is no definitive answer but we know from experience how *not* to do it. In their 1994 report, Abadi and Needham [10] defined a set of engineering principles for designing security protocols. The Abadi–Needham principles are all motivated by security failures and the principles were proposed as engineering practices. By adhering to the principles, one would eliminate some common mistakes and thereby prevent some of the more embarrassing security failures.

The requirements for the 3G security architecture also provide us with design input [11]. Based on the Abadi–Needham principles and 3G requirements, we now define some additional requirements and guidelines for the design of future mobile security protocols.

17.4.1   The Prudent Engineering Principles

The Abadi–Needham principles for security protocols are summarized below. A full account of the principles appears in DEC SRC Research Report 125, "Prudent Engineering Practice for Cryptographic Protocols" [10]. The principles represent sound engineering practices, and one is well advised to follow these principles. There is an additional unacknowledged principle, which states that one cannot trust unprotected information. This might seem self-evident but its still worth stating. We have taken the liberty of expressing this as the **0.th** principle. This principle can be traced to DEC SRC Research Report 39, "A Logic of Authentication" (the BAN-logic report) [12]. The principles (abridged) are:

0. Unprotected information cannot be trusted and cannot be used to develop security beliefs (the term "belief" has a special meaning in BAN logic).
1. Every message should say what it means. The interpretation of the message should depend only on its contents.
2. The conditions for a message to be acted upon should be clearly and unambiguously defined. A reviewer of the design should immediately see if the conditions are acceptable or not.
3. If the identity of the principal is essential to the meaning of the message, then the principal name should be explicitly given in the message.
4. Be clear about why encryption and cryptographic methods are being used. (That is, be clear on which security service is it that the cryptographic method is supposed to provide.)
5. When a principal signs information elements that are already encrypted, it should not be inferred that the principal knows the encrypted contents.
6. Be clear about why nonces are being used[3].
7. If a predictable quantity (such as a counter) is used to guarantee newness, then it must be protected to prevent an intruder from simulating the sequence.

---

[3] A nonce is a time-variant parameter (TVP) and is used to provide "freshness," "timeliness," and "uniqueness."

8. If timestamps in absolute time are used as nonces, then the difference between local clocks must be taken into account (both to limit the "replay window" and to prevent premature expiry). The integrity of the system clocks must be ensured, and proper synchronization between the clocks must be guaranteed.

9. It should be possible to deduce the protocol, which run of the protocol, and the message number within the sequence by inspecting a message. [*Extended interpretation:* The protocol state machine must be fully defined. The messages must be completely specified, and the encoding of all information elements must be fully defined.]

10. Trust relationships should be explicit. The reason for the relationships should be clear.

17.4.2   Mobile Security Requirements and Guidelines

The Abadi–Needham principles are not complete and do not cover the cellular system specific case. The following contains a set of additional requirements and guidelines for the mobile system setting.

Security is a system property, and the security architecture must be designed in a system context. Performance, in one way or another, is always an issue with real systems, and thus the security architecture must provide both effective and efficient security services. We are particularly concerned with the temporal performance, but should also consider the computational and communications performance. The requirements and guidelines presented here derive, in part, from the principles derived in Køien [13]. Before arriving at the requirements, we discuss and analyze some of the high-level requirements and mobile-specific aspects.

17.4.2.1   Home Control

The trust relationships between the principals are not fully bi-directional. The HE has security jurisdiction over the UE. This imbalance in the relationship is also reflected in the responsibility for charging and billing. The HE acts as a proxy for the UE; any charges that the UE incurs at the SN will be handled by the HE. While the UE unconditionally must trust the HE, the HE cannot afford quite the same level of trust.

In 2G and 3G systems, the HE delegates the execution of the challenge-response part of the AKA protocol to the SN. Operationally, this works fine but the HE must then abandon all control over the UE to the SN. While this was permissible for early 2G systems, it has become an increasingly naive approach over time. For a future system, we require the HE to enforce a measure of home control to protect the UE and its own interests.

17.4.2.2 Online Three-Way Authentication and Key Agreement Protocol

The 2G(3G) AKA protocols are two-staged offline protocols [1,2,14] (Figure 17.3). The strategy of using an offline delegated AKA protocol was permissible in the early years of 2G systems. However, delegating authentication to the SN implies delegating control over the UE to the SN. This is a naive approach and the amount of trust the HE then must place in the SN cannot really be justified.

For the 3G-WLAN case [3,15], one devised an adaptation of the basic 3G AKA scheme. Here, the challenge-response of the AKA protocol is executed between the UE and the corresponding HE entity. In this case, the SN is reduced to a passive party that just receives instructions and session key material. This is not acceptable from the SN point of view. The SN, as the provider of access services, is entitled to a certain amount of control to protect its assets.

We therefore contend that the two-party schemes used in 2G(3G) systems are inadequate. One must therefore design a solution where all three principals are active parties and where the jurisdiction characteristics of the trust relationships are respected. To be active means to respond. The AKA context must therefore be established online. The three principals are not equals, and physically the interactions between HE and UE are necessarily forwarded through the SN. We therefore do not have a straightforward three-party case. So, solutions that would have been appropriate for a tripartite case [16] do not automatically apply. Instead, we have a special three-way case.

17.4.2.3 Security Context Hierarchy

There is a natural hierarchy to security contexts [13]. The hierarchy has simultaneously both a temporal and a spatial dimension. The following is a suggestion as to how to structure the hierarchy. Figure 17.5 depicts the proposed three-way security context hierarchy and Figure 17.6 illustrates the dependencies between the levels.

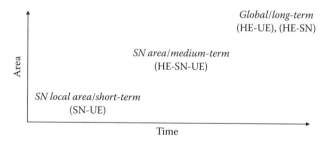

Figure 17.5   Security context hierarchy (UE centric).

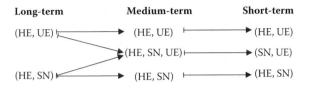

Figure 17.6   Security context level dependencies.

1. Long-Term Contexts:
   a. $LTC_{(HE,SN)}$: This context is based on roaming agreements. It will be long-lived and will normally apply for all HE subscribers and to the full coverage of the SN network. This context includes long-term security credentials for setting up protected communication tunnels between the principals.
   b. $LTC_{(HE,UE)}$: This context is defined by the UE subscription contract. It contains long-term security credentials, including the permanent UE identity *UEID*. The context exists for the duration of the subscription and is normally valid for all HE provided areas, which in principle is the sum of all HE roaming agreement areas.

   The HE will have a number of HE-SN roaming agreements (typically greater than 100). The SN can likewise have agreements with a large number of HE operators. The HE can serve millions of subscribers (UEs). A UE will have exactly one HE association.

2. Medium-Term Contexts:
   a. $MTC_{(HE,SN,UE)}$: This context derives from $LTC_{(HE,UE)}$ and $LTC_{(HE,SN)}$. It is established dynamically by the three-way AKA protocol in conjunction with the UE registration procedure. The validity of the context is determined by an SN area (server area or location area) and by a validity period.
   b. $MTC_{(HE,SN)}$: This context derives from the $LTC_{(HE,SN)}$ context and is independent of any specific UE. The context is therefore established and renewed independently of any UE activity. The validity is determined by usage (traffic volume) and a validity period.
   c. $MTC_{(HE,UE)}$: This context is derived from the $LTC_{(HE,UE)}$ context and is established by the three-way AKA protocol. The validity of the $MTC_{(HE,UE)}$ context is equal to the three-way $MTC_{(HE,SN,UE)}$ context.

3. Short-Term Contexts:
   a. $STC_{(SN,UE)}$: This context derives from $MTC_{(HE,SN,UE)}$ and secures the access link at the A-interface. It consists of symmetric-key key matrical for protection of user-related communication and

user-related control signaling. The validity of the context is determined by usage (traffic volume).

b. $STC_{(HE,SN)}$: This context derives from $MTC_{(HE,SN)}$ and secures the B-interface. The (HE-SN) context is established between security gateways. The HE and the SN can have multiple security gateways.

c. $STC_{(HE,UE)}$: This is an optional context between the UE and the HE. It can be used for HE security management of the UE and for UE notifications to the HE. Among the usages may be negotiation of new HE-UE authentication algorithms.

Note that when a context expires, all dependent context should be invalidated. The medium-term and short-term contexts have explicit termination conditions. Additionally, the principals can at will terminate the contexts prior to exceeding those conditions. For example, a UD power-off should cause a re-registration event. This event should automatically terminate all medium-term contexts (and thereby the short-term contexts).

17.4.2.4 Securing the B-Interface

Security for the B-interface is not mobile specific, and we assume that the B-interface channel will be based on the TCP/IP stack. This suggests that one should use a native IP security protocol for the B-interface. We therefore advocate the use of the IPSec [17] protocol over the B-interface. The $MTC_{(HE,SN)}$ security context can then be established with the Internet Key Exchange (IKEv2) Protocol [18] or the mobility extended IKEv2 Mobility and Multihoning Protocol (MOBIKE) [38]. The $STC_{(HE,SN)}$ context will consist of an IPSec Security Association (SA) set.[4]

A similar solution is found in 3G networks. For example, IPSec is used in securing the 3GPP core network control plane. This is captured in the Network Domain Security specification TS 33.210 NDS/IP [19], which specifies how to use IPSec/IKE when securing the inter-operator interface.

17.4.2.5 Exposure Control and Spatio-Temporal Context Binding

All things must end, and this is also the case for mobile security contexts. Over time, the security context will gradually be exposed and secrets may leak. One must therefore make sure that the security context expires before the exposure becomes critical. In traditional cryptosystems, one has two exposure dimensions and, correspondingly, two expiry conditions (IPSec is one example [17]). In a mobile setting, one can add a third exposure dimension with spatially based expiry [20,21]. The expiry conditions can be applied to the medium-term and short-term security contexts.

---

[4] The IPSec SA is unidirectional, and so at least two SAs will be needed.

1. *Temporally based expiry.* Cryptographic secrets/keys are susceptible to leakage over time. It is therefore sensible to treat key material as real-time data and to limit the lifetime. For example, the *default lifetime* for IPSec security associations is 28,800 seconds (8 hours).
2. *Usage (traffic volume) based expiry.* Cryptographic keys have limited entropy, and one must limit the amount of data that is protected with a particular key.
3. *Spatially based expiry.* Spatial expiry is implicit and inherent in mobile security contexts in the sense that the security context is bound to a node that covers a specific geographic area. One can refine this and specify the spatial validity explicitly.

17.4.2.5.1 Expressing the Expiry Conditions

The validity of the $MTC_{(HE,SN,UE)}$ context is confined by the spatial validity (an Area Code [*AC*], related to the registration area) and a validity period (*VP*). The geographical coverage of *AC* would depend on the network topology. We advocate that the area should be aligned with an aggregate server area (analogous to VLR/SGSN areas in 3G), and the *AC* may then consist of a set of location (paging) areas. The *AC* parameter should be published on a broadcast channel. There will also be local area codes for the location (routing) area (*LAC*), and this parameter can be used to further restrict the validity of the $STC_{(SN,UE)}$ context. The spatial expiry can only be controlled by the SN and the UE. There is therefore no compelling reason for the HE to know the actual *AC* or *LAC*; and for privacy reasons, it is best if the *AC/LAC* values are not known by the HE.

The validity period, *VP*, should be long enough to avoid excessive context invalidation, yet not be substantially longer than normal UE presence in an area. The actual mobility patterns can vary over time, and it should be possible to configure the *VP* parameter. We advocate that the SN determines a maximum period for the validity period (VP_{SN}) and publishes it on a broadcast channel. The UE should additionally have an HE determined maximum (VP_{HE}) value, and the context should be canceled when one of the validity periods expires. That is, $VP = Min(VP_{HE}, VP_{SN})$. Synchronization can be a problem for the temporal expiry condition. If one lets the *VP* parameter be expressed as a number of time units, then one can let the expiry time itself be expressed in absolute time ($t_{expiry} = t_{establish} + VP$).

17.4.2.6 Enhanced Home Control

Under some circumstances and for some systems, the HE may need stronger home control. By setting the $MTC_{(HE,SN,UE)}$ expiry interval to be short (i.e., frequent AKA invocations) and by tailoring the usage expiry setting, one can achieve tighter home control. There may also be circumstances where the HE needs to know where the UE is. The HE can then dictate the use of mobile phones with a protected location measurement module. Some 3G phones already have built-in GPS (NAVSTAR) receivers, and future mobile phones will probably be capable of receiving both NAVSTAR and Galileo[5] signals (conceivably also the Russian Glonass system). The SN will also be capable of measuring the UE position, and dedicated equipment will be deployed. The primary requirement has been to comply with the regulatory emergency requirements, the so-called E911/E112 requirements. In addition to the E911/E112 requirements, location measuring can be used for location-based services (these are higher-layer services and have no direct bearing on the link layer per se).

Spatial information is a real-time quantity. The measurement rate must therefore be adjusted according to UE velocity, etc. [20]. If UE location privacy is not an issue, one can order UE location reporting to the HE. Instead of solving the problem of enhanced home control during authentication, one can also implement a more generic access control mechanism. One scheme that allows for this is the so-called Spatial Role-Based Access Control (SRBAC) mechanism [22], and we generally advocate to use this or similar mechanisms to provide enhanced home control.

One can also envision scenarios where UE location privacy is an issue and where the HE nevertheless has a legitimate need for spatial control. One can then either use a trusted third party to be the location arbiter, or one can use Secure Multi-party Computation (SMC) methods to solve the contradictory and apparently irreconcilable requirements for spatial home control and for credible privacy protection. The SMC method can be used by having a protocol that solves the *point inclusion problem* [23,24]. One then has an HE defined area, a polygon P, and a UE determined point (x, y); and the trick is to find out whether (x, y) is inside P without divulging any information about (x, y) or P. The standard solutions have high computational and communications complexity but there exist solutions that are more efficient and adapted to a wireless access setting [25].

17.4.2.7 Performance and Integration Issues

Security procedures must be designed to meet the overall system performance requirements. We are particularly concerned with the accumulated delays during initial registration and user session establishment. An inspection of the 3G system call setup signaling scheme reveals that the procedures are merely evolved 2G procedures. For example, the 3GPP/UMTS TS 24.008 "Mobile Radio Interface Layer 3 Specification; Core Network

---

[5] Galileo is a EU/ESA initiative (www.esa.int). The first Galileo satellite was launched in late 2005.

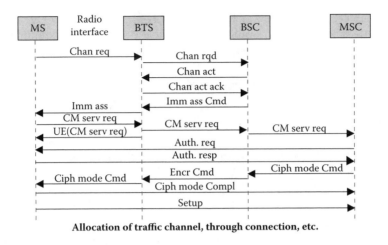

Allocation of traffic channel, through connection, etc.

Figure 17.7   Simplified GSM setup signaling.

Protocols;" the Stage 3 specification is largely a transposed version of
the corresponding GSM 04.08 specification. The 2G procedures were not
bad for the 2G systems, but the 2G service model focused on providing
telephony-type connections, that is, the system had a distinctively circuit-
switched connection-oriented service model. The channels were explicitly
set up in a highly ISDN-influenced sequential procedure (Figure 17.7 shows
an example). In addition, there were many technical limitations due to the
age of the 2G system architecture. New systems should not need to suffer from the 2G/3G history. The control model for future mobile systems
should therefore be designed from the start, and the main input should be
the system service model.

Integration is possible in many areas. We note that the Mobility Management *MM Location Registration* procedure will coincide with invocation of
the AKA protocol. Furthermore, we see that *MM Identity Presentation* must
necessarily precede execution of the AKA protocol. We therefore propose to
combine selected mobility management procedures with the security setup.
We argue that the total number of round-trips can be reduced, even when
deploying an online privacy enhanced three-way AKA protocol. There is an
important consequence of integrating *MM Location Registration, MM Identity Presentation*, and the AKA protocol; namely, that the sequence must
necessarily be initiated by the UE. This is because it is the task of the UE to
initiate *MM Location Registration*. In the 2G/3G systems, the network will
always be the initiator of the AKA protocol, but now we advise to change
the AKA initiator-responder scheme to improve the setup performance.

There is also integration potential for the establishment and renewal
of the short-term context, $STC_{(SN, UE)}$. For example, channel reallocation

may trigger $STC_{(SN,UE)}$ negotiation. The $STC_{(SN,UE)}$ negotiation can then be provided as a functional command, where the information elements are parameters to the channel reallocation command.

17.4.2.8 Security Context Separation

An SN can deploy multiple access networks, and in future mobile systems it is reasonable to expect that the user device will be able to connect to multiple access networks simultaneously. The different access types may all have different levels of security, and one is therefore ill-advised to use the same security context for different access types. It can therefore be argued that the UE and the SN should insist on establishing separate security contexts for each access type being used. One can conceivably have one common medium-term context ($MTC_{(HE,SN,UE)}$) that is used as a basis for derivation of separate short-term contexts ($STC_{(SN,UE)}$) for each of the access types. This is not without risks, and there will be technical issues with context bindings, expiry conditions, etc. We therefore argue against sharing medium-term contexts between different access types. A user device can also engage in communication with multiple SN operators. These SN operators can deploy the same access network types but the security context is tied to the SN and separate security contexts will consequently be needed. There are notable risks associated with allowing a user device to connect to multiple networks (operators) simultaneously (the risk of bridging secure and unsecured accesses is one).

17.4.2.9 Security Algorithm Separation

As demonstrated by Barkan, Biham, and Keller (BBK) in their attack on GSM [8], there can be severe consequences for not having the proper binding between the cryptographic algorithms and the session keys. In GSM, the session key (K_C) is not algorithm specific. Furthermore, the key can be used for multiple sessions, and the sessions can use different encryption algorithms. Due to earlier export restriction, one encryption algorithm (A5/2) was deliberately weakened to allow export of GSM systems to all countries[6]. Due to an unfortunate early decision in GSM design, the error protection is executed *before* encryption. The output data from the error protection stage is highly structured and, by design, contains a lot of redundancy. This is **not** a good characteristic for input to encryption algorithms. The result is that breaking the already weak A5/2 algorithm was made even easier. The upshot is that if an adversary can trick a user into using the A5/2 algorithm, even if it is only for a short period, the adversary will be able to

---

[6] The A5/2 algorithm has now been deprecated, and new handsets are required **not** to support it.

compute the secret key K_C. All communication with the compromised key, both prior to the A5/2 usage and subsequent to it, can now be read by the adversary. The adversary can also impersonate the user and initiate calls, etc. As long as the key has been compromised, it does not matter if the subscriber uses a secure algorithm thereafter. The lesson to be learned is that keys must be bounded to the algorithm (dynamically or statically). This is not a new thing. An example is IPSec, where the Security Association includes static algorithm binding.

17.4.2.10   Subscriber Privacy Aspects

The 3GPP security architecture [14] defines the following user identity confidentiality requirements (abridged):

- *User identity confidentiality*: the property that the permanent user identity cannot be eavesdropped on the radio access link.
- *User location confidentiality*: the property that the presence or arrival of a user in an area cannot be determined by eavesdropping on the radio link.
- *User untraceability*: the property that an intruder cannot deduce whether different services are delivered to the same user.

These requirements are reasonable and necessary, but not complete. The 3GPP security architecture does not capture the case where one requires identity or location privacy from internal system entities. Today, it is increasingly the case that one needs privacy protection from the networks. One cannot realistically expect the SN (there may be several hundred different SNs) to adhere to the HE/UE privacy policy. Furthermore, the UE should also expect some measure of privacy from HE monitoring.

The *user untraceability* requirement is a complex one. It not only requires protection against tracking of identified subscribers, but it must also provide protection against cases where the adversary does not know the UE identity.

With respect to protection against an external intruder, the 3GPP specifications do cover the requirements. Still, the 3GPP specifications fail to provide suitable protection mechanisms. In 3GPP networks, we have the following: the initial identity presentation must necessarily precede the identity verification (authentication). This means that the identity presentation takes place before one has negotiated encryption keys. Thus, the permanent subscriber identity (IMSI) will be transferred in cleartext on the over-the-air interface. This is unacceptable because it allows for subscriber location tracking. To mitigate the problem, the 3GPP SN will normally issue a local temporary identity called TMSI, which is to be used for subsequent subscriber identification. However, the scheme is totally inadequate with

Table 17.1   Knowledge of UE Privacy-Sensitive Data by Domain

Privacy Dimension	HE	SN	UE
UE Location	No	Yes	Yes
UE Identity	Yes	No	Yes

respect to active attacks and also suffers from vulnerabilities that may allow passive attacks to succeed (the allocated TMSI *may* be predictable) [1].

There have been many proposals on how to provide and handle subscriber privacy for mobile systems. The discussions in some of the articles are interesting [26–29], but the solutions are not particularly well fitted to meet the needs of future mobile systems.

17.4.2.11   Subscriber Privacy; Domain Separation

Table 17.1 gives an overview of the knowledge of privacy-sensitive subscriber identity and location information the principal will necessarily be able to acquire. The other principals cannot deny the principal this knowledge. Table 17.1 also indicates that we can provide *UE Identity Privacy* from SN monitoring and *UE Location Privacy* from HE monitoring. To attain subscriber privacy, we must make sure that the system signaling does not leak subscriber privacy-sensitive information.

The identity management scheme and the AKA protocol must therefore ensure that the HE cannot learn the UE location and that the SN will not learn the permanent UE identity. We note that the subscriber may choose to divulge the UE position to the HE or other parties to obtain location-based services. We conclude that this is an application layer decision. We also note that requirements for emergency calls (E112/E911) may dictate that the UE position be made available to the emergency control center. We additionally note that *Lawful Interception* (LI) requirements may also apply. To be able to provide disclosure of UE position to the *Law Enforcing Agency* (LEA) can then be a mandatory requirement.

17.4.2.12   Random Identities

In this clause we investigate the possibility of using random identities and propose a scheme that allows enhanced subscriber privacy to be provided by the AKA protocol.

Location privacy from SN monitoring hinges on the SN not knowing the permanent UE identity. This indicates that the UE should not use the permanent identity when registering with the SN. One solution is for the UE to choose a random reference identity for registration with the SN. This identity, which we refer to as the context identity (*CID*), is used as the primary identification of the medium-term context $MTC_{HE,SN,UE}$. The *CID* identity will act as a pseudo-anonymous UE identity. The *CID* must be constructed such that there is no apparent correlation between the *UEID* and the *CID*. More on the use of pseudo-random context identities can be found in References [25,30]. Because we want the HE to be able to forward data to the UE, we must allow the HE to learn the *UEID-CID* association. This must be done such that the SN cannot learn the *UEID-CID* association or the *UEID*. In terms of cryptographic functionality, we can solve this problem using public key encryption. The UE will then use the HE public key to encrypt the *UEID* and the *CID*. The UE must include a reference to the HE and the public key such that the SN can forward the data to the HE and such that the HE can find the corresponding private key.

Given a regime with randomly generated identities, there will be a certain risk of identity collision. A collision event is quite undesirable, and the SN must reject the access attempt with the duplicate *CID*. If we assume that there is no bias to the *CID* choices, we can use the approximation $p = k/m$, where p is collision probability, k the maximum number of users within the SN server area, and m is the range of the *CID*. To be on the safe side, we assume that an SN server will have, at most, one billion simultaneous users ($k = 10^9$). If we conservatively require the collision to occur for, at most, every 100 million AKA occurrences ($p = 1/10^8$), we have that *CID* must have a range of $m = 10^{17}$. The *CID* variable must therefore have a range of at least 57 bits. This is clearly not a problem, and we therefore conclude that the *CID* can be generated by a pseudo-random function (*prf*). Cryptographic *prf*s often provide us with sequences that are guaranteed not to repeat for the outcome space. This is not strictly necessary for production of *CID*, but there are other reasons that make us want nonrepeatable sequences.

The provision of a pseudo-random identity does solve our problem with identity privacy for the AKA protocol, and the use of asymmetric encryption makes sure that the attained identity privacy is effective against both external adversaries and the SN. However, it is insufficient with respect to tracking protection. We therefore introduce yet another anonymous UE identity. This identity is a session identity (*SID*), and it will only be used locally between the UE and the SN. The *SID* will then be a reference to the $STC_{(SN,UE)}$ context, and, as such, the *SID* will be used in cleartext over shared channels during paging and UE access requests. The UE and the SN will maintain an association between the *CID* and the *SID* for the duration of the $STC_{(UE,SN)}$ context. The HE need not know the *SID*, and we advise that the *SID* not be revealed to the HE. The *SID* can be generated by the UE, but this can complicate integration of radio resource management routines and establishment of the $STC_{(SN,UE)}$ context. We therefore advocate that it be the SN that allocates this identity. The use of the *SID* does not fully prevent subscriber tracking because it clearly is possible to link sessions using the same *SID*, but we can consider the untraceability goal to be adequately satisfied provided that the short-term sessions are sufficiently short. The lifetime of active sessions must also be contained to fulfill the untraceability goal. For a packet-switched architecture, one can consider each packet transfer as a session. This is the extreme case, but then the lifetime of a session will be very short. There is computational and communications overhead associated with context establishment, and the duration of the session lifetime must be balanced against the needs for performance and for security (privacy). We note that from an information theoretic perspective, one must prove the absence of any kind of correlation between each access (including usage patterns, etc.) to prevent tracking. This would include means to decorrelate physical radio-related information. This is outside the scope of our discussion.

The above scenario and the justification for tracking protection with random session identities are made with the assumption that the system uses common channels for paging and access requests. This assumption holds true for the current 2G/3G system, but it may not necessarily be true for future mobile systems. In fact, it is easy to envision systems where paging and access requests take place on dedicated control channels only. For ultra-wideband systems with orthogonal frequency division multiplex (OFDM/OFDMA) scheduling, one can, for example, allocate a permanent dedicated control channel as part of an integrated AKA protocol. Then there will be no need for session identities because session-oriented control signaling will be conducted on the secured dedicated channel. Instead, the problem will be to decorrelate the channel association because it will now be possible to track a subscriber by tracking the usage of the control channel. This problem might be more difficult to solve, but it might be possible to solve it using SN-controlled "pseudo-random channel hopping" or other methods.

17.4.2.13 Computational Balance

The UE must be able to compute the AKA functions on the fly. Modern mobile devices are powerful processing platforms and are capable of executing all the required cryptographic functions with ease. This is true even if we assume usage of computationally expensive algorithms, including Identity-Based Encryption (IBE) transformations, Diffie-Hellman (DH) exchanges, and Secure Multi-party Computations (SMC). The SM can be implemented on a smartcard. The smartcard is a computing platform in its own right; and while it does not have the computational power of the UD, it will still be sufficiently capable for most purposes. Modern smartcards also commonly have hardware support for crypto primitives. While there may be implementation challenges, we still expect the smartcard to have sufficient computational power to execute all necessary crypto-processing. Still, we must ensure that the processing load on the SM is reasonable, and that the AKA protocol is balanced such that the SM is not overloaded. We must also protect the SM against computational denial-of-service (DoS) attacks.

The SN must participate actively in the future AKA protocol. We note that the SN nodes will serve a comparatively large number of users and must be able to execute the AKA protocol on-demand. The instantaneous processing requirements may therefore be demanding, but not all sequences require real-time processing. For example, temporally induced security context renewals can be executed prior to context expiry and therefore need not be executed in real-time. Precomputation may also be possible for some operations (e.g., by precomputing DH parameters).

The HE will serve a large number of subscribers. To instantly compute context credentials may create a substantial load. AKA events are normally time critical; and for HE operators with distinct *busy hour* conditions, it may be a problem to serve all users. The HE must therefore be dimensioned for a high instantaneous crypto-processing load. Still, with optimized crypto-primitives implemented in hardware, we postulate that the capacity required need not be excessive.

17.4.2.14 Communication Balance

The radio channel is a shared physically restricted resource and there will necessarily be capacity limitations. Control signaling is not very demanding and, under most circumstances, we foresee no problems here. However, there may be restrictions on the MTU size during initial access signaling. For performance reasons, we must try to avoid message segmentation, and the MTU size may be a limiting factor in the AKA protocol design. MTU size restriction may preclude and complicate support for primitives that require large information elements, (e.g., Diffie-Hellman-based key generation and asymmetric cryptosystems with large blocksize). Support for data expanding ($E_K(M) \rightarrow C$, where $|C| \gg |M|$) primitives may likewise be problematic. So there will be some design restrictions on the A-interface.

The B-interface, on the other hand, is a fixed network interface. There should be no capacity problems for the B-interface.

17.4.2.15 Denial-of-Service (DoS) Prevention

It is difficult to provide DoS protection in a wireless environment. The AKA protocol must operate in an environment where it is very easy for an adversary to carry out access-denial DoS attacks simply by disrupting the radio transmission. These attacks cannot be prevented by the AKA protocol. An adversary may try to block channel access by issuing "invalid" requests as was the case with the TCP-syn attack [6], but this is significantly more complex than simply disrupting the channel by radio noise. However, these access-denial attacks are local in nature and, in contrast to the

TCP-syn attack, they do not scale well. It is therefore questionable if one should make an effort to try to avert this type of attack at the AKA protocol layer.

The exception is that we must make sure that the attacks cannot block specific users from accessing the system. An adversary should not be able to impersonate a legitimate subscriber and then block that subscriber from the system by executing too many false or failed AKA attempts. To limit the effect of computational DoS attacks, we suggest that the SN restrict the arrival rate of AKA invocations per access point. The HE, likewise, can limit the number of simultaneous AKA sessions from any given SN. Together, this will prevent a computational DoS attack from scaling. Needless to say, the SN and HE must monitor the number of unsuccessful access attempts due to authentication failure as this may be a sign that the system is under attack.

17.4.2.16 Security Termination Point and Key Distribution

In a real network, there will be many interfaces and a number of physical nodes. Our model with two interfaces and three principals (Figure 17.2) is too limited when its comes to discussing the issue of security termination points. In Figure 17.3 we see a more realistic scenario. We see that it is the MT, and not the UE, that terminates the confidentiality and integrity services at the subscriber end. This makes sense because the MT would process all data transfers, including the error correction codes, etc.

In the Universal Mobile Telecommunications System (UMTS), the network termination is at the radio network controller (RNC). The RNC is a UTRAN node and hence the termination is in the access network (AN). In GSM, one also terminates security in the AN; but for GSM, the termination point is the base station (BS). The problem with this is that the link between the BS and the base station controller (BSC) is often an unprotected radio link. Furthermore, in future mobile systems, the AP will likely be an inexpensive node with little physical protection. The APs will also be highly distributed and will generally not enjoy the safety of a protected location.

So we must conclude that the AP is not a good termination point. In general packet radio service (GPRS), one tried to fix the problems with security termination in the BTS, and thus moved the termination point to the serving GPRS support node (SGSN) core network (CN) node. However, we have now learned that this was too far into the network. The problem is that from a system design perspective, the CN should *not* need to know about the specifics of the ANs. When terminating security at the CN, one violates this principle. There are also other problems with quality measurements and AN monitoring that will then have to be routed via the CN. So we conclude that access security should not terminate in the CN but, rather, somewhere in the AN. Figure 17.8 gives an outline of the scenario and alludes to a solution.

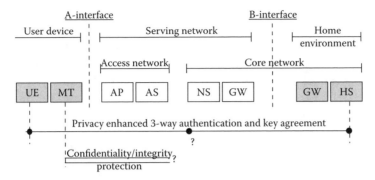

AP-Access Point; AS-Access Server; NS-Network Server; GW-Gateway; HS-Home Server

Figure 17.8   Security termination points.

Key generation also influences the termination point decision. For the subscriber and HE cases, the termination points are unproblematic. The SN case is more complex. Security termination and key generation must be seen in conjunction. Key generation is part of AKA processing, and the keys will therefore be produced at the AKA termination points. It is possible for the SN to derive keys at a CN node and then (securely) distribute the keys to the AN. This is the way the 3G systems do it. The immediate benefit is that the AN now is independent of the AKA protocol. Thus, it will be easier to integrate new access network types during the lifetime of the system. On the other hand, the system must then provide secure key distribution. The alternative solution is to terminate the AKA protocol in the AN, but we foresee problems regarding system evolution with this alternative. We therefore generally advise to terminate the AKA protocol at the edge of the CN, as is done in 3G systems.

17.4.2.17   Summary

We have identified the following high-level requirements for an access security architecture and an AKA protocol. The statements and requirements may not always apply, and the designer should judge their applicability and be selective when necessary.

17.4.2.17.1   High-Level Access Security Design Guidelines

1. *Home control.* The HE is entitled to have enforceable home control with respect to chargeable events.
2. *Online security context establishment.* The security context must be established online to guarantee freshness, and thereby provide

real-time home control for the mobility management registration event.

3. *Active three-way security context establishment.* All principal parties must be *active* participants in the establishment of the security context; all principals should have a measure of influence on the established security context. We therefore require an online three-way AKA protocol.

4. *Security context hierarchy.* To provide the necessary performance and flexibility, the security architecture must provide a security context hierarchy. This hierarchy must reflect the trust assumptions.

5. *Security context exposure control.* The security contexts must have sufficient exposure control mechanisms. The control dimensions include expiry conditions for usage, time, and area.

6. *Spatio-temporal security context binding.* The spatial and temporal control dimensions should be bounded to the security context explicitly. Usage-based expiry will additionally be needed for session key expiry.

7. *Identity management, registration, and security setup.* The procedures for identity management, registration, and security setup are logically connected and should be designed in conjunction.

8. *Channel allocation and session key agreement.* The procedures for channel allocation and session key agreement can be executed in conjunction and should be designed to allow this.

9. *Security context separation.* Short-term security contexts should never be shared over different access networks.

10. *Security algorithm separation.* Key material in a security context should never be used for more than one algorithm.

11. *Subscriber privacy — identity privacy.* Only the HE and the UE should ever be allowed to learn the permanent UE identity.

12. *Subscriber privacy — location privacy.* Only the UE and the SN should be able to learn the UE location as part of normal link layer operation. The HE should never be allowed to learn the UE location unless explicitly required.

13. *Subscriber privacy – untraceability.* No entity should be allowed to monitor UE movement over time. This implies that the UE must never use any visible identity or reference for a prolonged time.

14. *Subscriber privacy override.* For better or worse, regulatory requirements may dictate that the UE position and identity be revealed. Given HE and SN cooperation, it must therefore be possible to recover the UE identity and position.

15. *SN security termination.* Access security should terminate in the access network at an access server (AS). The three-way AKA protocol should terminate in a network server (NS) in the core network. The SN must secure the link between the AS and the NS.

We now have the input for designing a Privacy Enhanced three-Way Authentication and Key Agreement (PE3WAKA) protocol. The PE3WAKA protocol is an essential component in our proposal toward an improved access security architecture for future mobile systems.

17.5 Cryptographic Basis

17.5.1 Cryptographic Aspects

We now briefly discuss cryptographic issues for access link protection. Our need here is to provide adequate confidentiality and integrity/message authentication services for both user data and user-related signaling data. For these services and the intended context, we strongly recommend the use of symmetric-key crypto-primitives. These primitives are fast and efficient. Because our target system will likely be in operation for years to come, we must make sure that protection remains adequate for the intended lifespan of the systems. A time horizon of at least 10 to 20 years seems a realistic assumption. For the purpose of link layer protection, we do not require that data protection be effective for this time span, but we must ensure that the algorithms in place can be in operation for a substantial time period. To phase out an aging algorithm and phase in a new, improved algorithm takes considerable time, often several years. So we must be somewhat conservative when choosing our algorithms.

For confidentiality protection, the key size is the most important factor, but not the only important factor. One must also ensure that the internal state (block size) has sufficient size. For cryptographic reasons, one should not normally use a block cipher key for more than $2^{b/2}$ blocks (where b is the block size). For example, with a 64-bit block size, one should limit the lifetime of the keys to processing approximately 32 Gbytes of data. For future mobile systems, this limit is likely to be unacceptable. The standard for symmetric encryption is the Advanced Encryption Standard (AES) [31]. The AES is a block cipher with a block length of 128-bits and the "normal" case is to use AES with 128-bit keys. Thus, we recommend to use a cipher with 128-bit keys and 128-bit internal state for access security. A cipher must be used in a mode-of-operation to provide the desired confidentiality services. The mode-of-operation is essentially a specification on how the crypto-primitive should be used. For high-performance systems, there is a strong argument in favor of counter-based modes, because these modes will allow parallel processing of the crypto-blocks. The IEEE 802.11i standard [32] uses AES in this mode to provide WPA-2 protection for WLANs. The above discussion only covers block cipher algorithms, and historically this has been for a good reason. Stream ciphers have not enjoyed a good reputation. This is changing, however, and stream ciphers are equally well suited for link layer protection[7]. An example is the SNOW-3G stream cipher that is being developed by the 3GPP SA3 Working Group for use in UMTS access security[8].

Integrity/message authentication protection is different from encryption, in that it produces additional data, the integrity checksum value (*ICV*), that must be appended to the payload. The purpose of the ICV is to provide cryptographic error detection. We use a *MAC()* function to generate the *ICV*. The *MAC()* functions are constructed in different ways, the most common being from a cryptographic hash function (HMAC) or from a block cipher (CBC-MAC, etc.). A primary characteristic of a *MAC()* function is that it should be computationally infeasible to derive the secret *key* from knowledge of the *message* and the *ICV*:

$$MAC_{key}(message) \rightarrow ICV$$

The *message* is of variable length and is compressed to a fixed length *ICV*. Both the *key* length and the *ICV* length are important parameters. How large should the ICV be? The short answer is that it must be long enough to avoid problems with falsified data. However, it must also be short enough to be affordable. The *ICV* is, after all, communications overhead. The "standard" cryptographic hash functions have output digests with typical lengths of 128 to 256 bits. The upper end of this scale is fine if long-term protection is essential (digital document signing, etc.). Here, the protection time period will be in years (30 years is a common expectation), but this would be overkill for access link protection. If we look at IPSec [17], the transmitted checksum is 96 bit wide, truncated down from the computed checksum that is at least 128 bit wide. The 96 bit checksum is then used to protect a full IP packet. We observe that the protection goal for IPSec is for the network layer, and it is potentially used as an end-to-end service. For the link layer, the goals will be more modest. While we assume a powerful adversary, we note that our adversary will be in a hurry. The data being transferred will rapidly expire and successful falsifications will have little value after expiry (unless the key was revealed). In UMTS, the *ICV* is only 32 bit long, but this is dangerously short. The 32 bit UMTS *ICV* length can be traced to design restrictions arising from the evolved history of UMTS, and we do not suggest that 32 bits is enough for future systems.

---

[7] Although block cipher primitives are used in UMTS, the mode-of-operation used (modified Output-Feedback Mode) does actually provide a stream cipher service.

[8] The KASUMI algorithm and the corresponding f8/f9 functions have **not** been broken, but it is prudent practice to have an alternative ready. In this respect, it is a clear advantage that the cryptographic basis of the KASUMI and the SNOW-3G algorithms are constructed according to different principles.

When it comes to key size and the size of the internal state, one normally takes into account the *birthday paradox*. The birthday paradox indicates that to provide *collision resistance*, one must use roughly twice the "normal" key size [33]. That is, to provide 128 bit security, one must use a 256 bit key. However, one is *not* particularly concerned with collisions for link layer access security. Given a message m and a corresponding *ICV*, then an adversary can cause a collision if she can find a substitute message m' that also generates *ICV*. That is, $MAC_{key}(m) = MAC_{key}(m')$. The point here is that the adversary does not know the *key* but is still able to find a message m' that produces the same *ICV* as the original message m. However, a birthday paradox collision does not give the adversary any control whatsoever over the produced (false) message m'. So, in all likelihood, the collision m' is completely meaningless. The adversary will therefore only be able to produce a piece of apparently randomly corrupted data (which may be seen as a DoS-type of attack). Note that the effects of the attack are merely the equivalent of radio signal disruption.

With this in mind, we suggest using at least 64 bit checksum for signaling data, and we propose that it be computed with at least 128 bit keys and at least 128 bit internal state. For user data, one *can* use the same lengths, but we advocate longer checksums here. Because future systems will likely have adaptive coding and frame sizes, the actual checksum length may be adaptive as well. We note that there is an opportunity to combine the transmission protection integrity service and the cryptographic integrity protection, and this may save both processing time and bit overhead.

There are schemes that go even further and integrate integrity and confidentially protection directly with data encoding. As demonstrated by the GSM case [8] one must then be very careful with exactly how one designs such a scheme. It may also be advisable to provide a user data QoS service class with no user data integrity protection. This will make sense for applications and services that will provide full integrity protection at higher layers.

17.5.2 Pseudo-Random Functions

We use pseudo-random functions extensively in the AKA protocol. For our purpose, we define a generic pseudo-random function $prf(\cdot) \rightarrow rnd$ with the following characteristics:

- The function $prf_w(\cdot)$ will output *rnd* values in the output range $[0..w]$.
- The output will be uniformly distributed over the output range.
- The output of $prf(\cdot)$ will be unpredictable. That is, even after having observed n consecutive *rnd* values, the observer will not be able to guess output $n+1$.

■ The output of $prf_w(\cdot)$ will be nonrepetitive. That is, for any sequence n of *rnd* values, the value of *rnd* will never be repeated, provided the sequence length is shorter than the size of the output space, $(len(n) \leq |w|)$.

With our intended use it is important that the range w is chosen such that it is never exhausted.

17.5.3 Diffie-Hellman (DH) Exchange

The DH exchange is described and discussed extensively in Menezes et al. [33]. Suffice here to say that in a basic DH exchange, Alice and Bob must first agree on an appropriate prime p and a suitable generator g $(2 \leq g \leq p-2)$. Alice chooses a suitable random secret a such that $A = g^a \bmod p$. Bob chooses a suitable random secret b such that $B = g^b \bmod p$. Alice and Bob then exchange, A and B, and compute the DH secret s as shown below. The basic DH exchange is not authenticated and is therefore susceptible to man-in-the-middle attacks.

$$s = B^a \bmod p$$
$$s = A^b \bmod p$$

The medium-term security context will have as its basis a DH secret s. We propose to use an ordinary two-party DH exchange and then configure the usage of the derived secret such that no party will gain a definitive advantage in deciding s. There will be two distinct cases:

1. Execute the DH exchange over the over-the-air A-interface.
2. Execute the DH exchange over the fixed B-interface.

Case 1 is problematic. We will want to have a shared secret s with an entropy on the order of 256 bits. With standard DH methods, this will amount to an exchange of at least two 15-Kbit information elements over the A-interface [34]. During setup, this may not be acceptable. However, if we use an elliptic curve cryptography based DH method, we might be able to reduce the information element size considerably [34].

Case 2 is simpler. We do not have have to worry about information element sizes, as the bandwidth between SN and HE will be so large as to make the issue insignificant. The problem with case 2 is that the UE needs access to the secret. So, the HE must be able to forward the secret to the UE in a confidentiality and integrity protected manner.

17.5.4 Identity-Based Encryption (IBE)

For the purpose of the proposed AKA protocol, any public key encryption scheme can be used. The property that we require is that the UE be able to forward confidentiality-protected data to the HE. Because the SN will forward the data, we cannot include information that would allow the SN to learn the UE identity. This effectively prevents symmetric secret key methods because the UE is unable to include any reference as to who the initiator is, and this means that the HE will be unable to decide which secret key to use for decryption (we do *not* want to use symmetric group keys).

We must therefore use asymmetric methods in this case, that is, cryptosystems with encryption $E(\cdot)$, decryption $D(\cdot)$, and where $D_{k^{-1}}(E_k(M)) = M$ for the key k that satisfies $k \neq k^{-1}$. An essential property of asymmetric methods is that the public key is *public*. Therefore, the use of asymmetric methods will provide data confidentiality protection for transfers to the holder of the private key[9], but there will be no guarantees on who sent the encrypted data (i.e., no *data origin authentication*). We must therefore provide data authentication/data integrity in other ways.

IBE, which belongs to the public key methods, has the interesting property that one can use a public identity, or any other public information, as the encryption key. The idea of IBE dates back to 1984 when Shamir [35] asked for a public key encryption scheme in which the public key is an arbitrary text string. Shamir proposed a scheme in which the e-mail address could serve as the public key. So when Alice needs to send an encrypted message to Bob, she simply uses Bob's e-mail address as the public key to encrypt the message. There is no need for Alice to obtain Bob's certificate before proceeding. When Bob receives the encrypted message, he contacts a third party, called the private key generator (PKG), to retrieve the private key. The retrieval process is analogous to private key retrieval from a certificate authority (CA), and it is therefore a prerequisite that Bob and the PKG have an *a priori* arrangement with respect to authentication and key protection. Note that the private key can be generated subsequent to the use of the corresponding public key, and that the PKG will necessarily have access to the private key. The IBE problem was first solved by Boneh and Franklin [36], and the presentation given here is with the Boneh-Franklin IBE scheme in mind.

Because the identity string can be an arbitrary string one may construct it to contain expiry conditions. For example, using the e-mail example, one can construct an identity key in the following manner: $K_{ID} :=$ "bob@pe3waka.net"||"2007.01.01". The K_{ID} is then constructed to be valid until the date "2007.01.01". The inclusion of the expiry condition in the key

---

[9] This is the key corresponding to the public key. The private key itself is secret.

is useful, and it will be used in the proposed AKA scheme. The IBE functions include:

■ *The Setup() function.* The $Setup(k) \rightarrow sp, s$ function is computed by the PKG. The function takes a parameter, k, as input, and this parameter is used to generate the prime order groups $\mathbb{G}_1$, $\mathbb{G}_2$, and a bilinear map $\hat{e} : \mathbb{G}_1 \times \mathbb{G}_2 \rightarrow \mathbb{G}_2$. The output is system parameters sp and a master key, s. The sp identifier defines the groups $\mathbb{G}_1$, $\mathbb{G}_2$, the map $\hat{e}$, the finite message space $\mathcal{M}$, the finite ciphertext space $\mathcal{C}$, the hash functions H_1, H_2 and two generators P, P_{pub}. The P_{pub} generator is derived from P using the master key s. The system parameters sp need not be secret. The master key s will only be known to the PKG.

■ *The Extract() function.* The $Extract(sp, s, ID) \rightarrow d$ function is computed by the PKG. The function takes the system parameters, the master key, and an identity string (ID) as input. The identity string is viewed as an arbitrary-length binary string $ID \in \{0, 1\}^*$. The hash function $H_1 : \{0, 1\}^* \rightarrow \mathbb{G}_1^*$ is used on ID to produce the fixed-length identifier Q_{ID}. The output d is the (secret) private key. We then have a public-key key pair (ID, d).

■ *The Encrypt() function.* The $Encrypt_{ID,sp}(M) \rightarrow C$ function is computed by Alice. The function takes as input sp, ID, and the message $M \in \mathcal{M}$. The output is the ciphertext $C \in \mathcal{C}$.

■ *The Decrypt() function.* The $Decrypt_{d,sp}(C) \rightarrow M$ function is computed by Bob. The function takes as input sp, d, and the ciphertext $C \in \mathcal{C}$. The output is the plaintext message $M \in \mathcal{M}$.

In our use of IBE, the HE will play the role of both PKG and Bob. The UE will play the role of Alice. We will assume that the system parameter, sp, is preloaded onto the SM for use by the UE. It will be part of the long-term security context ($LTC_{(HE,UE)}$) credentials. Our motivation for using IBE instead of conventional asymmetrical cryptographic methods is largely due to the effortless and immediate context binding IBE will permit us to create.

17.5.4.1 Mutual Challenge-Response Schemes

We suggest the use of symmetric-key challenge-response schemes to provide entity authentication between the UE and the HE. These methods are computationally inexpensive. In contrast to the 3G AKA challenge-response method, we want to avoid the complication of a sequence number scheme. The 3G AKA protocol is executed in a single round-trip, and one then needs protection against replays (to provide the UE with a freshness guarantee). However, sequence numbers are a common source of configuration errors, which can both trigger unnecessary synchronization events and fail to protect against the use of aging security credentials. We will instead rely on a much simpler online, double challenge-response scheme to provide mutual entity authentication with guaranteed freshness. As for the 3G AKA scheme, we will use MAC signed challenges.

17.6 Privacy-Enhanced AKA Protocols

17.6.1 The PE3WAKA Example Protocol

We now present a Privacy Enhanced Authentication and Key Agreement (PE3WAKA) protocol. The protocol should be taken as an example and a suggestion on how to construct a privacy-enhanced AKA protocol. Different objectives and different system architectures may lead to radically different solutions. We assume that protection for the B-interface is established prior to PE3WAKA execution. For the sake of simplicity, we have constructed the PE3WAKA protocol such that it only generates the common medium-term security context $MTC_{(HE,SN,UE)}$ and the A-interface short-term context $STC_{(SN,UE)}$. The relevant information elements (IEs) are presented in Table 17.2.

Table 17.2   Information Elements

IE	Explanation/Comment
HEID	The public HE Identity; the *HEID* can be used to derive a HE address
SNID	The public SN Identity; the *SNID* can be used to derive an SN address
CH	The random challenge; subscripts denote the sender
RES	The corresponding response to a challenge; subscript matches the challenge
ID	Public IBE Key
d_{ID}	The corresponding private key; subscripts indicates the public key
Q_{ID}	IBE hashed ID key; used by the *Encrypt*() and *Extract*() functions
VP	The Validity Period
AC, LAC	Area Code and Local Area Code; these are SN areas (broadcast information)
DH	Diffie-Hellman public key; subscripts denote the creator
dhs	Diffie-Hellman secret
KC	(HE,UE) shared key belonging to the $LTC_{(HE,UE)}$ context
KA	(HE,UE) shared key belonging to the $LTC_{(HE,UE)}$ context
mtk	Keyset; belonging to A-interface medium-term context
stk	Keyset; belonging to A-interface short-term context
bkey	Keyset for B-interface, included in $STC_{(HE,SN)}$
sp, s	IBE parameters

17.6.2 Functions

Function 17.1 shows how to construct the pseudo-random UE identity *CID*. The *CID* identity is generated by the UE and is used for the $MTC_{(HE, SN, UE)}$ medium-term context.

$$UE : prf(\cdot) \rightarrow CID \qquad (17.1)$$

Function 17.2 shows how to construct the UE session identity (*SID*). The *SID* is generated by the SN and used for the $STC_{(SN, UE)}$ short-term context.

$$SN : prf(\cdot) \rightarrow SID \qquad (17.2)$$

Symmetric keys for the A-interface are constructed as shown in Functions 17.3 and 17.4. The function *KeyDeriv*() is a dedicated key derivation function. The shared secret *dbs* must contain sufficient entropy to allow for several key derivation runs. The *CID* and *SID* identifiers also serve to explicitly bind the keys to the context. We require keys for both confidentiality and integrity protection, and this means that we need two *independent* keys. The *KeyDeriv*() function does not depict this, and care must be taken to ensure key independence. The keyset is only to be used over the A-interface. We have therefore added the area codes to the derivation. This explicitly binds the keys to the areas. If the HE also needs to do the key derivation, we must make sure that there is no conflict between privacy goals and the area code knowledge.

$$SN, UE : KeyDeriv_{dbs}(CID, AC) \rightarrow mtk \qquad (17.3)$$

$$SN, UE : KeyDeriv_{dbs}(SID, LAC) \rightarrow stk \qquad (17.4)$$

The IBE public key pair is constructed as shown in Functions 17.5 through 17.7. Here we include key binding to both the area and the validity period. The derivation used is slightly different from the normal IBE construction, in that we use a modified *Extract'*() function. The HE will need to derive d_{ID} and thus it appears that it will learn the *AC*; but when we let the SN compute Q_{ID}, we circumvent this problem. We can therefore include the area code *AC* and yet not violate the UE location privacy. The radio information, *RI*, is a radio-related parameter that the HE cannot guess; that is, the *RI* should be nonpredictable as seen from the HE. A locally valid short-lived frame number may be a good *RI* choice.

$$ID := HEID||SNID||AC||VP||RI \qquad (17.5)$$

$$H_1(ID) \rightarrow Q_{ID} \qquad (17.6)$$

$$Extract'(sp, s, Q_{ID}) \rightarrow d_{ID} \qquad (17.7)$$

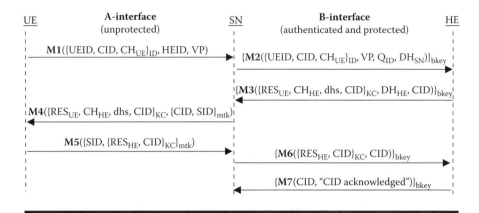

Figure 17.9 The PE3WAKA example protocol.

Functions 17.8 and 17.9 demonstrate how to derive the challenge-response data for use between the HE and the UE. The challenge data is generated using a pseudo-random function. When used, the challenge will have a subscript to indicate the generating principal. The response to the challenge is computed over the UE context identity and the challenge data. Function 17.9 must be a one-way function. To demonstrate authenticity, a preshared secret authentication key *KA* is used. The *KA* key belongs to the $LTC_{(HE, UE)}$ context.

$$prf(\cdot) \rightarrow CH \qquad\qquad (17.8)$$

$$Resp_{KA}(CH, CID) \rightarrow RES \qquad\qquad (17.9)$$

17.6.3 Protocol Outline

In the following we assume that the access channel has severe bandwidth restrictions. This would be similar to the 2G/3G case. In this scenario, we cannot afford to execute a Diffie-Hellman exchange over the A-interface. We therefore propose to let the HE execute the DH exchange on behalf of the UE. This is permissible because the HE has security jurisdiction over the UE, yet it creates a problem in that the HE must now transfer the derived DH secret to the UE such that neither the SN nor anybody else can read or modify the DH secret. Figure 17.9 provides a graphical presentation of the PE3WAKA protocol.

PE3WAKA Protocol: Bandwidth-Restricted A-Interface

1. $UE \rightarrow SN : A, HEID, VP$
   where: $A := \{UEID, CID, CH_{UE}\}_{ID}$
2. $SN \rightarrow HE : \{A, VP, Q_{ID}, DH_{SN}\}_{bkey}$

3. $HE \rightarrow SN : \{B, DH_{HE}, CID\}_{bkey}$
   where: $B := \{RES_{UE}, CH_{HE}, dhs, CID)\}_{KC}$
4. $SN \rightarrow UE : B, \{CID, SID\}_{mtk}$
5. $UE \rightarrow SN : \{SID, \{RES_{HE}, CID\}_{KC}\}_{mtk}$
6. $SN \rightarrow HE : \{\{RES_{HE}, CID\}_{KC}, CID\}_{bkey}$
7. $HE \rightarrow SN : \{CID, "acknowledge"\}_{bkey}$

17.6.4 Discussion and Analysis

17.6.4.1 Step-by-Step Discussion

When the UE needs to execute the PE3WAKA protocol, it begins by computing *ID*, *CID*, and the HE challenge CH_{UE}. It then protects this data with *ID* and forwards the encrypted data together with the *VP* value and the *HEID*.

The SN receives message M1 and looks up the HE address. SN then computes (or retrieves a precomputed) DH value DH_{SN}. It also computes Q_{ID}. It then adds *A* and *VP* to the message, before encrypting the whole message with *bkey*. The SN does not yet know the *CID* identity.

The HE receives message M2 and decrypts it using *bkey*. It then uses Q_{ID} to derive a corresponding private key to *ID*, and this key, d_{ID}, is used to decrypt *A*. The HE now *sees* UEID, CID, and CH_{UE}. It then tentatively accepts *CID* and computes a response RES_{UE}. The HE also computes a new challenge-response pair: (CH_{HE}, RES_{HE}). It then generates (or retrieves a precomputed) a DH value, and then computes the shared DH secret *dhs*. It then constructs block *B* and encrypts it with *KC*. Block *B*, the tentatively accepted *CID* and the DH value DH_{HE}, is then encrypted with *bkey* and transferred to the SN.

The SN receives message M3 from the HE and decrypts it. It now *sees* *CID* and tentatively accepts it. The SN then computes the DH secret *dhs*. From *dhs*, the SN generates the *mtk* keys. The *mtk* keys can *only* be derived by a party that has access to *dhs*. The SN also generates *SID*, and uses *mtk* to protect (*CID*, *SID*) and then sends the message to the UE.

When the UE receives message M4, it must first process *B*. The UE uses *KC* to decrypt *B*. It then confirms that *B* contains *CID* and that the response RES_{UE} from the HE is correct. The UE has now successfully authenticated the HE. It proceeds to compute its own response RES_{HE} to the HE challenge. Given that the UE accepts HE jurisdiction, the UE will also accept the HE-provided *dhs*. So the UE can use *dhs* to produce the *mtk* keys. It then uses the *mtk* keys to decrypt and verify the (*CID*, *SID*) block. The UE now knows that the SN has received *CID* from the HE. The UE also has confirmation that the HE has acknowledged the SN. The UE will therefore accept

SID as an authenticated reference. The UE now responds to the SN with a demonstration that it has received and accepted *SID*.

The SN receives message M5 and decrypts it to find *SID* and the response (RES_{HE}, *CID*) block. Because the presence of *SID* demonstrates that the UE has *dbs* possession, the SN will tentatively accept the *CID* as an authenticated reference. That is, from now on, the short-term context can be activated. The SN will, nevertheless, want HE confirmation. The SN forwards the (RES_{HE}, *CID*) block to the HE, together with the *CID*.

The HE receives message M6 and confirms the response RES_{HE} and the *CID* binding (both with SN and UE). It then accepts that UE is authenticated to be present at SN and that *CID* is the context reference. It then sends a message to the SN to confirm this. Upon receiving the final message M7, the SN now knows that the HE accepts liability for *CID* as an authenticated reference identity.

17.6.4.2   Efficiency Analysis

We now compare the round-trip efficiency of the PE3WAKA protocol to the corresponding 3G scheme. This amounts to assessing the cost of a 3G location updating sequence, including a 3G AKA sequence, and comparing it to the PE3WAKA round-trip cost (ref.: 3G TS 23.108 (ch. 7.3.1)):

■ UE→SN: *Location Updating Request*
■ SN→UE: *Authentication Request*
■ UE→SN: *Authentication Response*
■ SN→UE: *Location Updating Accept*

The 3G TS 23.108 specification only covers the UE-SN communication. The SN-HE part can be found in 3G TS 29.002, and it constitutes two separate request-reply sequences, where the SN first fetches the subscriber information and then the security credentials. In PE3WAKA, the subscriber information will be forwarded in parallel with message M3. The PE3WAKA protocol performs slightly better than the 3G scheme on the UE-SN interface. Note that message M4 serves as location registration confirmation. On the SN-HE interface, the PE3WAKA protocol requires two passes. The 3G scheme also requires two passes. We note that the UE-SN context is potentially operative after SN reception of message M5. The SN still wants HE confirmation, but it now has indirect *CID* confirmation. It is therefore safe for the SN to activate the short-term context. With this in mind, we can defend the claim that PE3WAKA is at least as efficient as the comparable 3G sequences.

17.6.5 Summary

We have demonstrated how one can design effective and efficient privacy enhanced three-way authentication and key agrement protocols for future mobile systems. The design does include elements of mobility management, and the example PE3WAKA protocol will, in fact, provide identity management and security context setup. It also effectively contains the registration phase. With other requirements or other system configurations, one can design the PE3WAKA protocol differently. For example, in Køien [37], is an example of how one can construct a PE3WAKA protocol for the case when one can afford to execute the DH exchange over the A-interface. We must also mention that it is very difficult to construct AKA protocols and that one should remain skeptical about new protocols until they have been subjected to formal analysis and verification. The current state-of-the-art in security protocol validation will catch a set of well-known security problems; and while successful validation does not imply correctness, it should at least demonstrate that the protocol is free from the more embarrassing errors. We must still note that the validation proof will be no better than the validation model, and that faithful and reasonably comprehensive modeling of advanced protocols is a very difficult task.

17.7 Conclusion

In this chapter we discussed various security architecture aspects for future mobile systems. We have attempted to define what we mean by access security and what security services we need in a mobile access security architecture. We have defined a simplified model of the principals and the mobile system. Subscriber privacy remains a growing concern, and future mobile systems must be able to offer effective subscriber privacy as a basic service.

We have presented and analyzed the requirements for a privacy-enhanced authentication and key agreement protocol for use in future mobile systems. This culminated in an example AKA protocol. The presented PE3WAKA protocol is capable of substantially improved user location/identity privacy compared to the 3G scheme. The PE3WAKA protocol provides enhanced privacy from eavesdropping and manipulation by an outside adversary, and it provides a measure of user location/identity privacy with respect to the serving network and the home operator. To achieve this, the user identity presentation scheme and the initial registration procedures had to be modified from the scheme used in 2G/3G systems. However, this change also allowed performance improvements by integration of signaling procedures that is anyway triggered by the same physical events. The PE3WAKA protocol also recognizes that all three principals should participate in the security context, and it provides online establishment of a three-way security context. Finally, the PE3WAKA protocol provides enhanced key distribution relative to 2G/3G schemes. In the PE3WAKA protocol, the serving network participates actively in creating security credentials for session key derivation. Contrast this with the 2G/3G scheme in which the home entity unilaterally decides the session keys.

All in all, we believe that the analysis and derived requirements presented herein represent a solid foundation on which to build future mobile access security architectures.

References

[1] G.M. Køien, An Introduction to Access Security in UMTS. *IEEE Wireless Communications Mag.*, Vol. 11, No. 1, pp. 8–18, February 2004.

[2] G. Rose and G.M. Køien, Access Security in CDMA2000, Including a Comparison with UMTS Access Security, *IEEE Wireless Communications Mag.*, Vol. 11, No. 1, pp. 19–25, February 2004.

[3] G.M. Køien and T. Haslestad, Security Aspects of 3G-WLAN Interworking. *IEEE Communications Mag.*, Vol. 41, No. 11, pp. 82–88, November 2003.

[4] K. Nyberg and V. Niemi, *UMTS Security.* ISBN 0-470-84794-8, Wiley, 2003.

[5] ITU-T, *SECURITY ARCHITECTURE FOR OPEN SYSTEMS INTERCONNECTION FOR CCITT APPLICATIONS*, Recommendation X.800, Geneva, 1991.

[6] CERT, *TCP SYN Flooding and IP Spoofing Attacks*, CERT Advisory CA-1996-21 1996.

[7] D. Dolev and A. Yao, On the Security of Public-Key Protocols, *IEEE Transactions on Information Theory*, Vol. 29, No. 2, pp. 198–208, 1983.

[8] E. Barkan, E. Biham, and N. Keller, Instant Ciphertext-Only Cryptanalysis of GSM Encrypted Communication, in *Proceedings of Crypto 2003*, Santa Barbara, CA, 2003.

[9] J. Walker, Unsafe at Any Key Size: An Analysis of the WEP encapsulation, Technical Report 03628E, IEEE 802.11 Committee, March 2000.

[10] M. Abadi and R. Needham, Prudent Engineering Practice for Cryptographic Protocols, DEC SRC Research Report 125, California, June 1994.

[11] 3GPP, emph3G Security; Security Threats and Requirements, Sophia Antipolis, France, 2001.

[12] M. Burrows, M. Abadi, and R. Needham, A Logic of Authentication, DEC SRC Research Report 39, California, 1990.

[13] G.M. Køien, Principles for Cellular Access Security, *NORDSEC 2004*, pp. 65–72, Espoo, Finland, November 2004.

[14] 3GPP, 3G TS 33.102: 3G Security; Security Architecture, Sophia Antipolis, France, 2004.

[15] 3GPP, 3G TS 33.234: 3G Security; Wireless Local Area Network (WLAN) Interworking Security, Sophia Antipolis, France, 2004.

[16] A. Joux, A One Round Protocol for Tripartite Diffie-Hellman, in *Proc. ANTS-IV 2000*, LNCS 1838, Springer-Verlag, 2000.

[17] S. Kent and K. Seo, Security Architecture for the Internet Protocol, RFC 4301, December 2005.

[18] C. Kaufman (Ed.), Internet Key Exchange (IKEv2), Protocol, RFC 4306, December 2005.

[19] 3GPP, 3G TS 33.210: 3G Security; Network Domain Security (NDS); IP Network Layer Security, Sophia Antipolis, France, 2005.

[20] G.M. Køien and V.A. Oleshchuk, Spatio-Temporal Exposure Control; An Investigation of Spatial Home Control and Location Privacy Issues, *IEEE PIMRC 2003*, pp. 2760–2764, Beijing, China, September 2003.

[21] G.M. Køien and V.A. Oleshchuk, Privacy-Preserving Spatially Aware Authentication Protocols: Analysis and Solutions, *NORDSEC 2003*, pp. 161–173, Gjøvik, Norway, 2003.

[22] F. Hansen and V.A. Oleshchuk, Spatial Role-Based Access Control Model for Wireless Networks, in *Proc. of IEEE Vehicular Technology Conference (VTC2003)*, pp. 2093–2097, 2003.

[23] M.J. Atallah and W. Du, Secure Multi-Party Computational Geometry, in *WADS2001: 7th International Workshop on Algorithms and Data Structures*, pp. 165–179, USA, August 8–10, 2001.

[24] W. Du and M.J. Atallah, Secure Multi-Party Computation Problems and Their Applications: A Review and Open Problems, *NSPW'01*, pp. 13–21, September 10–13, 2002.

[25] G.M. Køien and V.A. Oleshchuk, Location Privacy for Cellular Systems; Analysis and Solution, in *Proc. of Privacy Enhancing Technologies Workshop 2005*, Springer LNCS 3856, Cavtat, Croatia, 2005.

[26] H. Federrath, A. Jerichow, and A. Pfitzmann, MIXes in Mobile Communication Systems: Location Management with Privacy, in *Proc. First Int. Workshop on Information Hiding*, Cambridge, U.K., LNCS 1174, Springer, May/June 1996.

[27] G. Ateniese, A. Herzberg, H. Krawczyk, and G. Tsudik, Untraceable Mobility or How to Travel Incognito, *Computer Networks journal*, Vol. 31, No. 8, pp. 785–899, Elsevier B.V., April 1999.

[28] D. Samfat, R. Molva, and N. Asokan, Untraceability in Mobile Networks, *The First International Conference on Mobile Computing and Networking (ACM MOBICOM 95)*, Berkely, CA, November 1995.

[29] J. Go and K. Jim, Wireless Authentication Protocol Preserving User Anonymity, *The 2001 Symposium on Cryptography and Information Security (SCIS 2001)*, Oiso, Japan, January 2001.

[30] G.M. Køien, Privacy Enhanced Cellular Access Security, in *Proc. of the 2005 ACM Workshop on Wireless Security (WiSe 2005)*, pp. 57–66, Cologne, Germany, September 2005.

[31] FIPS 197, Advanced Encryption Standard (AES), NIST, November 2001.

[32] IEEE 802.11i, Amendment 6: Medium Access Control (MAC) Security Enhancements, IEEE, New York, 2004.

[33] A.J. Menezes, P.C. van Oorschot, and S.A. Vanstone, *Handbook of Applied Cryptography (5th printing)*, ISBN 0-8493-8523-7, CRC Press, Boca Raton, FL, June 2001.

[34] K. Lauter, The Advantages of Elliptic Curve Cryptography for Wireless Security, *IEEE Wireless Comm. Mag.*, Vol. 11, No. 1, pp. 62–67, February 2004.

[35] A. Shamir, *Identity-based Cryptosystems and Signature Schemes*, in *Proc. CRYPTO 1984*, pp. 47–53, LNCS 196, Springer-Verlag, 1984.

[36] D. Boneh and M. Franklin, Identity-Based Encryption from the Weil Pairing, *SIAM Journal of Computing*, Vol. 32, No. 3, pp. 586–615, 2003. (extended abstract in *Proceedings of Crypto '2001*, LNCS, Vol. 2139, Springer-Verlag, pp. 213–229, 2001.)

[37] G.M. Køien, *Privacy Enhanced Mobile Authentication*, in *Proc. of Wireless Personal Multimedia Communications (WPMC)*, Aalborg, Denmark, September 2005.

[38] P. Eronen (Ed.), IKEv2 Mobility and Multihoming Protocol (MOBIKE), RFC 4555, June 2006.

Chapter 18

The Effects of Authentication on Quality-of-Service in Wireless Networks

Wenye Wang, Wei Liang, and Avesh K. Agarwal

Contents

18.1 Overview

The emergence of public access wireless networks enables ubiquitous Internet services, whereas inducing more challenges on security concerns due to a shared transmission medium. As one of the most widely used security mechanisms, authentication is used to provide secure communications by preventing unauthorized usage and negotiating credentials for verification. However, authentication protocols can cause large signaling overhead and end-to-end delay of communications, further deteriorating overall system performance. Therefore, we study the effects of authentication on quality-of-service (QoS) of roaming mobile users with Internet protocol (IP) mobility because, ultimately, the goal of using security protocols is to provide users with reliable services. In this chapter we first provide an introduction to authentication mechanisms in mobile wireless networks. Then we describe a simple classification of security levels based on security functions, that is, information secrecy, data integrity, and resource availability. More importantly, we focus on the network security protocols that are applicable to wireless local area networks (WLANs) because WLANs are currently being used widely for wireless access to the Internet. After that, we present an analytical model to evaluate the effects of authentications on QoS in different mobile environments. Finally, we provide real-time measurements of authentication delays and overhead through an experimental study to manifest the significant effects of authentication protocols in wireless networks.

18.2 Introduction

Tremendous advances in wireless communication technologies have facilitated the ubiquitous Internet service, while involving more challenges to security concerns due to the shared open medium [3]. To provide security services in wireless networks, *authentication* is used as an initial process to authorize a mobile node (MN) for communication through secret credentials [33]. In an authentication process, an MN is required to submit secret materials such as certificates and challenge-response values for verification [7,18]. Verification is performed using a security association (SA), which is a relationship that affords security services with parameters such as session keys between the MN and its authenticator, etc. With an authentication process, the network resource can be protected by authenticating legitimate users. Information secrecy and data integrity can also be guaranteed using the negotiated secret credentials for encryption and message authentication. Therefore, the authentication service is directly related to network security in terms of network resource, information secrecy, and data integrity.

Meanwhile, authentication also has a large impact on the QoS in wireless networks. When a certificate-based authentication mechanism (i.e., public/private–key based authentication mechanism) is applied, the computation complexity of encrypting and decrypting data with public/private keys consumes more time and power [14]. Therefore, secret key–based authentication mechanisms, such as challenge-response authentication, are widely used in wireless networks [20,34], and are able to reduce the cryptographic load of authentication while inducing other challenges in both single-hop and multi-hop wireless networks.

In single-hop wireless networks, such as cellular and Mobile IP networks, the challenges of authentication lie in obtaining the credentials, such as keys, for MN authentication when MNs are roaming among wireless networks. For intra-domain roaming MNs (i.e., when MNs move within a network domain), the problem of obtaining the credentials, such as keys, for MNs has been solved by setting up a central authentication server for storing, verifying, and delivering the credentials. However, for inter-domain roaming MNs (i.e., the MNs that are moving among heterogeneous wireless systems managed by different service providers), the credentials of the MNs cannot be identified locally because the local authentication server has no information about them. Although the local authentication server may have an indirect trust relationship with the server that stores the credentials for the MNs, the credentials of the MN are encrypted and transmitted for remote verification hop-by-hop between authentication servers, due to lack of end-to-end SA. The transmission and encryption/decryption of credentials affect many QoS parameters, such as authentication cost in terms of signaling and encryption/decryption cost and authentication delay, which further affect other parameters such as call dropping probability.

To provide efficient and secure authentication in wireless networks with IP mobility support, such as mobile IP networks, two major issues should be considered: (1) authentication architecture and (2) authentication scheme. The objective of authentication architectures is to provide a secure interconnection between wireless networks. To this end, the manageability of networks, which is measured by the number of SAs between networks, has been identified as a requirement in mobile environments [1]. The authentication scheme is designed to verify the user and generate credentials with mutual trust. Because the mutual trust aims to protect the communication between networks and MNs, an authentication process is necessary to provide security. On the other hand, the efficiency of authentication with regard to authentication latency and bandwidth efficiency is also important because an authentication process introduces an overhead of communications and radio links may be idle in authentication waiting time.

Current research on authentication architecture is not sufficient to meet security and QoS requirements [11,13,20,24]. The authentication architecture is either based on a central authentication server, which is unrealistic for mass network environments [13], or based on chaining authentication servers with hop-by-hop static SAs between them, which is not suitable for large-scale and distributed networks [1,11]. Furthermore, authentication with hop-by-hop SAs can induce security problems such as man-in-middle attacks. A heavy burden of signaling cost and delay may be added to wireless networks by requiring hop-by-hop secure transmission for authentication. This overhead can degrade system performance in terms of bandwidth efficiency and call dropping probability.

Therefore, in this chapter we propose an analytical model to evaluate the effects of authentication mechanisms in mobile wireless networks to fill the blank field in which there is no quantitative analysis of authentication impacts on QoS and security simultaneously. In addition, we provide real-time measurements of authentication delays and overhead through experimental studies. The chapter is organized as follows. We introduce background knowledge of authentication in wireless networks in Section 18.3. In Section 18.5, we analyze the authentication impacts on security and QoS in mobile IP networks with the example of challenge/response authentication. We further provide real-time measurements of QoS parameters in an IP-based wireless local area network (WLAN) testbed in which a variety of authentication protocols are installed and evaluated in Section 18.6. Finally, we draw conclusions and make remarks on the future work of authentication mechanisms in wireless networks.

18.3 Authentication in Wireless Networks

Authentication is a process to identify a user with legitimate secret credentials. Based on the types of keys used for the authentication, authentication mechanisms, in general, can be categorized into two types: (1) *secret key–based* authentication and (2) *public/private key–based* authentication. In this section we first describe the above two basic authentication mechanisms used in computer networks. Then we move on to the authentication mechanisms in wireless networks, especially Mobile IP networks [27].

18.3.1 Secret Key-Based and Public Key–Based Authentication

Secret key–based authentication is a process to identify a user by encrypting and decrypting the credentials with a *secret key* shared between communicators. In this type of authentication, a *secret key*, also called a symmetric key, is preconfigured between two communicators. When an authentication process is initiated, the credentials, such as password or nonce, are encrypted and exchanged between the communicators with secret key–based cryptographic techniques, such as DES. Depending on the type of materials exchanged for the authentication, the secret key–based authentication mechanisms can be categorized into several types:

- *Password authentication.* Password authentication requires that a user inputs user name and password for the authentication. The user name and password are encrypted and transferred to a central server. The encrypted materials will be decrypted and verified at the central server [25].
- *Hash chain authentication.* Hash chain authentication requires that before a user applies for authentication, a chain of hash values must be derived from a preshared value (e.g., secret key). When an authentication occurs, a hash value derived at the end of the hash chain is used and removed after successful authentication [29].
- *Challenge-response authentication.* Challenge-response authentication requires that an authenticator generates a challenge value, a random number, and transmits it to the user who needs authentication. The user encrypts the challenge value and sends the result, called response value, back to the authenticator for verification [30].

From the description above, we can see that the secret key–based authentication mechanisms all depend on secret key–based cryptographic techniques, such as DES. When applied in wireless networks, because password authentication is not transparent to users, hash chain authentication and challenge-response authentication are widely used in wireless networks.

Public/private key–based authentication is a process to identify a user using the uniqueness of a *public/private key set*. In this type of authentication, a pair of asymmetric keys, public and private keys, is generated with a certain algorithm. A user who generates the pair of keys needs to transmit its public key to the authenticator before the authentication and keeps the private key for its own use. The private key matches the public key uniquely and is computationally infeasible to be derived from the public key. Only the data encrypted by the public key can be decrypted by the private key of the user. When a user needs to be authenticated, it encrypts a message with its private key, and sends it to its communication partner. The receiver is supposed to know the corresponding public key transmitted here previously. Because of the uniqueness of public/private keys, the user can be identified.

In the public/private key system, to fulfill the uniqueness requirement for public/private keys, complicated algorithms are developed to derive the key pair and encryption/decryption methods based on it. According to the algorithms and the entire procedure of authentication, the classical public/private key–based authentication mechanisms can be categorized into two types:

1. *Encryption-aided authentication.* Encryption-aided authentication with public/private key works by encrypting each message with the user's private key. The process to identify a user depends on the unique matching of the private and public keys. The algorithms in this category include RSA and Ellipse Curve Cryptogram (ECC) [21,35].

2. *Session key–aimed authentication.* Session key–aimed authentication with public/private key works by exchanging the public key with a user's communication partner. The final purpose is to construct a session key with a user's private key and the public key transmitted from its communication partner. The typical algorithms and procedures of authentication in this category include Diffie-Hellman and its variants [35].

Therefore, we can see that the security of public/private key authentication relies on the unique match of the key pair, which is guaranteed by the complex algorithms to derive the keys. On the other hand, when using public/private key–based authentication, either encryption-aided authentication or session key–aided authentication must consume more energy and delay for computation, which has proved to be greater than secret key–based authentication [14]. For more details on cryptosystems, see Menezes et al. [28].

18.3.2   Authentication Protocols in Wireless Networks

There are many authentication protocols in single-hop wireless networks, such as for WLANs with IP mobility support [9,19,20,24,34]. All of these protocols either focus on establishing an SA between two communication partners or try to design efficient schemes to improve the authentication efficiency in terms of reduced signaling messages and cryptographic load. As an example, we introduce an authentication protocol, the four-way handshake protocol, in the standard of transport layer security (TLS) that is able to establish static SA between two communication partners.

A four-way handshake protocol in TLS allows an MN and the authentication server to negotiate encryption algorithms and exchange keys to set up an SA before any application data is transmitted. The procedure of the four-way handshake protocol is illustrated in Figure 18.1 [14], which is divided into four phases. The first phase is to initiate a logical connection. At the second phase, a server sends a public key with its certificate to the client. The third phase is performed by the client to provide its public key and certificate to the server after successfully receiving and verifying the public key of the server. The fourth phase is to confirm cipher specification parameters, such as keys, algorithms, and lifetime of the SA, based on the shared secrets. After these steps, an SA is established between the client and the server. However, all of the algorithms applied in this protocol are time-consuming, especially when the client is an MN with limited computation capability and power [14]. In addition, the establishment of the SA cannot adapt to the QoS requirements of communications; for example,

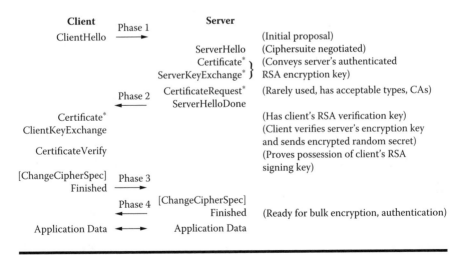

Figure 18.1   SSL/TLS four-way handshake protocol.

the cost and delay of authentication cannot be controlled and adjusted in the wireless environments.

Like the four-way handshake authentication protocol above, all other authentication protocols in single-hop wireless networks such as mobile IP networks have similar problems; that is, they only consider the security and efficiency of authentication and cannot adapt to the system performance.

Next we propose an analytical model to quantitatively evaluate the impacts of authentication on the security and system performance in single-hop wireless networks with the example of challenge response, which is the first piece of work that builds a direct numerical relationship between security and QoS. To understand the analysis, we specify the effect of authentication on security and QoS in the following section. Then we elaborate on the analysis of authentication in the subsequent section.

18.4 Effects of Authentication on Security and QoS

Challenge/response authentication is widely used in wireless networks and has significant effects on the security and QoS in single-hop wireless network. Thus, we provide an overview of challenge-response authentication first. The effects of the authentication are described thereafter.

18.4.1 Overview of Challenge-Response Authentication

Authentication in wireless networks is defined as a process in which the MN needs to send out the secret credentials for verification and negotiate SAs for communications. An SA is a trust relationship with many parameters, such as keys and algorithms, for secure service with cryptographic techniques [35]. To model an authentication process by considering signaling, we take challenge-response authentication as a prototype in this work, primarily because of its dominant applications in wireless networks, including WLANs and mobile IP. Even GSM/UMTS/CDMA200 employs challenge response with symmetric signed response using a message authentication code.

In challenge response–based authentication, a user is identified with shared SA by an authentication server that sends a *challenge value*, a random number, to the user for encryption, and verifies the returned value, called a *response value*, with decryption [22]. In a foreign network, a visiting MN sends out an authentication request to an access point (AP), which is a function unit for transmitting data. The request is forwarded by the AP to a local authentication server (LAS), which only takes charge of authentication for the visiting MNs that are roaming from foreign networks. If the LAS does not have enough information to verify the MN, it contacts the home authentication server (HAS) of the MN through an authentication architecture. An HAS is an authentication server that takes charge of the authentication for the MNs that subscribe the service in its network. And, an authentication architecture is composed of many authentication servers that share SAs with the LAS and HAS. Thus, when an HAS receives a request from the authentication architecture, it verifies the request using an SA shared with the MN. If the request is an inter-domain authentication request, the HAS sends a registration request to the MN's home agent (HA), which is a router in the home network that maintains the current location of the MN, to update the MN's location.

Note: Throughout this chapter, we assume that an MN is roaming in a foreign network domain.

Then, the challenge-response authentication for an MN in a foreign network domain can be categorized into three types: (1) intra-domain handoff authentication (2) session authentication and (3) inter-domain handoff authentication, with the signaling diagrams shown in Figure 18.2.

18.4.1.1 Intra-Domain Handoff Authentication

When an MN crosses the boundary of subnets in the foreign network domain with an ongoing service, an intra-domain handoff authentication is initiated. Because there is an ongoing communication session between the MN and an AP, one session SA exists between the MN and the LAS in the visiting network domain. Therefore, it is unnecessary to contact the HAS of the MN for authentication. In the case shown in Figure 18.2(A), the LAS that receives the authentication request from an MN sends a challenge value (i.e., a random value) to the MN. The MN encrypts the challenge value using the shared SA with the LAS. The encrypted number (i.e., a response value) is replied to the LAS. After decrypting the replied value and comparing it with the original challenge value by the LAS, the LAS then can authenticate the MN.

18.4.1.2 Session Authentication

When an MN starts a communication session in a subnet of a foreign network, a session authentication is initiated. Because there is no ongoing communication session between the MN and the AP, session SA does not exist between the MN and the AP, and it is necessary to contact the HAS of the MN for authentication. In the case shown in Figure 18.2(B), when the LAS receives the authentication request forwarded from the AP, it sends a challenge value to the MN. The MN encrypts the challenge value with the SA shared with the HAS, and replies with the response value to the LAS. The LAS must forward the challenge and response values to the HAS

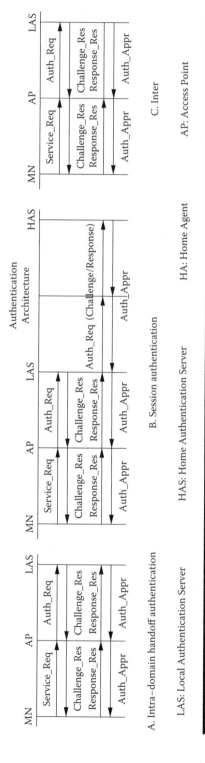

Figure 18.2   Challenge-response authentication in public wireless access networks.

of the MN for verification because the LAS does not share an SA with the visiting MN, and cannot decrypt the response value without the SA. After authentication at the HAS, the secret credentials such as keys to protect the communication can be generated and sent to the LAS.

18.4.1.3   Inter-domain Handoff Authentication

When an MN crosses the boundaries of different foreign network domains with an ongoing service, an inter-domain handoff authentication occurs. Because the session SA attached with the ongoing communication session is between the MN and the other AP, no session SA exists between the MN and the new AP, and it is necessary to contact the HAS of the MN for authentication. In the case shown in Figure 18.2(C), the signaling diagram is similar to that in the case of session authentication, except that the MN needs registration to its HA through the HAS because we assume that the MN needs registration only if it is crossing the boundaries of different network domains.

18.4.2   Effects of Authentication on Security and QoS Metrics

Security services are to provide information secrecy, data integrity, and resource availability for users. Information secrecy means to prevent improper disclosure of information in the communication, while data integrity aims to prevent improper modification of data and resource availability is considered to preventing improper denial of services [35].

To provide security services in wireless networks, challenge-response based authentication adopts several techniques to meet the requirements. First, challenge-response authentication enables the MN to share an SA with its HAS. The SA is unique and secret to other users. Therefore, the identification of the MN is unique, which can prevent unauthorized MNs from accessing the network resource. Thus, the resource availability for authorized users can be guaranteed. Second, new secret credentials such as session keys are generated and sent to communication partners during authentication. The distributed secret credentials are used to encrypt the data of communication and provide message authentication code for data integrity checking. Therefore, the authentication mechanism becomes a critical part in protecting information secrecy and data integrity because new secret credentials such as session keys are generated and transferred during this period. A well-designed authentication protocol then can provide great security by defeating well-known threats such as replay attacks and man-in-middle attacks.

In addition to the effect on the security, authentication also affects QoS metrics, such as authentication delay, cost, call dropping probability, and throughput of communication due to the generation of the overhead of communications. *Authentication delay* is defined as the time between when the MN sends out the authentication request and when the MN receives the authentication reply. During this authentication delay, no data for ongoing service can be transmitted, which may interrupt the connections. Therefore, the call dropping probability increases with increasing of authentication delay.

Authentication cost is defined as the signaling cost and processing load for cryptographic techniques. In challenge-response authentication, the challenge-response values must be transmitted back to the HAS of the MN for verification when the LAS has no SA shared with the roaming MN. Then, the signaling messages are transmitted between different LASs. The total number of signaling messages from the LAS to the HAS of the MN can be large if the authentication distance between them is long. Furthermore, the signaling messages need to be encrypted and decrypted hop-by-hop for protection due to the lack of a direct trust relationship between the LAS and the HAS. These multiple encryptions and decryptions increase the processing load of the networks. Moreover, the mobility and traffic patterns of MNs make authentication occur frequently in different scenarios because the authentication is initiated when an MN starts a communication session or crosses the boundaries of subnets with an ongoing service, which may cause an unbalanced distribution of authentication cost.

Compared to the effects of authentication on delay and cost, the *throughput* is affected by the authentication throughout the entire communication service. The throughput of the data communication is defined as the effective data transmitted in unit time. It can be greatly decreased due to authentication for several reasons. First, when authentication occurs, the authentication delay causes a temporary pause in data transmission, which decreases the throughput. Second, the cryptographic technique that is negotiated in authentication and protects the latter communication is changed from time required to time during authentication, which depends on the mobility of the MN and different security requirements of the networks that the MN is visiting. The key size and algorithms to encrypt and decrypt the data affect the time required to process the data. They will further reduce the effective data transmission rate due to the attachment of message authentication code for the data integrity check.

18.5 Analytical Model and Performance Evaluation

To analyze the performance of challenge-response authentication in wireless networks, we discuss the effect of authentication on security and QoS based on challenge-response authentication first. Then we describe a system model and define metrics used for performance evaluation. We analyze these metrics at different security levels based on mobility and traffic patterns, and provide the numerical results of our analysis on authentication cost, delay, and call dropping probability [27].

18.5.1 System Model

A system model is used to describe the authentication interaction between interconnected wireless networks. That means there are a number of *n* autonomous wireless networks. Each network domain has an *LAS* and an *HAS*. The LAS and HAS are central authentication servers in a network domain. However, an LAS only takes charge of authentication for visiting MNs, while an HAS is only responsible for the authentication of the MNs that subscribe services in the current network domain. The trust relationships between these LASs and HASs are maintained through an *authentication architecture*, which is an infrastructure composed of many proxy authentication servers and is designed to securely deliver the authentication messages between authentication servers [11]. It is assumed that the LAS and HAS are integrated together, and that the authentication architecture shares an SA with the LAS/HAS of a network domain.

Moreover, we need to describe the scenario and make assumptions on the mobility and traffic models.

18.5.1.1 Scenario

Assume that challenge-response authentication is implemented on the generic system model with signaling diagrams shown in Figure 18.2 because *our initial assumption is that an MN is roaming in foreign network domains.* For the intra-domain handoff authentication in foreign networks, Figure 18.2(A) shows the detailed process to realize this type of authentication. Similarly, the detailed processes on session and inter-domain handoff authentication in foreign networks are displayed in Figure 18.2(B) and Figure 18.2(C), respectively.

18.5.1.2 Mobility Pattern

The mobility pattern of an MN in our analysis is represented with the residence time of the MN in one subnet, denoted as T_r. We assume that T_r is a random variable and the probability density function (PDF) of T_r, denoted as $f_{T_r}(t)$, is a Gamma distribution with mean $1/\mu_r$ and variance V. Then, the Laplace transform of $f_{T_r}(t)$, $F_r(s)$ is:

$$F_r(s) = (\frac{\mu_r \gamma}{s + \mu_r \gamma})^\gamma, \quad \text{where} \quad \gamma = \frac{1}{V \mu_r^2} \tag{18.1}$$

Furthermore, if the number of subnets passed by an MN is assumed to be uniformly distributed between [1, M], the PDF of the residence time in a network domain, denoted as $f_{T_M}(t)$, can be expressed with Laplace transform $F_M(s)$ as follows [37]:

$$F_M(s) = \frac{1}{M} \left(\frac{\mu_r \gamma}{s + \mu_r \gamma} \right)^{\gamma} \frac{1 - \left(\frac{\mu_r \gamma}{s + \mu_r \gamma} \right)^{\gamma M}}{1 - \left(\frac{\mu_r \gamma}{s + \mu_r \gamma} \right)^{\gamma}} \tag{18.2}$$

Then, the mean value of residence time in this network domain, denoted as $\overline{T}_M$, can be expressed as:

$$\overline{T}_M = -\frac{\partial F_M(s)}{\partial s} \Big|_{s=0} = \frac{M+1}{2\mu_r} \tag{18.3}$$

18.5.1.3 Traffic Pattern

In the analysis, we consider the call arrival rate and call duration time of the MN as the traffic patterns of the MN. First we assume that the call arrival rate of the MN, which includes the incoming calls and outgoing calls, is a Poisson process with average rate λ_u; then the PDF of the call inter-arrival time, denoted as $f_{T_A}(t)$, can be determined by:

$$f_{T_A}(t) = \lambda_u e^{-\lambda_u t} \tag{18.4}$$

Moreover, we assume that a call duration time, denoted as T_D, has an exponential distribution with mean value $1/\eta$. Then, the PDF of call duration time, denoted as $f_{T_D}(t)$ can be written as:

$$f_{T_D}(t) = \eta e^{-\eta t} \tag{18.5}$$

Based on these assumptions for the mobility and traffic patterns of the MN, we evaluate the security and QoS metrics of authentication when the MN is roaming in our generic system model. The security and QoS metrics needed for evaluation are defined in the next subsection.

18.5.2 Performance Metrics

We categorize the performance metrics into security and QoS parameters. The security parameter is represented by security levels at which different levels of protection are provided. Meanwhile, we consider authentication cost and delay as the system performance for evaluation.

Table 18.1   Security Level Classification

Security Level i	Security Service			
	Integrity	Secrecy	Confidentiality	Availability Protection
1	No	No	No	No
2	No	No	Low	Low
3	No	No	Medium	Medium
4	Yes	Yes	High	High

18.5.2.1   Security Levels

There is much quantitative analysis of QoS in networks [8], whereas less analysis of security exists. This gap between QoS and security analysis demands quantization of security for engineering research. Therefore, the concept of *security level* becomes widely used for security evaluation [5,32,36]. However, all of them do not consider the nature of security (i.e., data integrity, secrecy, and availability). Therefore, we argue that the nature of security should become the standard to classify the security levels.

In our analysis, the *security level* is to indicate the level of protection provided by the authentication for quantitative analysis of security. The classification of security levels is shown in Table 18.1 according to the security functions described in Section 18.4.2, that is, protection in terms of integrity, secrecy, and resource availability. Because of different actions in challenge-response authentication, the protection of data integrity, secrecy, and availability may be different at different security levels.

- *Security Level 1.* Any MN can send data through an AP without authentication.
- *Security Level 2.* Authentication is implemented with medium access control (MAC) address and no keys are generated for subsequent communications.
- *Security Level 3.* Authentication is implemented with shared SA, and no keys are generated for the MN's communication.
- *Security Level 4.* Authentication is implemented with shared SA, and keys are generated for data encryption and message integrity check.

18.5.2.2   Average Authentication Cost

In this context, we define *authentication cost* as the sum of the signaling load and processing load for cryptographic techniques during one authentication operation. And, the *average authentication cost*, $C(i)$, is defined as

Table 18.2   Authentication Cost Symbols

Symbol	Description
c_s	Transmission cost on one hop
c_p	Encryption/decryption cost on one hop
c_v	Verification cost at an authentication server
c_{us}	Encryption/decryption cost for a session key
c_g	Key generation cost
c_{ts}	Transmission cost for a session key to other communication identities
c_{rg}	Registration cost

the sum of the authentication cost over a number of authentication requests in unit time at security level i, which can be written as:

$$C(i) = \sum_{\beta=1}^{3} \lambda_\beta [C_\beta^{(s)}(i) + C_\beta^{(p)}(i)], \qquad (18.6)$$

where β is the index of authentication type. $\beta = 1$ represents an intra-domain handoff authentication, $\beta = 2$ means a session authentication, and $\beta = 3$ is an inter-domain handoff authentication. We denote $C_\beta^{(s)}(i)$ and $C_\beta^{(p)}(i)$ as the signaling load and processing load of cryptographic techniques, respectively, of an authentication with type β at security level i. The arrival rate of requests for the authentication type β is defined as λ_β, which is related to the mobility and traffic patterns of MNs.

The authentication cost, $C_\beta(i)$ ($\beta = 1, 2, 3$ and $i = 1, 2, 3, 4$), is composed of $C_\beta^{(s)}(i)$ and $C_\beta^{(p)}(i)$, which depend on the authentication type β and security level i. For convenient analysis, we define a set of cost parameters in Table 18.2.

18.5.2.3   Average Authentication Delay

We define *authentication delay* as the time between when the MN sends out an authentication request and when the MN receives the authentication reply. The *average authentication delay*, $T(i)$, is defined as the sum of an authentication delay over a number of authentication requests in unit time at security level i. Then, $T(i)$ can be written as:

$$T(i) = \sum_{\beta=1}^{3} \lambda_\beta T_\beta(i), \qquad (18.7)$$

where $T_\beta(i)$ is the authentication delay per operation at security level i for authentication type β, and λ_β is the arrival rate of authentication requests with type β.

Table 18.3 Authentication Time Symbols

Symbol	Description
T_{pr}	Message propagation time on one hop
T_{tr}	Message transmission time on one hop
T_{ed}	Message encryption/decryption time on one hop
T_a	Authentication request service and waiting time at the AP
T_{sg}	Authentication request service and waiting time at the proxy authentication server
T_v	Authentication request service and waiting time at the HAS
T_{us}	Key encryption and decryption time
T_g	Key generation time at the HAS
T_{ts}	Transmission time for the session key to the other communication identities such as HA
T_{rg}	Registration request service and waiting time at the HA

To derive the delay for different types of authentications in different security levels, we use the same signaling diagram shown in Figure 18.2. We also define a set of time parameters shown in Table 18.3 for convenient description.

In summary, to evaluate $C(i)$ and $T(i)$ in Equations 18.6 and 18.7, we need to analyze λ_β, $C_\beta^{(s)}(i)$, $C_\beta^{(p)}(i)$, and $T_\beta(i)$. Based on the system models, assumptions, and definitions of the performance metrics described in Section 18.5.2, we can obtain the numerical results of our analysis on the effects of authentication in mobile IP networks. More details of the derivation procedures can be found in Liang and Wang [27].

18.5.3   Numerical Results

In this subsection, we evaluate the effects of mobility and traffic patterns on authentication cost $C(i)$ and authentication delay $T(i)$ at different security levels.

18.5.3.1   Assumptions and Parameters

The numerical results are presented based on the assumptions introduced in Section 18.5.1. Of the assumptions in that section, we consider an MN roaming within a foreign network. The mobility pattern of the MN is represented with the residence time in a subnet of the network domain, which is assumed to be a Gamma distribution with mean value $1/\mu_r$. The traffic patterns of an MN are represented by the call arrival rate and call duration time. The call arrival rate is assumed to be a Poisson process with mean value $1/\lambda_u$, and the call duration time is assumed to be an exponential distribution with mean value $1/\eta$.

Table 18.4   Parameters for Evaluation of QoS Metrics

Parameters for Authentication Cost					
c_s	c_p	c_v	c_g	c_{ts}	N_h
10	1	20	1	110	10
Parameters for Authentication Delay					
T_{th}	T_{pr}	T_{tr}	T_{ed}	T_g	M
3 sec	40 μ sec	20 msec	2 msec	2 msec	120
Parameters for Random Variables					
λ_u	η	γ	μ_r	ξ	
0.1 min^{-1}	0.3 min^{-1}	225	1/15 min^{-1}	15 sec^{-1}	

In the derivation of call dropping probabilities, we further assume that $M/M/1$ queues are used at APs, authentication servers such as LAS and HAS, and HAs with service rate μ_s and arrival rate of authentication requests λ_s. Let $\xi = \mu_s - \lambda_s$. The parameters to evaluate the authentication cost and delay are shown in Table 18.4.

There are many ways to determine values for the authentication costs. For example, the authentication cost for signaling can be measured by the number of messages, and the authentication cost for encryption can be measured by the number of CPU cycles. However, the most important problem here is how to make them consistent, that is, the values of the costs can be compared with each other on the same scale. To solve this problem, we assume that the encryption/decryption cost on one hop, c_p, and the key generation cost, c_g, are normalized to a cost unit because they are the lightest load compared with other costs and they have the similar operation in cryptographic techniques [17]. The values of other costs are determined by comparing to c_p and c_g with the time to finish the operation; that is, we use the ratio of processing time to represent the authentication cost instead of the actual processing time. The reason is that the time needed to finish an operation represents the load of the server to complete it. However, we do not use the processing time to represent the cost directly because we do not want to confuse the authentication cost with the authentication delay, and the authentication cost can be evaluated in various other ways.

When the maximum authentication message size is 4096 bytes [7], the transmission delay is about 20 milliseconds with the assumption of 2-Mbps link capacity [17]. The values of T_{ed} and T_g are obtained from existing research [17,23]. By assuming one network domain is about 100 square kilometers with a radius of 6 kilometers, the value of the propagation time, T_{pr}, can be determined by the distance between two LASs as shown in Table 18.4.

18.5.3.2   Effects of Mobility Pattern at Different Security Levels

The effects of mobility pattern on the authentication cost and delay are shown in Figure 18.3 and Figure 18.4. In these figures, we illustrate the relationships between the residence time of an MN in a subnet and authentication cost and delay, respectively.

In Figure 18.3, authentication costs at different security levels decrease with increasing residence time of an MN in a subnet because the longer an MN stays in the subnets, the less the intra-domain handoff authentication requests. And, if the residence time of an MN approaches infinity, the authentication cost will be stable on the session authentication cost because only session authentication exists in this case. Moreover, we can see that the security levels have different effects on the cost at the same residence time in a subnet — the higher the security level, the more the authentication cost because higher security levels impose more operations to provide secure services. For example, if we degrade the security level from 4 to 3, the authentication cost can be reduced up to 32 percent.

Figure 18.4 reveals the effect of residence time on the authentication delay. As we can see, authentication delay decreases with increasing residence time of an MN in a subnet. Similar to the authentication cost, this trend is due to the decrease in intra-domain handoff authentication requests. And, the higher security levels cause more authentication delay because of more operations needed for more secure services. The improvement in authentication delay by changing security levels from 4 to 3 is approximately 0.1 seconds, which is around 18.2 percent of the authentication delay at security level 3 when the residence time of an MN in a subnet is 27 minutes.

18.5.3.3   Effect of Traffic Load at Different Security Levels

The effects of traffic pattern on the authentication cost and delay at different security levels are demonstrated in Figure 18.5 and Figure 18.6.

Figures 18.5 and 18.6 show that the authentication cost and delay increase with increasing call arrival rate of an MN. As shown in Equations 18.6 and 18.7, authentication cost and delay are proportional to the call arrival rate λ_u, because variables λ_β, ($\beta = 1, 2, 3$), which denote the arrival rates of intra-domain authentication, session authentication, and inter-domain authentication, are proportional to λ_u. Moreover, a higher security level causes more cost and delay than a lower one. For example, if the security level is changed from 1 to 2, the authentication cost is about 7.5 times higher and 29 percent more time than those at security level 1.

In this section we investigate the impact of authentication on security and QoS in a combination of mobility and traffic patterns, which is critical to delivering secure and efficient services in wireless networks such as a WLAN. We analyzed the authentication cost and delay at different security

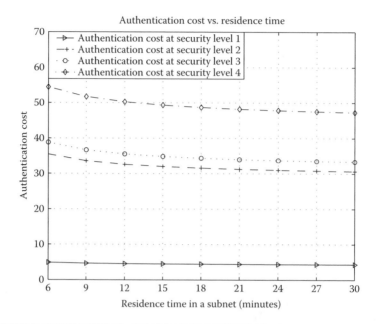

Figure 18.3   Authentication cost versus residence time in a subnet.

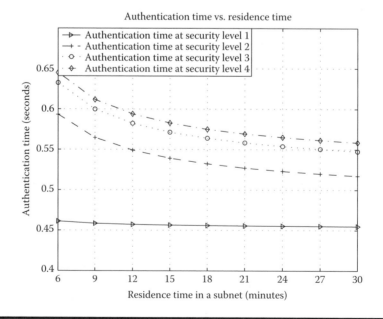

Figure 18.4   Authentication time versus residence time in a subnet.

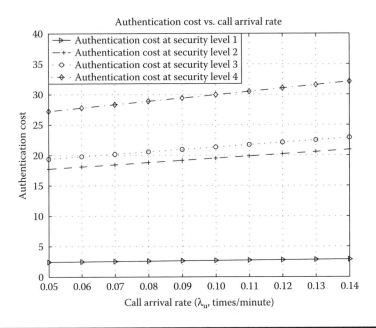

Figure 18.5   Authentication cost versus call arrival rate.

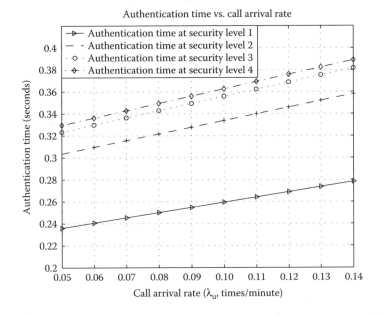

Figure 18.6   Authentication time versus call arrival rate.

levels in wireless networks based on a system model with a challenge-response mechanism. In the analysis, the mobility and traffic patterns are taken into account for QoS. Therefore, our analytical models and results can be used to obtain a fundamental understanding of the effects of authentication mechanisms in wireless networks because they provide a quantitative connection between the security and system performance with concern for adaptation to various mobile environments, which further proposes solid groundwork for an in-depth understanding of authentication impact, and demonstrates a framework for the future design of efficient authentication schemes in wireless networks.

Note that there are many security protocols developed for wireless networks in recent years [6,16]. Several security protocols, such as Wired Equivalent Privacy (WEP) and 802.1x port access control with Extensible Authentication Protocol (EAP) support, are proposed to address security issues [6,16]. Moreover, due to the strong security provided by the IP Security Protocol (IPSec) in wired networks, it is considered an alternative for wireless networks as well [31]. However, these protocols require extra network bandwidth in performing their functionalities regarding the configuration and transmission of signaling messages. Many details of implementation issues, such as cross-layer impacts, the processing and computing capability of mobile devices, and the effects of transmission channels, are very difficult to consider in analytical models or even in simulation studies. To present a more realistic view of the effects of authentication mechanisms in existing security protocols, in the next section we provide our experimental studies over a mobile IP-based WLAN testbed.

18.6 Experimental Studies on Authentication Mechanisms in WLANs with IP Mobility

To estimate the performance impact on a system, measurements are considered very important because they are helpful in obtaining a realistic view of the performance overhead associated with the system. Therefore, to gain a fundamental understanding of the performance impact due to security protocols, experimental studies were carried out in the past in various network environments [4, 10, 12, 15, 31]. However, these studies explored the advantages and disadvantages of security protocols in stand-alone mode. With the availability of several security protocols at different layers, it is intuitive to explore the possibilities and advantages associated with integrated security services at different layers, which would help in understanding the applicability of a particular *service* or *application* to real-time networks while at the same time maintaining the required QoS. With the increasing demand for better QoS by many real-time applications [26], it becomes critical to understanding the performance impact of security protocols for determining their applicability to real-time networks.

Therefore, our objectives for performing experimental study are manifold. We aim to study the cross-layer integration of security protocols to gain a deeper understanding of the trade-offs between performance overhead and security strength in wireless networks with heterogeneous devices. We focus on providing comprehensive real-time measurements of the performance overhead associated with security protocols at the various layers in wireless LANs with mobility support.

To achieve these goals, we set up a real-time experimental testbed, which is a miniature of existing wireless networks, to ensure that our experimental scenarios can represent typical deployment of WLANs. We use iPAQ, SharpZaurus, laptop, and desktop machines, each equipped with wireless cards to create heterogeneous environments. Moreover, stand-alone and integrated security protocols are configured at various layers and classified as individual and hybrid security policies, respectively. Authentication time and authentication cost are the performance metrics evaluated in our testbed. We configure roaming scenarios using Mobile IP open source code to support mobility by creating two subnets. TCP and UDP traffic generators are used to measure the impact of security protocols on different types of data streams [2].

Network scenarios are classified into nonroaming ($\mathcal{N}$) and roaming ($\mathcal{R}$), based on the user's current location (i.e., whether a user is in his home domain or a foreign domain, respectively). To make the description of scenarios clear, we assume that subnet A is the home domain for mobile nodes $A1$ and $A2$, and subnet B is the home domain for mobile nodes $B1$ and $B2$. All scenarios are demonstrated in Figure 18.7. Nonroaming scenarios, represented as $\mathcal{N}$, are defined as the scenarios when both communicating mobile users are in their home domain. Following are the details of various nonroaming scenarios configured in the testbed:

- *Scenario N1*: deals with the situation when both MNs are in the same subnet, which is also their home domain. For example, when communication occurs between $A1$ and $A2$ as shown in Figure 18.7(a).
- *Scenario N2*: MNs communicate with their HA that is acting as an application server providing services to mobile clients in the network. Here, part of the communication path is wired, which is not the case in scenario $N1$. As shown in Figure 18.7(b), this scenario occurs when HA A-HA communicates with $A1$ or $A2$.
- *Scenario N3*: to capture the impact of security services when participating MNs are in different domains. For example, when $A1$ or $A2$ communicates with $B1$ or $B2$, as shown in Figure 18.7(c).

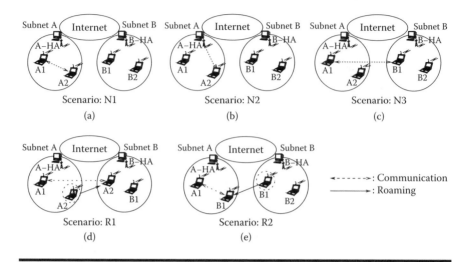

Figure 18.7   Nonroaming and roaming scenarios.

When at least one of two communicating mobile users is in a foreign domain, we refer to it as a roaming scenario, represented as $\mathcal{R}$. The following roaming scenarios are configured in our experimental testbed:

■ *Scenario R*1: specifies when one end node, which is in a foreign domain, is communicating with the other node, which is in its home domain, but the two nodes are in different domains. It aims to analyze the effect of security services on data streams when one node is roaming. As shown in Figure 18.7(d), this scenario occurs when node *A*2 roams to subnet B and communicates with *A*1.
■ *Scenario R*2: occurs when both nodes are in the same domain but one node is roaming. Therefore, the current network is the foreign domain for one node, whereas it is the home domain for the other node. It helps us in analyzing the performance impact on data streams when the roaming node is communicating with a nonroaming node in the same domain. For example, when node *B*1 roams to subnet A and communicates with *A*1 or *A*2, as shown in Figure 18.7(e).

18.6.1   Security Policies

Security policies (Table 18.5) are designed to demonstrate potential security services provided by the integration of security protocols at different layers. Each security protocol uses key management protocols, and various

Table 18.5 Features of Security Policies

Policy No.	Security Policies	Authentication	Confidentiality	Data Integrity	Non-Repudiation	Mutual Auth.
P_1	No Security					
P_2	WEP-128 bit key	Y	Y			
P_3	IPSec-3DES-SHA	Y	Y	Y	Y	Y
P_4	IPSec-3DES-SHA-WEP-128	Y	Y	Y	Y	Y
P_5	8021x-EAP-MD5	Y				
P_6	8021x-EAP-TLS	Y	Y		Y	Y
P_7	8021X-EAP-MD5-WEP-128	Y	Y			
P_8	8021X-EAP-TLS-WEP-128	Y	Y		Y	Y
P_9	8021X-EAP-MD5-WEP-128-IPSec-3DES-MD5	Y	Y	Y	Y	Y
P_{10}	8021X-EAP-TLS-WEP-128- IPSec-3DES-MD5	Y	Y	Y	Y	Y
P_{11}	8021X-EAP-MD5-WEP-128-IPSec-3DES-SHA	Y	Y	Y	Y	Y
P_{12}	8021X-EAP-TLS-WEP-128-IPSec-3DES-SHA	Y	Y	Y	Y	Y

authentication and cryptographic mechanisms. Therefore, a variety of security policies are configured in our experiments by combining the various mechanisms of security protocols.

18.6.2 Experimental Results

Now we analyze the performance cost associated with various security policies in terms of authentication time and authentication cost. Metrics (authentication time and authentication cost) are associated with the authentication phase and encryption/decryption process of a security policy, respectively.

18.6.2.1 Authentication Time

Authentication time is defined as the total time consumed in an authentication phase of a security policy. We consider authentication time, represented as (C_A), as the cost associated with the authentication phase of a security policy. It is due to the fact that the time involved in an authentication phase is one of the important factors contributing toward performance impact in a network. Here we describe steps to calculate the authentication time (C_A) as shown in Table 18.6.

Authentication time is associated with the initial phase of a security policy as defined in Section 18.6.1. During this period, an MN provides its credentials to the authentication server, such as an HA or foreign agent in the testbed, to access a network. Messages exchanged during the initial phase of a security policy vary with the security mechanisms involved in the policy. Moreover, the authentication time for various policies is obtained for nonroaming and roaming mobility scenarios, respectively. Table 18.6 demonstrates the individual components of authentication time associated with each security policy in various scenarios. Table 18.7 shows authentication time for individual protocols, whereas Table 18.8 shows authentication time $(C_A$ in sec) for IPsec and 802.1x policies. Since WEP does not involve exchange of control messages, there is no authentication time involved with it. Since Mobile IP is used for enabling mobility in the testbed, authentication time (C_A) for IPsec and 802.1x involves Mobile IP authentication time as well.

18.6.2.2 Authentication Cost

Authentication cost represents the performance overhead associated with a security policy. Because we compute the cost of the authentication phase of a security policy separately in terms of authentication time, authentication cost involves overhead due to other security features, such as encryption/decryption, data integrity, etc. Now we discuss the authentication cost associated with security policies in roaming and nonroaming scenarios.

Table 18.6 Authentication Time Computation

Scenario Policy	IPSec	802.1x-EAP(MD5/TLS) without IPSec	802.1x-EAP(MD5/TLS) with IPSec
Nonroaming ($\mathcal{N}$)	IPSec Tunnel Establishment Time in HN + MN registration time to HA	802.1x-EAP(MD5/TLS) authentication time in HN + MN registration time to HA	IPSec Tunnel Establishment Time in HN + 802.1x-EAP(MD5/TLS) authentication time in HN + MN registration time to HA
Roaming ($\mathcal{R}$)	IPSec Tunnel Establishment Time in HN + MN registration time to HA + MN registration time to FA	802.1x-EAP(MD5/TLS) authentication time in HN + MN registration time to HA + 802.1x-EAP(MD5/TLS) authentication time in FN + MN registration time to FA	IPSec Tunnel Establishment Time in HN + 802.1x-EAP(MD5/TLS) authentication time in HN + MN registration time to HA + 802.1x-EAP(MD5/TLS) authentication time in FN + MN registration time to FA

Table 18.7   Individual Authentication Time

Protocol	Time (sec)
Mobile IP(HA)	0.11
Mobile IP(FA)	1.432
IPSec	1.295
802.1x-EAP-MD5	0.317
802.1x-EAP-TLS	1.712

By analyzing the authentication cost, we capture the encryption and decryption time associated with security policies during data transmission. In addition, we have normalized experimental data for comparing results in various scenarios. Table 18.9 and Table 18.10 list authentication costs in nonroaming and roaming scenarios for TCP and UDP traffic streams, respectively. The values presented in boldface in Table 18.9 and Table 18.10 indicate the overall recommended security policy for a particular network scenario.

We notice from Table 18.9 that the authentication costs associated with policies P_4, P_9, P_{11}, P_{10}, and P_{12} in nonroaming scenarios are very close to each other, showing little variation. This is due to the fact that these policies use the same IPSec and WEP mechanisms, which are the dominating factors contributing toward their authentication costs. In general, the policies P_4, P_9, P_{11}, P_{10}, and P_{12} exhibit 16 percent higher authentication costs than P_3, and around 3.5 times higher than that of P_2, P_7, and P_8. The reason is that policies P_4, P_9, P_{11}, P_{10} and P_{12} have more than one level of encryption and decryption mechanisms associated with them. Further, we observe that P_5 and P_6 exhibit negligible authentication costs, which is due to the fact that these policies do not consist of any encryption/decryption mechanisms associated with them. Although, theoretically, the authentication costs of policies P_5 and P_6 should be zero, the small values obtained are due to some external factors in real-time environments.

Table 18.8   Authentication Time in Different Network Scenarios

Scenario Policy	IPSec (sec)	802.1x-EAP (MD5) without IPSec(sec)	802.1x-EAP (MD5) with IPSec(sec)	802.1x-EAP (TLS) without IPSec(sec)	802.1x-EAP with (TLS) IPSec(sec)
Nonroaming	1.405	0.427	1.722	1.822	3.117
Roaming	2.837	2.176	3.471	4.966	6.281

Table 18.9 TCP Authentication Cost (kbits/sec) under Various Network Scenarios

Network Scenarios		P_1	P_2	P_3	P_4	P_5	P_6	P_7	P_8	P_9	P_{10}	P_{11}	P_{12}
Non-Roaming ($\mathcal{N}$)	N1	0	71.10	264.90	318.32	11.88	11.47	75.88	**77.21**	286.89	291.48	313.94	301.77
	N2	0	70.90	273.45	302.8	2.09	3.92	101.88	**57.15**	347.33	299.25	298.78	296.13
	N3	0	108.78	304.59	331.68	7.29	4.33	105.11	**118.71**	343.84	382.87.48	378.10	343.52
Roaming ($\mathcal{R}$)	R1	0	90.43	209.54	216.27	1.97	6.15	104.81	92.19	246.49	259.97	251.56	**260.81**
	R2	0	208.04	318.32	367.53	25.60	1.79	232.66	230.78	393.13	381.70	391.12	**395.29**

Security Policies

Table 18.10 UDP Authentication Cost (kbits/sec) under Various Network Scenarios

Network Scenarios		P_1	P_2	P_3	P_4	P_5	P_6	P_7	P_8	P_9	P_{10}	P_{11}	P_{12}
								Security Policies					
Non-Roaming (N)	N1	0	97.68	174.50	186.58	0.54	5.70	111.59	95.98	163.15	164.66	185.81	**198.75**
	N2	0	51.77	101.20	173.18	6.21	7.26	42.74	55.28	177.83	170.30	194.41	**175.01**
	N3	0	139.14	193.05	289.85	41.23	53.36	162.31	168.99	284.47	289.38	304.68	**286.03**
Roaming (R)	R1	0	64.84	164.96	227.47	20.90	5.84	69.59	**73.87**	384.51	354.55	399.53	375.46
	R2	0	72.71	172.47	184.49	3.47	10.73	88.14	**99.27**	241.26	241.26	241.26	241.26

18.7 Conclusions

In this chapter we presented an overview of authentication mechanisms and their effects on QoS and security functions. Through both analytical studies and experimental measurements, we observed that authentication mechanisms can greatly affect system performance due to authentication delay and authentication overhead. In our analytical studies, we considered a mobility and traffic model in the evaluation of authentication delay and cost. However, many realistic issues such as processing capability, protocol implementation, and the cross-layer interactions are not taken into account in the model; our analytical results are much more optimistic than in the real world. Considering that the experimental measurements are very important to our understanding of the effects of authentication, we developed a WLAN testbed with IP mobility support. By examining different mobility scenarios, UDP/TCP traffic, and security policies, we observed that there is always a trade-off between security and system performance. Based on our studies, we conclude that authentication time or delay is one of the most critical factors in real-time applications. None of the current security protocols can support seamless roaming because of the large authentication delays. Therefore, we suggest that reducing authentication delay is the most important issue in designing new authentication protocols for mobile wireless networks.

References

[1] B. Aboba and J. Vollbrecht. Proxy Chaining and Policy Implementation in Roaming. RFC2607, June 1999.

[2] A. Agarwal and W. Wang. Measuring Performance Impact of Security Protocols in Wireless Local Area Networks. In *Proceedings of IEEE Broadnets*, October 2005.

[3] A.K. Arumugam, A. Doufexi, A.R. Nix, and P.N. Fletcher. An Investigation of the Coexistence of 802.11g WLAN and High Data Rate Bluetooth Enabled Consumer Electronic Devices in Indoor Home and Office Environments. *IEEE Transactions on Consumer Electronics*, 49(3):587–596, August 2003.

[4] N. Baghaei and R. Hunt. Security Performance of Loaded IEEE 802.11b Wireless Networks. *Computer Communications, Elsevier*, 27(17):1746–1756, November 2004.

[5] E. Bertino, S. Jajodia, L. Mancini, and I. Ray. Advanced Transaction Processing in Multilevel Secure File Stores. *IEEE Transactions on Knowledge and Data Engineering*, 10(1):120–135, February 1998.

[6] N. Borisov, I. Goldberg, and D. Wagner. Intercepting Mobile Communications: The Insecurity of 802.11. In *Proceedings of the Seventh Annual International Conference on Mobile Computing and Networking*, July 2001.

[7] P.R. Calhoun, J. Loughney, E. Guttman, G. Zorn, and J. Arkko. Diameter Base Protocol. draft-ietf-aaa-diameter-17.txt, December 2002.

[8] C. Chien, M.B. Srivastava, R. Jain, P. Lettieri, V. Aggarwal, and R. Sternowski. Adaptive Radio for Multimedia Wireless Links. *IEEE Transactions on Selected Areas in Communications*, 17(5):793–813, May 1999.

[9] T. Dierks and C. Allen. The TLS Protocol. RFC2246, January 1999.

[10] O. Elkeelany, M.M. Matalgah, K.P. Sheikh, M. Thaker, G. Chaudhary, D. Medhi, and J. Qaddour. Perfomance Analysis of IPSEC Protocol: Encryption and Authentication. In *Proceedings of IEEE Communication Conference (ICC)*, pp. 1164–1168, May 2002.

[11] S. Glass, T. Hiller, S. Jacobs, and C. Perkins. Mobile IP Authentication, Authorization and Accounting Requirements. RFC2977, October 2000.

[12] A. Godber and P. Dasgupta. Secure Wireless Gateway. In *Proceedings of ACM Workshop on Wireless Security (WiSe)*, pp. 41–46, September 2002.

[13] C.F. Grecas, S.I. Maniatis, and I.S. Venieris. Towards the Introduction of the Asymmetric Cryptography. In *Proc. of Sixth IEEE Symposium on Computers and Communications, 2001*, July 2001.

[14] V. Gupta, S. Gupta, and S. Chang. Performance Analysis of Elliptic Curve Cryptography for SSL. In *WiSe'02-ACM Workshop on Wireless Security*, September 2002.

[15] G.C. Hadjichristofi, N.J. Davis IV, and S.F. Midkiff. IPSec Overhead in Wireline and Wireless Networks for Web and Email Applications. In *Proceedings of IEEE International Performance, Computing, and Communications Conference, 2003*, pp. 543–547, April 2003.

[16] A. Hecker and A.H. Laboid. Pre-authenticated Signaling in Wireless LANs using 802.1X Access Control. In *Proceedings of IEEE Global Telecommunications Conference, GLOBECOM '04*, Vol. 4:2180–2184, November–December 2004.

[17] A. Hess and G. Schafer. Performance Evaluation of AAA/Mobile IP Authentication. In http://www-tkn.ee.tu-berlin.de/publications/papers/pgts2002.pdf, 2002.

[18] IETF. IEEE 802.11 Working Group. http://grouper.ieee.org/groups/802/11/index.html.

[19] IETF. http://standards.ieee.org/getieee802/download/802.1X-2001.pdf.

[20] H. Kim and H. Afifi. Improving Mobile Authentication with New AAA Protocols. In *IEEE International Conference on Communications*, 1:497–501, 2003.

[21] N. Koblitz. Elliptic Curve Cryptography. *Mathematics of Computation*, 48:203–209, 2002.

[22] H.P. Konigs. Cryptographic Identification Methods for Smart Cards in the Process of Standardization. *IEEE Communications Magazine*, 29(6):42–48, June 1991.

[23] M. Kurdziel, R. Clements, and G. Dennis. Harris 2G Encryption Engine Performance Measurements. Technical Report RFCD MD04, Harris Corporation, RF Communications Division, October 2004.

[24] B. Lee, T.Y. Kim, and S.S. Kang. Ticket-based Authentication and Payment Protocol for Mobile Telecommunications Systems. In *International*

Symposium on Dependable Computing, 2001. Proceedings, pp. 218–221, 2001.

[25] M. Leech. Username/Password Authentication for SOCKS V5. RFC9129, March 1996.

[26] M. Li, H. Zhu, S. Sathyamurthy, I. Chlamtac, and B. Prabhakaran. End-to-End Framework for QoS Guarantee in Heterogeneous Wired-cum-Wireless Networks. In *Proceedings of the First International Conference on Quality of Service in Heterogeneous Wired/Wireless Networks, 2004*, pp. 140–147, October 2004.

[27] W. Liang and W. Wang. A Quantitative Study of Authentication and QoS in Wireless IP Networks. In *Proc. of IEEE INFOCOM'05*, March 2005.

[28] A. J. Menezes, P.C. v. Oorschot, and S.A. Vanstone. *Handbook of Applied Cryptography*. CRC Press, Boca Raton, FL. 1996.

[29] A. Niemi, J. Arkko, and V. Torvinen. Hypertext Transfer Protocol (HTTP) Digest Authentication Using Authentication and Key Agreement (AKA). RFC3310, September 2002.

[30] C.E. Perkins and P.R. Calhoun. Mobile IP Challenge/Response Extensions. draft-ietf-mobileip-challenge-09.txt, February 2000.

[31] W. Qu and S. Srinivas. IPSEC-Based Secure Wireless Virtual Private Networks. In *Proceedings of IEEE MILCOM*, pp. 1107–1112, October 2002.

[32] D. Rosenthal and F. Fung. A Test for Non-disclosure in Security Level Translations. In *Proceedings of the 1999 IEEE Symposium on Security and Privacy*, pp. 196–206, May 1999.

[33] L. Salgarelli, M. Buddhikot, J. Garay, S. Patel, and S. Miller. The Evolution of Wireless LANs and PANs—Efficient Authentication and Key Distribution in Wireless IP Networks. *IEEE Personal Communications on Wireless Communications*, 10(6):52–61, December 2003.

[34] S. Shieh, F. Ho, and Y. Huang. An Efficient Authentication Protocol for Mobile Networks. *Journal of Information Science and Engineering*, 15:505–520, 1999.

[35] W. Stallings. Network Security Essentials. *Applications and Standards*, 2000.

[36] S. Sutikno and A. Surya. An Architecture of $F(2^{2N})$ Multiplier for Elliptic Curves Cryptosystem. In *Proceedings ISCAS 2000 Geneva on Circuits and Systems*, 1:279–282, May 2000.

[37] W. Wang and I.F. Akyildiz. Intersystem Location Update and Paging Schemes for Multitier Wireless Networks. In *Proc. of ACM/IEEE MobiCom'2000*, pp. 99–109, August 2000.

Index